Hot-Wire Anemometry

Hot-Wire Anemometry
Principles and Signal
Analysis

H. H. BRUUN

*Department of Mechanical and Manufacturing
Engineering, University of Bradford*

OXFORD   NEW YORK   TOKYO
OXFORD UNIVERSITY PRESS
1995

This book has been printed digitally in order to ensure its continuing availability

OXFORD
UNIVERSITY PRESS

Great Clarendon Street, Oxford OX2 6DP

Oxford University Press is a department of the University of Oxford.
It furthers the University's objective of excellence in research, scholarship,
and education by publishing worldwide in

Oxford New York

Auckland Bangkok Buenos Aires Cape Town Chennai
Dar es Salaam Delhi Hong Kong Istanbul Karachi Kolkata
Kuala Lumpur Madrid Melbourne Mexico City Mumbai Nairobi
São Paulo Shanghai Singapore Taipei Tokyo Toronto

with an associated company in Berlin

Oxford is a registered trade mark of Oxford University Press
in the UK and in certain other countries

Published in the United States
by Oxford University Press Inc., New York

A catalogue record for this book is available from the British Library

Library of Congress Cataloging in Publication Data
Bruun, H. H.
Hot wire anemometry : principles and signal analysis / H. H. Bruun.
1. Fluid dynamic measurements. 2. Hot-wire anemometer.
I. Title.
TA357.M43B78   1995   620.1'0287—dc20   94-46236

ISBN 0-19-856342-6

To my wife Ann and daughter Emma

PREFACE

The objective of this book is to provide a comprehensive, easy to access reference book for students, engineers, and researchers engaged in the practical use of any aspects of thermal anemometry. To emphasize the most common application of this type of measurement technique the title Hot-Wire Anemometry (*HWA*) has been selected for the book.

HWA is based on the convective heat transfer from a heated wire or film element placed in a fluid flow. Consequently any change in the fluid flow condition which affects the heat transfer from the heated element can be studied with *HWA*. Typical measured quantities are the components of the velocity vector and the temperature of both gases and liquids. More special applications include wall shear stress, gas concentration, and flow event detection.

HWA is used extensively for experimental fluid mechanics within the areas of Mechanical Engineering, Aeronautical Engineering, Chemical Engineering, Civil Engineering, Oceanography, and Experimental Physics.

The book is structured in such a way that it is of direct value to people with no or very little practical knowledge of *HWA* as well as to researchers with considerable experience in *HWA*. To achieve this dual objective the content of the book has been formulated within a logical framework, and wherever possible the necessary concepts are presented in the simplest possible way, without sacrificing the methodology and principles of the *HWA* application described.

The early chapters are in particular aimed at the inexperienced *HWA* user, and they represent the building blocks on which the later more advanced chapters are built.

The format of the book is as follows: Chapter 1 outlines the main advantages and disadvantages of *HWA* and formulates objectives of current *HWA* work in the context of Computational Fluid Dynamics (*CFD*). Chapter 2 describes the basic principles of *HWA*, such as heat transfer, frequency response, and spatial resolution. By necessity the presentation is somewhat mathematical and readers who are inexperienced in *HWA* should read Chapter 3 first. This chapter is an introduction to practical velocity measurements with *HWA*. It introduces the logical framework for a computer-based *HWA* system and identifies the individual steps in the complete experimental procedure ranging from probe selection through to presentation of analysed data. Chapters 4 to 6 describe velocity measurements in incompressible flows using the framework identified in Chapter 3. The practical aspects of single-component velocity measurements in both air and water flows are covered in Chapter 4. Chapter 5 contains an in-depth discussion of *HWA* probe types and signal analysis methods for two-component velocity measurements. Finally, Chapter 6 gives an account of *HWA* probes and techniques used to measure either simultaneously the three components of the velocity vector at a point or the corresponding three mean velocity components and six Reynolds stresses.

More advanced *HWA* applications are described in Chapters 7 to 11. Chapter 7 deals with the effect of variations in the ambient fluid temperature T_a on *HWA* signals, both in terms of compensation for slow variations in T_a as well as the measurement of fluctuations in T_a. *HWA* techniques for reversing flows and for the near-wall region are presented in Chapter 8, which includes a description of the pulsed hot-wire technique (*PWA*), the flying hot-wire technique (*FWA*), the necessary *HWA* correction procedure in the near-wall region and measurements of mean and time-dependent wall shear stresses. Chapter 9 covers measurements in two-phase flows, gas mixtures, and flows with significant compressibility effects. Chapter 10 gives an account of probe types and signal analysis methods used to measure the streamwise- or cross-stream vorticity component as well as all three components of vorticity measured simultaneously. The use of conditional sampling techniques in fluid-flow studies are discussed in Chapter 11. The phase-locked averaging technique is described with reference to the semi-regular phenomena which occur in turbomachinery and internal combustion engines. The identification of coherent structures in turbulent shear layers is examined with reference to the turbulent boundary layer on a flat plate and free-shear layers.

Time-series analysis is presented in a separate Chapter 12 as the principles involved can be applied to the time-history record of any physical process. The chapter considers common statistical properties for either a single time-history record or two simultaneous time-history records. Particular emphasis has been placed on the formulation of digital evaluation procedures, error estimates, and the specification of optimum sampling rate criteria.

Bradford   H. H. B.
June 1994

ACKNOWLEDGEMENTS

Full bibliographical references to all material reproduced are to be found at the end of the book.

The author wishes to thank the editors and authors for permission to reproduce the following figures and tables: *Journal of Fluid Mechanics* published by Cambridge University Press: Fig. 2.11 from Paranthoen *et al.* (1982*b*); Figs 2.23 and 2.24 from Gessner and Moller (1971); Fig. 4.15 from Perry and Morrison (1971*c*); Figs 4.19, 4.20, and 4.21 from Swaminathan *et al.* (1986*b*); Fig. 5.4 from Morrison *et al.* (1972); Fig. 5.21 from Kawall *et al.* (1983); Fig. 8.6 from Bradbury and Castro (1971); Fig. 9.12 from Owen *et al.* (1975); Fig. 10.11 from George and Hussein (1991); Figs 10.12 and 10.13 from Vukoslavčevic *et al.* (1991); Fig. 10.4 from Balint *et al.* (1991); Fig. 10.15 from Tsinober *et al.* (1992); Fig. 11.14 from Murlis *et al.* (1982); Figs 11.15 and 11.17 from Kovasznay *et al.* (1970); Fig. 11.18 from Chen and Blackwelder (1978); Fig. 11.19 from Subramanian *et al.* (1982); Fig. 11.21 from Corino and Brodkey (1969); Fig. 11.22 from Brodkey *et al.* (1974); Fig. 11.23 from Lu and Willmarth (1973); Fig. 11.24 from Alfredsson and Johannson (1984*b*); Fig. 11.25 from Blackwelder and Kaplan (1976); Figs 11.26 and 11.27 from Johannson and Alfredsson (1982); Fig. 11.28 from Luchik and Tiederman (1987); Fig. 11.29 from Alfredsson and Johansson (1984*b*); Fig. 11.30 from Johansson *et al.* (1987*b*); Fig. 11.31 from Randolph *et al.* (1987); Fig. 11.32 from Weisbrot and Wygnanski (1988) and Fig. 12.15 from Comte-Bellot and Corrsin (1971). Table 11.1 from Hedley and Keffer (1974) and Table 11.3 from Bogard and Tiederman (1986). *International Journal of Heat and Mass Transfer* published by Elsevier Science Ltd, the Boulevard, Langford Lane, Kidlington, OX5 1GB, UK: Fig. 7.1 from Koppius and Trines (1976) and Fig. 9.6 from McQuad and Wright (1973). Springer-Verlag: *Advances in Turbulence* (1987): Figs 8.22 and 8.23 from Janke; Fig. 11.31 from Randolph *et al.*; and *Advances in Turbulence* 2 (1989): Figs 5.15 and 5.16 from Hartmann and Dengel; Fig. 10.6 from Haw *et al.* (1989) and Fig. 10.9 from Kim and Fiedler. *AIAA Journal* published by the American Institute of Aeronautics and Astronautics: Figs. 9.7 and 9.8 from Way and Libby (1970) and Fig. 9.13 from Walker *et al.* (1989). John Wiley & Sons Ltd: Figs 12.10, 12.14, 12.17, and 12.21 from Bendat and Piersol (1966); Fig. 12.2 from Bendat and Piersol (1980) and Fig. 12.20 from Bendat and Piersol (1986).

Although every effort has been made to trace and contact copyright holders, in a few instances this has not been possible. If notified the publishers will be pleased to rectify any omission in future editions.

My research work and book has benefited from the contributions by many people. I would in particular like to thank:

(a) My past and current research students: H. H. Al-Kayiem, J. Ali, M. F. N. Al-Deen, O. O. Badran, A. A. Fardad, B. Farrar, A. Fitouri, A. R. R. Jaju, F. A. Hamad, M. A. Khan, M. K. Khan, Z. Mahmood, N. Nabhani, and A. L. Samways.

(b) My colleagues at the yearly advanced course at UMIST, Manchester, England: P. L. Betts, F. Durst, B. E. Launder, D. C. Jackson, M. A. Leschziner, C. Tropea, J. T. Turner, and A. Yule.

(c) A large number of people have very kindly provided comments on parts of this book, supplied pictures or copies of papers, and for their help I would like to express my appreciation to: A. A. Abdel-Rahman, M. Acrivlellis, R. J. Adrian, R. A. Antonia, C. Arcoumanis, R. S. Azad, J. Bataille, C. Beguier, M. F. Blair, J. P. Bonnet, F. V. Bracco, P. Bradshaw, K. Bremhorst, K. J. Bullock, I. P. Castro, R. Cheesewright, R. Chevray, Y. T. Chew, N. Collings, G. Comte-Bellot, N. Cumpsty, Dantec Measurement Technology, M. R. Davies, K. Döbbeling, H. Eckelmann, J. W. Elsner, S. Einav, T. D. Fansler, H. H. Fernholz, H. E. Fiedler, L. M. Fingerson, J. F. Foss, P. Freymuth, T. Hotta, H. J. Hussein, G. Janke, P. A. Libby, J. P. Johnston, F. E. Jørgensen, G. Kawall, J. K. Keffer, W. Kühn, B. Laksminaryana, P. Sheel Larsen, B. J. Legg, B. E. Launder, M. A. Leschziner, P. M. Ligrani, Y. Nagano, P. Paranthoen, S. O. Park, A. G. Piersol, J. Podzimek, R. Rask, K. Rehme, F. J. Resch, B. Sammler, V. A. Sandborn, W. J. Seale, A. Serizawa, S. A. Sherif, A. J. Smits, E. P. Sutton, T. Tagaki, W. G. Tiederman, T. S. I. Inc., J. M. Wallace, J. H. Whitelaw, P. O. Witze, D. H. Wood, N. B. Wood, and I. Wygnanski.

Finally I would like to give a special thanks to Anita Knight for her kindness and inspiration during the writing of Section 11.2 and to Mrs C. Noble for superb typing of the many versions of the manuscript for this book.

CONTENTS

NOTATION

Numbers indicate sections in which quantities are defined or first used. Occasional notation is defined where it is used.

A	calibration constant
A_w	wire cross-sectional area
$A^*(\theta)$	calibration function
A_1-A_6	calibration parameters
a	overheat ratio, R_w/R_a (2.2); speed of sound (2.1)
B	calibration constant
$\hat{B}(\alpha)$	calibration function, constant calibration exponent, n
$B^*(\theta)$	calibration function, variable calibration exponent, $n(\theta)$
B_1-B_6	calibration parameters
b	yaw coefficient
$b[\ ]$	bias error of [ ]
C	calibration constant (4.4); threshold level (11.2)
C_w	thermal capacity of a wire element, $\rho_w c_w (\pi/4) d^2 \ell$
C_1-C_{15}	calibration parameters
c_p	specific heat of fluid at a constant pressure
c_w	specific heat of wire material
c_s	specific heat of the substrate, hot-film probe
$D(t)$	detection function
D_1-D_6	calibration parameters
$\dfrac{\text{D}}{\text{D}t}$	$\equiv \bar{U}\partial/\partial x + \bar{V}\partial/\partial y + \bar{W}\partial/\partial z$; mean substantial derivative
d	wire (or sensor) diameter
E	anemometer output voltage
E_C	calculated voltage, calibration
E_G	output voltage from signal-conditioning unit
E_R	measured calibration reference anemometer voltage
E_T	theoretical wire voltage, resistance wire
E_w	wire voltage
E_0	anemometer output voltage at zero velocity
E_{off}	offset voltage, signal-conditioning unit
$E[\ ]$	expected value of [ ]
$E_u(f)$	one-dimensional (velocity) frequency spectrum
$E_u(k_1)$	one-dimensional (velocity) wave-number spectrum
$E_\theta(f)$	one-dimensional (temperature) frequency spectrum
$E_\theta(k_1)$	one-dimensional (temperature) wave number spectrum
e	fluctuating anemometer voltage
e_{off}	amplifier offset voltage

e_t	electrical pertubation signal
e_w	fluctuating wire voltage
f	frequency (Hz)
f_c	cut-off frequency, frequency where amplitude has decreased by 3 dB (2.3); Nyquist folding frequency (12.5)
$f(r_1)$	longitudinal correlation coefficient
$f(\alpha)$	yaw function
Δf	bandwidth resolution
G	gain of amplifier (in anemometer (2.3), or signal-conditioning unit (4.11))
Gr	Grasshof number $(g\rho^2 d^3\beta(T_w - T_a)/\mu^2)$
$G_x(f)$	autospectral density function (one-sided)
g	gravitational acceleration
$g(r_2)$	lateral correlation coefficient
$g(\alpha)$	yaw function
$H(f)$	amplitude transfer function for velocity fluctuations
$H_\theta(f)$	amplitude transfer function for temperature fluctuations
H_p	plateau level of $H_\theta(f)$
h	heat-transfer coefficient (2.1); wire separation, multi-wire probes (2.4); pitch coefficient (3)
I	electrical current in the sensor (2.1); intermittency function (11.2)
i	fluctuating electrical current in sensor
Kn	Knudsen number (λ/d)
k	turbulent kinetic energy $(\frac{1}{2}\overline{u_i u_i})$ (1.6); thermal conductivity of fluid (2.1); yaw coefficient (3); threshold (11.2)
k_w	thermal conductivity of wire material (2.1); thermal conductivity of wall material (8.8)
k_1	wave number $(2\pi f/\bar{U})$
L	longitudinal integral length scale (2.4); total wire length (5.2)
L_g	transverse integral length scale
L_p	prong length
ℓ	(active) length of wire (or sensor)
ℓ_c	cold length
ℓ_m	mixing length
ℓ_t	turbulence length scale
M	Mach number (U/a)
M	time constant of wire (or sensor)
M_p	time constant of prongs (2.3); time constant of pulsed wire, *PWA* (8.5)
M_s	time constant of sensor wire, *PWA*
m	yaw coefficient

N	signal-to-noise ratio (2.3); number of samples in a time-history record (3.5)
n	calibration exponent
n_{opt}	optimum calibration exponent
$n(\theta)$	calibration-exponent function
n_d	number of subrecords
Nu	Nusselt number (hd/k)
P	pressure
P_a	pressure of ambient fluid
P_{atm}	atmospheric pressure
Pe	Peclet number (Uh/k)
P_o	total pressure
Pr	Prandtl number ($c_p\mu/k$)
$P(x)$	probability distribution function
p	fluctuating pressure
$p(x)$	probability density function
$p(x, y)$	joint probability density function
$\dot{Q}_{fc}$	forced convective heat transfer rate
$\dot{Q}_{cp}$	conductive heat transfer rate to prongs
R	resistance of wire (or sensor)
R_a	resistance of wire at the ambient temperature, T_a
R_f	resistance of heated film sensor at the temperature, T_f
R_L	resistance of probe lead and cables
R_s	resistance of sensor wire, PWA
R_w	resistance of heated wire at temperature, T_w
R_0	resistance of wire at 0 °C
R_{20}	resistance of wire at 20 °C
R_1, R_2, R_3	resistors in Wheatstone bridge (Figs 2.6 and 2.12)
$R_x(\tau)$	autocorrelation function
$R_{xy}(\tau)$	cross-correlation function
Re	Reynolds number ($\rho\bar{U}d/\mu$)
Re_c	critical Reynolds number for forced convective heat transfer
Re_θ	momentum thickness Reynolds number ($\rho\bar{U}\theta/\mu$)
r	fluctuating wire (or sensor) resistance
S	shear parameter
S	Strouhal number ($fD/\bar{U}$)
S^*	shear parameter
SR	sampling rate
S_m	mass flow sensitivity, longitudinal component
S_{mv}	mass flow sensitivity, transverse component
S_{T_o}	total temperature sensitivity
S_u	CT-mode velocity sensitivity, longitudinal component

$S_{u,cc}$	CC-mode velocity sensitivity, longitudinal component
S_v	CT-mode velocity sensitivity, transverse component
$S_x(f)$	autospectral density function (two sided)
S_θ	CT-mode temperature sensitivity
$S_{\theta,s}$	CT-mode temperature sensitivity obtained by static calibration
$S_{\theta,cc}$	CC-mode temperature sensitivity
S_ρ	density sensitivity
s	prong spacing
s, n	directions in a streamwise coordinate system
T	temperature; sampling time or record length (3.5); time of flight, PWA (8.5); periodicity (11.1); averaging time (11.2)
T_a	temperature of ambient fluid
T_d	diffusion time, PWA
T_e	recovery temperature
T_f	film temperature $(T_w + T_a)/2$ (2.1); temperature of heated hot-film sensor (5.7)
T_o	stagnation temperature
T_p	prong temperature (2.3); Temperature of pulsed wire, PWA (8.5)
T_r	reference temperature (7.2); total record length (12.5)
T_w	temperature of heated wire
T_I	integral time scale
T_0	0 °C reference temperature
Tu	turbulence intensity $((\overline{u^2})^{1/2}/\bar{U})$
T_{in}	inner time scale (v/U_τ^2)
T_{out}	outer time scale (δ/U_∞)
t	time
Δt	sampling interval
$\hat{t}$	duration of heating pulse, PWA
Δt_d	time delay, PWA circuit
U, V, W	velocity components in x, y, z directions of a Cartesian space-fixed coordinate system
U_N, U_T, U_B	normal, tangential, and binormal velocity components in x', y', z' directions of a Cartesian wire-fixed coordinate system
U_o, V_o, W_o	velocity components at probe centre
U_s, V_s, W_s	velocity components in x_s, y_s, z_s directions of a Cartesian probe-stem coordinate system
U_p, V_p	velocity components of a moving probe (FHA) in a space-fixed coordinate system
U_r, V_r	relative velocity components measured with a moving probe (FHA) in a space-fixed coordinate system

U_r', V_r'	relative velocity components measured with a moving probe (*FHA*) in a probe-stem coordinate system
U_s, U_n	velocity components in an (s, n) streamwise coordinate system
U_a	area-averaged velocity
U_C	calculated calibration velocity
$U_{\mathscr{C}}$	centre-line velocity, pipe flow
$\bar{U}_m$	measured time mean velocity
U_R	measured calibration reference velocity
U_τ	friction velocity $(\tau_w/\rho)^{1/2}$
U^+	normalized mean velocity $(\bar{U}/U_\tau)$
U_∞	free-stream velocity
$\bar{U}_{i,E}(\theta)$	ensemble-averaged mean velocity components
u_i	fluctuating velocity components
u, v, w	fluctuating velocity components (index notation as for U, V, W)
u', v', w'	rms values of fluctuating velocity components
$u'_{i,LF}(\theta)$	rms values of low-pass filtered, fluctuating velocity components
$u'_{i,HF}(\theta)$	rms values of high-pass filtered, fluctuating velocity components
V	velocity vector
$\tilde{V}$	magnitude of the velocity vector
V_e	effective velocity
V_p	probe velocity vector
V_r	relative velocity vector measured by a moving probe (*FHA*)
V_t	turbulence velocity scale
var[ ]	variance of [ ]
$w(t)$	time window
$W(f)$	spectral window
X_N	ensemble average of X_n over N realizations
$X(f)$	Fourier transform of $x(t)$
$X(f, T)$	Fourier transform of $x(t)$ over a record length T
x_i	Cartesian space-fixed coordinate system
x, y, z	Cartesian space-fixed coordinate system, usually with x in the general direction of the flow and with y normal to the surface or the plane of the shear layer
x', y', z'	Cartesian wire-fixed coordinate system
x_s, y_s, z_s	Cartesian probe-stem coordinate system
x_p, y_p	coordinates of a moving probe (*FHA*)
$x(t), y(t)$	time-dependent variables
y	distance from wall
y^+	normalized distance from wall (yU_τ/ν)

z	lateral position
α	yaw angle (3); void fraction (9.1)
α_e	effective yaw angle
$\bar{\alpha}$	mean yaw angle
α_0	temperature coefficient of resistivity at 0 °C
α_{20}	temperature coefficient of resistivity at 20 °C
β	volume coefficient of expansion (2.1); pitch angle (3)
$\bar{\beta}$	mean pitch angle
Γ	concentration
γ	ratio of specific heats (c_p/c_v) (2.1); roll angle (6.1); signal dead time (9.1); fluctuating concentration (9.2); intermittency factor (11.2)
γ_p	forward-flow fraction
Δ	small change in value
δ	boundary layer thickness
ε	dissipation rate (1.6); normalized rms error (12.3)
ε^2	normalized mean squared error
ε_θ	temperature dissipation rate
η	Kolmogorov length scale (2.4); recovery factor (T_e/T_o) (9.3)
θ	fluctuating fluid temperature (1.4); flow angle (5.4); momentum thickness (10.2)
κ	von Kármán constant
λ	molecular mean free path (2.1); Taylor's microscale (2.4)
λ_θ	temperature microscale
μ	molecular viscosity of fluid
μ_t	turbulent viscosity
ν	kinematic viscosity (μ/ρ)
ν_t	eddy viscosity (μ_t/ρ)
ρ	density of fluid
ρ_s	density of substrate, hot-film probe
ρ_w	density of wire material
$\rho_{ij}(r_1, r_2, r_3)$	spatial correlation coefficient function
$\rho_x(\tau)$	autocorrelation coefficient function
$\rho_{xy}(\tau)$	cross-correlation coefficient function
σ_x	standard deviation of $x(t)$
$\sigma[\ ]$	standard deviation of [ ]
σ^2	variance
τ	time delay; decay time square wave test (2.3); temperature loading factor, $(T_w - T_e)/T_o$ (9.3)
τ_H	hold time; intermittency circuit
τ_w	mean wall shear stress (skin-friction)
φ	angle
χ	resistivity

ψ	angle between probe-stem and space-fixed coordinate system
ψ^2	mean square error
ω	angular frequency $(2\pi f)$
$\omega_x, \omega_y, \omega_z$	vorticity components in a Cartesian (x, y, z) space-fixed coordinate system

Subscripts and superscript

a	temperature reference condition, T_a
e	effective
f	film
i	component
i, j, k	denotes Cartesian coordinate directions
m	measured
n	non-turbulent
o	stagnation
off	offset
r	temperature reference condition, T_r
t	turbulent
w	temperature reference condition, T_w
0	value at 0 °C
1, 2	wires (or sensors) 1 and 2
∞	free-stream conditions
+	denotes quantities nondimensionalized by v, τ_w, and ρ
'	rms value
$\langle\ \rangle$	conditional average
$[\hat{\ }]$	estimate of $[\ ]$
$\overline{}$	(e.g. $\bar{U}$ or $\overline{u^2}$) denotes time-mean value.

ABBREVIATIONS

CC   Constant current
CFD   Computational fluid dynamics
CT   Constant temperature
FHA   Flying hot-wire anemometer
HWA   Hot-wire anemometer
LDA   Laser doppler anemometer
PWA   Pulsed wire anemometer
SN   Single normal
SY   Single yawed
3W   Three wire

1

INTRODUCTION

The primary objective in writing this book on practical Hot-Wire Anemometry (HWA) was to provide a reference textbook for the selection and use of the most appropriate hot-wire/film anemometry technique for different turbulent flow studies. This book has a logical framework, with simple principles and single-wire techniques preceding more advanced multi-sensor methods. It includes a relatively detailed discussion of the more common hot-wire/film probes and the related signal-analysis techniques, while shorter descriptions are given of more specialized HWA applications. A historical review of HWA has not been included since one has been written by Freymuth (1983a). Previous HWA contributions are summarized in a number of excellent hot-wire textbooks and reference papers including those by Kovasznay (1954), Hinze (1959, 1975), Corrsin (1963), Bradshaw (1971), Sandborn (1972), Comte-Bellot (1976), Vagt (1979), Bruun (1979a) Blackwelder (1981), Perry (1982), Fingerson and Freymuth (1983), Lomas (1986) and Müller (1987). The detailed bibliographies of thermal anemometry given by Freymuth (1978a, 1983b, 1992) are also valuable reference sources for HWA users.

HWA is based on convective heat transfer from a heated wire or film element placed in a fluid flow. Any change in the fluid flow condition that affects the heat transfer from the heated element will be detected virtually instantaneously by a constant-temperature HWA system. HWA can therefore be used to provide information related to for example, the velocity and temperature of the flow, concentration changes in gas mixtures, and phase changes in multi-phase flows.

1.1 ADVANTAGES OF HOT-WIRE/FILM ANEMOMETRY

HWA is likely to remain the principal research tool for most turbulent air/gas flow studies. For measurements in low and moderate turbulence intensity flows (less than ~ 25%) the main advantages of conventional HWA are:

1. *Cost* HWA systems are relatively cheap, in comparison with their main competitors, Laser Doppler Anemometers ($LDAs$).

2. *Frequency response* A standard hot-wire probe operated in conjunction with a modern Constant-Temperature (CT) anemometer at optimum conditions has, except at low velocities, a flat frequency response from 0 to 20–50 kHz. Measurements up to several hundred kilohertz are therefore

easy to obtain. In contrast, measurements with an *LDA* system are normally restricted to less than 30 kHz.

3. *Size* A typical hot-wire sensor is about 5 μm in diameter and 1.25 mm long, although wire elements as small as 1 μm by 0.25 mm have been used. In contrast, a typical *LDA* measuring volume is 50 μm by 0.25 mm.

4. *Velocity measurements* Hot-wire probes with one or more sensors are commercially available, which allow measurement of one, two, or three components of the velocity vector at specified points in the flow field. Both *HWA*s and *LDA*s have a very wide velocity range, from very low velocity to high-speed (compressible) flows.

5. *Temperature measurements* Simultaneous measurements of the fluctuating velocity and temperature field can be obtained using a multi-sensor probe, which usually contains one sensor operated in the 'cold-wire' mode.

6. *Two-phase flow* Hot-film probes can be used for measurements in flows containing a continuous turbulent phase and distributed bubbles (liquid/gas or liquid/liquid flow). When a bubble hits the sensor, an interaction will take place between the probe and the interface between the bubble and the continuous phase. However, provided that the bubble is larger than the size of the sensing element, this interaction process can be used for signal-analysis purposes, as discussed in Chapter 9. In applying this technique, there is no significant restriction on the bubble concentration (void fraction). In contrasts, if an *LDA* technique is applied to a bubbly flow, then a low void fraction must be used due to the requirement for a clear optical path.

7. *Accuracy* Both *HWA* and *LDA* can give similar, very accurate results (0.1–0.2%) in carefully controlled experiments. However, in many practical applications, a 1% accuracy is more likely for both systems.

8. *Signal-to-noise ratio* Hot-wire anemometers are clearly superior since they can have very low noise levels. A resolution of one part in 10 000 is easily accomplished, while with an *LDA* one part in 1000 is difficult to achieve with present technology (Fingerson and Freymuth 1983).

9. *Probe and analysis selection* A *HWA* system is relatively simple to operate, and as shown in this book, the selection and use of probes and the related calibration, data acquisition, and analysis can be carried out within a logical framework.

10. *Signal analysis* The output from a *HWA* system is a continuous analogue signal. Consequently, both conventionally/conditionally-sampled time-domain and frequency-domain analysis can be carried out.

11. *Spatial information* The use of two or more spatially separated probes enables the measurement of spatial/temporal correlations of turbulent fluctuations. Hot-wire rakes in conjunction with conditional sampling can iden-

tify the spatial/temporal development of large-scale structures within turbulent flows.

12. *Special probes* Special *HWA* probes, and the related signal analysis, can be used to evaluate turbulent quantities such as intermittency, dissipation rate, vorticity, etc.

1.2 SIGNIFICANT *HWA* PROBLEMS AND PROPOSED SOLUTIONS

High-turbulence-intensity flows Conventional *HWA* is restricted to low and moderate turbulence intensity flows. At high turbulence intensities errors can occur from two sources. For single-sensor and *X*-configuration probes, errors can be caused by neglecting higher-order terms in the series expansion for the effective velocity (Chapters 4–6). The second type of error is usually referred to as a rectification error. Due to its rotational symmetry, the wire element is insensitive to a reversal of the flow direction, which may occur in high turbulence intensity flows. Also, for *X*- and triple-wire (3*W*) probes, signal ambiguity will occur when the velocity vector falls outside the approach-acceptance angle (Chapters 5 and 6). There are no satisfactory compensation techniques for these errors for stationary *HWA* probes. A Flying Hot-wire Anemometer (*FHA*) system (Chapter 8) may be used to resolve the rectification problem. This method is equivalent to the frequency shift in *LDA* systems. An alternative anemometer system is the Pulsed-wire Anemometer (*PWA*) technique (Chapter 8).

Probe-disturbance Placing a probe in the flow will modify the local flow field. However, for a well-designed probe, the corresponding errors will often be very small (see Chapters 4 and 5), and the related flow disturbances are usually incorporated into the calibration procedure. However, a stationary probe may significantly modify any disturbance-sensitive flow phenomenon such as, for example, flow separation. In this case, an *FHA* or *LDA* system should be used.

Liquid flow Several problems may occur when *HWA* is used for the study of single-phase liquid flow, as discussed in Section 4.10.2. Accumulation of fouling material on the sensor can be a major problem. To minimize this contamination problem, a filtration unit is recommended if a recirculating test facility is used. Probe fouling is less of a problem for probes with large sensors, and hot-film probes are normally used in liquid flow studies. A second problem in liquid flows is 'temperature' contamination of the probe calibration due to the small overheat ratio used. Even small changes in the temperature of the liquid will result in substantial changes in the probe calibration, necessitating frequent probe recalibration. Due to these *HWA* problems, the *LDA* technique is also used for liquid flow studies since naturally occurring particles can produce the required seeding.

Probe breakage Although a hot-wire/film probe is a very delicate instrument, it can, if handled carefully, last for many months or years. For low-speed air-flow studies the most common cause of breakage is improper handling by inexperienced operators. To avoid time-consuming and, for multi-sensor probes, also expensive repairs, new researchers should be carefully instructed in the safe way of removing (and replacing) probes from their storage container. Also, the placing, of a hot-wire probe in a narrow passage or through a mounting hole should *always* be done using a specially designed guide system and *not* by hand. By insisting on these simple instructions, the author has reduced manual probe breakage to less than one breakage per student per year. Probe breakage may also occur due to 'burn out'. This will take place if the overheat ratio is accidently set too high, or it will take place by an ageing process if the probe is operated at a high overheat ratio setting. In high-speed flow, probes may be damaged/broken due to the impact of relatively fine particles and all research wind-tunnels should therefore be fitted with an air filter.

Contamination The deposition of impurities in the flow on the sensor can dramatically alter the calibration characteristic and reduce the frequency response. When this occurs, the probe must be recalibrated and, if necessary, cleaned and then recalibrated. This can be very time-consuming, and it is strongly recommended that a flow filter is incorporated into the experimental flow facility wherever possible. A common method of removing dust from a sensor is to brush it very carefully with an artist's brush. However, contamination can also be caused by oil droplets in air flows, or by scale/slime in water flows. Depending on the extent of the contamination, it may be possible to remove such deposits by soaking in a liquid with some solvent properties. Swirling of the probe or acoustic agitation of the liquid will speed up this process (Wyatt 1953). Commonly used solvents include methylated spirit, acetone, and carbon tetrachloride. Jimenez *et al.* (1981) cleaned hot-film probes used in water by wiping them with a cotton swab dipped in acetic acid. After cleaning, the probe should be washed in distilled water and allowed to dry before recalibration.

1.3 *HWA* RESTRICTIONS/*LDA* APPLICATIONS

HWA and *LDA* are complementary techniques. Each has its own exclusive advantages and limitations, and a significant number of flow situations exist where both techniques can be used successfully. The main *HWA* exclusion is in hostile environments, such as combustion, where probe damage will occur. *LDA* is the practical choice for such applications. Turbulent gas or liquid flows with solid particles or fine droplets are also normally studied by *LDA* techniques. For liquid flow studies, *LDA* is often preferred since the seeding requirement is usually satisfied by naturally occurring fine particles, and because of the existence of the related probe-fouling problem for *HWA*. In high turbulence intensity flow, a doppler-shifted *LDA* system can provide unambiguous results, which is not the case for a stationary hot-wire probe system. The *FHA* or *PWA* methods can be used to solve this problem.

However, *LDA* methods have several restrictions. *LDA* systems are costly, and bias effects and seeding problems may occur (see for example, Durão *et al*. 1980; Dring 1982; Stock and Fadeff 1983; Johnson *et al*. 1984; Stevenson *et al*. 1984; Haghgooie *et al*. 1986; Winter *et al*. 1991a,b,c, and Ruck 1991). Simultaneous velocity (*LDA*) and temperature ('cold-wire' probe) measurements may not be reliable due to contamination of the 'cold wire' probe by naturally occurring particles (water) or by seeding particles (air). *LDA* does not give a continuous signal, and the spectral information is much more difficult to evaluate (Gaster and Bradbury 1976; George *et al*. 1978; and Roberts and Gaster 1980). Finally, *LDA* cannot provide some of the more advanced turbulence quantity information obtainable by *HWA*.

1.4 THE NEED FOR FLOW MEASUREMENTS

Fluid flow plays an important role in processes occurring in our environment, in hydraulic machines, in manufacturing industries, and in many other areas. Most practical flows are turbulent, and turbulent processes contribute significantly to the transport of momentum, heat, and mass. Turbulence is also the factor responsible for most fluid friction losses which can be observed as pressure losses in fluid-transport systems.

In the past, the main objective of experimental flow and heat-transfer investigations was overall performance; that is, integral information. As an example, consider a heated circular cylinder (or a water-cooled pipe) placed in a cross-flow as shown in Fig. 1.1. The flow parameters considered were the mean velocity, U_∞, and the temperature, T_∞, of the approach flow, and for the heated cylinder the quantity measured was the electrical heat

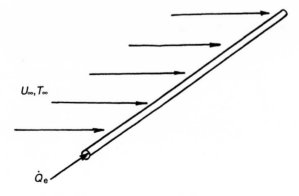

Fig. 1.1. Heat transfer from a heated cylinder in a cross flow.

flux, $\dot{Q}_e$. For design and performance evaluations the results were usually correlated in a nondimensional format

$$Nu = f(Re, Pr) \qquad (1.1)$$

linking the heat transfer via a Nusselt number, Nu, to the flow field in terms of the Reynolds and Prandtl numbers, Re and Pr, respectively. From detailed measurements of these quantities, and by evaluating the fluid properties at a suitable average temperature, empirical correlations in the form of eqn (1.1) were obtained. In many applications the heated cylinder (or the water-cooled pipe) illustrated in Fig. 1.1 formed part of a large heat-exchanger unit, in which case many additional measurements were required to cover the geometrical variation in the positioning of the individual elements. Such experiments were repetitive and time-consuming, and this approach only involved the empirical correlation of time-mean quantities with little or no attempt to determine the effect of turbulence in the flow.

For most practical fluid-dynamic and heat-transfer applications, the aim is to develop accurate predictive methods, and the rapid growth of Computational Fluid Dynamics (CFD) has therefore had a major impact on the objectives of many contemporary experimental flow studies. CFD is an evolving science in which turbulence modelling (see Section 1.6) is the pacing issue. At the moment, the accuracy and physical realism achieved often leave a lot to be desired, but improvements are constantly being achieved and the generality is being extended. State of the art CFD can give credible solutions for very complex conditions, including three-dimensional recirculation in complex geometries, transient flows, combustion, and two-phase flow features. CFD is now being used as an integral part of design procedures. For example, the flows in turbine/compressor passages and combustion chambers are being analysed in the design context, flows over wings and entire aircraft are being computed by aircraft manufacturers, flows in diesel engines are being analysed by various engine manufacturers, flows around ships are being computed, and quite accurate 7–10-day weather forecasting is now possible. The main limitation in all these applications arises from turbulence modelling.

CFD can also be used as an aid in understanding the physics of complex flow/heat transfer processes. In general, a turbulence model is derived from rational general principles (such as a second-moment closure) and the 'calibration' coefficients are determined by comparison with experimental data in well-defined simple shear flows. Applying such a model (with constant values of the 'calibration' coefficients) to complex flow situations can often provide valuable information. Although it may not be possible to achieve accurate numerical values, very often the principal aerodynamic features are correct and they enhance the understanding of complex conditions for which experimental data may be lacking.

With the rapid growth of *CFD*, a new role has therefore developed for *HWA*. To validate existing *CFD* computer codes containing turbulence models, high-quality experimental flow data are required. The format of the required experimental information has changed from the data needed for much of the earlier statistical turbulence work. For experimentalists to make a significant input to *CFD*, it is necessary that they should become familiar with the basic principles and assumptions/simplifications made in *CFD*. To illustrate the kind of experimental flow information required for the development of turbulence models, the principles of some of the *simpler* turbulence models will be outlined. However, the reader should also be aware that significant additional *CFD* problems relate to: (i) simplification of the full Navier–Stokes equations; (ii) the accuracy with which the derivatives in eqns (1.4) and (1.5) are approximated in terms of algebraic equivalents; (iii) selection of grid points; and (iv) sensitivity of the solution to the boundary conditions, including the inlet conditions. This last problem has been highlighted by George (1988, 1990). Using momentum-integral considerations for an axisymmetric buoyant plume and for an axisymmetric jet, George showed that the spreading rate would be different if the experiments take place in either a very large room or in an infinite environment. Consequently, for such flow types, unless the boundary conditions are well-understood and specified, any measurements, however high their quality, will be of limited value for *CFD* comparison.

HWA studies are also being carried out for a number of different reasons. Since *CFD* is not yet an operational tool, there is still a strong need for detailed, accurate flow and temperature measurements to ensure the satisfactory performance of new complex flow systems and fluid machinery. *HWA* is often the ideal tool for providing the required information related both to the mean velocity and temperature field and to the magnitude and distribution of the turbulence. *HWA* also remains the principal tool for basic studies of the physics of turbulent flows. It can also be used to identify specific flow phenomena such as the transition from laminar to turbulent flow, flow separation, or relaminarization.

1.4.1 Time-mean governing equations

Fluid flow and any associated heat-transfer processes are governed by the equations of continuity, momentum (Navier–Stokes), and energy. Due to the complexity of these equations, most practical *CFD* work will, for the forseeable future, be centred around solutions to the related time-averaged equations. For a statistically steady, isothermal, flow of a Newtonian fluid with constant density and viscosity these equations can be written in a Cartesian (x, y, z) coordinate system with velocity components (U, V, W) as the *continuity* equation

$$\frac{\partial \bar{U}}{\partial x} + \frac{\partial \bar{V}}{\partial y} + \frac{\partial \bar{W}}{\partial z} = 0 \tag{1.2}$$

and the *momentum (Navier–Stokes)* equations

$$\rho \left[\bar{U} \frac{\partial \bar{U}}{\partial x} + \bar{V} \frac{\partial \bar{U}}{\partial y} + \bar{W} \frac{\partial \bar{U}}{\partial z} \right] = - \frac{\partial \bar{P}}{\partial x} + \mu \nabla^2 \bar{U} - \rho \left[\frac{\partial \overline{u^2}}{\partial x} + \frac{\partial \overline{uv}}{\partial y} + \frac{\partial \overline{uw}}{\partial z} \right], \tag{1.3a}$$

$$\rho \left[\bar{U} \frac{\partial \bar{V}}{\partial x} + \bar{V} \frac{\partial \bar{V}}{\partial y} + \bar{W} \frac{\partial \bar{V}}{\partial z} \right] = - \frac{\partial \bar{P}}{\partial y} + \mu \nabla^2 \bar{V} - \rho \left[\frac{\partial \overline{uv}}{\partial x} + \frac{\partial \overline{v^2}}{\partial y} + \frac{\partial \overline{vw}}{\partial z} \right], \tag{1.3b}$$

$$\rho \left[\bar{U} \frac{\partial \bar{W}}{\partial x} + \bar{V} \frac{\partial \bar{W}}{\partial y} + \bar{W} \frac{\partial \bar{W}}{\partial z} \right] = - \frac{\partial \bar{P}}{\partial z} + \mu \nabla^2 \bar{W} - \rho \left[\frac{\partial \overline{uw}}{\partial x} + \frac{\partial \overline{vw}}{\partial y} + \frac{\partial \overline{w^2}}{\partial z} \right]. \tag{1.3c}$$

In these equations, ∇^2 denotes Laplace's operator $(\partial^2/\partial x^2 + \partial^2/\partial y^2 + \partial^2/\partial z^2)$. In deriving these equations, each quantity was separated into its time-mean and fluctuating parts; for example $U = \bar{U} + u$. The overbars denote the time-mean value. This notation is retained throughout this book unless otherwise stated. However, to explain the basic principles of turbulence modelling, the time-mean governing equations can also be expressed in tensor format, with the continuity equation given by

$$\frac{\partial \bar{U}_j}{\partial x_j} = 0 \tag{1.4}$$

and the momentum (Navier–Stokes) equation given by

$$\rho \bar{U}_j \frac{\partial \bar{U}_i}{\partial x_j} = - \frac{\partial \bar{P}}{\partial x_i} + \frac{\partial}{\partial x_j} \left[\mu \frac{\partial \bar{U}_i}{\partial x_j} - \rho \overline{u_i u_j} \right] \tag{1.5a}$$

which can also be expressed as

$$\rho \frac{\partial \bar{U}_j \bar{U}_i}{\partial x_j} = - \frac{\partial \bar{P}}{\partial x_i} + \frac{\partial}{\partial x_j} \left[\mu \left(\frac{\partial \bar{U}_i}{\partial x_j} + \frac{\partial \bar{U}_j}{\partial x_i} \right) - \rho \overline{u_i u_j} \right]. \tag{1.5b}$$

This formulation facilitates the explanation of the principles and the assumptions made in some of the simpler turbulence models (see Section 1.6).

The measured flow-field data can be classified according to their use in *CFD* work. Considering the general case of both heat and mass transfers and denoting the time-mean and fluctuating fluid temperature respectively, as $\bar{T}_a$ and θ we have:

(1) mean field variables, $\bar{U}_i$, $\bar{T}_a$, $\bar{P}$;

(2) turbulent second moments appearing in the time-mean Navier–Stokes and energy equations, $\overline{u_i u_j}$, $\overline{\theta u_i}$, $\overline{\theta^2}$;

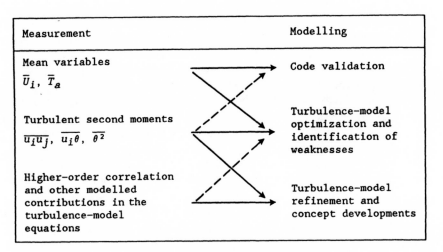

Fig. 1.2. Use of measured data in *CFD*.

(3) higher-order correlations and other modelled contributions, which occur in turbulence-model equations;

(4) turbulent quantities which can advance the understanding of the turbulent flow structure (dissipation of turbulence energy, intermittency, conditioned variables, pressure–velocity correlations, etc.).

The use of these types of flow-field data is illustrated in Fig. 1.2.

For isothermal flow, the direct output of any *CFD* analysis is the mean-velocity components at discrete points throughout the flow region investigated. Therefore, as shown in Fig. 1.2, to validate *CFD* code in terms of its accuracy and its range of applicability, the primary requirement is accurate *mean* velocity and pressure data. A *CFD* code can, within its range of application, be used for design purposes as a fast, reliable, interpolation tool. As demonstrated in this book, mean-velocity information can be obtained by a number of different *HWA* methods. The effect of a turbulence model within a *CFD* code is often observed as a loss in the mean pressure, $\bar{P}$, between the flow inlet and the exit. To optimize a turbulence model contained within a *CFD* code, measurements of both the mean-velocity components, $\bar{U}_i$, and the Reynolds stresses, $-\rho \overline{u_i u_j}$, are required, as shown in Fig. 1.2, and as discussed in Section 1.6. Type 3 and 4 measurements are used primarily for model and concept developments.

1.5 MEAN VELOCITY MEASUREMENTS

If *only* mean-velocity information is required then pressure probes are often used, provided that the turbulence intensity is so low that flow reversal does not take place. For flow fields with a known mean-flow direction, the mean velocity is usually measured with a conventional pitot-static probe. A review of pressure probes for fluid measurements has been given by Chue (1975).

Five-hole probes have been used extensively in flow fields containing three mean-velocity components. These probes are usually made up of an aerodynamically shaped body with a central 'total pressure' tapping and four symmetrically arranged 'static' pressure tappings. The probes are employed in either a stationary or a nulling mode. In the nulling mode, the probe is rotated so that each side port reads the same pressure. The corresponding probe orientation identifies the mean-flow direction. However, this method is cumbersome, and most pressure probes are employed in the stationary mode, in which the top-to-bottom and side-to-side pressure differences are measured. Calibration functions are then used to find the flow direction and the magnitude of the mean velocity. Many different geometries have been investigated. Studies of spherical pressure probes have been reported by Morrison *et al.* (1967), Nowack (1970), Judd (1975), and Dixon (1978), and a conically shaped probe was reported by Huffman *et al.* (1980). The probe used and studied most extensively is the five-tube pressure probe (Bryer and Pankhurst 1971; Treaster and Yocum 1979; Samet and Einav 1984; and Ligrani *et al.* (1989a). Miniature probes of this type, as shown in Fig. 1.3, have been manufactured and tested in order to reduce spatial-resolution errors and blockage effects (Richard and Johnson 1988, and Ligrani *et al.* 1989b). The use of a related four-hole probe has been proposed by Shepherd (1981) and Sitaram and Treaster (1985). Zilliac

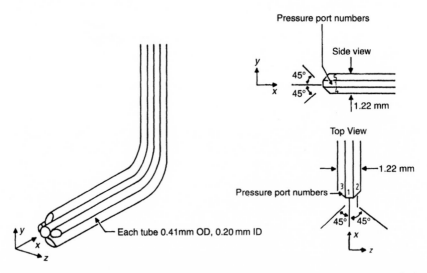

Fig. 1.3. A miniature five-hole pressure probe for measurement of the three mean-velocity components in low-speed flow. (From Ligrani *et al.* 1989a.)

(1993) has proposed the use of a seven-hole probe, the advantage being that the useful flow-angle range of the probe is greatly extended over the range of a five-hole probe with a similar geometry, but the calibration complexity is increased. The pressure readings are influenced by the turbulent-flow structure (Becker and Brown 1974; Christiansen and Bradshaw 1981; and Cho and Becker 1985), but most comparative studies have found that *HWA* and pressure probes give a similar accuracy for the measured mean-velocity vector.

Some flow fields have large variations in the mean-flow direction. A pressure probe may be used initially to identify this direction at each measuring point, followed by measurements with an aligned hot-wire probe (Johnston 1970). However, for most *HWA* studies involving measurements of turbulent quantities, the use of an additional pressure probe will add to the complexity of the experimental procedure without any gain in the accuracy of the measured quantities.

1.6 SOME SIMPLE TURBULENCE MODELS

Equation (1.5) shows that in order to obtain solutions for the mean-velocity field it is necessary to provide information about the Reynolds stresses, $-\rho\overline{u_i u_j}$. Consequently, many turbulent flow studies have investigated the relationships between the Reynolds stresses and the mean-velocity field under different flow conditions. The development of useful mathematical approximations for these relationships is known as turbulence modelling. Most models are of the following types:

1. *Turbulent-viscosity models* In these models the Reynolds stresses are related, by a Boussinesq-type hypothesis, to the mean-velocity field.

2. *Reynolds-stress models* These models contain mathematical approximations to the transport equations for the Reynolds stresses, $-\rho\overline{u_i u_j}$.

3. *Large-eddy simulation* The solution procedure is separated into an instantaneous motion for the large-scale structure and a statistical sub-grid model for the small-scale turbulence.

1.6.1 Turbulent-viscosity, μ_t, models

In turbulent-viscosity models it is assumed that the Reynolds stresses, $-\rho\overline{u_i u_j}$, are related to the mean rate of strain, $S_{ij} = \frac{1}{2}[\partial\bar{U}_i/\partial x_j + \partial\bar{U}_j/\partial x_i]$ of the mean-velocity field. This is similar to the relationship for the viscous stresses, and it is known as the Boussinesq hypothesis. Introducing a turbulent viscosity, μ_t, the Reynolds stresses can be expressed as

$$-\rho\overline{u_i u_j} + \frac{2}{3}\rho k \delta_{ij} = \mu_t\left(\frac{\partial\bar{U}_i}{\partial x_j} + \frac{\partial\bar{U}_j}{\partial x_i}\right), \tag{1.6}$$

where $k = \frac{1}{2}\overline{u_i u_i}$ denotes the turbulent kinetic energy and δ_{ij} is the Kronecker-delta tensor.

From dimensional analysis, μ_t may expressed as

$$\mu_t = \rho \ell_t V_t, \tag{1.7}$$

where ℓ_t and V_t are suitable length and velocity scales for the turbulence.

1.6.1.1   Zero-equation (algebraic) models

The zero-equation models were developed and applied to thin shear layers, including wakes, jets, and ducted flows, where the strain field is dominated by a single shear strain. Using the notation given by eqns (1.2) and (1.3), and aligning the x-coordinate with the mean-flow direction, $\bar{U}$, the only significant velocity gradient will be $\partial \bar{U}/\partial y$, and of the Reynolds stresses only the term $-\rho \overline{uv}$ needs to be considered. Equations (1.2) and (1.3) can therefore be reduced to

$$\frac{\partial \bar{U}}{\partial x} + \frac{\partial \bar{V}}{\partial y} = 0 \tag{1.8}$$

$$\rho \left(\bar{U} \frac{\partial \bar{U}}{\partial x} + \bar{V} \frac{\partial \bar{U}}{\partial y} \right) = -\frac{d\bar{P}}{dx} + \frac{\partial}{\partial y} \left(\mu \frac{\partial \bar{U}}{\partial y} - \rho \overline{uv} \right). \tag{1.9}$$

In a duct flow the complete flow field is usually computed by a thin-shear solver. In an (external) boundary-layer flow the pressure gradient must be prescribed or is given by the Bernoulli relationship

$$-\frac{d\bar{P}}{dx} = \rho U_\infty \frac{dU_\infty}{dx}$$

and a known variation in $U_\infty(x)$. If $U_\infty(x)$ is not known (say when the boundary layer is in a large inviscid stream which accelerates or decelerates in an unknown manner) then one would opt for a viscous/inviscid inter-action calculation; this is often done in aerofoil flows.

The first turbulence model developed for thin shear layers is known as the mixing-length model (Prandtl 1925). Setting ℓ_t in eqn (1.7) equal to a mixing length, ℓ_m, and expressing the corresponding turbulent velocity scale, V_t, as

$$V_t = \ell_m \left| \frac{\partial \bar{U}}{\partial y} \right| \tag{1.10}$$

the relationship for $-\rho \overline{uv}$ becomes

$$-\rho \overline{uv} = \rho \ell_m^2 \left| \frac{\partial \bar{U}}{\partial y} \right| \frac{\partial \bar{U}}{\partial y}. \tag{1.11}$$

Provided that an algebraic equation is given for the spatial distribution of ℓ_m, then a solution can be obtained from eqns (1.8) and (1.9) for the mean-velocity field in terms of $\bar{U}$ and $\bar{V}$. To optimize the turbulence model corresponding to eqn (1.11), it is necessary to measure the spatial distribution of $\bar{U}$ and $\overline{uv}$ in order to determine the most appropriate algebraic equation for ℓ_m.

The mixing-length model is relatively simple to use since the specification of the Reynolds stress, $-\rho\overline{uv}$, only involves an algebraic equation and not the solution of an additional differential equation. This method permits realistic predictions of the mean-velocity and shear-stress distribution in simple boundary-layer flows. However, it takes no account of convection and diffusion of turbulence and it is not suitable for flows containing recirculating flow regions.

1.6.1.2 One-equation models

In one-equation models the turbulence length scale, ℓ_t, in eqn (1.7) is set equal to the mixing length, ℓ_m, (or a closely related length scale), which is prescribed by an algebraic equation. However, as proposed independently by Kolmogorov (1942) and Prandtl (1945) the reference velocity for the turbulence is usually selected as $V_t \simeq k^{1/2}$, where k is the turbulent kinetic energy, and the expression for the turbulent viscosity becomes

$$\mu_t = C_{\mu 1}\ell_m k^{0.5}. \tag{1.12}$$

The turbulent kinetic energy, k, is determined from a transport equation derived from the Navier–Stokes equations. Using tensor notation, the exact equation can be expressed as (Hinze 1959):

$$\frac{Dk}{Dt} = -\frac{\partial}{\partial x_j}\overline{u_j\left(\frac{p}{\rho} + \frac{u_iu_i}{2}\right)} - \overline{u_iu_j}\frac{\partial \bar{U}_i}{\partial x_j} + v\frac{\partial}{\partial x_j}\overline{u_i\left(\frac{\partial u_j}{\partial x_i} + \frac{\partial u_i}{\partial x_j}\right)}$$
$$\quad\text{I}\qquad\qquad\quad\text{II}\qquad\qquad\qquad\text{III}\qquad\qquad\qquad\text{IV}$$

$$-v\overline{\left(\frac{\partial u_j}{\partial x_i} + \frac{\partial u_i}{\partial x_j}\right)\frac{\partial u_i}{\partial x_j}}. \tag{1.13}$$
$$\quad\text{V}$$

This is the *turbulence-energy equation*, which states that the rate of change in the kinetic energy of turbulence per unit of mass of the fluid (I) is equal to the convective diffusion by turbulence of the total turbulence energy (II), plus the energy transferred from the mean motion through the turbulence shear stresses (III), or the production of turbulence energy, plus the work done per unit of mass and of time by the viscous shear stresses of the turbulent motion (IV), plus the dissipation per unit of mass by the turbulent motion (V). This equation can be rewritten in the form

$$\frac{Dk}{Dt} = -\frac{\partial}{\partial x_j} \overline{u_j \left(\frac{p}{\rho} + \frac{u_i u_i}{2} \right)} - \overline{u_i u_j} \frac{\partial \bar{U}_i}{\partial x_j} + v \frac{\overline{\partial^2 k}}{\partial x_j \partial x_j} - v \overline{\frac{\partial u_i}{\partial x_j} \frac{\partial u_i}{\partial x_j}}. \qquad (1.14)$$

Denoting the viscous dissipation of the turbulent motion per unit mass by ε, and assuming that the turbulence is locally isotropic at the small scales of turbulence, we have

$$\varepsilon = v \overline{\left(\frac{\partial u_j}{\partial x_i} + \frac{\partial u_i}{\partial x_j} \right) \frac{\partial u_i}{\partial x_j}} = v \overline{\frac{\partial u_i}{\partial x_j} \frac{\partial u_i}{\partial x_j}}. \qquad (1.15)$$

For a plane two-dimensional boundary-layer flow at a high Reynolds number, so that viscous diffusion of turbulent energy can be ignored, eqn (1.14) can be expressed as

$$\frac{Dk}{Dt} = -\overline{uv} \frac{\partial \bar{U}}{\partial y} - \frac{\partial}{\partial y} \left(\frac{\overline{v^3 + vu^2 + vw^2}}{2} + \frac{\overline{pv}}{\rho} \right) - \varepsilon, \qquad (1.16)$$

with ε given by the last expression in eqn (1.15). For the purpose of turbulence modelling, each term in this equation must be expressed in terms of known or determinable quantities. As reported by Launder and Spalding (1972), eqn (1.16) may be expressed in the following approximate form

$$\rho \frac{Dk}{Dt} = \mu_t \left(\frac{\partial \bar{U}}{\partial y} \right)^2 + \frac{\partial}{\partial y} \left(\frac{\mu_t}{\sigma_k} \frac{\partial k}{\partial y} \right) - C_D \frac{\rho k^{3/2}}{\ell_m}, \qquad (1.17)$$

where σ_k and C_D are coefficients determined from calibration by reference to experimental data from simple flow configurations. If the flow contains recirculating flow regions then additional terms will appear in the equation.

Optimization or improvement of a turbulence model based on eqns (1.14) or (1.16) requires, in principle, accurate measurements of the spatial distribution of the mean velocity, $\bar{U}$, the shear stress, $-\rho\overline{uv}$, the turbulent kinetic energy, k, third-order velocity correlations (such as $\overline{v^3}$, $\overline{vu^2}$, $\overline{vw^2}$), the pressure–velocity correlation, $\overline{pv}$, and the dissipation rate, ε. Some of these terms may be difficult to measure directly, and ε is often estimated from an energy balance.

It is generally accepted that $k^{1/2}$ provides a better representation of V_t than $\ell_m |\partial \bar{U}/\partial y|$. However, because ℓ_m is still prescribed by an algebraic equation, the results obtained by the use of a one-equation model are usually not much better than Prandtl's mixing-length model.

1.6.1.3 Two-equation models

These models utilize two transport equations to determine μ_t. The turbulent reference velocity, V_t, is usually selected as $k^{1/2}$ as in the one-equation models. In a number of investigations, the reference length scale ℓ_t has been defined in terms of $z = k^m \ell_t^n$ (m and n constants) with z being a quantity for which an exact equation can be derived from the Navier–Stokes equations. The earliest proposal was that of Kolmogorov (1942), who chose the quantity $k^{1/2}/\ell_t$, for z, which is proportional to the mean frequency of the most energetic motion. However, the dissipation rate, ε, has been favoured by most workers using the two-equation approach. There are specific reasons for selecting this approach: an exact equation can be derived for ε with relative ease, and ε appears directly as an unknown quantity in the transport equation, eqn (1.14), for k. Another reason for favouring ε is that this variable gives rise to an equation with the minimum number of terms; specifically, wall-related fragments are absent (Leschziner, personal communication).

Davidov (1961) proposed the use of a transport equation for ε. An exact equation for ε can be obtained by taking the derivative of the Navier–Stokes equation, eqn (1.5), with respect to x_j, multiplying each side by $2\nu \partial u_i / \partial x_j$ and averaging (Launder 1984). For a constant density, steady flow at high Reynolds numbers where local isotropy is assumed to prevail for the small-scale turbulence, Launder *et al.* (1975) have shown that the following form of the equation for ε is appropriate

$$\frac{D\varepsilon}{Dt} = -\frac{\partial}{\partial x_k}\left[\overline{\nu u_k \left(\frac{\partial u_i}{\partial x_\ell}\right)^2} + \frac{\nu}{\rho}\overline{\frac{\partial p}{\partial x_i}\frac{\partial u_k}{\partial x_i}}\right] - 2\overline{\left(\nu \frac{\partial^2 u_i}{\partial x_k \partial x_\ell}\right)^2} - 2\nu \overline{\frac{\partial u_i}{\partial x_k}\frac{\partial u_i}{\partial x_\ell}\frac{\partial u_k}{\partial x_\ell}}.$$

$$\text{I} \qquad\qquad\qquad \text{II} \qquad\qquad\qquad \text{III} \qquad\qquad \text{IV}$$

$$(1.18)$$

This ε-equation contains terms representing convection by the mean motion (I), diffusion by the turbulent motion (II), and destruction of fluctuating vorticity by viscous forces (III). In addition, this equation contains a term (term IV) representing the generation of vorticity due to vortex stretching connected with the energy cascade. For fully developed high Reynolds number turbulence, the transport equations for k and ε, for two- or three-dimensional flow calculations, can be modelled and expressed in tensor notation (Launder and Spalding 1974) as

$$\frac{Dk}{Dt} = \frac{1}{\rho}\frac{\partial}{\partial x_j}\left(\frac{\mu_t}{\sigma_k}\frac{\partial k}{\partial x_j}\right) + \frac{\mu_t}{\rho}\left(\frac{\partial \bar{U}_i}{\partial x_j} + \frac{\partial \bar{U}_j}{\partial x_i}\right)\frac{\partial \bar{U}_i}{\partial x_j} - \varepsilon \qquad (1.19)$$

$$\frac{D\varepsilon}{Dt} = \frac{1}{\rho}\frac{\partial}{\partial x_j}\left(\frac{\mu_t}{\sigma_\varepsilon}\frac{\partial \varepsilon}{\partial x_j}\right) + C_{\varepsilon 1}\frac{\mu_t}{\rho}\frac{\varepsilon}{k}\left(\frac{\partial \bar{U}_i}{\partial x_j} + \frac{\partial \bar{U}_j}{\partial x_i}\right)\frac{\partial \bar{U}_i}{\partial x_j} - \frac{C_{\varepsilon 2}\varepsilon^2}{k}, \qquad (1.20)$$

and the turbulent viscosity, μ_t, can be expressed as

$$\mu_t = \frac{C_\mu \rho k^2}{\varepsilon}. \qquad (1.21)$$

Here, σ_k, σ_ε, $C_{\varepsilon1}$, $C_{\varepsilon2}$ and C_μ are calibration coefficients determined by reference to simple flow configurations (for example grid turbulence, equilibrium wall shear layer, etc.) The models are applied to such flows and the coefficients are determined to make the model simulate the experimental behaviour. The coefficients are kept invariant once calibrated. The values of these coefficients, as recommended by Launder and Spalding (1974), are given in the following table.

C_μ	σ_k	σ_ε	$C_{\varepsilon1}$	$C_{\varepsilon2}$
0.09	1.0	1.3	1.44	1.92

It is now common practice to compare the measured and computed values of the turbulent kinetic energy, k, and the turbulence length scales. Measured integral length scales (see Section 2.4.2) are often assumed to be comparable to the mixing length, ℓ_m, discussed in Section 1.6.1.2. An expression for ℓ_m can be obtained by applying the k-ε model to an equilibrium, near-wall boundary layer.

Introducing eqn (1.21) into eqn (1.6) gives

$$-\rho\overline{uv} = \mu_t \frac{d\bar{U}}{dy} = \rho C_\mu \frac{k^2}{\varepsilon} \frac{d\bar{U}}{dy}, \tag{1.22}$$

and for equilibrium conditions eqn (1.16) reduces to

$$-\overline{uv} \frac{d\bar{U}}{dy} = \varepsilon. \tag{1.23}$$

Combining eqns (1.22) and (1.23) and solving for $-\overline{uv}$ gives

$$-\overline{uv} = C_\mu^{1/2} k. \tag{1.24}$$

The logarithmic velocity distribution in the turbulent part of the flow can be expressed as

$$\bar{U} = \frac{U_\tau}{\kappa} \ln\left(\frac{U_\tau y}{\nu}\right) + A,$$

where U_τ is the friction velocity. Introducing the mixing length $\ell_m = \kappa y$ we get

$$\frac{d\bar{U}}{dy} = \frac{U_\tau}{\kappa y} = \frac{(-\overline{uv})^{1/2}}{\ell_m}. \tag{1.25}$$

Finally, combining eqns (1.23), (1.24), and (1.25) and solving for ℓ_m gives

$$\ell_m = C_\mu^{3/4} \frac{k^{3/2}}{\varepsilon} \tag{1.26}$$

In the standard k-ε model, the viscous near-wall region was bridged by wall functions, where the velocity at the first grid node is linked to the wall shear stress by the logarithmic law of the wall.

Jones and Launder (1972, 1973) extended the original k-ε model to the low Reynolds number form, which allows calculations right up to a solid wall. There are now about 12 different low Reynolds number models, some are better than others in some circumstances. The more recent versions are described by Lam and Bremhorst (1981), Rodi and Scheuerer (1986), Nagano and Hishida (1987), Nagano and Kim (1988), and Nagano and Tagawa (1990). In general, the modified k- and ε-equations can, setting $v = \mu/\rho$ and $v_t = \mu_t/\rho$, be expressed as

$$\frac{Dk}{Dt} = \frac{\partial}{\partial x_j}\left\{\left(v + \frac{v_t}{\sigma_k}\right)\frac{\partial k}{\partial x_j}\right\} - \overline{u_i u_j}\frac{\partial \bar{U}_i}{\partial x_j} - \varepsilon + D, \tag{1.27}$$

$$\frac{D\varepsilon}{Dt} = \frac{\partial}{\partial x_j}\left\{\left(v + \frac{v_t}{\sigma_\varepsilon}\right)\frac{\partial \varepsilon}{\partial x_j}\right\} - C_{\varepsilon 1}f_1\frac{\varepsilon}{k}\overline{u_i u_j}\frac{\partial \bar{U}_i}{\partial x_j} - C_{\varepsilon 2}f_2\frac{\varepsilon^2}{k} + E. \tag{1.28}$$

The Reynolds stress, $-\rho\overline{u_i u_j}$, can be expressed by eqn (1.6) with

$$v_t = C_\mu f_\mu k^{1/2}\ell_t = \frac{C_\mu f_\mu k^2}{\varepsilon}, \tag{1.29}$$

where $\ell_t = k^{3/2}/\varepsilon$ is the eddy length scale.

To extend the k-ε model to low Reynolds numbers it should be noted that the empirical constants C_μ, $C_{\varepsilon 1}$, and $C_{\varepsilon 2}$ in eqns (1.28) and (1.29) have been multiplied by functions f_μ, f_1, f_2 which involve the molecular viscosity. Some of the proposed models also include additional terms, denoted by D and E. These terms are based on an analysis of the limiting behaviour of turbulence as the wall is approached (for example, it is known that k decays quadratically close to the wall and, based on a Taylor series expansion, ε will approach a constant value related to the gradient of k (Patel et al. 1981). However, recent direct simulations show that ε peaks very close to the wall (Leschziner, personal communication). Summaries of various proposed k-ε models are given by Patel et al. (1985) and Nagano and Hishida (1987).

The one- and two-equation models are related in that both are based on identical arguments of dimensional homogeneity between viscosity, length, and velocity scales. The principal difference between the two methods is that in the one-equation model ℓ_t is prescribed by an algebraic equation, while in the two-equation model the length scale, ℓ_t, is determined indirectly from transport equations for k and ε. The two-equation k-ε model is now widely used, and it can be applied to flows containing recirculating regions. However, this model is still a very simplified representation of turbulent flow processes, and the limitations of the eddy viscosity models show up as soon as the conditions of the computed flow depart from simple unidirectional shear (Leschziner 1990). For example, the skin-friction produced by a curved boundary layer, the spreading rate of a plane curved wall or a free-shear layer, or the structure of a weakly swirling round jet cannot be computed adequately by such models unless flow-specific *ad hoc* modifications are introduced (Rodi 1979; Rodi and Scheuerer 1983). These related defects are all rooted in the model's fundamental inability to account for selective amplification or attenuation of different Reynolds stresses by *curvature-related* strain components (Leschziner 1990).

1.6.2 Advanced turbulence modelling

The time-averaged Navier–Stokes equations contain the Reynolds stresses, $-\rho\overline{u_i u_j}$, and exact transport equations for $\overline{u_i u_j}$ can be derived as shown by, for example, Hinze (1975, p. 323). A substantial effort has been directed towards second-moment closures containing modelling of the individual terms in these transport equations. Comparing computational results for many complex flow types obtained by both k–ε and by second-moment closure methods, Leschziner (1990) concluded that significant improvements in the predictive accuracy can be obtained by the latter method in flows containing large recirculation regions and involving swirl. In all cases, such improvements appear to be rooted in the ability of the second-moment closure models to mimic the strong response of the turbulent structure to streamline curvature.

A discussion of complex turbulence models is beyond the scope of this book. An introduction to and a summary of the topic of turbulence modelling have been given by Launder and Spalding (1972, 1974), Bradshaw (1972), Launder *et al.* (1975), Reynolds (1976), Bradshaw *et al.* (1981), Lumley (1983), Rogallo and Moin (1984), Patel *et al.* (1985), Kutler (1985), Schumann and Friedrich (1987), Launder (1989), Leschziner (1990), and So *et al.* (1991).

To develop more realistic turbulence models it is necessary to carry out fundamental turbulence studies in order to obtain a better understanding of both the large- and fine-scale nature of turbulence and of their interactions. *HWA* will remain the principal tool for obtaining such information.

2

BASIC PRINCIPLES OF
HOT-WIRE ANEMOMETRY

This chapter describes the basic principles, which apply to most hot-wire anemometer (*HWA*) applications. The topics discussed include heat transfer, the sensitivity coefficients of the hot-wire signal, the frequency response of a sensor operated in the Constant-Current (*CC*) and Constant-Temperature (*CT*) modes, and the spatial resolution. These aspects contribute significantly to the performance of a *HWA* system. To give an adequate description of these topics the presentation is somewhat mathematical, and it is recommended that readers who are new to *HWA* read Chapter 3, Introduction to Velocity Measurements, first.

Hot-wire anemometry (*HWA*) is based on convective heat transfer from a heated sensing element. The most common sensor configurations are cylindrical hot-wires and hot-films deposited on cylindrical fibres. The theory presented in this chapter applies specifically to these two types of sensors. However, other types of hot-film probes are also used, and the theory can be modified to cover such sensor configurations. Many of the heat-transfer aspects are similar for both hot-wire and hot-film probes, and the term hot-wire will cover both of these probe types, unless otherwise stated. Features which apply specifically to either hot-wire sensors or to hot-film sensors are identified in the text.

2.1 HEAT TRANSFER

The heat transfer from a heated wire placed in a fluid flow depends on both the properties of the ambient fluid (density, ρ, viscosity, μ, thermal conductivity, k, and specific heat, c_p, etc.) and the parameters of the flow (velocity vector, V, fluid temperature, T_a, pressure, P, etc.).

2.1.1 Infinitely long wire element

Early heat-transfer relationships were expressed in terms of the nondimensional groups of Nusselt (Nu), Reynolds (Re), Prandtl (Pr), Grashof (Gr), and Mach (M):

$$\text{Nu} = \frac{hd}{k}, \qquad \text{Pr} = \frac{c_p\mu}{k}, \qquad \text{Re} = \frac{\rho U d}{\mu},$$

$$\text{Gr} = \frac{g\rho^2 d^3 \beta (T_w - T_a)}{\mu^2}, \qquad \text{M} = \frac{U}{a},$$

where h is the heat-transfer coefficient,
k is the thermal conductivity of the fluid,
μ is the molecular viscosity of the fluid,
c_p is the specific heat of the fluid at a constant pressure,
ρ is the density of the fluid,
g is the gravitational acceleration,
β is the volume coefficient of expansion,
T_w is the temperature of the heated wire,
T_a is the temperature of the ambient fluid
a is the speed of sound,
U is the flow velocity, and
d is the diameter of a cylindrical element.

Following the pioneering experimental and theoretical work by King (1914), the convective heat transfer is often expressed in the form

$$Nu = A + BRe^{1/2}, \qquad (2.1)$$

where A and B are empirical calibration constants for each fluid. Kramers (1946) considered the results of heat-transfer experiments for wires placed in air, water, and oil. Selecting the 'film' temperature $T_f = (T_w + T_a)/2$ as the reference temperature for the fluid properties μ, ρ, and k, he obtained satisfactory results in the ranges $0.01 < Re < 10\,000$ and $0.71 < Pr < 1000$ using

$$Nu = 0.42Pr^{0.2} + 0.57Pr^{0.33}Re^{0.50}. \qquad (2.2)$$

At high flow velocities in a gas (more than $100\,m\,s^{-1}$ in air), compressibility effects in the flow around the wire become significant, and the Mach number, M, and the specific heat, c_p, must be considered as variables. In low-density flows the relevant parameter is the Knudsen number, $Kn = \lambda/d$, where λ is the molecular mean free path. The Knudsen number is related to the Mach and Reynolds numbers by

$$Kn = \left(\frac{\gamma\pi}{2}\right)^{1/2}\frac{M}{Re},$$

where γ is the ratio of the specific heats. These flow types are considered in Chapter 9.

At very low velocities, natural convection from the hot-wire probe becomes important. Its effect depends on the value of the Grashof number, Gr, and Collis and Williams (1959) concluded from their experiments in air with hot-wire probes with large values of the length-to-diameter aspect ratio, ℓ/d, that the buoyance effect can be neglected when $Re > Gr^{1/3}$. The effect of natural convection on practical HWA is discussed in Section 4.8.

In the forced-convective heat-transfer range, Collis and Williams found, by evaluating the fluid properties at the film temperature, T_f, that their experimental data collapsed in the range $0.02 < \text{Re} < 44$ on the curve

$$\text{Nu}\left[\frac{T_f}{T_a}\right]^{-0.17} = 0.24 + 0.56\text{Re}^{0.45}. \tag{2.3}$$

This relationship contains a temperature-loading factor $(T_f/T_a)^{-0.17}$, in which the corresponding temperatures are absolute temperatures (K).

The fluid properties ρ, μ, and k are all temperature dependent, and the values of Nu and Re therefore depend on the reference temperature used, as discussed by Kovasznay (1950a), Grant and Kronauer (1962), Bradshaw (1971), Koch and Gartshore (1972), Morrison (1974), and Bruun (1975a). This heat-transfer aspect is described in Chapter 7.

2.1.2 Finite length hot-wire sensors

The heat transfer from a hot-wire probe containing a finite aspect ratio sensor (see Fig. 2.1) deviates from that of an infinitely long wire. The sensing element may extend to the prongs, as shown in Fig. 2.1, or the active part of the wire may be removed from the prongs by a plating technique, as described in Section 2.2.1. In comparison with the wire element, the prongs are massive, and the prong temperature, T_p, will therefore remain at a temperature close to the time-mean ambient fluid temperature, $\bar{T}_a$. Since the wire is operated at an elevated temperature, conductive heat transfer will take place towards the prongs, resulting in a temperature distribution within the wire element. This temperature distribution can be determined from the heat-rate balance equation for an incremental wire element, dx, (see Fig. 2.1):

$$d\dot{Q}_e = d\dot{Q}_{fc} + d\dot{Q}_c + d\dot{Q}_r + d\dot{Q}_s, \tag{2.4}$$

where $d\dot{Q}_e$ is the electrical heat-generation rate,
$d\dot{Q}_{fc}$ is the forced-convective heat-transfer rate,
$d\dot{Q}_c$ is the conductive heat-transfer rate,
$d\dot{Q}_r$ is the radiation heat-transfer rate, and
$d\dot{Q}_s$ is the heat storage rate.
The individual terms in eqn (2.4) can be expressed as: The heat-generation rate by an electrical current, I,

$$d\dot{Q}_e = \frac{I^2 \chi_w}{A_w}\, dx, \tag{2.5}$$

where χ_w is the electrical resistivity of the wire material at the local wire temperature, T_w, and A_w is the cross-sectional area of the wire. The forced-convection heat-transfer rate, $d\dot{Q}_{fc}$, to the fluid can be expressed in terms of the heat-transfer coefficient h, as

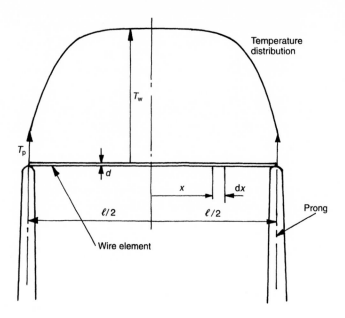

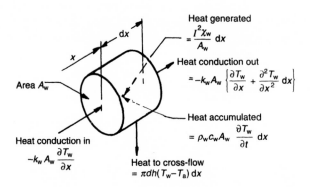

Fig. 2.1. The hot-wire geometry and heat balance for an incremental element.

$$d\dot{Q}_{fc} = \pi dh (T_w - T_a) dx. \tag{2.6}$$

Figure 2.1 shows that the total conduction heat-transfer rate out of the element is

$$d\dot{Q}_c = -k_w A_w \frac{\partial^2 T_w}{\partial x^2} dx, \tag{2.7}$$

where k_w is the thermal conductivity of the wire material at the temperature T_w. The radiation heat-transfer rate is

$$dQ_r = \pi \, d\sigma \, \varepsilon (T_w^4 - T_s^4) dx, \tag{2.8}$$

where σ is the Stefan–Boltzmann constant, ε is the emissivity of the sensor, and T_s is the temperature of the surroundings. In most HWA applications this term is very small, and it is omitted in Fig. 2.1 and in the following discussion. The heat-storage rate is

$$dQ_s = \rho_w c_w A_w \frac{\partial T_w}{\partial t} dx, \tag{2.9}$$

where ρ_w is the density of the wire material and c_w is the specific heat of the wire material per unit mass.

By inserting these relationships in eqn (2.4) the following equation can be obtained

$$k_w A_w \frac{\partial^2 T_w}{\partial x^2} + \frac{I^2 \chi_w}{A_w} - \pi \, dh (T_w - T_a) - \rho_w c_w A_w \frac{\partial T_w}{\partial t} = 0. \tag{2.10}$$

2.1.2.1 Steady-state temperature distribution in sensor elements

Under steady conditions $\partial T_w / \partial t = 0$. As will be shown in Section 2.1.2.2, χ_w can be expressed as $\chi_w = \chi_a + \chi_0 \alpha_0 (T_w - T_a)$, where χ_a and χ_0 are the values of the resistivity at the ambient fluid temperature, T_a, and at $0\,°C$, and α_0 is the temperature coefficient of resistivity at $0\,°C$. Equation (2.10) can therefore be rewritten as

$$k_w A_w \frac{d^2 T_w}{dx^2} + \left(\frac{I^2 \chi_0 \alpha_0}{A_w} - \pi \, dh \right) (T_w - T_a) + \frac{I^2 \chi_a}{A_w} = 0. \tag{2.11}$$

Since the fluid temperature, T_a, is constant along the wire, this equation is of the form

$$\frac{d^2 T_1}{dx^2} + K_1 T_1 + K_2 = 0,$$

where

$$T_1 = T_w - T_a$$

and

$$K_1 = \frac{I^2 \chi_0 \alpha_0}{k_w A_w^2} - \frac{\pi dh}{k_w A_w},$$

$$K_2 = \frac{I^2 \chi_a}{k_w A_w^2}.$$

Solutions to eqn (2.11) and related heat-transfer equations have been considered by a number of investigators including King (1914), Corrsin (1963),

Davies and Fisher (1964), and Champagne *et al.* (1967). In the study by Davies and Fisher (1964), a constant mean value was assumed for h to keep K_1 constant. Depending on the magnitude of h, the value of K_1 may be positive or negative. For most hot-wire applications, K_1 will be negative, and in this case the solution for a wire of length ℓ becomes

$$T_w = \frac{K_2}{|K_1|}\left[1 - \frac{\cosh(|K_1|^{1/2}x)}{\cosh(|K_1|^{1/2}\ell/2)}\right] + T_a. \tag{2.12}$$

The mean wire temperature, $T_{w,m}$ is obtained by integrating the temperature distribution $T_w(x)$ in eqn (2.12) along the length, ℓ, of the wire

$$T_{w,m} = \frac{K_2}{|K_1|}\left[1 - \frac{\tanh(|K_1|^{1/2}\ell/2)}{|K_1|^{1/2}\ell/2}\right] + T_a. \tag{2.13}$$

The conductive heat loss, $\dot{Q}_{cp}$, to the two prongs is $2k_w A_w |dT_w/dx|_{x=\ell/2}$, and the temperature gradient at the end of the wires is determined from eqn (2.12) as

$$\left|\frac{dT_w}{dx}\right|_{x=\ell/2} = \frac{K_2}{|K_1|^{1/2}}\tanh(|K_1|^{1/2}\ell/2). \tag{2.14}$$

The temperature distribution $T_w(x)$ given by eqn (2.12) can be compared with the temperature $T_{w,\infty}$ of a similar infinitely long wire also heated by a current, I. Using the heat-transfer analysis given in Section 2.1.2.3 it can be shown that $K_2/|K_1| = T_{w,\infty} - T_a$ and

$$\frac{1}{|K_1|^{1/2}} = \ell_c = \frac{d}{2}\left(\frac{k_w}{k}\frac{R_w}{R_a}\frac{1}{Nu}\right)^{1/2}, \tag{2.15}$$

where R_w and R_a are the resistances of the sensor at the temperatures T_w and T_a respectively. Equation (2.15) is the 'cold length', ℓ_c, introduced by Betchov (1948a, b) and discussed by Corrsin (1963). Equation (2.12) can therefore be expressed as

$$\frac{T_w - T_a}{T_{w,\infty} - T_a} = 1 - \frac{\cosh(x/\ell_c)}{\cosh(\ell/2\ell_c)}, \tag{2.16}$$

and similarly for the mean wire temperature, $T_{w,m}$,

$$\frac{T_{w,m} - T_a}{T_{w,\infty} - T_a} = 1 - \frac{\tanh(\ell/2\ell_c)}{\ell/2\ell_c}. \tag{2.17}$$

The variation in the temperature profile, in the form $(T_w - T_a)/(T_{w,m} - T_a)$, as a function of $\ell/2\ell_c$, is shown in Fig. 2.2. Hinze (1959) has shown that the relationship between the maximum temperature, $T_{w,max}$ (at $x = 0$), and the mean wire temperature, $T_{w,m}$, is given by

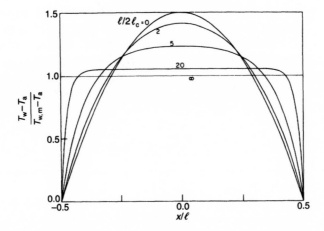

Fig. 2.2. The temperature distribution along a hot-wire for various values of $\ell/2\ell_c$. (From Freymuth 1979.)

$$\frac{T_{w,\,max} - T_a}{T_{w,\,m} - T_a} = \frac{(\ell/2\ell_c)\cosh(\ell/2\ell_c) - \ell/2\ell_c}{(\ell/2\ell_c)\cosh(\ell/2\ell_c) - \sinh(\ell/2\ell_c)}. \tag{2.18}$$

It can be observed from eqn (2.15) that the 'cold length,' ℓ_c, does not depend on the wire length, ℓ. For a $5\,\mu m$ tungsten wire operated at an overheat ratio R_w/R_a of 1.8, ℓ_c will be about $30d$. Standard hot-wire probes often have an active wire length of $\ell = 1.25\,mm$, giving a value for $\ell/2\ell_c$ of about 4. For such probes, eqn (2.18) predicts that the value of $(T_{w,\,max} - T_a)/(T_{w,\,m} - T_a)$ is 1.3.

For hot-wire anemometer applications it is usually advantageous to minimize the rate of the conductive end losses, $\dot{Q}_{cp}$, relative to the forced-convective heat-transfer rate, $\dot{Q}_{fc}$. Using eqns (2.14) and (2.15), $\dot{Q}_{cp}$ can be expressed as

$$\dot{Q}_{cp} = 2k_w A_w \frac{(T_{w,\,\infty} - T_a)}{\ell_c} \tanh(\ell/2\ell_c), \tag{2.19}$$

and the forced-convective heat-transfer rate from a wire element of length ℓ is obtained by integrating eqn (2.6)

$$\dot{Q}_{fc} = \int_{-\ell/2}^{\ell/2} \pi \, dh (T_w - T_a) dx.$$

Introducing eqn (2.16) for $T_w - T_a$ and assuming that h is constant, the equation for $\dot{Q}_{fc}$ can be expressed as

$$\dot{Q}_{fc} = \pi \, d\ell h (T_{w,\,\infty} - T_a)\left[1 - \frac{2\ell_c}{\ell}\tanh\frac{\ell}{2\ell_c}\right]. \tag{2.20}$$

It should be noted from eqn (2.19) that the conductive end losses only vary slowly with increasing wire length when $\ell > \sim 2\ell_c$. In contrast, the forced-convective heat-transfer, eqn (2.20), increases nearly linearly with ℓ. Therefore, to minimize the effect of the conductive end losses, the wire should be as long as possible and the wire material selected should have a low value of k_w. Values of k_w for common hot-wire materials are shown in Table 2.1. However, to satisfy spatial resolution criteria, as discussed in Section 2.4.2, a standard 4–5 μm diameter tungsten wire usually has an active wire length, ℓ, of about 1.25 mm. For such probes, the conductive end losses are about 15 per cent of the total heat transfer from the heated-wire element.

The larger diameters ($d = 25$–70 μm) of cylindrical hot-film probes make the corresponding values of ℓ/d much smaller (20–50), but as the thickness of the film element is very thin the end losses will be relatively small. As shown by Fingerson and Freymuth (1983), the effect of conduction losses to the prongs is similar for a tungsten wire and a film sensor whose ℓ/d ratio is approximately one-sixteenth that of the tungsten wire. However, the most important difference is the additional heat loss to the substrate on which the film is deposited. This heat-loss mechanism has a significant influence on the performance of hot-film probes under unsteady-flow conditions, as discussed in Section 2.3.2.2.

2.1.2.2   The resistance of sensor elements

The resistance, R, of a wire element at a uniform temperature is given by

$$R = \frac{\chi \ell}{A_w},   \quad (2.21)$$

where ℓ is the wire length, A_w is its cross-sectional area and χ is the resistivity defined as the resistance per unit length and per unit cross-sectional area. Typical values of χ at room temperature (20 °C) for common hot-wire materials are quoted in Table 2.1. As discussed by, for example, Sandborn (1972) and Bruun (1975a), measured values of χ for finely drawn wires may deviate substantially from the values quoted in International Standard Tables for materials in bulk. Manufacturers of hot-wire probes usually provide this information in the form of the sensor resistance at ambient room temperature (20 °C), R_{20}.

As described in Section 2.1.2.1, the temperature $T_w(x)$ of a heated wire varies along its length. Provided that $T_w(x)$ is known, the mean wire temperature, $T_{w,m}$ can be evaluated from

$$T_{w,m} = \frac{1}{\ell} \int_{-\ell/2}^{\ell/2} T_w(x)\,dx.   \quad (2.22)$$

Also, $T_{w,m}$ can be related to the corresponding hot resistance, R_w, of the

TABLE 2.1 *Physical properties of common hot-wire materials*

Material	Ultimate tensile strength (N cm^{-2})	Temperature coefficient of resistivity α_{20} (°C^{-1})	Resistivity χ_{20} ($\mu\Omega$ cm)	Thermal conductivity, k_w (W cm^{-1} °C^{-1})	Density, ρ_w (kg m^{-3})	Specific heat, c_w (kJ kg^{-1} °C^{-1})	Available as Wollaston wire?	Melting point (°C)	Comments
Tungsten	250 000	0.0036	5.5	1.9	19 300	0.14	No	3410	Oxidizes above 350 °C, cannot be soldered
Platinum	35 000	0.0038	9.8	0.7	21 500	0.13	Yes	1770	Soft and weak
Platinum–rhodium (90–10%)	70 000	0.0016	19	0.4	19 900	0.15	Yes	1830	Stronger than Pt
Platinum–iridium (80–20%)	140 000	0.0008	32	0.17	21 600	0.13	Yes	1840	Stronger than Pt

sensor by applying eqn (2.21) to an incremental wire element, and integrating over the wire length

$$R_w = \int_{-\ell/2}^{\ell/2} \frac{\chi_w}{A_w} \, dx. \tag{2.23}$$

The temperature dependence of the resistivity for hot-wire materials has been shown in many studies (for example, Laufer and McClellan 1956; Hinze 1959; Bruun 1975a) to be of the form

$$\chi_w = \chi_0[1 + \alpha_0(T_w - T_0) + \beta_0(T_w - T_0)^2], \tag{2.24}$$

where T_0 (0 °C) is the reference temperature for χ_0, α_0, and β_0. For an accurate evaluation of the mean wire temperature, $T_{w,m}$, it may be necessary to include the second-order term in eqn (2.24). However, for velocity measurements, a sufficiently accurate estimate of the mean wire temperature can be obtained by using the linear approximation of eqn (2.24):

$$\chi_w = \chi_0[1 + \alpha_0(T_w - T_0)]. \tag{2.25}$$

Inserting eqn (2.25) into eqn (2.23) gives

$$R_w = \int_{-\ell/2}^{\ell/2} \frac{\chi_0[1 + \alpha_0(T_w - T_0)]}{A_w} \, dx. \tag{2.26}$$

Integrating eqn (2.26) and *denoting, unless stated otherwise, the mean wire temperature* (see eqn 2.22) *by* T_w, we obtain

$$R_w = R_0[1 + \alpha_0(T_w - T_0)], \tag{2.27}$$

which provides the relationship between the hot resistance, R_w, and the mean temperature, T_w, of the wire element. Typical values for the temperature coefficient, α_{20} of common hot-wire materials are given in Table 2.1. In practical *HWA* applications the reference temperature is often selected as room temperature (20 °C). In this case, eqn (2.27) can be rewritten as

$$R_w = R_{20}[1 + \alpha_{20}(T_w - T_{20})], \tag{2.28}$$

and it should be noted that as the value of the reference temperature is changed so will the corresponding value of α, with

$$\alpha_{20} = \frac{R_0}{R_{20}} \alpha_0. \tag{2.29}$$

2.1.2.3   The heat-transfer relationship

Infinitely long wire A relatively simple heat-transfer relationship can be developed for an infinitely long wire, as the conductive end losses can be ignored. The heat balance for a wire segment, of length ℓ, can, using eqn (2.10), be written as

$$I^2 R_{w,\infty} = \pi\, dh\ell (T_{w,\infty} - T_a) = \pi\ell k (T_{w,\infty} - T_a)\mathrm{Nu}. \tag{2.30}$$

From eqn (2.27) it follows that

$$T_{w,\infty} - T_a = \frac{R_{w,\infty} - R_a}{\alpha_0 R_0}. \tag{2.31}$$

Introducing eqns (2.2) and (2.31) into eqn (2.30) gives

$$I^2 R_{w,\infty} = \frac{\pi\ell k}{\alpha_0} \frac{R_{w,\infty} - R_a}{R_0} (0.42\mathrm{Pr}^{0.2} + 0.57\mathrm{Pr}^{0.33}\mathrm{Re}^{0.50}). \tag{2.32}$$

For *HWA* applications this equation may be written in the form

$$\frac{I^2 R_{w,\infty}}{R_{w,\infty} - R_a} = A + BU^{0.5}, \tag{2.33}$$

where

$$A = 0.42\, \frac{\pi k\ell}{\alpha_0 R_0}\, \mathrm{Pr}^{0.20},$$

$$B = 0.57\, \frac{\pi k\ell}{\alpha_0 R_0}\, \mathrm{Pr}^{0.33} \left[\frac{\rho d}{\mu}\right]^{0.50}.$$

Finite length wire For a hot-wire probe with a finite-length active wire element, the conductive end losses must be taken into account. In practice this is often achieved by modifying eqn (2.33) as

$$\frac{I^2 R_w}{R_w - R_a} = A + BU^n, \tag{2.34}$$

in which R_w is the real hot resistance of the wire element. It is related to the real, mean, wire temperature, T_w, by

$$T_w - T_a = \frac{R_w - R_a}{\alpha_0 R_0}. \tag{2.35}$$

The exponent n introduced into eqn (2.34) also accounts for the fact that the square-root heat-transfer relationship is not very accurate (see Section 4.4.1). For any given *HWA* application, the values of A, B, and n can be determined by a suitable calibration procedure, as described in Chapter 4.

Hot-wire probes are used primarily for the measurement of velocity and temperature fluctuations, and the operational conditions and the sensor geometry are determined to a large extent by velocity- and temperature-sensitivity considerations. These are discussed in the following section for both the Constant-Temperature (*CT*) and Constant-Current (*CC*) hot-wire-anemometer modes. Another very important aspect is the frequency response of a hot-wire probe, when operated in both the *CC* and the *CT* modes, as discussed in Section 2.3.

2.1.2.4   Velocity and temperature sensitivities

Introducing the wire voltage $E_w = IR_w$ and using eqn (2.35) for the temperature difference $T_w - T_a$ in eqn (2.34) gives

$$\frac{E_w^2}{R_w} = (A + BU^n)(T_w - T_a),\tag{2.36a}$$

in which $\alpha_0 R_0$, has been included in the calibration constants A and B. For measurement purposes, the response equation is often expressed in terms of the anemometer output, E. For a balanced anemometer bridge, the relationship between E and E_w is

$$E = \frac{R_1 + R_L + R_w}{R_w} E_w,$$

where R_1 and R_L, (the probe and the cable resistance) are resistances defined in Figs 2.6 and 2.12. Expressing eqn (2.36a) in terms of E gives

$$\frac{E^2 R_w}{(R_1 + R_L + R_w)^2} = (A + BU^n)(T_w - T_a).\tag{2.36b}$$

CT mode In the CT mode the hot resistance, R_w, remains virtually constant, independent of the flow conditions. It is also often assumed that A, B, and n in eqn (2.36a) have constant values which are independent of the velocity and the temperature. The accuracy of these assumptions is discussed in Chapters 4 and 7.

The CT mode velocity and temperature sensitivities corresponding to eqn (2.36a) are (Elsner 1972; Bruun 1979a)

$$S_u = \frac{\partial E_w}{\partial U} = \frac{nBU^{n-1}}{2}\left[\frac{R_w(T_w - T_a)}{A + BU^n}\right]^{1/2},\tag{2.37a}$$

$$S_\theta = \frac{\partial E_w}{\partial \theta} = -\frac{1}{2}\left[\frac{R_w(A + BU^n)}{T_w - T_a}\right]^{1/2},\tag{2.37b}$$

where θ represents a small fluctuation in the fluid temperature, T_a. The fluctuating wire voltage signal, e_w, is related to the velocity, u, and temperature, θ, fluctuations by

$$e_w = S_u u + S_\theta \theta.\tag{2.38a}$$

For practical applications eqn (2.38a) is often expressed in terms of the fluctuating anemometer voltage, e, by

$$e = S_u u + S_\theta \theta,\tag{2.38b}$$

with the sensitivities defined as $S_u = \partial E/\partial U$ and $S_\theta = \partial E/\partial \theta$.

Typical values for S_u and S_θ (see eqns (2.37a–b)) are shown in Fig. 2.3

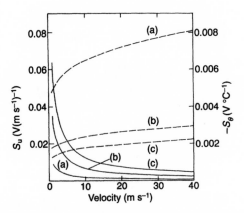

Fig. 2.3. The variation in (---) the velocity and (—) the temperature sensitivity with velocity and for temperature differences $T_w - T_a$ of: (a) 10 °C, (b) 100 °C and (c) 250 °C.

for a 2 mm long 5 μm diameter, tungsten hot-wire probe operated by a *CT* anemometer. For a given velocity U, the sensitivities S_u and S_θ can be seen to vary inversely with $T_w - T_a$, with the value of S_u increasing and the value of S_θ decreasing with increasing values of $T_w - T_a$. A high overheat ratio is therefore recommended for the measurement of velocity fluctuations. To obtain a high temperature sensitivity, a low overheat ratio must be used.

CC mode For temperature measurements, a hot-wire is often operated in the *CC* mode as a resistance thermometer using a very low overheat ratio. When a resistance-wire probe is placed in a non-isothermal turbulent flow, both R_a and R_w will vary with time, and the *CC* mode velocity and temperature sensitivities, corresponding to eqn (2.34), can be expressed in terms of the corresponding time-mean values $\bar{R}_a$ and $\bar{R}_w$ (see, for example, Hinze 1959; Wyngaard 1971a; Fulachier 1978)

$$S_{u,cc} = -\frac{(\bar{R}_w - \bar{R}_a)^2 nB\bar{U}^{n-1}}{I\bar{R}_a} = -\frac{nB\bar{U}^{n-1}I^3\bar{R}_w^2}{\bar{R}_a(A + B\bar{U}^n)^2}, \quad (2.39a)$$

$$S_{\theta,cc} = \frac{\alpha_0 I\bar{R}_w R_0}{\bar{R}_a}. \quad (2.39b)$$

The fluctuating voltage signal, e_w, from the resistance-wire is related to the temperature and velocity fluctuations by

$$e_w = S_{\theta,cc}\theta + S_{u,cc}u, \quad (2.40)$$

and to avoid contamination of the temperature signal by the velocity fluctuations the ratio

$$\frac{S_{u,\,cc}}{S_{\theta,\,cc}} = -\frac{nB\bar{U}^{n-1}\bar{R}_w I^2}{(A + BU^n)^2 R_0 \alpha_0} \qquad (2.41)$$

should be minimized by operating the resistance-wire with as low a current, I, as possible, retaining a sufficiently large signal-to-noise ratio of the *CC* anemometer system (LaRue *et al.* 1975; Mestayer and Chambaud 1979).

2.2 *HWA* PROBES

Several different heat-transfer probes have been developed for *HWA* applications. The most common are the hot-wire and hot-film probes.

2.2.1 Hot-wire probes

The primary role of *HWA* is as a research tool for turbulent flow studies. It is therefore important that the sensing element and its associated electronic circuit (see Section 2.3) should provide a virtually instantaneous response to the fastest occurring flow changes. Also, to obtain accurate turbulence measurements, the dimensions of the hot-wire sensor should not be much larger than the Kolmogorov length scale of the smallest eddies, as discussed in Section 2.4.2. For measurements of statistical quantities (such as time-mean velocity components, Reynolds stresses, etc.) standard probe sizes ($d \simeq 4$–$5\,\mu$m, $\ell = 1.25$ mm) will often give satisfactory results, since the spatial resolution error is mainly observed as an attenuation of the energy of the smallest eddies. However, accurate spectral measurements related to the dissipating eddies, the dissipation rate, ε, etc., require special techniques and much smaller probes.

A single-sensor hot-wire probe consists of a short length of a fine-diameter wire attached to two prongs which are usually made of stainless steel or nickel. The active part of the wire element may extend to the prongs, or it may be restricted to a central part by using plated wire ends near the prongs. For most practical *HWA* applications, the wire materials are tungsten, platinum or platinum alloys (usually platinum–rhodium 90–10 per cent or platinum–iridium (80–20 per cent). For very fine wires, platinum and its alloys are usually selected since it is available with diameters as small as $0.25\,\mu$m in the form of Wollaston wires. In the Wollaston process, a thin platinum rod is usually covered by a thick sheath of silver. The assembly is then drawn through a die to give a small outer diameter. By first soldering or welding the Wollaston wire to the prongs, the small, fragile, platinum wire can subsequently be exposed by etching the silver sheath from the wire. Wollaston wires are also available for platinum alloys.

There is no suitable sheathing material for tungsten, so the Wollaston process cannot be applied to tungsten wires. Instead a tungsten wire is first drawn to the minimum diameter possible (10–$12\,\mu$m). Final reduction is

obtained by electrical etching in an acid bath, producing wires with diameters as small as $2\,\mu$m. A spot-welding technique is required to attach the
tungsten wire to the prongs. A plating technique can be applied to remove
the active element from the prongs.

2.2.1.1 Applications of hot-wire probes

Hot-wire probes are normally used in gas flows due to their small size and
well-defined calibration characteristics. To obtain a good frequency response (see Section 2.3.2.1) most hot-wire sensors have diameters of $5\,\mu$m
or less. Such thin wires are very fragile, and the tensile strength of the
material is an important operational parameter. As shown in Table 2.1,
tungsten has a much higher tensile strength than platinum and its alloys.
Consequently, for velocity measurements in air and gas flows with temperatures less than about 150 °C, probes with tungsten wires are normally used.
Although tungsten has a high melting point ($\sim$3400 °C) it cannot be used
at high temperatures since it oxidizes at about 350 °C. It is therefore usually
operated at a mean wire temperature, T_w, of less than 250 °C. A platinum
coating is often applied to the tungsten wire element to provide oxidization
resistance and long-term stability. For measurements in high-temperature
flows (150 °C $< T_a <$ 750 °C), wires of platinum and its alloys are used.
Very fine platinum wires, operated in the 'cold-wire' mode are also used
for temperature measurements, as discussed in Chapter 7.

2.2.1.2 The overheat ratio

For most velocity measurements in air, it is not necessary to know the exact
value of the overheat ratio, R_w/R_a. The manufacturers of hot-wire probes
usually supply values of R_{20} and α_{20}. A simplified approach can be applied
to the setting of the hot resistance, R_w, when the flow temperature is close
to 20 °C. In this case, sufficiently accurate results may be obtained by
setting R_a equal to R_{20}, which is given by the manufacturers. For a
tungsten probe, the overheat ratio, R_w/R_a, is usually set to a value of less
than 2, with 1.8 being the recommended value. As described in Section
2.1.2.4, a high wire temperature should be selected in order to achieve a
high velocity sensitivity. However, to avoid oxidization it is essential that
the wire temperature at any point along the wire element is kept well below
350 °C. To illustrate these points, consider a 1.25 mm long, $5\,\mu$m diameter,
tungsten wire (DANTEC 55P01), for which typical values are $R_{20} = 3.5\,\Omega$
and $\alpha_{20} = 0.0036$ °C^{-1}. Using the approximation $R_a = R_{20}$ and selecting
an overheat ratio of 1.8, the operational resistance, R_w, of the wire element will be $1.8 \times 3.5\,\Omega = 6.3\,\Omega$. The corresponding mean wire temperature, T_w, can be evaluated from eqn (2.28) as being equal to about 230 °C,
and the maximum wire temperature, $T_{w,max}$, (see eqn (2.18)) as being

equal to about 290 °C, which is below the oxidization temperature. For more accurate applications, the use of home-made probes and flows with temperatures which deviate from 20 °C, it is necessary to determine R_a either indirectly from eqn (2.28) by using R_{20} and the temperature difference $T_a - T_{20}$ or by direct measurements. More importantly, for practical *HWA* applications it is necessary to take into account the probe and cable resistance, R_L, which includes:

(1) the lead resistance, R_p, of the probe, including the prongs but excluding the wire element itself (this resistance value is often supplied by the probe manufacturers);

(2) the resistances of the connection leads in the probe support, R_s, and of the probe cable, R_c, used to connect the probe to the anemometer.

The value of $R_s + R_c$ can be measured if the hot-wire probe is replaced by a shorting probe. Then, with $R_L = R_p + R_s + R_c$ and having selected an overheat ratio, R_w/R_a, the operational resistance of the anemometer can be set using the following equations for the resistance values measured at the end of the connecting cable

$$R_{a, m} = R_a + R_L, \qquad (2.42a)$$

$$R_{w, m} = R_w + R_L. \qquad (2.42b)$$

$R_{w, m}$ is the actual value to be set on the anemometer.

For the DANTEC 55P01 probe considered, $R_{20} \simeq 3.5\,\Omega$ and typically $R_L \simeq 1.6\,\Omega$. Consequently, $R_{a,m} \simeq 5.1\,\Omega$. With an overheat ratio of 1.8, $R_w = 6.3\,\Omega$, and the hot resistance set is $R_{w,m} = 6.3 + 1.6 = 7.9\,\Omega$. The consequences of ignoring R_L can be significant. If the value of R_L is assumed to be zero then the measured value $R_{a,m} = 5.1\,\Omega$ is interpreted as being equal to R_a. Having selected an overheat ratio of 1.8 and assuming that $R_w = R_{w,m}$, the hot resistance will be set to $R_{w,m} = 1.8 \times 5.1\,\Omega \simeq 9.2\,\Omega$. However, the true overheat ratio, R_w/R_a, will be $(9.2 - 1.6)/(5.1 - 1.6) \cong$ 2.15, corresponding to a mean wire temperature of 340 °C and a maximum wire temperature of about 425 °C. If this approach is used then the hot-wire probe will operate at a significantly higher real overheat ratio than expected, leading to a reduction in the probe life and possible wire burn out due to oxidization in the central region of the wire element.

2.2.2   Hot-film probes

Hot-film sensors are thin ($\sim 0.1\,\mu$m) platinum or nickel films, which are deposited on thermally insulating substrates (usually quartz). The most common substrate shapes are cylinders, wedges and cones, as shown in Fig. 2.4 (a–c). For surface measurements, hot-film sensors are available as flush-mounted probes, Fig. 2.4(d), or mounted on thin plastic foils which

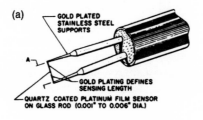

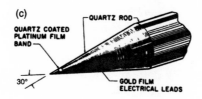

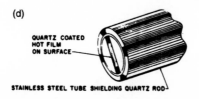

Fig. 2.4. Hot-film probe types: (a) cylindrical, (b) wedge, (c) cone, and (d) flush mounted. (Reprinted with the permission of TSI Inc.)

can be glued to the wall. The films are deposited by cathode sputtering to ensure a uniform thickness of the sensing element. A thick layer of conducting material is connected by sputtering to the ends of the film to supply the electrical heating current. The film is usually coated with a $1-2\,\mu m$ thick deposited layer of quartz or a similar insulating material. This coating protects the film material from abrasive particles and provides electrical insulation for hot-film probes used in liquids. For cylindrical hot-film probes, the active element is usually 25–70 μm in diameter and 1–2 mm long. The quartz fibre is gold-plated at the ends to remove the film element from the prongs (Fig. 2.4(a)). The size of the sensing element on the standard wedge and cone probes is usually 1–1.5 mm in the cross-stream direction and 0.1–0.2 mm in the stream-wise direction. The cold resistance of the film element is usually in the range 5–15 Ω.

2.2.2.1 Applications of hot-film probes

Air flows When accurate turbulence results are required in gas flows (including air), then hot-wire probes are normally used, as described in Section 2.2.1.1. However, there are applications where mean-velocity measurements and long-term calibration stability are the primary requirements. For such investigations, cylindrical hot-film probes can be considered. The related advantages and disadvantages are discussed in Section 4.10.1.

Liquid flows Hot-film probes are normally used for measurements in liquids due to their more sturdy construction. Also, in conducting liquids, such as water, probes with an impervious quartz coating must be used to electrically insulate the film element. Although it is possible in principle to coat wires, it is difficult to keep the insulation sufficiently thin and impervious.

The use of hot-film probes in water is complicated by several factors including probe fouling, bubble formation on film elements, and the sensitivity of the calibration relationship to small changes in the water temperature due to a low overheat ratio, R_w/R_a, of about 1.05–1.08. These aspects are discussed in Section 4.10.2.

2.3 MODES OF *HWA* OPERATION

There are two main modes of operating a hot-wire probe:

(1) the Constant-Current (*CC*) mode, in which the probe temperature varies,

(2) the Constant-Temperature (*CT*) mode, in which the probe resistance (and thereby its temperature) is kept virtually constant by varying the current.

2.3.1 Selection of anemometer types

The primary function of *HWA* is the measurement of fluctuating velocities and/or temperatures. For modern anemometers using low-drift integrated-circuit amplifiers, Freymuth (1968) has shown that, for the same operating conditions, both the *CC* and the *CT* anemometers are equivalent in their signal-to-noise ratios. The *CT* anemometer is much simpler to use than the

CC type, and most velocity measurements are now carried out with *CT* systems.

As discussed in Section 2.1.2.4, the temperature sensitivity increases when the overheat ratio is decreased and many measurements of fluctuating temperatures are carried out by operating a wire at a low overheat ratio as a resistance-wire by a *CC* anemometer, as discussed in Chapter 7.

In the selection and use of a *CC* or *CT* anemometer system it is necessary to consider the electronic noise, defined as the minimum detectable change that can be measured. The signal-to-noise ratio is important because:

1. Measurements of very low signal levels are often required.

2. Most turbulent flows have a significantly high frequency content and in both anemometer types the electronic compensation for the thermal lag of the sensor causes the electronic noise to increase as the frequency increases.

The signal-to-noise ratio of both *CC* and *CT* anemometers has been investigated by Kidron (1967), by Freymuth (1968, 1978*b*), and by Fingerson and Freymuth (1983). The electronic noise is primarily associated with the following aspects of the anemometer circuit: (i) the hot-wire or hot-film sensor, (ii) the resistors in the anemometer bridge, and (iii) the amplifier circuit. In *CT* anemometers, the amplifier is part of a feedback loop which compensates for the thermal lag of the hot-wire. To compensate a *CC* anemometer, the bridge output is often connected to an *RC* filter with a rising frequency response, and the resulting output signal is often boosted by an amplifier unit.

2.3.1.1 *The signal-to-noise ratio of hot-wire and hot-film sensors*

To maximize the signal-to-noise ratio, N_w, of a hot-wire probe, Fingerson and Freymuth (1983) concluded that one should: (i) operate the wire at a high overheat ratio, (ii) use a wire material with a high temperature coefficient of resistance, and (iii) use a thin wire to minimize its thermal capacity. For a $4\,\mu m$ diameter tungsten wire with a length $\ell = 1.25\,mm$, they found about 4% noise contribution when measuring 0.1% turbulence with a 0–50 kHz bandwidth. For measurement of very low turbulence intensities it is important to use a low-pass filter set to the maximum frequency of interest to minimize the electronic noise which primarily occurs at high frequencies. In such applications, the anemometer should still be tuned for the maximum frequency response.

For a hot-film probe, the larger diameter of the film sensor results in a decrease in the signal-to-noise ratio, N_f, between 10 and 10^3 Hz (Freymuth 1978*b*). This is illustrated in Fig. 2.5, which compares the signal-to-noise ratio for a $4\,\mu m$ diameter wire and a $50\,\mu m$ diameter film sensor. However,

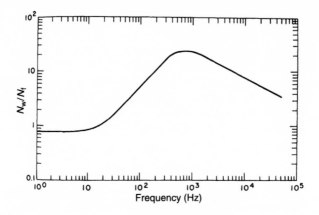

Fig. 2.5. The frequency dependence of N_w/N_f which compares the signal-to-noise ratios of hot-wire and hot-film anenometers. (From Freymuth 1978*b*.)

for most practical hot-film applications the larger noise value for the film at about 1 kHz is not a problem.

2.3.2 The constant-current mode

A typical *CC* circuit incorporating a Wheatstone bridge is shown in Fig. 2.6. Selecting, at a specified velocity, an overheat ratio R_w/R_a, the calculated value of R_w is first set by adjusting R_3 using the relationship

$$\frac{R_w + R_L}{R_1} = \frac{R_3}{R_2},\qquad(2.43)$$

which applies when the bridge is in balance, as observed with the galvanometer, G. The probe and cable resistance, R_L, is defined in Section

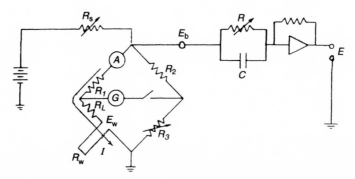

Fig. 2.6. A typical *CC* circuit incorporating a Wheatstone bridge and an *R–C* compensation circuit. (From Perry 1982.)

2.2.1.2. This condition is achieved by adjusting the resistance R_s, and the corresponding current, I, through the wire is measured by the ammeter, A. During calibration, the current, I, is kept constant for each velocity setting. The bridge is balanced by adjusting the resistances R_3 and R_s, and the corresponding value of R_w is determined from eqn (2.43). This procedure makes *CC* anemometers laborious to use. Knowing the value of I and assuming that $n = 0.5$, a least-squares curve-fitting technique can be applied to the calibration data to determine the calibration constants A and B in eqn (2.34). For measurement purposes, a first-order series-expansion technique can be applied to eqn (2.34), provided that the turbulent fluctuations are of small amplitude and low frequency (Hinze 1959). The corresponding response to large velocity fluctuations has been studied by Corrsin (1963) and Comte-Bellot and Schon (1969). The application of frequency compensation to the output signal from the Wheatstone bridge is described in the next section.

2.3.2.1 The frequency response of a hot-wire sensor

Due to the thermal inertia of the wire element, it will not respond instantaneously to changes in the flow condition when operated in the *CC* mode. The frequency response of the wire element itself can be estimated if it is assumed that it is long enough for the conductive end losses to be ignored, and that the temperature along the wire is uniform. The corresponding heat balance for the total wire element, of length ℓ can, using eqn (2.34) for the convective heat transfer, be written as

$$I^2 R_w = (R_w - R_a)(A + BU^n) + \rho_w c_w A_w \ell \frac{dT_w}{dt}. \tag{2.44}$$

By expressing T_w in terms of R_w this equation becomes

$$I^2 R_w = (R_w - R_a)(A + BU^n) + \frac{C_w}{\alpha_0 R_0} \frac{dR_w}{dt}, \tag{2.45}$$

where C_w $(= \rho_w c_w (\pi/4) d^2 \ell)$ is the thermal capacity of the wire element. Equation (2.45) can be expressed as

$$\frac{C_w}{\alpha_0 R_0 (A + BU^n - I^2)} \frac{dR_w}{dt} + R_w = \frac{(A + BU^n)}{(A + BU^n - I^2)} R_a. \tag{2.46}$$

This is a first-order equation, whose coefficients are not, in general, constant. The temporal variation of the forcing function $\varphi(t) = R_a(A + BU^n)/(A + BU^n - I^2)$ may be due to variations in the velocity, U, to fluctuations in the flow temperature, T_a, and thereby to fluctuations in R_a; or in a nonisothermal flow the temporal variation may be due to variations in both U and R_a. In all three cases, the response of R_w to these variations is characterized by the *same* single time constant, M,

$$M = \frac{C_w}{\alpha_0 R_0 (A + BU^n - I^2)}. \tag{2.47}$$

Using eqn (2.34) the equation for M can also be written as

$$M = \frac{C_w (R_w - R_a)}{\alpha_0 R_0 I^2 R_a}. \tag{2.48}$$

Equations (2.47) and (2.48) for M can be applied both to a standard hot-wire probe ($d = 4$–$5\,\mu m$) operated at a high overheat ratio and to a resistance-wire ($d \leqslant 1\,\mu m$) operated with a very low heating current, I. For both types of probes, eqn (2.47), shows that M is strongly dependent on U. If the fluctuations are small, and we apply a small-perturbation theory to eqn (2.46)

$$U = \bar{U} + u, \qquad R_w = \bar{R}_w + r_w, \qquad R_a = \bar{R}_a + r_a$$

then it can be transformed into a first-order differential equation with constant coefficients (Fulachier 1978). In this case, R_w and R_a should be replaced in eqn (2.48) by their corresponding time-mean values, $\bar{R}_w$ and $\bar{R}_a$, and the solution for the fluctuating resistance, r_w, will be of the form

$$r_w = \int_{-\infty}^{t} \exp\left(-\frac{t - \tau}{M}\right) \varphi(\tau)\, d\tau, \tag{2.49}$$

where $\varphi(\tau)$ is the forcing function. If, for example, a velocity step change, u_0, corresponding to a step function φ_0, is applied at $t = 0$ then the response $r_w(t)$ will be

$$r_w(t) = \begin{cases} 0, & t < 0 \\ \varphi_0(1 - e^{-t/M}), & t \geqslant 0, \end{cases} \tag{2.50}$$

as shown in Fig. 2.7. The resistance change is seen to lag the forcing function, and the time constant M is a measure of this lag.

The small-perturbation response of the hot-wire sensor in the frequency domain can be evaluated if it is assumed that u is a simple periodic function with an angular frequency, $\omega\ (= 2\pi f)$. The response characteristics are often specified in terms of the amplitude transfer function, $H(\omega)$, and the phase angle, ψ. As shown by, for example, Hinze (1959), the amplitude of the fluctuating hot-wire output signal is attenuated by a factor of $(1 + \omega^2 M^2)^{-1/2}$. The corresponding phase angle, ψ, is given by $\psi = \tan^{-1}(-\omega M)$. From these relationships, it follows that when $\omega = 1/M$ the amplitude is reduced by a factor $1/1.414$ (or -3 dB) and the corresponding phase angle is $-45°$.

The expression for M can be rewritten by introducing $C_w = \rho_w c_w (\pi/4) d^2 \ell$ and $R_0 = \chi_0 \ell/(d^2 \pi/4)$ in eqn (2.48):

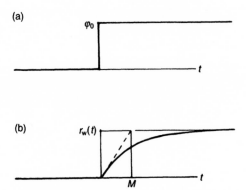

Fig. 2.7. The exponential rise in the resistance of a sensor (with a time constant M) when an instantaneous change in velocity is applied to a hot-wire sensor.

$$M = \left\{\frac{c_w \rho_w}{\chi_0 \alpha_0}\right\} \left\{\left[\frac{\pi}{4} d^2\right]^2\right\} \left\{\frac{\bar{R}_w/R_a - 1}{I^2}\right\}. \tag{2.51}$$

Corrsin (1963) has pointed out that these three properties represent the effects of the wire material, the geometry, and the operational conditions. It follows (Blackwelder 1981) that different wire materials may change the time constant by a factor of about 2, with platinum giving the smallest value. The time constant is seen to be independent of the length but strongly dependent on the wire diameter. However, for a fixed overheat ratio, if the diameter is reduced then the heating current must also be reduced. Corrsin (1963) has estimated that the combined effect of the geometry and operational groups is to cause the time constant to vary as $d^{3/2}$. As shown in Section 7.4.3, this diameter dependence is valid when $d > \sim 1\,\mu$m. Consequently, the most significant way of reducing the time constant is to use very fine wires. Because of the approximations used and the uncertainties in the values of the parameters, eqn (2.51) will only provide an estimate of the time constant. Hinze (1975) and Bremhorst *et al.* (1977) have presented improved-response evaluations, which include the effect of the conductive end losses. Their results do not deviate significantly from the simpler theory presented here.

   In practice, more reliable results can be obtained by direct testing. For a typical $5\,\mu$m diameter, $1.25\,$mm long, tungsten wire operated at a high overheat ratio the value of the time constant, M, is about $0.6\,$ms. The frequency response has been measured both by internal and external heating methods, and similar results have been reported for both types of technique. Kidron (1966) used a microwave simulator for controlled variation of the air-flow velocity. (A similar technique was also used by Yeh and Van Atta (1973) to measure the time constant of a $0.25\,\mu$m diameter platinum

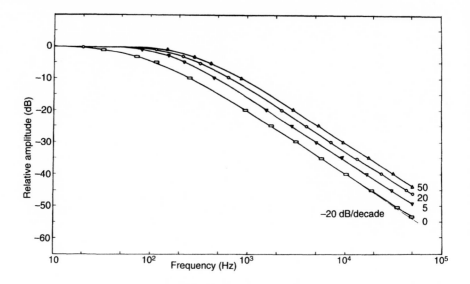

Fig. 2.8. Frequency response of a 5 μm diameter, tungsten hot-wire probe operated in the *CC* mode for various air-flow velocities (m s^{-1}). (From Kidron 1966.)

resistance-wire.) The transfer functions $H(f)$ measured by Kidron at different flow velocities are shown in Fig. 2.8. The results show that the value of *M* decreases with increasing velocity, as predicted by eqn (2.47). When the convective heat-transfer effect predominates, the response was found to approximate to that of a first-order system with a predicted $(1 + \omega^2 M^2)^{-1/2}$ (or -20 dB decade^{-1}) amplitude attenuation. However, a detailed examination of the $H(f)$ curve at zero velocity revealed that the system's response is not that of a first-order system, due to natural convective heat transfer and large conductive end losses. The flat part of the transfer functions only extends to about 100 Hz, which is far too low for turbulence measurements without frequency compensation. An extension of the applicable frequency range can be obtained by using a simple *RC* compensation technique for *CC* operation, as shown in Fig. 2.6. Dryden and Kuethe (1929) were the first to introduce electronic compensation for the thermal lag of the wire sensor, thereby laying the foundation for modern hot-wire anemometers as accurate instruments for turbulent flow studies. However, for velocity measurements, the *CC* compensation technique has now been superseded by the *CT* technique described in Section 2.3.3.

2.3.2.2 The frequency response of a cylindrical hot-film sensor

The evaluation of the frequency response of a hot-film is more complicated than for a hot-wire probe since the heat loss from the sensor includes conduction losses to the substrate. In this case there are two main heat-loss mechanisms from the film element: (1) convective heat loss to the surrounding fluid, and (2) conduction into the substrate and thereby indirectly to the fluid. Each of these heat loss mechanisms has its own time constants.

At low frequencies, sufficient time is available for the heat to be conducted directly through the substrate and into the fluid. A cylindrical probe therefore behaves as a hot-wire sensor in this frequency range. The time constant of the convective heat transfer can be evaluated by an analysis similar to that for the hot-wire probe (Blackwelder 1981). However, since the heat capacity of the substrate is much greater than the film it must be used to evaluate the time constant. For a cylindrical hot-film of thickness t mounted on a substrate of diameter d, the time constant due to convective heat losses to the surrounding fluid is

$$M = \left\{ \frac{c_s \rho_s}{\chi_0 \alpha_0} \right\} \left\{ \frac{\pi^2 d^3 t}{4} \right\} \left\{ \frac{\bar{R}_w / R_a - 1}{I^2} \right\}, \qquad (2.52)$$

where c_s and ρ_s are the specific heat and density of the substrate material. As the diameters of hot-film probes are much larger than for hot-wires, the frequency at which roll-off occurs will be correspondingly lower. The amplitude transfer function, $H(f)$, of a 0.5 mm-long platinum film on a 25 μm diameter quartz fibre in an air flow at various velocities is shown in Fig. 2.9, demonstrating the velocity dependence of $H(f)$. The roll-off frequency is less than 10 Hz, and the initial amplitude attenuation is similar to that of the hot-wire probe with a slope of -20 dB decade^{-1} between 50 and 300 Hz. However, above 300 Hz, the slope changes to -10 dB decade^{-1}. This is

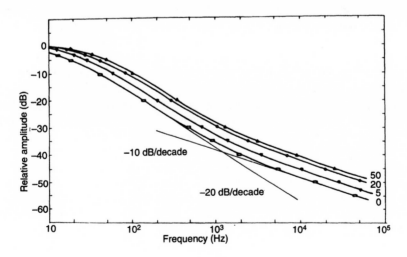

Fig. 2.9. The frequency response of a 25 μm diameter platinum hot-film probe operated in the *CC* mode for various flow velocities (m s^{-1}). (From Kidron 1966.)

caused by heat being conducted into the substrate and being stored during part of the cycle and then being returned to the heated film, when its temperature is lower than that of the substrate. As a consequence, the response is different from that of a hot-wire at higher frequencies. Due to the presence of massive substrates in cone and wedge probes, their frequency response deviates significantly from that of a cylindrical probe, as discussed in Section 2.3.3.2.

2.3.2.3   The frequency response of a resistance-wire probe

Temperature fluctuations are often measured by a thin platinum resistance-wire ($d \leqslant 1 \, \mu$m) operated at a low heating current, I, in the CC mode. This method was first proposed by Corrsin (1949), who calculated the time constant of the temperature-sensing wire and proposed the use of a compensation network similar to the one shown in Fig. 2.6 for CC anemometers. In the simple heat-transfer analysis presented in section 2.1.2.1 it was assumed that the prongs had a temperature equal to the ambient temperature. For isothermal flow this has been shown to be a good approximation. However, when the ambient fluid temperature, T_a, varies, then the time constant of the prongs, M_p, must be taken into account. Since the prongs are massive compared to the thin resistance-wire element, they will not be able to follow variations in T_a, except at very low frequencies, and the prong temperature, T_p, will remain close to the time-mean ambient fluid temperature, $\bar{T}_a$. This causes a dynamic conductive heat-transfer process between the wire and the prongs. In most theoretical analysis it is assumed that the response of the prongs to fluctuations in T_a is as a first-order system with a time constant M_p, which is much larger than the time constant M of the wire element itself (Fiedler 1978). The amplitude transfer function, $H_\theta(\omega)$, of a resistance-wire probe to temperature fluctuations will therefore be of the form shown in Fig. 2.10, which demonstrates the existence of both a low-frequency and a high-frequency attenuation (see, for example, Paranthoen et al. 1982b; Lecordier et al. 1984; Larsen et al. 1986; Tsuji et al. 1992). Fiedler (1978) and Paranthoen et al. (1982b) have also shown that the shape of $H_\theta(\omega)$ depends on the length of the Wollaston portion of the sensor element. Most resistance-wire probes and related theories correspond to a fully etched platinum wire.

The low-frequency attenuation phenomenon has been studied by a number of investigators including Maye (1970), Højstrup et al. (1976, 1977), Bremhorst (1978), Bremhorst and Gilmore (1978), Smits et al. (1978), Millon et al. (1978), Perry et al. (1979), Perry (1982), Paranthoen et al. (1982b), Paranthoen et al. (1984), Lecordier et al. (1984), Petit et al. (1985), and Larsen et al. (1986). Paranthoen et al. (1982b) have shown that an approximate value for the plateau level H_p (Fig. 2.10) can be obtained from

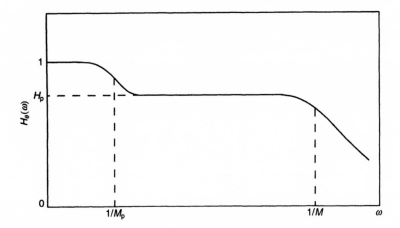

Fig. 2.10. The amplitude transfer function for temperature fluctuations, $H_\theta(\omega)$, of a resistance-wire probe. (From Lecordier *et al.* 1958.)

$$H_p = 1 - 2(\ell_c/\ell)\left(\frac{\eta^3 - 1}{\eta^2 - 1}\right), \tag{2.53}$$

where $\eta = \ell_b/\ell_c$, $\ell_c = (d/2)[k_w/(k\mathrm{Nu})]^{1/2}$ is the cold length (see eqn (2.15)), and ℓ_b is the thickness of the thermal boundary layer on the prongs. The effect of ℓ_b is usually only significant at low velocities.

To minimize the low-frequency amplitude attenuation, the value of ℓ/ℓ_c and therefore of ℓ/d should be as large as possible, as illustrated in Fig. 2.11. For practical purposes it is often possible to ignore the low-frequency attenuation when $\ell/d > \sim 1000$. Due to spatial-resolution criteria, a large value for ℓ/d is normally obtained by using very fine wires ($d \leqslant 1 \, \mu m$). This has also the benefit of decreasing the time constant M of the wire element itself. The principal drawback is that such wires are very fragile.

The analysis for the high-frequency attenuation caused by the thermal heat capacity of the wire element (see Section 2.3.2.1) can also be applied to the sensor element of a resistance-wire. A general discussion of the parameter dependence of the corresponding time constant M and its measurement is presented in Section 7.4.3.

2.3.3 The constant-temperature mode

There are major advantages in maintaining the hot-wire or hot-film at a constant operational temperature and thereby at a constant hot resistance, since the thermal inertia of the sensor element is automatically adjusted when the flow conditions vary. This mode of operation is achieved by incorporating a feedback differential amplifier into the *HWA* circuit to obtain

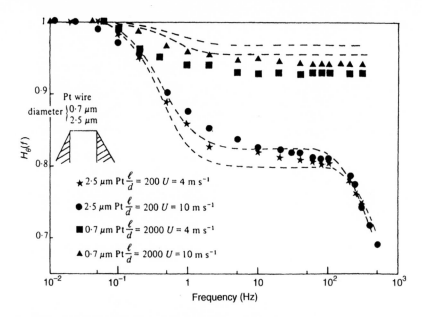

Fig. 2.11. The effect of the ℓ/d ratio on the transfer function of resistance-wires at low frequencies. (From Paranthoen *et al.* 1982*b*. Reproduced with the kind permission of Cambridge University Press from the *Journal of Fluid Mechanics*.)

a rapid variation in the heating current to compensate for instantaneous changes in the flow velocity.

The advantages of the *CT* operation were recognized at an early stage (Weske 1943; Ossofsky 1948). However, it was not until the mid 1960s that reliable *CT* anemometers became available with the development of highly stable, low-drift integrated-circuit amplifiers.

The principle of a *CT* circuit is illustrated in Fig. 2.12. The hot-wire probe is, as in the *CC* mode, placed in a Wheatstone bridge. As the flow conditions vary the error voltage $e_2 - e_1$ will be a measure of the corresponding change in the wire resistance. These two voltages form the input to the operational amplifier. The selected amplifier has an output current, i, which is inversely proportional to the resistance change of the hot-wire sensor. Feeding this current back to the top of the bridge will restore the sensor's resistance to its original value. Modern amplifiers have a very fast response, and in the *CT* mode the sensor can be maintained at a constant temperature except for very-high-frequency fluctuations. Various *CT* circuits have been reported in the literature (for example, Freymuth 1967; Wyngaard and Lumley 1967; Perry and Morrison 1971*a*; Perry 1982; Miller *et al.* 1987). Commercial high-quality *CT* anemometers are now also readily available. Figure 2.13 shows a computer-controlled *CT* anemometer and a nozzle-calibration facility.

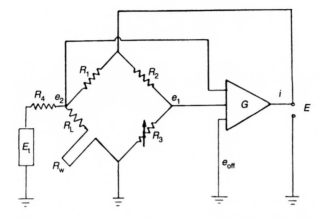

Fig. 2.12. A *CT* anemometer containing a Wheatstone bridge, a feedback amplifier, and an electronic-testing subcircuit.

Fig. 2.13. A computer-controlled *CT* anemometer and a nozzle calibration facility. (Reprinted with the permission of DANTEC Measurement Technology.)

Modern *CT* anemometers often contain a low-pass filter to decrease the electronic noise, a subcircuit for determining and setting the overheat ratio, a square-wave subcircuit for frequency testing and a minimum of two controls to optimize the frequency response to fast fluctuations. The most common controls are (Borgos 1980):

(1) cable trim, either a capacitor or inductor in the bridge, that is used to vary the bridge reactance;

(2) an offset voltage, e_{off}, in the amplifier (Fig. 2.12);

(3) the amplifier gain, G (Fig. 2.12);

(4) a stability circuit or a similar board with a limiting circuit in the amplifier.

The resistance ratio R_2/R_1 in Fig. 2.12 is called the bridge ratio. To effectively use the available current from the amplifier, the resistance on the right-hand-side (passive) of the bridge is normally larger than that on the left-hand-side which contains the hot-wire probe. The bridge ratio usually has a value between 5 and 20. For optimum frequency response and for the compensation of very long cables, a bridge ratio of 1:1 can be used.

The upper frequency limit of the *CC* anemometer is always higher than that of the corresponding uncompensated *CC* anemometer. The dynamic response is governed by the response equations for: (i) the bridge, (ii) the sensor element, and (3) the amplifier (Freymuth 1967). There are significant differences in the frequency responses obtained for hot-wire and hot-film probes, as discussed below.

2.3.3.1   The frequency response of a hot-wire probe operated in the CT mode

The frequency response of a hot-wire probe operated by a *CT* anemometer has been studied by many investigators including Grant and Kronauer (1962), Berger *et al.* (1963), Freymuth (1967, 1977*a*), Wyngaard and Lumley (1967), Wyngaard and Sheih (1968), Davis (1970), De Haan (1971), Perry and Morrison (1971*a*), Hinze (1975), Wood (1975), Freymuth and Fingerson (1977), Lu (1979), Smits and Perry (1980), Lord (1981, 1982), Perry (1982), and Sherlock (1984).

The response to large velocity fluctuations has been discussed by Freymuth (1969, 1977*b*, 1978*c*), and Comte-Bellot (1977). Fingerson and Freymuth (1983) concluded that errors due to nonlinearities in the dynamic response at large amplitudes are negligible for most measurement conditions. This is especially true if the upper frequency limit, f_c (-3 dB), of the *CT* anemometer system is maintained as large as possible so that the large amplitude fluctuations are at frequencies of less than 10% of f_c.

There are two different methods by which the dynamic response of the anemometer system can be determined. In the first, a small perturbation is added to the flow in which the probe is placed. The perturbation can be obtained by using, for example, a shock tube, an oscillatory flow, or an imposed temperature variation; alternatively, the probe itself may be shaken to create the velocity fluctuation. In the second approach an electronic disturbance signal, e_t, is applied to the anemometer, either at the top of the bridge or to the error signal e_2, as shown in Fig. 2.12. The electronic testing may take the form of either a square-wave or sine-wave test.

Comparisons of theory and experiments by Freymuth (1977a), Freymuth and Fingerson (1977), and Fingerson and Freymuth (1983) have demonstrated that the frequency response of a *CT* anemometer with two-parameter optimization can be accurately modelled by a third-order system. As explained by Freymuth (1977a), the third-order time constant plays a significant part in the high-frequency performance of a *CT* anemometer system. Other researchers (Perry and Morrison 1971a; Wood 1975; Blackwelder 1981) have explained the frequency response of a *CT* anemometer in terms of a second-order system, but this is consistent only with a one parameter system adjustment, such as a first-order damping. The differences in predicting the frequency optimization and the high-frequency performance are considerable. The high-frequency asymptotic amplitude response for a third-order system is f^{-3} whereas an f^{-2} response is predicted by a second-order system. The asymptotic phase angle is $-270°$ and not $-180°$; the dependence of the cut-off frequency, f_c, (-3 dB) on the overheat ratio $a = R_w/R_a$ is $(a-1)^{1/3}$ and not $(a-1)^{1/2}$. Also, for a sine-wave test, the high-frequency amplitude response is f^{-2} not f^{-1}, and the asymptotic phase angle is $-180°$ instead of $-90°$ (Freymuth, personal communication). A second order system approach may, however, be useful to illustrate the effect of system damping. As shown by Blackwelder (1981), the related response equation for the fluctuating bridge voltage e can be expressed as

$$\frac{d^2 e}{dt^2} + 2\zeta\omega_0 \frac{de}{dt} + \omega_0^2 e = \omega_0^2 \left\{ S_u u + S_t \left[M\frac{de_t}{dt} + \left(1 + \frac{2\bar{R}_w(\bar{R}_w - R_a)}{R_a(\bar{R}_w + R_1)}\right)e_t \right] \right\},$$
(2.54)

where ω_0 is the natural frequency of the electronic circuit, ζ is the damping coefficients, and S_u and S_t are the sensitivity coefficients of the velocity and electronic test signals. Assuming that the wire had an overheat ratio, R_w/R_a, of 1.5, the sensor had a time constant, M, of 0.4 ms, the amplifier had a time constant of 25 μs and that the value of the system gain parameter was 1250, Blackwelder (1981) evaluated the natural frequency $f_0 = (1/2\pi)\omega_0$ as 55 kHz. Figure 2.14 shows the related amplitude transfer functions for velocity fluctuations for values of ζ equal to 0.1, 1 and 10. The optimum response is seen to correspond to critical damping conditions ($\zeta \simeq 1$).

The upper frequency limit, f_c, of the amplitude transfer function, $H(f)$, for a *CT* system is velocity dependent. This is illustrated in Fig. 2.15 for

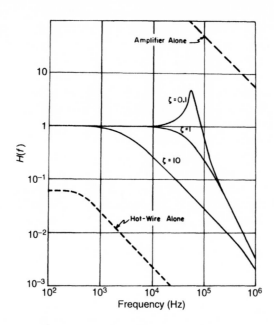

Fig. 2.14. The predicted amplitude transfer function for velocity fluctuations of a *CT* anemometer. (From Blackwelder 1981.)

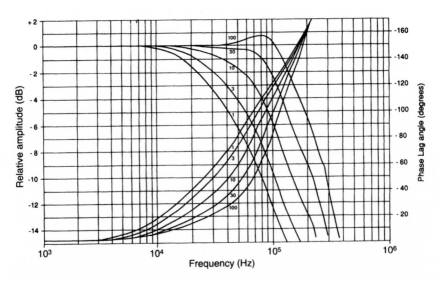

Fig. 2.15. The frequency response of a 5 μm diameter tungsten hot-wire probe operated by a DANTEC 55M10 *CT* anemometer, for air velocities of 1, 3, 10, 30, and 100 m s^{-1}. (Reprinted with the permission of DANTEC Measurement Technology.)

a 5 μm tungsten wire (DANTEC 55P11). The corresponding phase-lag angle is observed to increase rapidly with the frequency when $f > f_c$. To reduce this phase change, the amplifier of the (DANTEC 55M10) *CT* system selected for hot-wire-probe use has a frequency-dependent gain of -1.2 dB octave^{-1} (DISA 1971, 1977).

Direct testing by velocity perturbations of the dynamic response of a *CT* anemometer is rather cumbersome, and electronic testing is usually carried out instead by injecting a small electronic square-wave or sine-wave signal into the bridge and observing the response of the anemometer's output voltage, *E*. A sine-wave test is useful in analysing the dynamic response of a *CT* anemometer, as discussed by Freymuth (1977a) and Blackwelder (1981). It should be noted from eqn (2.54) that the forcing function includes a contribution from the derivative of the test signal. The consequence of this is that the response is flat up to $f = 1/(2\pi M)$, where M is the time constant of the uncompensated wire element. As the frequency is increased further the amplitude increases at 20 dB decade^{-1}. Both of these features are observed on the sine-wave result for a 4 μm tungsten hot-wire probe (TSI T1.5) shown in Fig. 2.16. Using a log–log plot, the slope of the increasing part of the curve can be seen to be equal to 1. The response curve also identifies the value of the cut-off frequency, f_c. Sine-wave tests are however, not practical for the two-parameter optimization of the *CT* anemometer discussed in Section 2.3.3. This optimization is most conveniently

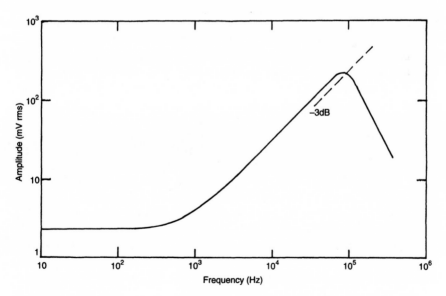

Fig. 2.16. A sine-wave test on a 4 μm diameter tungsten hot-wire probe operated by a *CT* anemometer for an air velocity of 10 m s^{-1} and a 5:1 bridge ratio. (Reprinted with the permission of TSI Inc.)

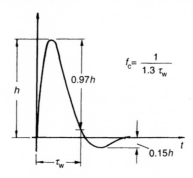

Fig. 2.17. The square-wave test response of a hot-wire probe operated by a CT anemometer.

carried out by a square-wave test. The optimum response to a square-wave test can be obtained (Freymuth 1977a) by adjusting the controls until the response signal in Fig. 2.17 is achieved, which has about a 15% undershoot relative to the maximum. For applications which require a truly flat response, a sine-wave test is recommended. Defining τ_w as the time from the start of the pulse until the response signal has decayed to 3% of its maximum value, the cut-off frequency, f_c, can be evaluated from

$$f_c = \frac{1}{1.3\tau_w}. \tag{2.55}$$

A comparison of a square-wave and a sine-wave test for a $4\,\mu m$ tungsten hot-wire probe operated in the CT mode with a $5:1$ bridge ratio at a flow velocity of $10\,\mathrm{m\,s^{-1}}$ gave similar values for f_c of about 90 kHz.

As discussed by Perry (1982), an increase in velocity usually results in an increase in the frequency response and a decrease in the value of the damping coefficient, ζ. Instability therefore tends to occur at higher velocities. Consequently, to avoid the possibility of instability occurring during a flow investigation, the system should be optimized at the highest occurring velocity. The effect of off-optimum operation on the frequency response of CT anemometers has been discussed by Freymuth (1982).

The low-frequency dynamic prong effect It has been suggested by Perry *et al.* (1979), Smits and Perry (1980) that a low-frequency dynamic prong effect may exist in CT hot-wire anemometry. However, this dynamic effect is insignificant in isothermal flows for standard hot-wire probes ($\ell/\ell_c > 8$) operated at a high overheat ratio and exposed to velocity fluctuations only (see Section 4.9). At very low overheat ratios, Freymuth (1979) has evaluated that this effect may result in an amplitude attenuation of about 5%. However, the low-frequency response of a CT hot-wire probe in a nonisothermal flow is considerably more complex, as discussed in Section 2.3.3.3.

2.3.3.2 *The frequency response of hot-film probes operated in the CT mode*

Cylindrical hot-film probes The relatively large diameter of the fibre substrate (25–70 μm) on which the film is deposited influences the frequency response of the uncompensated film probe, as discussed in section 2.3.2.2. This substrate effect can also be observed when the film sensor is operated in the *CT* mode. Results of a sine-wave test in an air flow for a 50 μm diameter, platinum hot-film probe (TSI-20) are shown in Fig. 2.18. Initially, the response curve is flat and the slope of the increasing part of the curve reaches a value of 1 as did the curve for the hot-wire probe (Fig. 2.16). However, the slope then rapidly reduces to a value of 1/2 due to heat retention in the substrate during part of the heating cycle. Due to the change in gradient from 1 to 1/2 the corresponding amplitude transfer function, $H(f)$, for velocity perturbations will *not* be flat at low and moderate frequencies in contrast to the hot-wire probe. For a properly adjusted anemometer system, this variation in $H(f)$ results in about a ±5–10% uncertainty in measured turbulent quantities. Figure 2.18 also identifies the upper frequency limit, f_c. This value has, for a system with an optimum setting, been compared with the value obtained from the response curve from a square-wave test (Freymuth 1981) using

$$f_c = \frac{1}{\tau_f},\qquad(2.56)$$

where τ_f is defined in Fig. 2.19. For a bridge-ratio setting of 5:1, both methods predicted, for an air flow velocity of 10 m s^{-1}, a similar value for f_c of about 75 kHz, which is sufficient for most turbulent-flow studies.

Wedge and cone hot-film probes Due to the heat retention in the relatively massive substrate and the related conductive 'side' losses in wedge

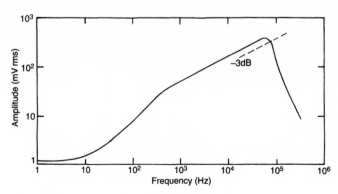

Fig. 2.18. A sine-wave test on a 50 μm diameter platinum hot-film probe operated by a *CT* anemometer for an air velocity of 10 m s^{-1} and a 5:1 bridge ratio. (Reprinted with the permission of TSI Inc.)

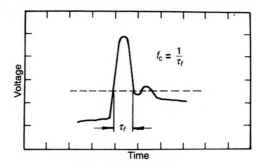

Fig. 2.19. The square-wave test response of a cylindrical hot-film probe operated by a *CT* anemometer. (Reprinted with the permission of TSI Inc.)

and cone hot-film probes, their frequency response deviates from that of a cylindrical film probe. Bellhouse and Schultz (1967) and Bellhouse and Rasmussen (1968) have demonstrated that the low-frequency characteristics of $H(f)$ for wedge-shaped hot-film probes deviated from the steady-state sensitivity value. This was particularly noticeable when they were used in air flows where the turbulent quantities measured with the wedge probe were about 50% smaller than the corresponding hot-wire-probe results. These workers developed a one-dimensional model to describe this low-frequency effect, and they measured the amplitude attenuation both by shaking the probe and by comparative measurements in which a hot-wire probe and a wedge probe were placed in a flow containing either sound waves or grid-generated turbulence. Similar low-frequency attenuations were observed in all cases. For measurements in water, Bellhouse and Schultz found that the low-frequency attenuation was considerably reduced. This one-dimensional model does not take into account the geometry of the substrate, and more recent results have shown that wedge- and cone-shaped

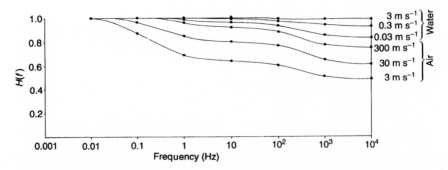

Fig. 2.20. Predicted amplitude transfer function of a wedge sensor (TSI-1232) at various velocities in air and water. (From Nelson and Borgos 1983.)

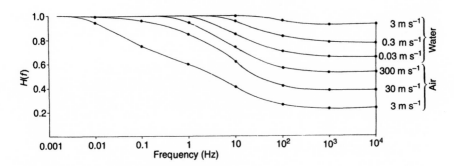

Fig. 2.21. Predicted amplitude transfer function of a cone sensor (TSI-1230) at various velocities in air and water. (From Nelson and Borgos 1983.)

probes have different frequency responses. This has been reported in a comparative experimental and theoretical study by Nelson and Borgos (1983). Typical predictions of the amplitude transfer functions, $H(f)$, for velocity perturbations are shown in Figs. 2.20 and 2.21 for, respectively, a wedge and a cone sensor. The cone probe exhibits the largest attenuation, but the wedge sensor is seen to have the most complex $H(f)$ function, with two separate attenuation regions. The attenuation for both probes is large in air. However, in water, the maximum attenuation of the wedge probe is 5% or less, except at very low velocities. This is consistent with the observations by Bellhouse and Schultz (1967). The corresponding attenuation for the cone probe is about three times larger than for the wedge probe.

The frequency response of a wedge or cone hot-film probe is optimized by varying the available controls until the response signal to a square-wave test is in the form shown in Fig. 2.19. The tail-off response will vary slightly with the geometry of the hot-film probe, but the upper frequency limit, f_c, can be estimated by eqn. (2.56). Another practical approach is to compensate for the attenuation due to the substrate by using an amplifier with a frequency-dependent gain. DISA (1971) describes such an amplifier with a response of $-10\,\text{dB decade}^{-1}$. The advantage of this approach is that the response of the system to a square-wave signal will approach that of a hot-wire probe; see Fig. 2.17. The disadvantage is that normally only one amplifier response setting is available to cover all types of hot-film probes, and the various hot-film probes have been shown to exhibit different amplitude attenuations. Also, this approach will not compensate for the deviation from a flat response at low frequencies.

2.3.3.3 The dynamic response of a CT hot-wire probe in a non-isothermal flow

Hot-wire anemometry is often used in flows with simultaneous velocity and temperature fluctuations. As shown in Section 2.1.2.4 a hot-wire sensor probe is sensitive to both velocity and temperature fluctuations. If it is assumed, for a single, normal, (SN) hot-wire probe, that the 'static' calibration relation in eqn (2.36a) is valid for the wire voltage, E_w,

$$\frac{E_w^2}{R_w} = (A + BU^n)(T_w - T_a),$$

then the fluctuating anemometer signal, e_w, is related to the velocity fluctuation, $u(=U - \bar{U})$, and to the ambient fluid temperature variation, $\theta(=T_a - \bar{T}_a)$, by eqn (2.38a)

$$e_w = S_u u + S_\theta \theta,$$

where S_u and S_θ are the velocity and temperature sensitivities. This section deals with corrections to eqn (2.38a) based on the frequency response of a hot-wire probe exposed to combined velocity and temperature fluctuations. A related correction method for both SN and X-probes is given by Smits and Perry (1981). Measurements of temperature fluctuations or corrections for temperature drift are discussed in Chapter 7.

For accurate velocity measurements, a hot-wire probe is usually operated at a high overheat ratio, making the ratio of the velocity sensitivity, S_u, to the temperature sensitivity, S_θ large. In most of the recent experimental methods the smaller temperature term in eqn (2.38a) is treated as a correction term, with the fluctuating temperature, θ, measured independently by the resistance-wire (or 'cold-wire') method described in Section 7.4.3. This is the approach considered in this section.

As discussed in Section 2.3.3.1, the low-frequency dynamic prong effect can be ignored when considering the response to velocity fluctuations of a standard hot-wire probe operated in the CT mode at a high overheat ratio. Consequently, in eqn (2.38a), the velocity sensitivity, S_u, can be obtained by static calibration methods using eqn (2.37a). However, a correction is required for the temperature term in eqn (2.38a) if S_θ is specified by the temperature sensitivity eqn (2.37b) obtained by a static calibration, $S_{\theta,s}$, (Freymuth 1979; Paranthoen et al. 1983; Larsen et al. 1986). The amplitude transfer function, $H_\theta(f)$, for temperature fluctuations exhibits a significant low-frequency dynamic effect, as shown in Fig. 2.22. To obtain accurate results, the temperature sensitivity in eqn (2.38a) must therefore be expressed as

$$S_\theta = S_{\theta,s} H_p', \tag{2.57}$$

where H_p' is lower than unity and it is the plateau level of the transfer function, $H_\theta(f)$, in Fig. 2.22. If this correction is not made then overcompensation of the temperature influence in eqn (2.38a) will occur. The measurement of θ is usually carried out with a thin resistance-wire operated in the CC mode, and the following relationship applies for the measured temperature fluctuation, θ_m:

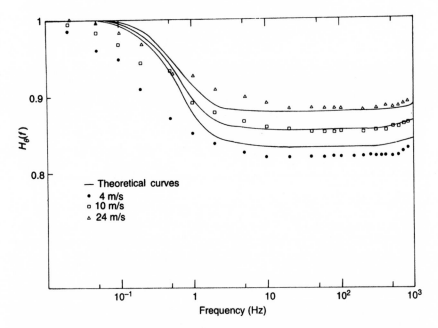

Fig. 2.22. The amplitude transfer function, $H_\theta(f)$, for a $4\,\mu$m diameter tungsten hot-wire probe (TSI 1210-T1.5) operated by a *CT* anemometer and exposed to temperature fluctuations (from Paranthoen *et al.* 1983).

$$\theta_m = H_p\theta, \qquad\qquad (2.58)$$

where H_p is the plateau level of the amplitude transfer function for temperature fluctuations of the resistance-wire (see Fig. 2.10). By combining eqns (2.38a), (2.57), and (2.58) we obtain

$$e_w = S_u u + S_{\theta,s}\left(\frac{H_p'}{H_p}\right)\theta. \qquad\qquad (2.59)$$

If $H_p'/H_p \sim 1$ then no correction is necessary. This can be achieved by using the same kind of prongs for the two sensors and by selecting the length of the wires so that the two transfer functions have the same plateau level (Paranthoen *et al.* 1983).

2.4 SPATIAL-RESOLUTION ERRORS

The spatial nonuniformity of the flow along the length, ℓ, of the sensing element has to be taken into account. There are two main sources of error. The first is due to changes in the mean velocity along the length of the wire. This condition exists in a boundary-layer flow, when the hot-wire element is perpendicular to or at an angle to the surface. The second occurs when the length, ℓ, of the wire is not small compared with the fine scales of the turbulence.

2.4.1 Non-uniform mean-velocity distributions

Single normal (SN) hot-wire probes Nonuniform mean-velocity distributions have been investigated by Mattioli (1956) and by Furth (1956), as quoted by Corrsin (1963), and more recently by Gessner and Moller (1971). These workers considered an *SN* probe placed perpendicularly to the flow with the wire aligned in the direction of the mean-velocity gradient, as shown in Fig. 2.23. The steady temperature distribution in the wire is determined by the heat balance, given by eqn (2.11), for an incremental element dx. The heat-transfer coefficient, h, will, in general, be a function of the local values of T_w and $\bar{U}$. When the velocity distribution is uniform along the wire length, ℓ, then the temperature distribution, $T_w(x)$, will be symmetrical as shown in Figs 2.2 and 2.24 ($S = 0$). However, when the wire is exposed to a mean-velocity profile $\bar{U} = \bar{U}(x)$, the analysis by Gessner and Moller predicts an asymmetric temperature profile, as shown in Fig. 2.24.

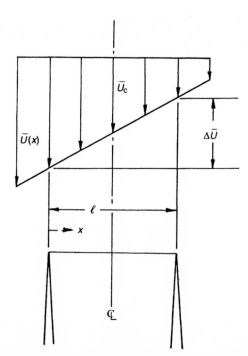

Fig. 2.23. An *SN* hot-wire probe in a mean shear flow. (From Gessner and Moller 1971.) Reproduced with kind permission of Cambridge University Press from the *Journal of Fluid Mechanics.*)

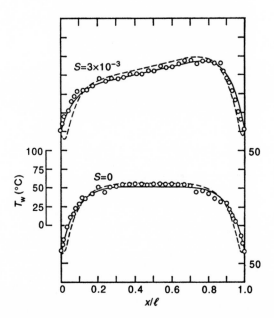

Fig. 2.24. A comparison between analytical and experimental wire-temperature distributions in uniform and shear flow. $\ell/d = 300$, (---) Collis and Williams (1959), (—) Kramers (1946). (From Gessner and Moller 1971. Reproduced with kind permission of Cambridge University Press from the *Journal of Fluid Mechanics*.)

This nonlinear heat-transfer effect was used to calculate the corresponding corrections to the measured values of $\bar{U}$ and $u'(=(\overline{u^2})^{1/2})$. To describe the parameter dependencies of these corrections, Gessner and Moller introduced a shear parameter

$$S = \frac{\Delta \bar{U}}{\bar{U}_c} \frac{d}{\ell}, \tag{2.60}$$

where $\Delta \bar{U}$ is the velocity variation across the length, ℓ, of the wire element, and $\bar{U}_c$ is the centre-line velocity, as shown in Fig. 2.23. For comparison, let us also introduce the shear parameter

$$S^* = \Delta \bar{U}/\bar{U}_c \tag{2.61}$$

and apply both shear parameters to a boundary layer, of thickness δ, with a linear mean-velocity profile. If a wire element, of length, ℓ and diameter d, is placed a distance y from a wall then it follows from simple trigonometry that

$$S = \frac{d}{y}$$

and

$$S^* = \frac{\ell}{y}$$

and therefore

$$S^* = S\frac{\ell}{d}. \tag{2.62}$$

The result by Gessner and Moller shows that the corrections increase with increasing values of either S or ℓ/d. This indicates that the primary parameter dependence is on S^*. It would probably have been simpler to apply their theory to a given shear flow if their results had been presented in terms of S^*, with ℓ/d as a second-order correction parameter. Gessner and Moller concluded—for a $5\,\mu$m diameter platinum wire with a relatively large ℓ/d ratio of 400 operated at an overheat ratio, R_w/R_a, of 1.8 in a shear flow with $S = 2 \times 10^{-3}$ (eqn (2.60)) in the velocity range $10\,\mathrm{m\,s^{-1}} \leqslant \bar{U}_c \leqslant 100\,\mathrm{m\,s^{-1}}$—that the error for $\bar{U}$ was -4.3% and it was -5.8% for u'. A significant reduction in the errors occurs if the ℓ/d ratio is reduced. For an ℓ/d ratio of 200, the above errors in $\bar{U}$ and u' were reduced to -1.3% and -1.4%, respectively.

These corrections apply to flows with low or moderate values of the shear flow parameter. Sandborn and Seegmiller (1975) and Sandborn (1976) have reported studies of flows with large values of the shear parameter, in which the velocity varied from 0 to about $80\,\mathrm{m\,s^{-1}}$ over a distance of 38 mm. They compared the outputs from an SN hot-wire probe placed with the wire element either parallel or perpendicular to, the surface. Substantial differences were observed in the values of $\overline{u^2}$ measured in the two wire orientations. Attempts to develop a reliable correction method were not successful. For measurements of the longitudinal velocity component it is, of course, recommended that the sensing element of the SN hot-wire probe is placed parallel to the surface. In this case, a variation in the mean velocity will occur across the diameter of the wire. This effect, which is usually assumed to be small, has been studied by So (1976).

Single inclined hot-wire probes The time-mean cross-velocity components and the related Reynolds stresses in, for example, a boundary-layer flow can be measured by placing an inclined SN or an SY probe at two or more positions in a plane either parallel or perpendicular to the surface. In the latter case the wire element of the probe will be exposed to a shear effect similar to that described for a SN probe placed perpendicularly to the surface.

X-hot-wire probes An X-probe placed in a flow with a mean-velocity profile may be exposed to two different types of shear effect. These effects can be explained by considering a boundary-layer flow over a flat surface. Introducing a Cartesian (x, y, z) coordinate system with the x-coordinate

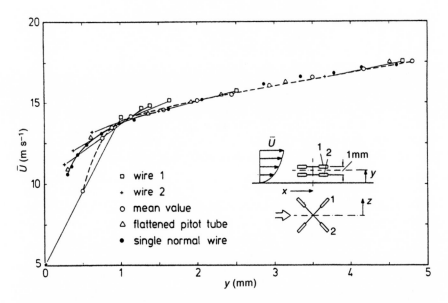

Fig. 2.25. The influence of the mean shear on X-wire measurements. (From Vagt 1979.)

aligned with the mean-flow direction, the velocity components will be denoted (U, V, W). The first shear type occurs when the plane of the X wires is perpendicular to the surface, exposing both wires to the same mean-velocity gradient. As described in Section 5.4.3.1, the values of U and V are often evaluated by a sum-and-difference method applied to the signal from the two wires. It therefore follows, to first order, that U will contain a mean shear error similar to that described for the SN probe, while the related error for V will be small. The second type can be observed when the X-probe is placed parallel to the surface. Due to the separation, h, between the two wires they will be exposed to two different mean velocities, as illustrated in Fig. 2.25. This shear effect has been studied by Hirota *et al.* (1988), who found the error in U to be small, but that a significant error occurred in W. This error can be corrected by the use of an additional X-probe which is a mirror image of the first X-probe, and then adding the results from the X-probes. However, when the distance of the probe from the surface becomes similar to, or smaller than, the separation between the two wires, then all X-probe techniques become unreliable, as shown in Fig. 2.25. A solution to this type of shear effect is to place the two wires in the same plane in a V-configuration (see, for example, Kreplin and Eckelmann 1979; Pandaya and Lakshminarayana 1983). However, the disadvantage of this approach is poor spatial resolution.

2.4.2 Non-uniform turbulent velocity distributions

Turbulent flows consist of eddies of different sizes. The energy exchange between the mean flow and the turbulence is governed by the dynamics of the large eddies. The energy extracted by the turbulence from the mean flow therefore enters the turbulence mainly at length scales which are comparable to the integral length scale L (which is defined by eqn (2.66)). Within the turbulent flow, an energy cascade takes place where energy is transferred from large eddies to small eddies by a process of vortex stretching. The large-scale motion is often nonisotropic with strongly preferred directions. However, for many turbulent flows, the fine-scale motion is locally isotropic (that is, the fine-scale motions which are superimposed on the large-scale motion have no preferred direction). Viscous dissipation of turbulent energy determines the scales of the smallest eddies. The size of the dissipating eddies depends only on the kinematic viscosity v and on the dissipation rate, ε. The Kolmogorov length scale, η, where

$$\eta = (v^3/\varepsilon)^{1/4}, \tag{2.63}$$

is usually taken as a measure of the size of the dissipating eddies. However, as stated by Bradshaw (1971, p. 39), 'It must be emphasized that these various length scales are *typical* scales for use in order of magnitude arguments. For instance, the main contribution to the energy may be spread over a range of length scales of much more than 10:1 with the position of the integral length scale L in this range depending strongly on the type of flow considered, and obviously even the dissipating eddies contain some energy: by "energy containing" and "dissipating" ranges of eddy size we mean those that contain *most* of the energy or effect *most* of the dissipation.'

The size of a typical large eddy can be determined from the spatial correlation between the velocity fluctuations at two adjacent points in the flow field. For homogeneous isotropic turbulence, the general spatial correlation coefficient function

$$\rho_{ij}(r_1, r_2, r_3) = \frac{\overline{(u_i)_A (u_j)_B}}{(u_i)_A^{2^{1/2}} (u_j)_B^{2^{1/2}}} \tag{2.64}$$

for the fluctuating velocity components (u_1, u_2, u_3) at two points A (x, y, z) and B $(x + r_1, y + r_2, z + r_3)$ can be described in terms of the longitudinal and lateral correlation coefficients $f(r_1)$ and $g(r_2)$ (Hinze 1959). Expressing these correlations in terms of the longitudinal velocity fluctuation and setting $u_1 = u$, we have

$$f(r_1) = \rho_{11}(r_1, 0, 0) = \frac{\overline{u(0) u(r_1)}}{\overline{u^2}} \tag{2.65a}$$

$$g(r_2) = \rho_{11}(0, r_2, 0) = \frac{\overline{u(0)\,u(r_2)}}{\overline{u^2}} \tag{2.65b}$$

Typical variations for $f(r_1)$ and $g(r_2)$ are shown in Fig. 2.26. It is usually observed that $f(r_1)$ approaches zero asymptotically as r_1 increases, while $g(r_2)$ becomes negative for some values of r_2.

The size of the large energy-containing eddies is usually described by the longitudinal integral length scale, L

$$L = \int_0^\infty f(r_1)\,dr_1. \tag{2.66}$$

The corresponding transverse integral length scale, L_g, is

$$L_g = \int_0^\infty g(r_2)\,dr_2. \tag{2.67}$$

When the separation r_1 is in the mean-flow direction then $f(r_1)$, and therefore L, is difficult to measure due to thermal wake interference between two parallel wires separated in the mean-flow direction (Ko and Davies 1971). Ko and Davies carried out measurements in the potential core of a 50 mm diameter jet, at a position with a low turbulence intensity of about 4%. The axial separation, Δx, between the two wires was varied over the range $\Delta x/d = 0\text{–}10^4$, where d is the wire diameter, and Ko and Davies observed a significant wake effect on both $\bar{U}$ and $\overline{u^2}$ measured by the downstream wire. Even the presence of a cold upstream wire caused a substantial wake effect, and for a heated hot-wire the wake disturbance can be observed in $\overline{u^2}$ for values of $\Delta x/d$ in excess of 10^4 (Fig. 2.27). Spatial correlations, the integral scale, and the velocity gradient in the mean-flow

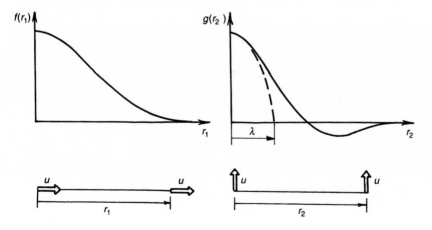

Fig. 2.26. Longitudinal, $f(r_1)$, and transverse, $g(r_2)$, correlation coefficients.

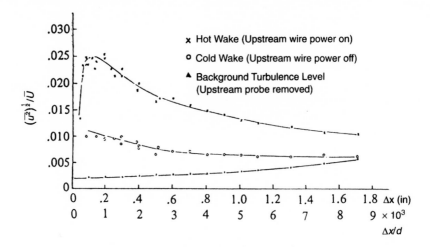

Fig. 2.27. Wake-turbulence intensity disturbance measured by a hot-wire probe placed in the hot and cold wake of a second upstream hot-wire probe. (From Ko and Davies 1971.)

direction are therefore not normally evaluated by the parallel-wires method. Instead, Taylor's convected, frozen-turbulence-pattern hypothesis

$$\frac{\partial}{\partial x} = -\frac{1}{\bar{U}}\frac{\partial}{\partial t} \qquad (2.68)$$

is usually applied. This procedure relates the longitudinal variation to the temporal variation at a point, by assuming the convection of a 'frozen' turbulent structure past the measurement point with the mean velocity. The value of the integral length scale, L, can therefore be calculated from the autocorrelation coefficient function, $\rho_u(\tau)$, obtainable by a SN probe:

$$\rho_u(\tau) = \frac{\overline{u(t)u(t+\tau)}}{\overline{u^2}}, \qquad (2.69)$$

where $u(t)$ is a measured time-history record of the longitudinal-velocity fluctuation and τ is a variable time delay. An integral time-scale T_I can be defined as

$$T_I = \int_0^\infty \rho_u(\tau)\,\mathrm{d}\tau \qquad (2.70)$$

and by using Taylor's hypothesis the value of L may be estimated from

$$L = \bar{U}T_I. \qquad (2.71)$$

To describe the turbulence of a flow, it is often convenient to introduce the Taylor microscale, λ, which is usually defined as

$$\lambda = \left[\frac{\overline{u^2}}{\overline{(\partial u/\partial x)^2}} \right]^{1/2}. \qquad (2.72)$$

Due to the thermal wake interference between two wires separated in the mean-flow direction, it is difficult to measure λ (defined by eqn (2.72)), but if Taylor's frozen-turbulence-pattern hypothesis is applied then the value of λ can be estimated from

$$\lambda = \bar{U} \left[\frac{\overline{u^2}}{\overline{(\partial u/\partial t)^2}} \right]^{1/2}. \qquad (2.73)$$

Alternatively, the microscale λ can be calculated from

$$\lambda = \left[\frac{\overline{u^2}}{\int_0^\infty k_1^2 E_u(k_1)\, \mathrm{d}k_1} \right]^{1/2}, \qquad (2.74)$$

where $k_1 = 2\pi f/\bar{U}$ is the wavenumber and $E_u(k_1)$ is the one-dimensional wavenumber spectra defined in such a way that

$$\overline{u^2} = \int_0^\infty E_u(k_1)\, \mathrm{d}k_1. \qquad (2.75)$$

$E_u(k_1)$ is related to the one-dimensional frequency spectrum $E_u(f)$ by

$$E_u(k_1) = \frac{\bar{U}}{2\pi} E_u(f). \qquad (2.76)$$

Taylor's microscale, λ, is not a direct measure of either the large energy-containing eddies nor of the smallest dissipating eddies. However, λ can be related to both $g(r_2)$, and thereby to L_g, and to η. As shown by Hinze (1959), and as illustrated in Fig. 2.26, the transverse correlation coefficient, $g(r_2)$, will, for small values of r_2, approach a parabolic function of r_2 given by

$$g(r_2) = 1 - \frac{r_2^2}{\lambda^2}. \qquad (2.77)$$

For isotropic turbulence, the dissipation rate in eqn (1.15), ε, can be expressed as

$$\varepsilon = 15\nu \,\overline{(\partial u/\partial x)^2}. \qquad (2.78)$$

Introducing the Taylor microscale, λ, from eqn (2.72), the expression for ε becomes

$$\varepsilon = 15\nu \frac{\overline{u^2}}{\lambda^2}. \qquad (2.79)$$

By inserting eqn (2.74) into eqn (2.79), the dissipation rate, ε, can also be expressed as

$$\varepsilon = 15\nu \int_0^\infty k_1^2 E_u(k_1) \, dk_1. \tag{2.80}$$

These formulations for ε are often used to evaluate the Kolmogorov length scale, $\eta = (\nu^3/\varepsilon)^{1/4}$.

The relationship between the longitudinal and transverse integral scales, L and L_g, and the Taylor microscale, λ, vary with the structure of the turbulent flow. As an example, for boundary layers developing inside a circular pipe, Favre *et al.* (1957) found that $L/\lambda \simeq 10$ while $L_g/\lambda \simeq 1$. The value of these ratios illustrate the anisotropic nature of the large-scale eddies in many turbulent flows. From a simple energy budget, it can be shown (see, for example, Hinze 1959; Tennekes and Lumley 1972) that the ratios L/λ and λ/η can be estimated from

$$\frac{L}{\lambda} = \frac{A}{15} \, \mathrm{Re}_\lambda, \tag{2.81}$$

$$\frac{\lambda}{\eta} = 15^{1/4} \, \mathrm{Re}_\lambda^{1/2}, \tag{2.82}$$

where A is an undetermined constant, which is presumably of the order of unity, and Re_λ is the microscale Reynolds number, which can be expressed for isotropic turbulence as

$$\mathrm{Re}_\lambda = \frac{u'\lambda}{\nu}, \tag{2.83}$$

where $u' = (\overline{u^2})^{1/2}$. In the above example $L/\lambda \simeq 10$, and, if A is assumed to be equal to 1, then the corresponding value of the ratio λ/η will be about 25. A length ratio of the largest energy-containing eddies to the smallest dissipating eddies of about 1000 is therefore common in turbulent flows; this imposes severe restrictions on the accuracies and simulations in Computational Fluid Dynamics (*CFD*) work.

To achieve a perfect spatial resolution, a hot-wire probe should have a wire length, ℓ, which is short compared with all the eddy size which occur. It should also have a wire length-to-diameter ratio, ℓ/d, which is greater than about 200 to minimize the degradation of the sensitivity and the frequency response due to the heat conduction to the prongs. Consequently, hot-wires, which are much shorter than the energy-containing eddies are normally used. This allows accurate resolution of the large-scale eddies. However, for standard hot-wire probes, with $\ell = 1.25$ mm and $d = 4$–$5\,\mu$m, imperfect spatial resolution of the fine-scale turbulence will occur in many laboratory flows, in which the Kolmogorov length scale, η, may typically be of the order of 0.1 mm. The magnitude of the resultant spatial-resolution error and the related probe-size criteria depend on whether only statistical quantities (such as Reynolds stresses, etc.) are needed, or whether spectra, dissipation, or vorticity information are also required.

SN hot-wire probes The early investigations of the spatial resolutions of hot-wire probes were largely theoretical and they were mostly based on an assumption of local isotropy. One of the earliest was by Skramstad, as reported by Dryden *et al.* (1937). Related studies have also been reported by Frenkiel (1949, 1954), Corrsin and Kovasznay (1949), and Uberoi and Kovasznay (1953). For an *SN* probe placed perpendicularly to the mean flow, $\bar{U}$ and aligned with the r_2-direction (see Fig. 2.26), the measured fluctuating velocity, u_m, will be

$$u_\mathrm{m} = \frac{1}{\ell} \int_0^\ell u(r_2)\,\mathrm{d}r_2.$$

For the case of homogeneous, isotropic, turbulent flow, Skramstad derived the following relationship for the measured normal Reynolds stress

$$\overline{u_\mathrm{m}^2} = \overline{u^2}\, \frac{2}{\ell^2} \int_0^\ell (\ell - r_2)g(r_2)\,\mathrm{d}r_2. \tag{2.84}$$

The signal attenuation is therefore a function of the ratio of the wire length, ℓ, and the transverse integral scale, L_g (see eqn (2.67)). If the wire length, ℓ, is smaller than L_g, and thereby $\ell < \lambda$, then eqn (2.77) applies, and eqn (2.84) can be expressed as

$$\overline{u^2} = \overline{u_\mathrm{m}^2}\left(1 - \frac{\ell^2}{6\lambda^2}\right)^{-1}. \tag{2.85}$$

This equation indicates that, provided ℓ is small compared with the size of the large energy-containing eddies, the error in $\overline{u^2}$ will be small. If $\ell = \frac{1}{2}\lambda$ then eqn (2.85) predicts that the error in $\overline{u^2}$ will be only 4%.

Spatial-resolution errors are most severe in the near-wall region of turbulent boundary-layer flows. In the near-wall region, the Kolmogorov length scale can be as small as $100\,\mu\mathrm{m}$, and even the fine-scale structure may be anisotropic. In this region, Ligrani and Bradshaw (1987*b*) found that the Kolmogorov length scale, η, was typically about $2\nu/U_\tau$, where U_τ ($=(\tau_w/\rho)^{1/2}$) is the friction velocity and τ_w is the wall shear stress. To study such fine-scale structure, Willmarth and Sharma (1984) developed and used subminiature hot-wire probes with wire lengths smaller than η. Spatial-resolution problems for 'burst' measurements in turbulent boundary layers (see Section 11.2.1.2) have also been reported by Blackwelder and Haritonidis (1983). Further developments of subminiature probes have been reported by Ligrani and Bradshaw (1987*a,b*), who investigated in detail how the wire geometry affects the measured value of $\overline{u^2}$. In their studies, the length, ℓ, of the sensing element was varied from $50\,\mu\mathrm{m}$ to $3\,\mathrm{mm}$. The size of the wire diameter was 0.625, 1.25, or $5\,\mu\mathrm{m}$. With the Kolmogorov length scale, η, being about $100 \mu m$, the ℓ- and d-range used enabled the ratio ℓ/η to be varied between 1.6 and 15 and the length-to-diameter ratio, ℓ/d to be varied from 70 to 600. Their results demonstrated that the largest 'true' value of $\overline{u^2}$ was obtained with a $0.625 \mu m$ diameter subminiature probe with an ℓ/d ratio of 360 and $\ell/\eta = 1.6$. From their experimental results, they concluded that comparatively few of the energy-containing eddies had a spanwise size smaller than ten Kolmogorov length scales. From this observation, they made the practical conclusion that the error in $\overline{u^2}$ would be less than -4% if $\ell/\eta < \sim 10$ and $\ell/d > 200$. Provided $\eta \geqslant 100 \mu m$, which is the case for most turbulent flow, then these criteria are satisfied for standard SN hot-wire probes with $\ell = 1.25 \text{ mm}$ and $d = 4\text{-}5 \mu m$.

The effect of incomplete spatial resolution is most noticeable in the high-frequency attenuation of measured velocity spectra. Wyngaard (1968) has extended the work of Uberoi and Kovasznay (1953) and calculated the spectral transfer functions for SN and X-hot-wire probes. In this analysis, it was assumed that the small-scale turbulence was isotropic and Pao's form (Pao 1965) was used for the three-dimensional spectrum.

Figure 2.28 shows the one-dimensional spectral response E_u^m/E_u versus $k_1\ell$ for various values of η/ℓ (for an SN hot-wire probe); $E_u^m(k_1)$ is the measured spectrum of the fluctuating longitudinal velocity component u, $k_1 = 2\pi f/\bar{U}$ is the corresponding wavenumber, $E_u(k_1)$ is the true spectrum, and ℓ is the wire length. A significant feature of these results is that the $k_1\ell$ value at which attenuation of $E_u(k_1)$ becomes important increases with η/ℓ. For $\eta/\ell = 1$, the 10% attenuation point is $k_1\ell = 4$, while for $\eta/\ell = 0$ it decreases to 0.5.

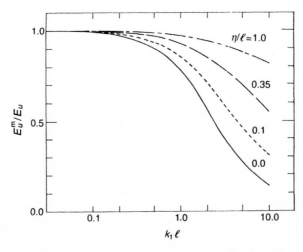

Fig. 2.28. The one-dimensional spectral response for an SN probe, for various values of η/ℓ. (From Wyngaard 1968.)

Ligrani and Bradshaw (1987a,b) have evaluated the effect of ℓ/η on measured one-dimensional spectra $E_u^m(k_1)$ in the near-wall region of a boundary-layer flow. Using wires with a $0.625\,\mu m$ diameter, they measured spectra at the position of the maximum value of $\overline{u^2}$, at $y^+ = 17$, and at $y^+ = 169$ ($y^+ = yU\tau/v$), and they compared their results with the theoretical corrections of Wyngaard (1968). At $y^+ = 169$ they obtained a good agreement, and it is therefore justified to use Wyngaard's correction (based on isotropy of the small-scale turbulent structure) to obtain the true spectra from such small sensors at this position in the boundary layer. This conclusion is also in agreement with measurements at $y^+ = 150$ by Derksen and Azad (1983) However, at $y^+ = 17$, Ligrani and Bradshaw found that the flow deviated considerably from isotropy, and the measured attenuations were significantly larger than Wyngaard's corrections.

The corresponding spatial correction for measurements of the (average) dissipation rate, ε, has been studied by Ligrani and Bradshaw (1987b), Turan and Azad (1989), Azad and Kassab (1989), Browne *et al.* (1991), Antonia and Mi (1993a), and Elsner *et al.* (1993). Hussein and George (1990), Hallbäck *et al.* (1991), and Antonia *et al.* (1993) have studied the effect of wire spacing on the direct measurement of spatial derivatives using parallel wire probes. The measurement of $\partial u/\partial y$ is of particular importance in turbulent boundary-layer flows. The quantity $\overline{(\partial u/\partial y)^2}$ represents a significant contribution to the average energy-dissipation rate, ε, (see eqn (1.15)) and the spanwise mean-square vorticity $\overline{\omega_z^2}$ (see eqn (10.3c)). A comparison between measured and direct numerical simulation (DNS) data indicates that reasonable statistics for $\partial u/\partial y$ can be obtained when the separation between the hot wires is in the range 2–4 Kolmogorov length scales. It should, however, be noted that as $\Delta y \rightarrow 0$ the gradient measurement becomes very sensitive to calibration accuracies and noise.

X-hot-wire probes The spatial resolution of X-hot-wire probes is more complex since it depends on ℓ/η, ℓ/d, and the spacing, h, between the two wires. Based on isotropy of the small-scale structure, Wyngaard (1968) has also developed theoretical corrections for the measured one-dimensional spectra of u and v. The spectral attenuation with ℓ/η was found to be similar to that of the *SN* probe, and it was concluded that the spectral responses were influenced by crosstalk terms between the two wires, resulting in $\overline{u^2}$ becoming larger and $\overline{v^2}$ becoming smaller with increasing values of h/ℓ. This is the opposite trend to the measured variations in $\overline{u^2}$ and, $\overline{v^2}$ with h/η found in a study by Browne *et al.* (1988), where a $5\,\mu m$ diameter, Pt-10% Rh wire was used with an active length, ℓ, of 0.75 mm. Measurements were carried out in a turbulent flow with a Kolmogorov length scale, η, of 0.45 mm. Consequently, $\ell/\eta = 1.7$ and $\ell/d = 150$; these values are within the sub-miniature *SN*-probe range studied by Ligrani and Bradshaw (1987a,b). The effect of the spatial separation, h, between the two wires has also been investigated in several numerical-simulation database investigations (Moin and Spalart 1989; Suzuki and Kasagi 1990. Tagawa *et al*. 1992). These investigations predict that the measured values of $\overline{u^2}$ will be too low and that the values of $\overline{v^2}$ will be too high. A comparative investigation has been carried out by Ligrani *et al*. (1989c) , using a standard X-probe, a subminiature X-probe, and a subminiature SN probe. Measurements with all three probes in a turbulent boundary layer showed that $\overline{u^2}$ was not strongly influenced by the probe size or type, but the standard X-probe measured lower values of $\overline{v^2}$ than the subminiature X-probe.

The spatial resolution of other probe types has also been investigated. Wyngaard (1969) studied a vorticity probe, and a number of investigators have reported work related to the spatial resolution of resistance-wires (Wyngaard 1971b; Larsen and Højstrup 1982; Lecordier *et al*. 1984; Browne and Antonia 1987).

3

INTRODUCTION TO
VELOCITY MEASUREMENTS

To explain the concepts and principles of practical velocity measurements by hot-wire anemometry (HWA), consider the case of a hot-wire probe containing one or more sensors placed in a fluid flow. The purpose of the sensors is to provide information related to the velocity vector at the centre point of the hot-wire element(s). For the discussion contained in Chapters 3 to 6, the flow will be assumed to be isothermal and the density ρ will be assumed to be constant. The velocity field will be described by the velocity vector, V, at the point of the sensing element(s), where V is a function of time for a turbulent flow. The direct output voltage from the anemometer is denoted by E, with indices of 1, 2, etc., in the case of a multi-sensor probe. In this chapter it will be assumed that the response equation for the output voltage, E, from an anemometer connected to a given hot-wire sensor can be expressed in the form of a simple power law

$$E^2 = A + BV_e^n, \tag{3.1}$$

where V_e is an effective velocity, which is related to the velocity vector, V.

It is generally accepted that a hot-wire sensor responds uniquely to the velocity V, provided that it is specified in a *wire-fixed* coordinate system (x', y', z'), as illustrated in Fig. 3.1. The x'- and y'-axes are normally contained in the plane of the prongs. The velocity vector, V, may be specified either in terms of its magnitude, $\tilde{V}$, the yaw angle, α, and the pitch angle, β, or in terms of the corresponding three velocity components, U_N (normal), U_T (tangential), and U_B (binormal). If the response of the hot-wire element to all three velocity components was identical, the effective velocity, V_e, would be equal to the magnitude, $\tilde{V}$, of the velocity vector. However, a hot-wire sensor has different responses to the three velocity components U_N, U_T, and U_B; and, in particular, the sensitivity to the tangential velocity component, U_T, is very small. One commonly used relationship, often referred to as Jørgensen's equation, which takes these factors into account, and thereby defines the concept of an effective velocity, is

$$V_e^2 = U_N^2 + k^2 U_T^2 + h^2 U_B^2, \tag{3.2}$$

where k and h are often referred to as the sensor's yaw and pitch coefficients. The notation yaw and pitch factor, or sensitivity coefficients, has also been used. Typical values for k and h for a standard plated hot-wire

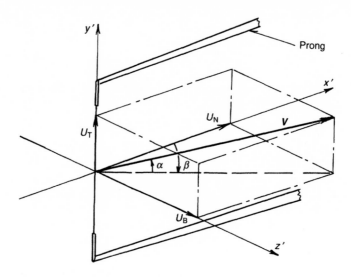

Fig. 3.1. A wire-fixed coordinate system and the corresponding velocity components.

probe are, respectively, 0.2 and 1.05. By inserting eqn (3.2) into eqn (3.1) we can obtain

$$E^2 = A + B(U_N^2 + k^2 U_T^2 + h^2 U_B^2)^{n/2}. \tag{3.3}$$

This equation represents one possible hot-wire response equation, and it enables the velocity information from hot-wire probes to be evaluated.

A logical analysis of a computer-based hot-wire anemometer system used for velocity measurements will identify a similar number of basic steps in all investigations. In some of the proposed methods it may be possible to combine two or more of these steps (as described in Chapters 4 to 6) , but for introductory purposes it is useful to identify all the individual steps:

(1) specification of the measurements required;
(2) selection of the anemometer type and the hot-wire probe;
(3) selection of the hot-wire response equation;
(4) calibration of a hot-wire probe;
(5) digital measurement requirements;
(6) measurements with a hot-wire probe;
    (i) identification of the mean-flow direction;
    (ii) data acquisition and storage;
(7) data analysis;
    (i) data conversion;
    (ii) time-series analysis;
(8) data presentation;
(9) uncertainty analysis.

3.1 SPECIFICATION OF THE MEASUREMENTS REQUIRED

As discussed in Chapter 1, measurements with hot-wire probes are carried out for a large number of reasons and in a wide range of flow situations. In certain confined spaces it may be necessary to restrict the probe choice and therefore the type of measurements. However, in most cases, the selection of the required velocity information is based on a balance between what is possible in principle and what can be achieved in practice, taking into account each investigator's constraints such as the available: time, hot-wire probes, anemometer equipment, and computer facilities. To highlight the available options, Chapters 4 to 6 have been divided into procedures for obtaining one, two, or three components of the velocity vector at the sensor position. As will be shown for the cases of two- or three-component measurements, more than one probe type can be used to obtain certain types of *statistical* velocity results. A major criterion in the selection of the probe type is therefore whether *only* statistical information of the velocity components (such as time-mean values, Reynolds stresses etc.) is required, or whether the investigator *needs* to measure and evaluate the time dependence of two or three components of the velocity vector at the measuring points selected.

3.2 SELECTION OF THE ANEMOMETER TYPE AND THE HOT-WIRE PROBE

A *HWA* probe operated by a Constant-Temperature (*CT*) anemometer is usually used for velocity measurements in isothermal flows due to the system's very good frequency response (see Section 2.3.3). In the case of a gas, for example, air, an upper-frequency limit, f_c ($-3\,$dB), in excess of 50 kHz can often be obtained for both hot-wire and hot-film probes. A relatively high overheat ratio, R_w/R_a, is normally selected for velocity measurements. As explained in Section 2.1.2.4, when the overheat ratio is increased the sensitivity to velocity fluctuations, S_u, of the hot-wire probe is also increased, and at the same time the sensitivity to temperature variations in the flow, S_θ, is reduced. For the typical case of a 4–5 μm diameter tungsten wire, an overheat ratio of 1.8 represents an acceptable compromise between obtaining a high value of S_u and avoiding oxidation of the tungsten wire element, which occurs at a temperature of about 350 °C. For measurements in a conducting liquid, such as water, quartz-coated hot-film probes are usually used, and the temperature of the sensor must be kept low to avoid the formation of air bubbles on the surface of the film element (Rasmussen 1967). The overheat ratio is usually about 1.05–1.08. In non-isothermal flows, the hot-wire probe signal will contain both a velocity and a temperature component. A Constant-Current (*CC*) anemometer is often used in conjunction with a resistance-wire to measure the temperature fluctuations and to compensate the *CT* hot-wire probe signal for its temperature content. The measurement and analysis procedures for combined tempera-ture and velocity measurements are presented in Chapter 7.

To clarify the selection procedure, only conventional hot-wire probes with one, two or three sensors will be considered at this stage. They can be classified as: (i) Single Normal (*SN*) probes (ii) Single Yawed (*SY*) probes (iii) *X*-probes, and (iv) triple-wire (3*W*) probes. Briefly, an *SN* probe can provide information about the velocity component in the mean-flow direction. An *SY* probe, *aligned* with the mean-flow direction, can (if placed at two yaw positions and used in conjunction with an *SN* probe) provide *statistical* information about the velocity in the mean-flow direction and about one of the transverse components. If it is feasible to rotate the *SY* probe 360° around its axis, then all three time-mean velocity compo-nents, as well as all six Reynolds stresses, can in principle be measured. An *X*-probe can evaluate two simultaneous velocity components in the plane of the *X*-probe, provided that the mean-flow direction is contained in this plane. Finally, by using a 3*W* probe, the time dependence of the complete velocity vector, *V*, can be determined. However, in the use of any of these probe types, there are restraints related to the permissible variations of the yaw and pitch angles of the velocity vector relative to the sensor element(s). These aspects are discussed in Chapters 4 to 6.

In selecting a probe type, an investigator must balance the capability of each probe type against the cost and time required for its use. Multi-wire probes (*X*-probes and 3*W* probes) have the distinct advantage that they can provide simultaneous information about two or three of the velocity com-ponents of the velocity vector, *V*, at the centre of the sensing elements. Also, as far as the signal analysis and interpretation are concerned, digital signal analysis enables multi-wire probes to be used with relative ease. In practice, there is relatively little difference between writing a programe for a single wire or a multi-wire probe, provided that the signal-analysis equa-tions are specified. (These aspects are discussed in Chapters 4 to 6.) Once such programs are operational then, on a modern computer, the difference in computer time between analysing one, two, or three input signals is usually not of major importance compared to the other time factors in the complete experiment. Furthermore, data analysis of simultaneous velocity-component records enables detailed flow information to be extracted which is unobtainable using a single-probe technique.

However, compared with a single-wire probe, considerable increases in *cost* and *calibration time* are required when multi-wire probes are used. Long-term calibration stability, particularly for multi-wire probes, can be a major criterion, in which case a hot-film probe may be the preferred choice, as discussed in Section 4.10. A multi-wire probe is more expensive than a single-sensor probe, and the commercial cost of repairing broken wires on such probes is significantly higher. Also, each sensor of a multi-wire probe requires a separate anemometer and signal-conditioning unit. The

analog-to-digital (A/D) converter unit must have a separate converter channel for each sensor, and the A/D unit must be capable of providing simultaneous data acquisition from all the sensors in the multi-wire probe. The
calibration procedure and the related experimental time increases significantly with the use of multi-wire probes. In the case of an SN probe, normally only a velocity calibration is required. For a two-position SY-probe
method, a velocity and a yaw calibration must be performed. An X-probe
requires a velocity and a yaw calibration for both sensors. Finally, each
wire of a $3W$-probe should be calibrated for velocity, yaw, and pitch, which
significantly increases the calibration procedure and time in comparison
with an X-probe. It should also be noted that in many signal-analysis
methods, accurate values for the mean yaw ($\bar{\alpha}$), and pitch ($\bar{\beta}$), angles are
required. For an X-probe the values of $\bar{\alpha}$ for each wire are needed, and in
the case of a $3W$ probe the values of $\bar{\alpha}$ and $\bar{\beta}$ should be measured for each
wire.

3.3 SELECTION OF THE HOT-WIRE RESPONSE EQUATION

A hot-wire sensor responds to both the magnitude, $\tilde{V}$, of the velocity vector
and its orientation relative to the wire element. The relationship for the
anemometer voltage, E, may therefore be specified as

$$E = F(\tilde{V}, \alpha, \beta). \tag{3.4}$$

Alternatively, E may be described in terms of the velocity components in
the wire-fixed coordinate system (see Fig. 3.1)

$$E = F(U_N, U_T, U_B). \tag{3.5}$$

The accuracy of the hot-wire relationship selected for the calibration and
the signal analysis should either have been proved previously, or if an
investigator chooses a new method then its accuracy should be demonstrated. The response equation is often expressed in the form of eqn (3.3).
As demonstrated in Chapters 4 to 6, this calibration equation forms the
basis for several different methods of analysis. The related advantages/
disadvantages and accuracies of these and other methods will be discussed,
and it will be shown that improved speed of analysis and/or improved
accuracy can be obtained by considering the features and capabilities of
modern computer-based systems.

3.4 CALIBRATION OF A HOT-WIRE PROBE

Equation (3.3) contains a number of calibration constants (A, B, n, k, and
h), and it is the specific aim of a probe calibration to determine the values
of these constants. Most calibrations and subsequent measurements are carried out with stationary hot-wire probes. (The concepts of dynamic calibra-

tion and the use of a moving hot-wire probe are described in Chapters 4 and 8 respectively.) Most calibrations are carried out in either a special calibration facility or in a flow field in which the magnitude and direction of the flow vector is known. In these experiments, the velocity field should be uniform over the space occupied by the sensors and prongs. The turbulence intensity, $Tu(= (\overline{u^2})^{1/2}/\bar{U})$, must also be low ($Tu < {\sim}0.5\%$), otherwise errors will be introduced into the calibration constants.

The accuracy of the hot-wire relationship selected should be evaluated as part of the calibration procedure. As a general rule, the selected response equation should, using the calibration constants evaluated, provide a good approximation to the calibration data over the *complete* velocity range, and, if required, the yaw and the pitch range.

Once the calibration constants have been obtained, then eqn (3.3) provides a specific relationship between (U_N, U_T, U_B) and E; that is, if U_N, U_T, and U_B are known, then E can be evaluated from eqn (3.3). However, for measurement purposes it is necessary to *invert* this relationship, and simplifications/assumptions are often introduced at this stage. Consequently, to test the *overall* accuracy of the selected calibration/interpretation method, it is strongly recommended that the initial measurements are carried out in a *known* flow situation; for example, in a fully developed pipe flow. By comparing the results obtained by the selected method with known velocity reference values the *minimum* uncertainty in the selected method can be established. These calibration aspects are discussed in Chapters 4 to 6.

3.5 DIGITAL MEASUREMENT REQUIREMENTS

The units contained in a typical digital measurement system are illustrated in Fig. 3.2. It is recommended first that the output voltage, E, is passed from a hot-wire anemometer through a low-pass filter, which is used either for spectral-analysis purposes (see Chapter 12) and/or to remove high-frequency electrical noise (see Section 2.3.1.1). This signal is then fed to a signal-conditioning unit. The purpose of this unit is to match the output signal E_G from this unit to the input voltage range of the A/D converter unit. This aspect is discussed in Section 4.11.2. The purpose of the A/D converter is to change the analog input signal, E_G, into the equivalent value in digital format. The primary A/D converter unit specifications are:

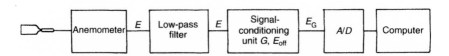

Fig. 3.2. Single-channel digital-measurement system.

voltage range, resolution, and sampling rate. Firstly, the voltage range of the hot-wire probe should be adjusted by the signal-conditioning unit to match the input-voltage range of the A/D converter, which is typically 0–10 V or ±5 V. The second aspect is the resolution of the A/D unit. All A/D converters work on the principle of discretization; that is, an n-bit converter will divide the input-voltage range into 2^n different output values. Consequently an eight-bit converter only provides $2^8 = 256$ different values, while a twelve-bit unit will subdivide the range into $2^{12} = 4096$ values. Clearly the rounding-off or quantization error on a twelve-bit unit is much smaller than on an eight-bit unit. Most modern A/D converters have twelve-bit (or higher) resolutions. A high resolution reduces the maximum sampling rate, but for practical purposes this is no longer a significant problem. Thirdly, a large range of available sampling rates is of primary importance for most investigations. The minimum requirement for a single-channel A/D unit used for general hot-wire/hot-film measurements would be a variable sampling rate from about 10 samples s^{-1} up to about 20 000–50 000 samples s^{-1}. For spectral analysis studies and high-velocity experiments in a gas, it may be necessary to have access to an A/D unit with a maximum sampling rate of about 100 000–250 000 samples s^{-1}. These figures refer to the sampling-rate requirement for the signal from a single sensor. In the case of a multi-wire probe, each sensor must be connected to its own A/D channel, and the same data-acquisition requirements apply to each channel. In particular, the signals from a multi-wire probe must be converted simultaneously. If this 'simultaneous' conversion is implemented inside the A/D unit, then it usually involves the addition of a sample-and-hold (S/H) device on the input side of each A/D channel. For most microcomputers/minicomputers, the data transfer to the computer from the different A/D channels is implemented using a multiplexer, with which the computer scans, in sequence, the various input channels. Modern electronics now permit a maximum sampling rate of 50 000 samples s^{-1} per channel on a four channel unit with sample-and-hold. Sampling rates in excess of 250 000 samples s^{-1} can be achieved by using direct parallel data acquisition for each input channel.

Digital measurements and related time-series techniques require the specification of several parameters. In a given experiment, it is necessary to specify the number of input channels, NC, the sampling rate, SR, and the number of samples, N, for each channel. If $NC > 1$ then the value of N and SR will be the same for each input channel. Once the values of N and SR have been specified then the related sampling time, T, is given by

$$N = SR \times T. \tag{3.6}$$

From eqn (3.6), note that it is only possible to specify two of the three parameters N, SR, and T. The selection of SR, N, and T for a given experiment, and the related time-series analysis, is described in Chapter 12.

Briefly, the selection of these parameters depends on: (i) the types of measurement required, (ii) specification of the level of uncertainty, (iii) the available computer-memory data space, AM, for the acquisition of continuous data records and their subsequent analysis, and (iv) the maximum data-acquisition and analysis time permitted for each measuring point. As will be demonstrated in Chapter 12 several of these conditions are interrelated. In general, the uncertainty of the evaluated results will be small when the number of statistically independent samples is large. Consequently, if only time-averaged values such as the mean velocity, the Reynolds stresses, etc., are required, then the most accurate results are obtained when the samples are statistically independent, which requires the sampling interval to be at least twice the integral time-scale, T_I, of the fluctuating signal (see Chapter 12), usually resulting in a relatively slow sampling rate, SR, and a relatively long sampling time, T. If the time interval between each sample is shorter than T_I, then most of the data acquired will be statistically redundant, and for a fixed value of N this results in an increased uncertainty in the evaluated time-averaged values. For spectral (and the related correlation) analysis the interactions between the data-acquisition parameters are stronger. In air-flow studies, for example, clearly defined energy spectra can be identified for frequencies up to and sometimes in excess of 10–20 kHz, and if the full energy spectrum is to be measured then a very high sampling rate, SR, must be used. Consequently, unless a large value of N can be used, the sampling time T will be rather short, and this may result in a relatively high uncertainty in the calculated spectra (see Chapter 12).

3.5.1 Digital measurement systems

The available types of digital measurement systems may broadly be divided into three groups. In the first method, the output signal(s) from one (or more) hot-wire sensors are initially recorded, instead of by direct acquisition to a computer, on either a frequency modulation (FM) tape recorder or a digital tape recorder which incorporates an A/D converter system. The stored data may be transferred to a minicomputer or a main-frame computer for subsequent analysis. The advantage of this method are that data acquisition can take place at positions far removed from the computer which is used for the data analysis. Also, by using a tape recorder, no practical constraint exists on the sample length of a single record. This may be useful for conditional-sampling techniques, which often require long sampling times (see Chapter 11). However, this method incurs the additional cost of a high-quality multichannel tape recorder, and recording data on a tape recorder is incompatible with online data checking and analysis. The second method relates to the approach adopted by many larger research groups. In this case a minicomputer or main-frame computer with an associated

A/D facility is located in a separate room, and data links are provided to individual experiments which are often placed in different laboratories. This method only involves a single computer system, but the initial cost of the computer and the subsequent running cost are often high, and it operates on a time-sharing basis.

Finally, in many investigations a smaller microcomputer/minicomputer system may be assigned to a particular experiment. During the period of the data acquisition and analysis this computer system is fully dedicated to one particular experiment, and it is available for no other purpose. However, once one particular set of experiments is completed, the system can be used in another experiment. The recent decrease in the cost of microcomputers and the related large increase in their central memory storage, in their incorporated hard-disc facility, and in their improved speed of analysis have now transformed microcomputers into sophisticated systems for experimental work. The author has successfully applied several microcomputer-based systems to a variety of research investigations. However, it should be emphasized that for most applications all of the three digital-measurement systems mentioned should be capable of providing the information required.

3.6 MEASUREMENTS WITH A HOT-WIRE PROBE

In a HWA investigation the steps outlined in Sections 3.1–3.5 must first be considered and then implemented. The hot-wire response equation is selected, and a calibration must be carried out to determine the relevant calibration constants. Also, a suitable A/D and computer system is assumed to be available. However, before reliable hot-wire results can be obtained, it is necessary to evaluate the mean-flow direction and set the parameters of the digital measuring system.

Identification of the mean-flow direction In the case of SN and SY probes, and for some of the methods applied to X-probes it is necessary for the probe-stem to be aligned with the mean-flow direction. Consequently, for these HWA applications, the mean-flow direction must be identified at each measuring point and the probe must be aligned before the measurements of the instantaneous or fluctuating velocity components. This adds an initial measurement procedure (see Section 4.11.1) to the velocity measurements.

Data acquisition and storage The parameters of the digital measuring system must first be set: the cut-off frequency of the low-pass filter, the offset and gain of the signal-conditioning unit for A/D input optimization (see Section 4.11.2), and the sampling rate, SR ,and sampling time, T (or the number of samples, N) for the A/D unit (as explained in Chapter 12). Data acquisition can now be initiated, and on completion a voltage record, $E_G(m)$, is available in digital form in the case of an SN or an SY probe, and two simultaneous time histories, $E_{G1}(m)$ and $E_{G2}(m)$, are available for an X-probe. These values may be stored in the computer using an array format, or in the form of an address location. As discussed in Section 4.12.2.2, the latter format facilitates data handling.

3.7 DATA ANALYSIS

3.7.1 Data conversion

The initial task in most data analysis is to convert the acquired anemometer voltage, $E_G(m)$, into the related velocity information, and $E_G(m)$ can be converted into the corresponding anemometer values, $E(m)$, by using the relationships given in Section 4.4. The form of the calibration equation, eqn (3.3), shows that it is a relatively simple and unique procedure to evaluate E from U_N, U_T, and U_B. However, for the data analysis, this procedure must be *inverted* to obtain U_N, U_T, and U_B from E. This is a highly nonlinear process and many approximate solutions have been proposed, as will be described in Chapters 4 to 6. The inversion and the subsequent data conversion can basically be separated into three individual steps. Firstly, by inverting eqn (3.1) the effective velocity can be obtained; that is,

$$V_e = [\,(E^2 - A)/B\,]^{1/n}. \tag{3.7}$$

It is possible, using eqn (3.7), to convert the digital voltage values, $E(m)$, into the corresponding effective-velocity values, $V_e(m)$. Originally this inversion process was carried out by a separate analog 'linearizer'. Today, digital data acquisition and conversion is normally used, and an analog 'linearization' is both unnecessary and obsolete. The second step involves expressing V_e in terms of velocity components in the *wire-fixed* coordinate system.

Wire-fixed coordinate system The effective velocity V_e, for a wire element in a hot-wire probe can be related by eqn (3.2),

$$V_e^2 = U_N^2 + k^2 U_T^2 + h^2 U_B^2,$$

to the velocity components (U_N, U_T, U_B) in a coordinate system (x', y', z') fixed to that particular wire element (see Fig. 3.1).

Probe-stem coordinate system It is usually convenient to convert the wire-fixed velocity components (U_N, U_T, U_B) for each of the wires in a hot-wire probe into the velocity components (U_s, V_s, W_s) in a common *probe-stem* coordinate system (x_s, y_s, z_s).

For single-element (SN or SY) hot-wire probes only one wire-fixed coordinate system is introduced, which simplifies the velocity transformation to the probe-stem coordinate system. For these probe types, it is often a requirement that the probe-stem is aligned with the mean-flow direction, which must be known (or be determined by an initial experimental procedure) at all measuring points.

Each wire in an X-probe has its own (x', y', z') coordinate system, resulting in two different, and often not orthogonal, coordinate systems. It is therefore useful to transform the (U_N, U_T, U_B) components from each wire into velocity components in a *common* probe-stem coordinate system. In general, it is not necessary to align an X-probe, provided the direction of the mean-flow does not deviate too much from the probe axis. The corresponding 'misalignment' problem can easily be resolved for X-probes and $3W$ probes using digital data analysis, as discussed in Chapters 5 and 6.

Space-fixed coordinate system The results are usually required in a space-fixed coordinate system (x, y, z); the corresponding velocity components are denoted by (U, V, W). In this book it will be assumed, unless otherwise stated, that the probe-stem is aligned with the x-axis of the space-fixed coordinate system which has been selected. This enables the output from the hot-wire probe to be evaluated directly in terms of the space-fixed velocity components (U, V, W). It is normally assumed that $\bar{U}$ is the predominant mean-velocity component.

Several of the single-wire-probe techniques require the probe to be aligned with the mean-velocity vector, which may have two or three significant components in the space-fixed coordinate system selected. Also, due to test-facility constraints, and for measurements close to a wall, it may be necessary to incline the probe-stem relative to the x-axis of the space-fixed coordinate system. In all these cases, it is necessary to convert the velocity components in the probe-stem coordinate system into the corresponding values in the space-fixed coordinate system. This conversion is illustrated in Fig. 3.3, and the corresponding (two-dimensional) transformation equations are

$$U = U_s \cos \psi - V_s \sin \psi \qquad (3.8a)$$

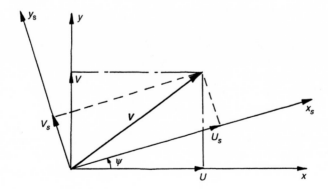

Fig. 3.3. Velocity-component transformation from the probe-stem (x_s, y_s), to the space-fixed (x, y) coordinate system.

$$V = U_s \sin \psi + V_s \cos \psi, \qquad (3.8b)$$

where ψ is the angle between x-axis and the x_s-axis.

3.7.2   Time-series analysis

The outcome of the above analysis steps is a digital time-history record(s) of one or more of the velocity components. The information for an SN probe is a time record of the instantaneous velocity component, U, (that is, $U(t)$) in the mean-flow direction. An X-probe will provide simultaneous records of, for example, $U(t)$ and $V(t)$, while a $3W$ probe can provide a complete description of the velocity vector $V(t)$, in the format $U(t)$, $V(t)$, and $W(t)$. If an SY probe is used then it is possible to obtain *time-mean statistical information* related to two or three of the velocity components. In all these methods, time-series analysis must be applied to obtain the statistical velocity information required. As time-series analysis can be applied to a digital record representing any physical parameter, including velocity and temperature, it will be described separately in Chapter 12.

3.8   DATA PRESENTATION

The output data from the time-series analysis is often in the form of mean-velocity components and Reynolds stresses, and for reference purposes it is common practice to normalize such data and, if possible, to compare them with some predetermined reference condition. As part of the experimental procedure, a data presentation should be selected which will highlight the main features of the experimental data. Some common methods of data presentation are outlined, using mean-velocity data for the flow behind a backward-facing step (measured with a flying hot-wire probe), with particular reference to the spatial development of the flow field. A common experimental procedure is measurement of the transverse variation in the longitudinal velocity component $\bar{U}(y)$, at different axial positions, and a related profile presentation is shown in Fig. 3.4(a) for various longitudinal positions, x/H, downstream of the step (H is the step height). The same velocity information can also be presented in the form of an isocontours plot, Fig. 3.4(b), which facilitates the interpretation of the data in terms of the magnitude and spatial development of the velocity field. Both of these data-presentation procedures can be applied to the mean velocity and to turbulent quantities. If both $\bar{U}$ and $\bar{V}$ are measured, the two mean-velocity components can be combined into a vector plot, as shown in Fig. 3.4(c). This presentation clearly identifies both the deflection of the outer potential flow and the extent and direction of the mean flow within the separation region.

With the development of two-dimensional and three-dimensional

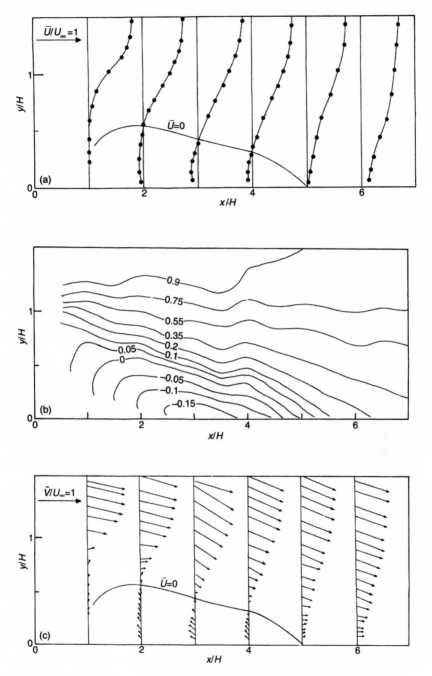

Fig. 3.4. The flow behind a backward facing step: (a) $\bar{U}/U_\infty$ profiles, (b) isocontours of $\bar{U}/U_\infty$, (c) a vector plot of ($\bar{U}/U_\infty$, $\bar{V}/U_\infty$). U_∞ is the free-stream velocity. (From Badran 1993.)

computer-based display techniques, data can now be presented in some very impressive ways. However, for scientific evaluation, the main criterion in selecting a data-presentation method, is still that it conveys the maximum amount of significant flow information in the simplest possible format.

3.9 UNCERTAINTY ANALYSIS

To obtain reliable results it is necessary to identify the various causes of uncertainty in *HWA* and to estimate the magnitude of the individual uncertainties. The limitations related to the frequency response of a *HWA* system and the spatial-resolution problems are discussed in Chapter 2. Chapters 4 to 7 contain a discussion of the following uncertainties: (i) probe disturbance; (ii) the curve-fit uncertainty of the selected response equation; (iii) errors caused by signal analysis simplifications; and (iv) signal interpretation errors in high turbulence intensity flows, including the reversed-flow ambiguity of hot-wire-probe signals. Chapter 12 describes time-series errors caused by restrictions in the sampling rate and the sampling time. For a general introduction to uncertainty analysis the reader is referred to papers by Moffat (1982, 1985) and by Kline (1985).

3.10 CONCLUSION

This introduction to velocity measurements by the *HWA* method has outlined the individual steps that should be considered in order to achieve the selected experimental objectives. It has been demonstrated that the formulation of the objectives for a *HWA* study involves decisions related both to the type of measurements required and to the identification of a logical sequence of steps which are necessary to obtain the chosen objectives. The detailed information contained in the following chapters has been presented in accordance with the framework outlined in this chapter, in order to enable the reader to select and implement a complete digital *HWA* system for velocity and, if necessary, for temperature measurements. Using the material contained in this book, the reader should acquire an overview of a computer-based *HWA* system, which involves *HWA* equipment, computer and *A/D* systems, calibration and response equations, time-series analysis, and estimation of the overall signal uncertainty. It is hoped that this will result in improved planning and execution of *HWA* investigations in terms of the reliability of the results obtained and in a reduction in the overall time of each experiment.

4

ONE-COMPONENT VELOCITY
MEASUREMENTS

4.1 INTRODUCTION

This chapter is concerned with the use of a single normal (SN) hot-wire or hot-film probe for measurements of the velocity component in the mean-flow direction. The presentation will follow the step-by-step procedure outlined in Chapter 3, but in the case of an SN-probe some of the options are predetermined. For an SN-probe, the wire-fixed (x', y', z') and the probe-stem (x_s, y_s, z_s) coordinate systems are often the same. If the probe-stem is aligned with the mean-flow direction, see Fig. 4.1, and if this orientation is selected as the space-fixed coordinate system then the analysis can be carried out directly in terms of the velocity components (U, V, W) in the selected space-fixed system. This probe-stem orientation and analysis procedure will be used below unless otherwise stated.

4.2 SINGLE NORMAL HOT-WIRE PROBE TYPES

The standard SN hot-wire probe types usually have one of the configurations shown in Fig. 4.2(a,b). The main difference between the two probe types is that in Fig. 4.2(b) the sensing element has been removed by a plating technique from the prongs to minimize the aerodynamic probe disturbance,

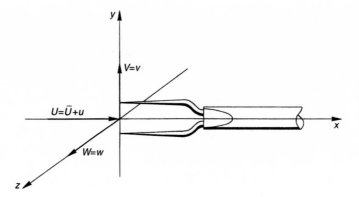

Fig. 4.1. An SN hot-wire probe placed with the probe-stem parallel to the mean-flow direction.

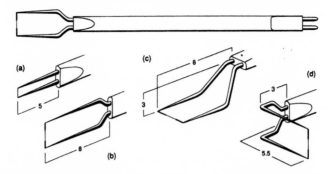

Fig. 4.2. Hot-wire-probe types: (a) unplated, (b) plated, (c) a boundary-layer probe, (d) a probe with bent prongs. The dimensions are in millimetres. (Reprinted with the permission of DANTEC Measurement Technology.)

(see Section 4.2.2). For most measurements, it is recommended that the plated-probe type, Fig. 4.2(b), is used in conjunction with a parallel probe-stem orientation. However, this probe orientation has disadvantages in boundary-layer studies since the probe-stem and the support prevent measurements near the wall. One solution is to use a specially adapted boundary-layer probe, in which the prongs are displaced from the probe-stem, as indicated in Fig. 4.2(c). This design enables the probe to be placed with the wire very close to the wall, with the prongs in the flow direction and the probe body also in the flow direction but out of the near-wall region. In this way, blocking effects and vortex shedding are avoided. Another solution is to place the probe-stem at an angle to the wall, since this orientation enables the wire element to be moved close to the wall. In some flow situations it may be necessary to use an SN-probe with the stem perpendicular to the mean-flow direction. However, in these cases, probe-interference effects caused by vortex shedding from the prongs and the stem may contaminate the hot-wire signal, as discussed in Section 4.2.3. To reduce the related prong interference, it is possible to bend the prongs at 90° to the probe-stem, as shown for the probe in Fig. 4.2(d). This reduces the vortex shedding from the prongs, but it does not significantly reduce the disturbances caused by vortex shedding from the stem. For flow measurements in a highly curved wall jet, strong interference effects were found even for probes which had been optimized with regard to aerodynamic interference in other regimes (Hartmann 1982). For accurate measurements in the near-wall region of wall-boundary-layer flows, Willmarth and Sharma (1984) and Ligrani and Bradshaw (1987a,b) have developed and used special subminiature probes (Fig. 4.3).

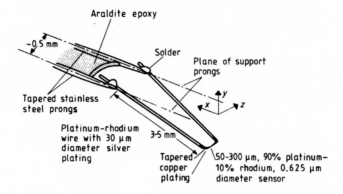

Fig. 4.3. A schematic diagram of a subminiature hot-wire sensor. (From Ligrani and Bradshaw 1987*a*.)

4.2.1 Hot-wire-probe design and pitch-angle effects

Hoole and Calvert (1967) and Gilmore (1967) presented the first results which demonstrated the effect of the probe geometry on the velocity field at the sensor location. An unplated *SN*-probe (DISA 55A25) was placed in a uniform flow with the sensing element perpendicular to the mean-flow direction, and the pitch, β, dependence (see Fig. 3.1) was measured by rotating the probe around the wire axis. Comparing the output voltages for a parallel- ($\beta = 0°$) and a perpendicular- ($\beta = 90°$) stem orientation they found a 15–20% variation in the evaluated velocity. Clearly, for the unplated probe type, the prong and stem configurations had a significant and undesirable effect on the flow field at the sensor position. The observation of the pitch effect resulted in more detailed investigations which aimed at identifying the effects of the probe geometry on the velocity field at the sensor position. Using a probe with a geometry similar to that illustrated in Fig. 4.2(b), Dahm and Rasmussen (1969) studied the change in the indicated velocity by varying the pitch angle, β, from 0° to 45° or 90°. The sensing element was a tungsten wire with an active length $\ell = 1$ mm and a diameter $d = 5\,\mu$m. The variable probe geometry was related to changes in the prong spacing, s, and prong length, L_p. Their results, shown in Fig. 4.4, confirmed the earlier pitch results for an unplated probe. If a plated probe with $s \geqslant 2$ mm is used, then the pitch dependence is seen to be reduced significantly. The length of the prong, L_p, was found to have only a small effect provided $L_p \geqslant 8$ mm. For a practical hot-wire probe, the minimization of the pitch dependence must be balanced with the normal requirement that the probe is relatively compact. The DANTEC *SN*-probe (55P01) with dimensions of $\ell = 1.25$ mm, $s = 3$ mm, and $L_p = 8$ mm appears to be a satisfactory compromise between these conflicting requirements.

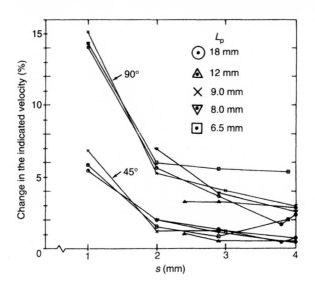

Fig. 4.4. The variation in the indicated velocity with the pitch angle, β, prong spacing, s, and prong length, L_p, for a plated *SN* hot-wire probe. (From Dahm and Rasmussen 1969.)

The pitch dependence may be evaluated by positioning the *SN*-probe so that the yaw angle, α, (Fig. 3.1), is zero. Selecting the parallel probe-stem orientation as a reference ($\beta = 0°$) and introducing the related space-fixed velocity components into eqn (3.2) gives

$$V_e^2 = U^2 + h^2 W^2 = \tilde{V}^2(\cos^2\beta + h^2\sin^2\beta), \qquad (4.1)$$

where $\tilde{V}$ is the magnitude of the velocity vector. If $E(\beta)$ is measured at a constant flow velocity, $\tilde{V}$, and it is assumed that the power-law relationship in eqn (3.1) applies, that is,

$$E^2 = A + BV_e^n,$$

then $h(\beta)$ can be obtained from

$$h = \frac{1}{\sin\beta}\left[\left(\frac{E^2(\beta) - A}{E^2(0) - A}\right)^{2/n} - \cos^2\beta\right]^{1/2}. \qquad (4.2)$$

The variation in the value of h with the pitch angle, β, and with the flow velocity has been studied in a number of investigations for both unplated and plated standard *SN*-probes (Jørgensen 1971a; Bruun and Tropea 1980; Samet and Einav 1985; Mobarak *et al.* 1986; Wagner and Kent 1988a). In general, it was found that the variation in h with β was small and that the value of h decreased slowly with increasing velocity. For an unplated, standard *SN*-probe, a typical value of h is about 1.10–1.12, and the corresponding value for a standard plated wire is about 1.05. Larger values for h for unplated probes have been quoted in two of the above investigations. In the study by Mobarak $et\,al.$ (1986) the value of the calibration constant A in the power-law relationship of eqn (3.1) was set equal to E_0^2, where E_0 is the voltage at zero velocity. As shown in Section 4.4.1 such an approach will result in a poor curve-fit. The values of h evaluated by Samet and Einav (1985) were based on an unusual nonsymmetric pitch dependence for $V_e(\beta)$. However, it is possible that a significant uncertainty occurred in their $\beta = 0°$ reference position, since an offset, $\Delta\beta$, of 3–4° will result in a nearly symmetrical pitch dependence for their $V_e(\beta)$ results.

4.2.2 Aerodynamic-disturbance effects

The above investigations demonstrated the variation in the indicated velocity with the pitch angle; however they did not provide any direct evidence of the individual effects caused by the prongs and the stem or how these effects may explain the response of the hot-wire probe to the three velocity components (U, V, W). These aspects have been investigated by Comte-Bellot $et\,al.$ (1971), Adrian $et\,al.$ (1984), and Merati and Adrian (1984).

In the investigation by Comte-Bellot $et\,al.$ (1971) the perturbations caused by the various parts of the probe were studied using a large probe, with a prong spacing of $s = 8\,\mathrm{mm}$ and a prong length of $L_p = 38\,\mathrm{mm}$ as a reference. The total variation in the indicated velocity with the pitch angle for the reference probe was only about 1.5%. The relative magnitudes of the individual stem and prong effects are shown in Fig. 4.5 for a modified DISA 55A25 probe. The three main conclusions were: (i) the total probe

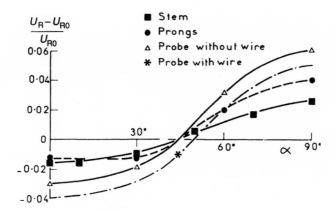

Fig. 4.5. Perturbations resulting from the stem, prongs, and wire in relative order of magnitude for a modified DISA 55A25 SN-probe. (From Comte-Bellot $et\,al.$ 1971.)

disturbance was approximately equal to the sum of the disturbances from the individual elements of the probe; (ii) the largest perturbation was usually caused by the prongs; and (iii) global pitch tests demonstrated that the observed variation in the indicated velocity was due to a modification of the flow pattern by the probe, and not to a modification of the temperature of the prongs. From their extensive measurements, and the use of a simple potential flow model, they concluded that the aerodynamic pitch dependence may be reduced to about 2.5% if a probe with the following dimensions is used:

stem diameter:	3 mm
prong length	20 mm
prong diameter	0.6 mm
prong spacing	6 mm.

However, long thin prongs are aerodynamically undesirable since they are prone to vibrations due to a combination of their small stiffness and their low natural frequency. As discussed by Vagt (1979), for a specified probe disturbance, it is possible to reduce the length of the prongs by a factor of two if the prong diameter is tapered from 0.5 mm at the base to about 0.05 mm at the tip.

The magnitude of the flow perturbation caused by the various parts of the probe have been evaluated theoretically by Adrian et al. (1984) and by Merati and Adrian (1984) using a potential-flow slender-body theory. However, the flow separation, which occurs at large values of the pitch angle, is not accounted for by this theory. The relationship between the undisturbed flow field, $V = (U, V, W)$, and the disturbed flow field, $V^* = (U^*, V^*, W^*)$ at the sensor position was derived using a small-perturbation theory

$$U^* = U(1 + \delta_x), \qquad V^* = V(1 + \delta_y), \qquad W^* = W(1 + \delta_z),$$

and expressions were presented for δ_x, δ_y, and δ_z in terms of the geometry of the prongs and stem and the orientation of the probe. To relate the evaluated flow perturbations to the response equation for V_e (eqn (3.1)), these investigators assumed that the perturbed effective velocity, $V_e = V_e^*/(1 + \delta_x)$, had the following component relationship

$$(V_e^*)^2 = (U^*)^2 + k^2(V^*)^2 + (W^*)^2.$$

The predicted pitch dependence was found (for probes with straight wires) to agree reasonably well both with their experimental results and with the earlier results of Comte-Bellot et al. (1971).

The information contained in these papers has resulted in increased confidence in the correct use of SN hot-wire probes. For a well-designed plated probe the value of the pitch coefficient, h in eqn (3.2) will typically be about

1.05. The related perturbation, δ_x, of the velocity component in the mean-flow direction is also small, and it is virtually independent of the flow velocity. It is accounted for, in practical *HWA*, by the calibration procedure.

4.2.3 Probe-interference effects caused by vibration

Prong vibration Another interference effect may occur, when the pitch angle, β, is in the range $60° < \beta < 120°$ and the wire is straight between the prongs (Vagt 1979). In this position, the prongs may vibrate due to vortex shedding. The vibration of the prongs can cause a periodic change of the electrical resistance of the wire, and a so-called strain-gauge signal is generated. The strain-gauge effect was demonstrated by measuring the rms (root mean square) values of the fluctuations. When the *SN*-probe was inserted into a flow field and rotated around the wire axis the result remained unchanged until a specific pitch angle was reached (in their experiment, $\beta \approx 55°$). At this position the rms signal increased sharply, demonstrating the onset of the strain-gauge effect. For this reason, the manufacturers of hot-wire probes usually mount the wire with a small amount of slack. This procedure is also recommended to users doing their own probe repairs.

Wire vibration Regular vortex shedding from a circular cylinder in a cross flow is observed when the Reynolds number, Re, is greater than ~40. This phenomenon was observed by Collis and Williams (1959) for a $53\,\mu\mathrm{m}$ diameter hot-wire probe in air. For a standard 4–$5\,\mu\mathrm{m}$ diameter tungsten wire probe, this condition corresponds to a flow velocity of greater than $100\,\mathrm{m\,s}^{-1}$. Also, in a laboratory flow, the Kolmogorov length scale, η, may be as small as $100\,\mu\mathrm{m}$, and viscous dissipation is likely to restrict the development of regular vortex shedding.

Cylindrical hot-film probes (with typical diameters of 50 or $70\,\mu\mathrm{m}$) are, however, prone to vortex shedding, as reported by Pitts and McCaffrey (1986). The onset of regular vortex shedding occurs at a relatively low velocity, of about $10\,\mathrm{m\,s}^{-1}$ in air flows; it can be observed in practical *HWA* by both a change in the shape of the heat-transfer relationship and by noise in the fluctuating signal.

Large vibrations of a hot-wire filament have been observed by Perry and Morrison (1971*b*, 1972) and by Perry (1972, 1973), who have developed a related theory for an elastically bowed wire due to thermal expansion. The phenomenon occurred during measurements of velocity-perturbation profiles across a number of Kármán-vortex streets shed from circular cylinders. Under such conditions, it is possible that the hot-wire skips into a nearly circular orbit with the same frequency as the Kármán-vortex street. This means that the wire will move with the perturbing force, thus causing a decrease in the anemometer output because of a decrease in the fluid velocity relative to the hot-wire. However, current evidence suggests that

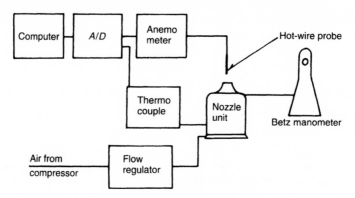

Fig. 4.6. A nozzle-calibration facility.

vibrations of hot-wire filaments will only be important when the spectrum of the velocity perturbation is narrow and the heated filament is not straight. Consequently, for measurements in broad-band turbulence, the effect (if any) of wire vibrations can usually be ignored.

4.3 CALIBRATION OF AN *SN* HOT-WIRE PROBE

In many hot-wire applications the anemometer equations selected and the corresponding calibration method are intrinsically linked; that is, they cannot be specified separately. This condition usually applies when the calibration involves more than one variable, such as for example, velocity and yaw. However, for an *SN*-probe the relationship is a one-dimensional one, that is

$$E = F(U), \tag{4.3}$$

where E is the output from the anemometer connected to the *SN*-probe and U is the velocity component in the mean-flow direction. In this case it is possible to consider the calibration procedure and the selection of the response equation separately. Due to the individual features of each probe, to variations in the anemometer setting, and to possible variations in the flow temperature, it is necessary to carry out a calibration each time a hot-wire probe is used for an experiment. This involves placing the probe in a flow with a known variable reference velocity, U_R, and a low turbulence intensity ($Tu < \sim 0.5\%$). The flow velocity must be perpendicular to the wire element, and the stem orientation must be specified (it is usually parallel to the flow). In a wind-tunnel the value of the reference velocity, U_R, can be measured by placing a pitot-static tube next to the hot-wire probe, provided the velocity profile is flat. If a special nozzle-calibration facility is used, as in Fig. 4.6, then U_R can be determined from the pressure drop across the nozzle. Placing the probe about one diameter down-stream from the nozzle exit, the corresponding static pressure, P, will be equal to the atmospheric pressure, P_{atm}. By measuring the corresponding total pressure, P_0 (which is equal to the stagnation pressure in the nozzle settling chamber), relative to P_{atm}, the flow velocity can be evaluated from $P_0 - P_{atm} = \frac{1}{2}\rho U^2$, where ρ is the density of the fluid. However, for a *short nozzle*, Bremhorst and Listijono (1987) have shown that the static pressure in the exit plane may deviate from P_{atm} due to the formation of a *vena contracta*.

The purpose of a calibration experiment is to obtain a set of calibration points (E_R, U_R), usually 10–30 points, spaced evenly over the selected velocity range ($U_{R,min}$ to $U_{R,max}$). Due to fluctuations occurring in even the best calibration facilities, the reference values (E_R, U_R) should be the time-mean values over a suitable time period of say 10–30 s. In the investigation by Bruun *et al.* (1988), it was shown that similar results were obtained using 10, 20, or 40 calibration points. This experimental result is supported by the numerical simulation experiment of Swaminathan *et al.* (1984b) which iden-tified, for practical applications, a sample size of 20–30 points. To reduce the uncertainty in the calibration data, great care should be exercised during the calibration. Since a *CT* anemometer has a very high frequency response, the output voltage, E_R, will usually correspond to the existing velocity U_R. However, since U_R is normally varied in a stepwise manner, a significant difference can occur between the true value of U_R and the value obtained using the ΔP reading of the pressure-measuring device. Due to the length of the connection tubes and the volume of air contained in the pressure-measuring device, its response time will be very slow, and a significant settl-ing time must be used for each calibration point. This problem can be minimized by using a high-quality, flush-mounted, pressure transducer.

4.3.1 *In situ* calibration versus the use of special calibration facilities

The most satisfactory arrangement is to calibrate the hot-wire probe in the facility in which the measurements are to be taken, since the disturbance caused by the probe and its holder and the influence of the geometry of the test facility will be the same during calibration and experiment. How-ever, in practice it may not always be possible to carry out such an *in-situ* calibration. The test flow may not contain a known potential flow part, constraints may exist in the variation of the velocity, and (if appropriate) in the yaw and pitch angles. Because of these restrictions, it is often neces-sary to calibrate the probe in either a small wind-tunnel reserved for this purpose or in a separate, special, calibration facility, which can be obtained commercially. This approach may have several disadvantages. If the test facility and the calibration unit are placed on different sites, it will neces-sitate moving the probe, anemometer, and computer system. Such a move

requires the anemometer to be switched off and in many cases the discon-
nection and reconnection of probe and cables. This is a highly undesirable
approach since a large number of uncertainties may be introduced between
the calibration and the experiment. A significant improvement to this problem has been implemented by the author and his students. The nozzle-
calibration facility (DANTEC 55D90) illustrated in Fig. 4.6 is relatively
small and therefore highly mobile. Consequently, the nozzle unit was incorporated as a (removable) part of the test facility by placing it in a special
holder just outside the test rig. The traversing mechanism and the related
side wall of each test rig were designed in such a manner that the probe
could be withdrawn from the test facility and exposed to the nozzle flow
without disconnecting the probe. Calibrations have also been carried out
by placing the nozzle-calibration facility inside a wind-tunnel. This approach
solved most of the problems caused by using a separate calibration facility.
However, this approach does necessitate an integrated design approach to
the calibration and testing.

In addition to the above problems, the author has found that small
calibration facilities have an additional disadvantage. The aerodynamic-
disturbance effects described in Section 4.2.2 correspond to the probe being
placed in an infinitely large uniform flow. This situation corresponds closely
to a probe being placed in a large wind-tunnel. However, the cross-sectional
area of most special calibration facilities is relatively small, and for the
DANTEC nozzle-calibration unit the probe represents a sizeable blockage
of the exit-jet flow, resulting in a deflection of the nozzle flow around the
probe. Khan *et al.* (1987) and Khan (1989) measured the corresponding
difference in the interpreted velocity and found a difference of about 3%
for the largest nozzle ($d = 12.4$ mm).

4.4 HOT-WIRE CALIBRATION EQUATIONS

In selecting a hot-wire response equation an investigator may be faced with
conflicting objectives, such as high accuracy or ease of use. It is the purpose
of this section to compare a number of methods in terms of their accuracy
and/or their ease of implementation on a computer system. To demonstrate that a selected relationship has a good accuracy, it is necessary to
minimize any calibration errors, to select an appropriate method of curve
fitting, and to specify the resultant uncertainty. Also, the accuracy of the
calibration facility itself should be better than that required for the *HWA*
equipment.

To optimize the calibration constants which appear in the selected hot-
wire relationship, $E = F(U)$, a least-squares curve-fitting method, based on
the sum of errors squared (*SES*) may be applied to, for example , the
voltage difference $E_R - E_C$; that is,

$$SES = \sum_{i=1}^{N} (E_R - E_C)^2. \tag{4.4a}$$

Alternatively the *SES* may be evaluated from

$$SES = \sum_{i=1}^{N} (E_R^2 - E_C^2)^2, \tag{4.4b}$$

which simplifies the curve-fitting procedure when the power-law relationship in eqn (4.8) is used (Pitts and McCaffrey 1986). E_R is the measured calibration voltage and E_C is the voltage calculated from the selected relationship, $E = F(U)$, using the measured calibration velocity, U_R; that is,

$$E_C = F(U_R). \tag{4.5}$$

The *SES* method may be applied to any analytical function, and it will determine the optimum value of the calibration constants involved. For comparison purposes, the *goodness of fit* of the selected hot-wire response relationship may be described by the related normalized standard deviation, ε_u,

$$\varepsilon_u = \left[\frac{1}{N} \sum_{i=1}^{N} (1 - U_R/U_C)^2 \right]^{1/2}, \tag{4.6}$$

where U_C can be obtained by inverting eqn (4.5) using the now-known calibration constants:

$$U_C = F^{-1}(E_R). \tag{4.7}$$

4.4.1 Power laws

Following the original hot-wire investigation by King (1914), the relationship $E = F(U)$ is often assumed to be of the form

$$E^2 = A + BU^n. \tag{4.8}$$

King's evaluations suggested that $n = 0.5$, but the results obtained by Collis and Williams (1959) have demonstrated that a better curve fit can be obtained using $n = 0.45$, provided $0.02 < \text{Re} < 44$. This is an unusually large velocity range since a constant-exponent power law is often only a good curve fit over a moderate velocity range, U_{max}/U_{min}, of 10–20. In later detailed investigations the value of n has been treated as a variable parameter. The value of $n = n_{opt}$ which gave the smallest value of *SES* in eqn (4.4a or b) was chosen as the optimum condition by van Thinh (1969), Bruun (1976a), Bruun and Tropea (1980, 1985), Swaminathan *et al.* (1983), and Pitts and McCaffrey (1986). A typical variation with n in the related normalized standard deviation, ε_u, is shown in Fig. 4.7. At moderate

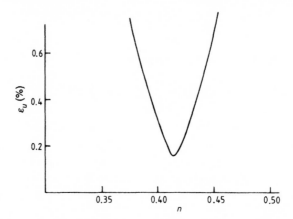

Fig. 4.7. Variation in ε_u with the exponent n in the power-law relationship given by eqn (4.8).

velocities, the optimum value of n for a typical $5\,\mu\text{m}$ tungsten hot-wire probe usually lies in the range 0.4–0.45. Figure 4.7 also shows that the uncertainty increases rapidly when n deviates from n_{opt}. A similar approach, keeping A as a variable parameter and solving for B and n, is reported in Bruun *et al.* (1988). Attempts have been made to use the power-law relationship with $A = E_0^2$, where E_0 is the voltage at zero velocity. However, several investigators (Kjellström and Hedberg 1970; Bruun 1971; Elsner and Gundlach 1973) have shown that this results in n being a strong function of the velocity, which greatly complicates its use. It is recommended instead that A is treated as a variable parameter ($A \approx 0.8E_0^2$). Finally, results have also been reported for the simultaneous solutions for A, B, and n (Swaminathan *et al.* 1986a; Bruun *et al.* 1988). The accuracy of these various methods is compared in Section 4.5.

4.4.2 Extended power laws

Van der Hegge Zijnen (1956) proposed, for an extended Reynolds number range, $0.01 < \text{Re} < 500\,000$, the use of the following nondimensional heat-transfer equation

$$\text{Nu} = 0.35 + 0.5\text{Re}^{0.5} + 0.001\text{Re}. \qquad (4.9)$$

Similar relationships have subsequently been suggested by Richardson (1965) and by Fand and Keswani (1972). This approach has been applied to hot-wire probes by Davies and Patrick (1972) and by Siddall and Davies (1972) in the form

$$E^2 = A + BU^{0.5} + CU, \qquad (4.10)$$

from which U can be obtained by a simple inversion process.

4.4.3 The universal-function principle

Due to the limitations imposed by various analytical hot-wire relationships, Bruun (1971) investigated whether a *universal function principle* could be established and implemented. The closest approximation to a universal-shape function was obtained using the equation form

$$E^2 - E_0^2 = CF(U), \qquad (4.11)$$

where E_0 is the measured voltage at zero velocity, $F(U)$ is the universal function, and C is an individual constant for each probe and test. The validity of this method was investigated by placing two SN-probes 1 and 2, simultaneously in a uniform calibration flow. Provided that eqn (4.11) is valid, the ratio $(E^2 - E_0^2)_1/(E^2 - E_0^2)_2$ should remain constant, that is, independent of the velocity. The typical results shown in Fig. 4.8 demonstrate that the principle applies with a high degree of accuracy at moderate and high velocities but some deviation is observed at low velocities. Also, this method was restrictive since it was necessary to use a specific probe type, a fixed probe orientation, and a nearly fixed overheat ratio. Based on measured data, Bruun (1971) evaluated a universal-shape function in a tabulated format. This function has been used to provide reference data for a number of subsequent calibration evaluations, in particular, those aimed at establishing the validity of calibration relationships over large velocity ranges (Siddall and Davies 1972; Freymuth 1972; Kinns 1973; Nishioka and Asai 1988). With the general availability of digital computers for both calibration and measurement purposes the author no longer recommends the use of this method.

4.4.4 Spline-fits

Curve fitting by splines is a recent mathematical tool. If one fits a polynomial through n points, then if n is large either solution instability occurs or the polynomial fit will generally have many wiggles in it. However, it is known that such wiggly behaviour is inconsistent with the calibration of

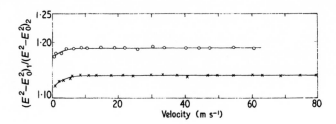

Fig. 4.8. The difference in the shape of the universal calibration law given by eqn (4.11), for two sets of two 2 mm long, hot-wire probes.

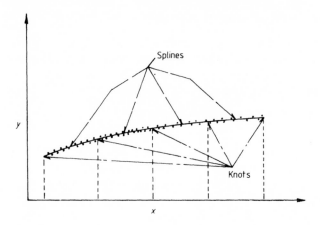

Fig. 4.9. The spline-fit notation and principle: (•) experimental points.

hot-wire probes. Consequently a curve-fitting method that guarantees a smooth fit through a sequence of points may have many advantages. A cubic-spline-fitting method satisfies this requirement. The spline method is the numerical analog of using a thin, uniform, flexible strip (a drafting spline) to draw a smooth curve through and between points on a graph. These points through which the curve is required to pass are called the knots of the spline. The data range is divided into n subintervals, see Fig. 4.9, and to obtain a smooth fit of the curve, each subinterval is approximated by a cubic function $S(x)$ with the following properties:

(1) within each subinterval, $x_j \leqslant x \leqslant x_{j+1}$, $S(x)$ is a cubic polynomial;
(2) the function $S(x)$ and its two derivatives S', S'' are continuous over the whole range $x_0 \leqslant x \leqslant x_n$.

4.4.5 Polynomial curve fits

Based on historical and physical considerations, the hot-wire equations described in the previous sections have been specified in the form $E = F(U)$ (eqn (4.3)). With this formulation, it is necessary to carry out an inversion process, $U = F^{-1}(E)$ (see Section 4.12.2), in order to obtain the required velocity information. However, mathematically there is no reason why the hot-wire relationship should not be expressed in the form $U = F(E)$.

George *et al.* (1981, 1989) introduced a polynomial equation of the form,

$$U = A + BE + CE^2 + DE^3 + \dots \qquad (4.12)$$

Expressing the velocity U in a polynomial form of E has computational advantages since the required velocity, U, can be obtained directly from eqn (4.12). This method was tested, as described in Section 4.5, by a computer program which could fit a polynomial of up to the tenth degree to raw calibration data. However, it was found that the solution of the polynomial became unstable above the fourth order.

The accuracy of a polynomial fit in terms of E^2 has been investigated by Swaminathan *et al.* (1986*a*) by applying a least-squares fitting method to the following second- and third-order E^2 models

$$U = A + BE^2 + C(E^2)^2, \tag{4.13}$$

$$U = A + BE^2 + C(E^2)^2 + D(E^2)^3. \tag{4.14}$$

The accuracy of these and the other methods is described in the next section.

4.5 THE ACCURACY OF HOT-WIRE CALIBRATION EQUATIONS

A comparative investigation was carried out to establish the relative merit of the above hot-wire equations (Bruun *et al.* 1988). Selected results are presented below for a *SN* plated probe (DANTEC 55P01) using a moderate velocity range of 5–50 m s^{-1}. Similar results apply to other *SN*-probe types and velocity ranges. The quoted results are also in broad agreement with the results obtained by Bruun and Tropea (1980, 1985) and by Swaminathan *et al.* (1986*a*). First, to establish the reliability of the raw calibration data, a power-law equation was applied.

Power law Typical results are shown in Table 4.1 using the three curve fitting methods proposed in Section 4.4.1. Table 4.1 demonstrates that a good accuracy ($\varepsilon_u \simeq 0.1$–0.15%) can be obtained in all three cases. The curve fit corresponding to simultaneous solutions for A, B, and n can be seen to give the best accuracy, but the complexity of this method may not justify the moderate increase in the accuracy, particularly if a microcomputer system is used.

Having proved the suitability of the calibration data by the power-law method, the other hot-wire relationships outlined in Section 4.4 were applied to the same calibration data.

TABLE 4.1 *Simple and extended power-law curve fits, for a velocity range of 5–50 m s^{-1}*

	A	B	C	n	ε_u (%)
$E^2 = A + BU^n$					
Simultaneous A, B, n solution	5.980	4.608	–	0.4137	0.11
Increment, n	6.018	4.587	–	0.4145	0.15
Increment, A	6.005	4.592	–	0.4143	0.15
$E^2 = A + BU^{1/2} + CU$	7.544	3.449	−0.005	0.5	0.46

TABLE 4.2 *Polynomial curve fits in the velocity range* $5-50 \, m \, s^{-1}$

Equation	ε_u (%)
$U = A + BE + CE^2$	2.5
$U = A + BE + CE^2 + DE^3$	0.24
$U = A + BE + CE^2 + DE^3 + FE^4$	0.15
$U = A + B(E^2) + C(E^2)^2$	0.80
$U = A + B(E^2) + C(E^2)^2 + D(E^2)^3$	0.15

Extended power laws The accuracy of eqn (4.10) was evaluated and the corresponding results are also presented in Table 4.1. The value obtained for $C(\sim -0.005)$ demonstrates that the term CU can be considered as a small correction to the simple power law. Since $n = 0.5$ a simple conversion is possible, but the value of ε_u is seen to have increased by a factor of three to four compared with a simple power law. This result is consistent with the variation in ε_u with n for the simple power-law shown in Fig. 4.7.

Polynomial fits The calibration data used in Table 4.1 were also used as the input data to a polynomial-fit program. Table 4.2 gives the results for the quadratic, the cubic, and the fourth degree in E, as well as for the two variations of the polynomial equation in E^2 suggested by Swaminathan *et al.* (1986*a*). The general trend shown in Table 4.2 was confirmed by different tests, and several conclusions can be drawn from these results. The curve fit is not as good as for the simultaneous solution of A, B, and n for the simple power law of eqn (4.8), but the best-polynomial-fit solutions compare favourably with the accuracy obtained for eqn (4.8) using a variable n for the optimization. To achieve this accuracy, it is necessary to use either a full fourth-order polynomial in E or a third-order polynomial in E^2. These conclusions are consistent with the majority of the results given by Swaminathan *et al.* (1986*a*).

Spline fits To evaluate the spline fits it was necessary to use two subroutines of a NAG Fortran library on the input data (from Table 4.1). One of these subroutines evaluated the coefficients of the cubic spline, whilst the other evaluated the spline from the coefficients. The other input, apart from the calibration data, was the number of subintervals selected for the data range. These results are presented in Table 4.3 with the data range divided into a number of equal subintervals (1 to 10). The results in Table 4.3 clearly show that, in terms of curve fitting accuracy, a spline-fit method with many subintervals is superior to any of the other methods considered. However, in computational terms, it is a rather cumbersome and time-consuming method. Consequently, based on current computer technology, it may not yet be suitable for implementation on a minicomputer/micro-

TABLE 4.3   *Spline-fits in the velocity range*
5–50 m s^{-1}

Number of intervals	ε_u (%)
1	0.23
2	0.14
4	0.14
6	0.071
8	0.053
10	0.055

computer. The results in Table 4.3 are, however, useful as a reference. It is likely that the value of ε_u ($\sim 0.05\%$), achieved when the number of subintervals is greater than seven, mainly relates to the measurement error and only contains a small contribution from curve-fitting approximations. Consequently, for the other methods, the amount by which ε_u exceeds 0.05% will mainly reflect errors due to curve-fitting approximations to the true hot-wire curve. If this is the case, the accuracies ($\varepsilon_u \sim 0.1$–0.15%) obtained by the simple power law and the best polynomial fits seem quite adequate for most purposes.

4.6 HOT-WIRE CALIBRATION EQUATIONS FOR EXTENDED VELOCITY RANGES

Several of the equations discussed in Sections 4.4 and 4.5 have been shown to give a good curve fit over a moderate velocity range on the basis of the ε_u (eqn (4.6)) criterion. However, ε_u only represents the 'average' error over the velocity range. The variation in the local error $(U_C - U_R)/U_R$ was calculated using a power-law curve fit for the calibration data from Table 4.1. The results are plotted in Fig. 4.10, and they show a good curve fit for n_{opt} and a poor approximation for $n = 0.5$. The related velocity sensitivity, S_u (the slope of the calibration curve dE/dU), can, for the power-law equation, be evaluated from

$$S_u = \frac{dE}{dU} = \frac{nBU^{n-1}}{2E}, \qquad (4.15)$$

and Fig. 4.11 contains the corresponding results for $n = n_{opt}$ and $n = 0.5$. The value of dE/dU can be seen to be strongly dependent on the value of n, as will be discussed further in Section 4.9.

The extended power law of eqn (4.10), was introduced by Siddall and Davies (1972) and Davies and Patrick (1972); they claimed that it gave a good curve fit over a large velocity range (0–160 m s^{-1}). However, sub-

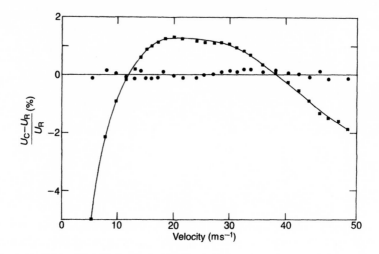

Fig. 4.10. The variation in the curve-fit error $(U_C - U_R)/U_R$ with velocity for: ($\bullet$) $n = n_{opt}$, and ($\blacksquare$) $n = 0.5$ (eqn (4.8)).

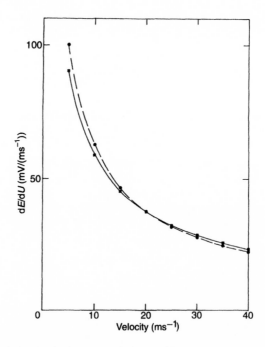

Fig. 4.11. The variation in the hot-wire velocity sensitivity dE/dU (eqn (4.15)) with velocity for: ($\bullet$) $n = n_{opt}$, and ($\blacksquare$) $n = 0.5$ (eqn 4.15)).

sequent computer-based calibration investigations by Bruun and Tropea (1985), Swaminathan *et al.* (1986*a*), and Bruun *et al.* (1988) have all shown that the extended power law of eqn (4.10) does not give a more accurate curve-fit than the single power law of eqn (4.8) even over an extended velocity range. Consequently, for measurements over an extended velocity range, it is not recommended that a single analytical function is used for the complete velocity range. Even if a low value of ε_u is obtained in such cases, it may not reflect the real uncertainty in the lower or upper part of the velocity range (Bruun *et al.* 1988). One solution to this problem would be to divide the velocity range into two or more subranges, and then to apply a separate optimization in each velocity subrange. By subsequent computer logic, the complete velocity range can be covered by a number of different hot-wire relationships, each applicable over a specific voltage and therefore over a specific velocity range. That this principle may lead to a closer approximation to the true hot-wire equation was clearly demonstrated by the results for the spline-fit method (Section 4.4.4), in which the total velocity range was divided into a number of subintervals.

4.7 CALIBRATION AT LOW VELOCITIES

The velocity-sensitivity curve, dE/dU, in Fig. 4.11 illustrates that *HWA* is, in principle, a very accurate method for low-velocity investigations. In the conventional calibration methods described in Sections 4.3 to 4.6 the velocity was evaluated from a pressure difference reading, ΔP, using either a pitot-static tube or the pressure drop across a calibration nozzle. However, when the velocity becomes smaller than about 3 m s^{-1}, the value of ΔP in air flows becomes so small ($\Delta P < 1 \text{ mm } H_2O$) that it is difficult to obtain accurate velocity results by these methods. Consequently, at low velocities, a number of special calibration methods have been proposed as outlined below.

Laminar pipe-flow calibration In several investigations (Almquist and Legath 1965; Andrews *et al.* 1972; Koppius and Trines 1976; Manca *et al.* 1988; and Lee and Budwig 1991), the calibration was carried out using a fully developed laminar flow in a circular pipe. For such a flow, $U_{\mathcal{C}} = 2U_a$, where $U_{\mathcal{C}}$ is the maximum velocity on the centre line and U_a is the area-averaged velocity. Consequently, by placing a probe on the centre line, the corresponding velocity can be evaluated by measuring the flow rate either by a flow meter or by slowly filling (or draining) a container with water. For small tube diameters, it is necessary to consider the blockage effect caused by the probe (see Section 4.3.1) and to evaluate the integration effect of the velocity profile, $U(r)$, over the wire length.

Conservation of mass principle For measurements in wind-tunnels, Bradshaw (1971) has proposed that the probe is placed in the settling chamber of the wind-tunnel upstream of the contraction and that the velocity is measured by means of a pitot-static tube placed in the working section of the wind-tunnel. By this method it may be possible to obtain a speed reduction of 1:10. However, it is necessary to check for the uniformity of the velocity profile across the cross-sectional areas where the two probes are placed, and to compensate for the growth of the boundary layer. The principle of the conservation of mass was also applied in an investigation by Seifert and Graichen (1982). Their calibration was based on measurements, with the hot-wire probe being calibrated, of the flow rate and of the corresponding anemometer output voltage profiles in the exit plane of a calibration nozzle.

Vortex shedding method This procedure utilizes the regular nature of the vortex shedding behind a circular cylinder at low Reynolds numbers. According to Roshko (1954), the relationship for the nondimensional Strouhal Frequency, F, can be expressed as

$$F = 0.212Re - 4.5, \quad (50 < Re < 150),$$
$$F = 0.212Re - 2.7, \quad (300 < Re < 2000). \tag{4.16}$$

Here, F, and Re are given by $F = fd^2/v$ and $Re = dU/v$, where f is the frequency of vortex shedding, d is the cylinder diameter, v is the kinematic viscosity, and U is the flow velocity. An anemometer probe is required to record the vortex-shedding phenomenon, and this technique is reported as having been used successfully by Perry (1982) by placing a 3 mm wire across the working section of a wind-tunnel. This method has also been used by Berger (1964), and it was applied by Lee and Budwig (1991), who used an empirical vortex-shedding relationship.

Moving probe calibration Calibration methods which rely on the known motion of the probe have also been used. If the fluid is a liquid then it is feasible to move the fluid (in a container) past the probe, as described by Dring and Gebhart (1969), but such a method is impractical for a gas. For gases (and also for liquids), several investigations have been carried out in which the probe is moving. As a probe only responds to the relative velocity of the fluid, such a method should, in principle, give a correct calibration of the anemometer probe. To ensure accurate calibration the following criteria must be satisfied: (i) the velocity of the probe and the related anemometer signal must be measured accurately, (ii) the fluid must be stationary (or move with a known velocity), and (iii) the motion of the probe must be virtually free of vibrations.

Several investigations, utilising the linear motion of a probe mounted on a sled, have been reported for the velocity range 0–3 m s⁻¹ (Baille 1973; Aydin and Leutheusser 1980; Tabatabai *et al.* 1986; Tsanis 1987; Heikal *et al.* 1988; Tewari and Jaluria 1990). (The calibration of a moving hot-film probe in water is described in Section 4.10.2.) The details of these investigations centre around procedures for obtaining a quiescent fluid, and the development of control systems to ensure a constant velocity of the sled during the calibration period for each calibration point. Consequently, all these methods require calibration runs at different calibration velocities to obtain a set of conventional calibration points (E_R, U_R) over the selected velocity range. For practical applications, the measured calibration data must be curve fitted, and in principle any of the methods described in Section 4.4 can be used. The results of Aydin and Leutheusser (1980) and Tsanis (1987) have shown that the (E_R, U_R) calibration data can, with a very high degree of accuracy, be approximated by the simple power law of eqn (4.8), provided that the value of U_{max}/U_{min} does not exceed 10–20. Their results demonstrated that the exponent n varies with the velocity range used: for $0.04\,\mathrm{m\,s^{-1}} < U < 0.2\,\mathrm{m\,s^{-1}}$, $n \cong 1.0$; for $0.05\,\mathrm{m\,s^{-1}} < U < 0.7\,\mathrm{m\,s^{-1}}$, $n \simeq 0.6$–0.8; and for $0.7\,\mathrm{m\,s^{-1}} < U < 3\,\mathrm{m\,s^{-1}}$, $n = 0.4$–0.41. The results of Heikal et al. (1988) indicate that, without much loss in accuracy, the applicable velocity range can be extended by using an extended power law.

To reduce the related calibration time and procedure, Bruun et al. (1989) and Haw and Foss (1990) have proposed the use of a variable-velocity calibration method. Such a method can be developed if modern digital techniques are applied to a single sweep with a variable probe velocity. Using a hot-wire probe placed at the tip of a swinging arm, of radius R, it is possible during one traverse to record simultaneously both the anemometer output, $E(t)$, and the angular position $\varphi(t)$ of the probe. The probe velocity $U(t) = R\omega(t)$ can be obtained by differentiating $\varphi(t)$ (Haw and Foss 1990). By applying a least-squares curve-fitting technique to selected data from the simultaneous $E(t)$ and $U(t)$ time histories, the calibration constants in the response equation can be determined. The disadvantage of this procedure is that each set of (E, U)-values corresponds to a single reading, instead of to conventional time-averaged values. A single-sweep method will therefore be sensitive to probe vibrations and to digital quantization errors caused by the n-bit nature of all digital A/D converters. To minimize these errors, Bruun et al. (1989) combined the variable-velocity single-sweep method with a time-integration procedure.

4.8 CALIBRATION AT VERY LOW VELOCITIES (MIXED-FLOW REGIME)

The methods described in Sections 4.3 to 4.7 refer to flow situations where the heat transfer from the sensor is dominated by forced convection. When the forced-convection velocity is reduced, a mixed-flow regime is encountered in which both forced- and natural-convection heat-transfer phenomena influence the hot-wire response equation. Collis and Williams (1959) carried out heat-transfer studies for long platinum wires in a horizontal air stream. The wire diameters used were in the range 3–53 μm and the minimum ℓ/d ratio was about 2000. The temperature loading, T_w/T_a where T_w is the wire temperature and T_a is the air temperature (both in Kelvin), was varied in the range 1.1–2.0. For forced convection, they demonstrated that the data from the three wires tested and all the overheat ratios could be expressed in terms of a general heat-transfer relationship (eqn (2.3)) which included a temperature loading factor

$$\text{Nu}_f \left(\frac{T_f}{T_a} \right)^{-0.17} = 0.24 + 0.56 \text{Re}^{0.45},$$

where $T_f = \frac{1}{2}(T_w + T_a)$ is the film temperature, Nu_f is the forced-convection Nusselt number, and Re is the wire Reynolds number. Their investigation also indicated that the lower Reynolds number limit, Re_c, for the forced convection flow region could be specified as

$$\text{Re}_c = \text{Gr}^{1/n}, \tag{4.17}$$

where Gr is the Grashof number (as defined in Section 2.1.1) and n is equal to about 3 for air. From their measurements, Collis and Williams (1959) estimated the value of Re_c to be 0.02. Mahajan and Gebhart (1980) indicated that the limit occurred at 0.04, while Ligrani and Bradshaw (1987a) observed deviations from their forced-convection heat-transfer relationship at a Reynolds number of 0.07.

Several authors have attempted to derive general equations for the heat transfer from a cylinder in the mixed-flow regime. The first study was by Van der Hegge Zijnen (1956), who developed separate response equations (in air) for natural and forced convection. He suggested the correlation of the data for the mixed-flow regime by vector addition of the natural- and forced-convection Nusselt numbers, but the comparison between theory and experiments was poor. Hatton et al. (1970) carried out a detailed investigation of forced- and natural-convection effects on the response of heated cylinders. Their study included flow directional effects by using horizontal flows, vertical-downwards flows and vertical upwards flows. The diameters of the cylinders were in the range 100–1260 μm (that is, they were substantially larger than a conventional hot-wire probe), and the length of 120 mm gave an ℓ/d ratio in the range 90–1200. The temperature difference, $T_w - T_a$, was varied from 30 to 200 °C. Their best-fit correlation for forced convection was

$$\text{Nu}_f \left(\frac{T_f}{T_a} \right)^{-0.154} = 0.384 + 0.581 \text{Re}^{0.439}, \tag{4.18}$$

which is of a similar format as eqn (2.3), proposed by Collis and Williams (1959). For natural convection they obtained

$$\text{Nu}_n \left(\frac{T_f}{T_a} \right)^{-0.154} = 0.384 + 0.59 \text{Ra}^{0.184}, \tag{4.19}$$

where $Ra = Gr \times Pr$ is the Rayleigh number (the subscript n denotes natural convection). As these last two equations have the same form, the authors argued that any natural convection can be expressed by an equivalent Reynolds number, Re_n, which they evaluated as

$$Re_n = 1.03Ra^{0.418}. \qquad (4.20)$$

For the mixed-flow regime they proposed the use of eqn (4.18) with Re being replaced by an effective Reynolds number, Re_e, obtained by vectorially adding the Reynolds numbers due to natural, Re_n, and forced convection, Re; that is

$$Re_e^2 = Re^2 + 2Re\,Re_n \cos \theta + Re_n^2, \qquad (4.21)$$

where θ is the angle between the forced-convection flow and the upward-vertical flow. However, significant differences were observed between their theoretical and experimental results. A similar approach was also adopted by Jackson and Yen (1971).

In general, heat-transfer investigations covering large variations in several parameters have proved inconclusive. Subsequent practical studies of the response behaviour of SN-probes in the mixed-flow region have revealed consistent hot-wire characteristics. The effect of the probe orientation in both vertical and horizontal flows was studied by Christman and Podzimek (1981) for an unplated SN-probe (DISA 55P11) placed in a laminar flow in a pipe. The flow direction could be changed by rotating the test facility. Their data for a single overheat ratio of 1.8 demonstrated the effect on the hot-wire calibration of both the wire orientation and the flow direction. For a horizon-

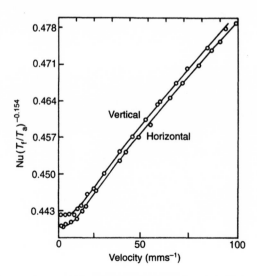

Fig. 4.12. Calibration curves for horizontal flow with the probe wire in the horizontal and vertical positions. (From Christman and Podzimek 1981.)

tal flow, Fig. 4.12, a consistent difference was observed in the hot-wire reading with the probe orientation. It is therefore important that the probe is mounted with the same orientation during both the calibration and the measurements. In the range 3–7 mm s^{-1}, multivalent velocity readings were observed. Due to the consistency of the individual calibration curves, it appears safe to accept 10 mm s^{-1} as the lower reliable region for measurements in horizontal flows. The results for the vertical flow in Fig. 4.13 show a clear difference in the calibration curves for upwards and downwards flow. This is caused by the natural-convection velocity being, respectively, added to or subtracted from the forced-convection velocity. The magnitude of the natural convection velocity was evaluated as being 6.8 mm s^{-1}. The lowest reliable velocity for measurements in vertical flows also appears to be about 10–15 mm s^{-1}. The difference in the hot-wire response between vertical upwards and downwards flows has been confirmed by Cowell and Heikal (1988). In their investigations, the probe was moved in a quiescent fluid, and their results also demonstrate the influence of the flow direction on the hot-wire output in the mixed-flow regime. However, in many practical flows, where the effects of flow direction and probe orientation are not known, the lowest reliable velocity may be as high as 20 cm s^{-1} (Paul and Steimle 1977).

The *SN* subminiature probe developed by Ligrani and Bradshaw (1987*a*) (Fig. 4.3) was calibrated in air for a velocity range covering natural to forced convection. They plotted their calibration data in the form

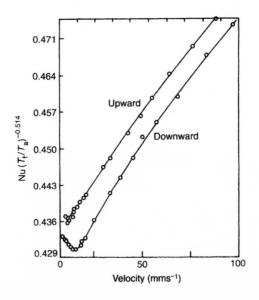

Fig. 4.13. Calibration curves for an upward and downward orientated flow. (From Christman and Podzimek 1981.)

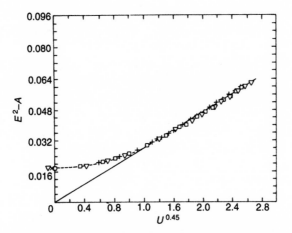

Fig. 4.14. Three sets of subminiature *SN* hot-wire sensor calibration data in the forced- and natural-convection flow regimes. (From Ligrani and Bradshaw 1987*a*.)

$$E^2 = A + BU^{0.45},$$

and they observed good agreement between data from different calibrations (Fig. 4.14). The lower limit for forced convection was evaluated as $Re_c = 0.07$. In the mixed-flow regime, the calibration data was approximated by a third-order polynomial of the form

$$\alpha = a_0 + a_1\beta + a_2\beta^2 + a_3\beta^3, \tag{4.22}$$

where $\alpha = U^{0.45}$ and $\beta = E^2 - A$.

4.9 DYNAMIC CALIBRATION OF AN *SN* HOT-WIRE PROBE

Sections 4.3 to 4.8 have demonstrated that the hot-wire relationship must be selected carefully in order to obtain a good curve-fit to the calibration data. However, before digital computers became generally available, considerable uncertainty existed in the evaluation of the velocity sensitivity, dE/dU, of the hot-wire equation selected. A dynamic calibration procedure has been proposed by Perry and Morrison (1971*c*) to overcome this problem. Similar methods were previously applied by Dryden and Kuethe (1929) and by Schubauer and Klebanoff (1946), but Perry and Morrison have developed this method into a standard calibration procedure which has been used extensively at Melbourne University. The method involves shaking the hot-wire probe at low frequencies in a uniform flow of known velocity, U, thereby subjecting the wire to a known velocity fluctuation, u. The corresponding anemometer output will consist of a mean value, $\bar{E}$, and a voltage fluctuation, e. The value at dE/dU obtained by this dynamic calibration can be defined as e/u, or as the ratio of the respective rms values e' and u' (Perry and Morrison 1971c)

$$\left(\frac{dE}{dU}\right)_{dyn} = \frac{e'}{u'}. \tag{4.23}$$

One of the main reasons for developing this dynamic calibration technique was the uncertainty of *analog* calibration and analysis methods, which often involved analysis of the direct (nonlinear) voltage signal, E. Separating the instantaneous values of $E(t)$ and $U(t)$ into the corresponding time-mean and fluctuating components, that is, $E(t) = \bar{E} + e$ and $U(t) = \bar{U} + u$, and applying the simple power law of eqn (4.8), gives the following relationship between $(\bar{E}, \bar{U})$ and (e, u) for a low turbulence intensity flow

$$\bar{E}^2 = A + B\bar{U}^n \tag{4.24}$$

and

$$e = \frac{dE}{dU}\, u, \tag{4.25}$$

where dE/dU is given by eqn (4.15)

$$\frac{dE}{dU} = \frac{nB\bar{U}^{n-1}}{2\bar{E}}.$$

Before computers became readily available, the value of dE/dU was often obtained by selecting the best curve fit from a graphical differentiation of a plot of $\bar{E}^2$ versus $\bar{U}^n$ (Perry and Morrison 1971c). The corresponding variations in dE/dU with n are shown in Fig. 4.15, and Perry and Morrison (1971c) claimed that the uncertainty in the local value of dE/dU may be as high as 20%. However, to identify the best curve-fit from such a procedure is very difficult and this *graphical* curve-fit method is *not* recommended. Also, the variation in dE/dU with n which is shown in Fig. 4.15 is incorrect since the sensitivity curves for different n will intersect, as shown previously in Fig. 4.11. Perry and Morrison's dynamic calibration results for dE/dU are also shown in Fig. 4.15. However, since the static calibration results they presented were not of a sufficiently high quality, any reliable conclusion concerning the relative merits of the dynamic and the static calibration of hot-wire probes cannot be made. This has been the topic of further comparative investigations. Kirchhoff and Safarik (1974) also reported different results obtained by these two methods, but the static calibration results that they presented were evaluated using $n = 0.5$, which has been shown to give a poor curve-fit to the measured calibration data.

In principle, the same results should be obtained by static and dynamic calibration methods, provided that a thermal equilibrium exists between the

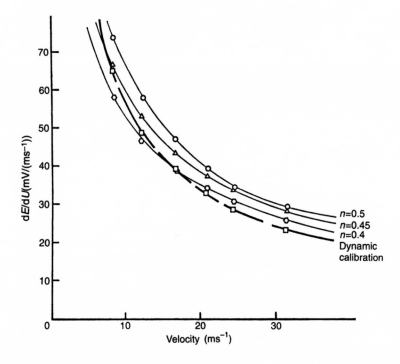

Fig. 4.15. The hot-wire probe velocity sensitivity, dE/dU, obtained by static and dynamic techniques. (From Perry and Morrison 1971c. Reproduced with kind permission of Cambridge University Press from the *Journal of Fluid Mechanics*.)

wire element and the prongs during the perturbation cycle. In this case, the following relationship applies

$$\frac{\mathrm{d}E}{\mathrm{d}U} = \lim_{\substack{u' \to 0 \\ f \to 0}} \frac{e'}{u'}, \tag{4.26}$$

where f is the perturbation frequency. The phenomenon of a low-frequency, dynamic, thermal inequilibrium at the wire/prong interface was discussed in Section 2.3.2.3. It was demonstrated that a significant dynamic prong effect exists in nonisothermal turbulent flows for short resistance-wires operated at a low overheat ratio by a constant-current (*CC*) anemometer. However, for a standard hot-wire probe placed in an isothermal turbulent flow and operated at a high overheat ratio by a constant-temperature (*CT*) anemometer, there is no evidence of a significant dynamic prong effect (see Section 2.3.3.1). This is supported by the numerical evaluations by Olivari (1976), whose results show that the prongs have a considerable influence on the frequency response and on the transient response of a hot-wire anemometer used in the *CC* mode. However, operating the wire in the *CT* mode completely changes the balance between the various terms of the unsteady heat transfer equation, and the effect of the prongs becomes negligible. Bremhorst and Gilmore (1976*a*) carried out a detailed experimental comparison of the static and the dynamic calibration of *SN*-probes operated in the *CT* mode, using a method related to eqn (4.26). They obtained a good agreement between the two methods, except for the power-law relationship with $n = 0.5$, which was shown in Section 4.4.1 to be a poor approximation. A similar agreement between static and dynamic calibration was also demonstrated by Bruun (1976*a*) and Soria and Norton (1990).

4.10 SINGLE NORMAL HOT-FILM PROBES

Most of the principles for *SN* hot-wire probes also apply to *SN* hot-film probes. Specific features related to their construction and frequency response are discussed in Chapter 2. Some additional aspects must be considered when using hot-film probes for practical *HWA* applications.

4.10.1 Measurements in gas flows

When accurate turbulence results are required in a gas flow (including air), then hot-wire probes are normally used, as described in Section 2.2.1.1. However, there are applications where very accurate turbulence results may not be the primary requirement. For such investigations, cylindrical hot-film probes can be considered, taking the following advantages/disadvantages into account:

1. A hot-film probe is much more robust than a hot-wire probe. Probe breakage is therefore less likely to occur, but their replacement is more expensive than for a hot-wire probe.

2. Film probes have a much more stable geometry than hot-wire probes, and (due to their much bigger surface) they are less susceptible to contamination, giving long-term calibration stability.

3. Hot-film probes with the sensor(s) coated with quartz or aluminium oxide can be used in dirty gas flows, since the coating prevents the film from being eroded.

4. The rigid construction of hot-film probes ensures that the sensor angles (yaw and pitch) remain virtually constant.

5. The diameter of cylindrical hot-film probes (25–70 μm) is much larger than a typical hot-wire probe ($\sim 5\,\mu$m), and the larger-diameter film sensors generate their own turbulence (~ 1–2%) due to regular vortex shedding even at moderate velocities (Re > ~ 40). The onset of vortex shedding also changes the shape of the hot-film calibration relationship (Collis and Williams 1959; Pitts and McCaffrey 1986).

6. Hot-film probes have a more complex frequency response, primarily due to the substrate on which the film is deposited. As a consequence, for a cylindrical film-probe operated by a CT anemometer, the amplitude transfer function at low and moderate frequencies will not be flat (see Section 2.3.3.2). For a properly adjusted CT hot-film anemometer system the corresponding amplitude variation is about ±5-10%, introducing a related uncertainty in the turbulence quantities evaluated.

7. The recommended overheat ratio, R_w/R_a, and the temperature of the film, T_w, depend on the material used for the sensing element. For a platinum film probe ($\alpha_{20} = 0.0024\,°\mathrm{C}^{-1}$) a typical overheat ratio is 1.4, giving a value for $T_w - T_a$ of about 160 °C. For a nickel film probe ($\alpha_{20} = 0.0040\,°\mathrm{C}^{-1}$) with a normal overheat ratio of 1.8 the value of $T_w - T_a$ is about 200 °C.

4.10.2 Measurements in water flows

Hot-film probes are normally used for the study of water flows, and the following HWA aspects must be considered in order to obtain reliable measurements.

Electrolysis In the early HWA investigations attempts were made to use hot-wire probes in a conducting medium such as water. However, electrolysis has proved to be a major problem for an unprotected hot-wire element. The potential drop across the wire causes electrolysis of the water, and this was originally thought to be the cause of the often-observed formation of bubbles on the wire or film element. More importantly, it resulted in a slight, but continuous, reduction in the diameter of the wire, and therefore in its cold resistance. The resulting continuously shifting calibration makes an unprotected hot-wire probe useless for measurements in a conducting medium such as water. To overcome the problem of electrolysis, coated hot-wire probes have been tried, but it is difficult to make a thin coating impervious to water and, if too thick a coating is applied, the probe becomes very insensitive to velocity fluctuations. Measurements in liquids by HWA first became a practical option with the development of the hot-film probe (Ling and Hubbard 1956). The use of a thin (1–2 μm) quartz coating on hot-film probes has, to a large extent, solved the electrolysis problem. The thin protective layer has a negligible effect on the frequency response, except for frequencies above about 50 kHz.

Cracking of quartz coating Sensor failures due to cracking of the quartz coating have been reported by Jones and Zuber (1978). This failure was traced to a forced resonant vibration of the sensor on its support legs, caused by vortex shedding from the cylindrical element. This problem was particularly acute for a 25 μm diameter probe (TSI) for velocities of about 1.5 m s^{-1}. Using a 50 μm probe significantly reduced, but did not entirely eliminate, the problem. Cracking can also occur if a large potential drop exists between the water surrounding the probe and the film element. This problem can be minimized by connecting the ground reference of the anemometer to the water in the test facility.

Bubble formation on the probe The early hot-wire and hot-film studies revealed that under certain conditions gas bubbles formed on the probe element, leading to significant calibration shifts. Several different theories were proposed, and the work of Rasmussen (1967) demonstrated that this was due to the thermal formation of bubbles on a heated surface immersed in air-saturated water. It was observed that the bubble problem mainly occurred if bubbles of a certain size were already suspended in the water flow. To minimize this problem, it is recommended that the water is allowed to stand before use and that cascading in air is avoided by using a submerged return/filler pipe in a water tank. The bubble-formation problem can usually be eliminated by restricting the temperature difference, $T_w - T_a$, between the film and water to about 20 °C. The corresponding overheat ratio, R_w/R_a, is about 1.05–1.1, the value depends on the film material, (platinum or nickel) used.

At sufficiently high velocities, *cavitation* can occur. Jones and Zuber (1978) found cavitation occurred for a 25 μm diameter probe (TSI) above 9 m s^{-1}. For a 50 μm diameter probe, cavitation was observed in the 4.5–6 m s^{-1} range.

Temperature drift and probe contamination Temperature drift and probe contamination often occur simultaneously, and the effects can be difficult to separate. The temperature difference, $T_w - T_a$, for a hot-film probe in water is only about 20 °C whereas it is 250 °C for a hot-wire probe in air. It therefore follows from the response equation (eqn (2.36a)

$$\frac{E_w^2}{R_w} = (A + BU^n)(T_w - T_a),$$

that even a change of a fraction of a degree in the water temperature, T_a, can result in a large shift in the probe calibration. Tan-atichat *et al.* (1973) have shown that a 3 °C shift in the water temperature can give a 100% error in the mean-velocity reading. This problem can be minimized by (i) the use of a recirculating test facility with a large storage tank, and therefore a large heat capacity; or (ii) by monitoring the temperature of the water and either applying a temperature correction (see Section 7.3) or, for a recirculating test facility, include a temperature-controlled heating/cooling system in the storage tank (Resch 1970). A more practical option may be frequent calibrations of the hot-film probe. This is often preferred due to the additional problem of probe contamination.

Probe fouling can be a major problem due to the gradual build up of scale, algae, and minerals on the probe, causing a shift in the calibration and a loss of sensitivity. It has been found that a conical probe is significantly less affected by contamination than a wedge probe (Resch 1970; Warschauer *et al.* 1974). Cylindrical hot-film probes are also prone to contamination by particles sticking to the fibre element. In a recirculating test facility these problems can be minimized by using de-ionized water, a by pass filtration unit and possible algae inhibitors. However, such procedures cannot be applied to experiments using nonreturn tap water. DISA (1965) and Jimenez *et al.* (1981) found that if a probe was cleaned just before the test (with a soft brush or with a cotton swab dipped in acetic acid), then the surface contamination was so fast that data taken at the end of the run could not be made to correspond with those taken at the beginning. Instead, it was recommended that the probe should be 'aged' in running water for at least 30 min after being cleaned. This precontamination ageing slowed the drift considerably, and allowed each test run to be considered as a unit with a constant calibration characteristic. The same principle was also studied by Wu and Bose (1993) for a V-shaped hot-film probe operated in tap water in a towing tank. They found that the probe needed to be 'preaged' for 80–100 h to obtain nearly constant calibration conditions over a 7 h period.

Minimization of probe contamination and temperature drift The author has, with co-workers A.L. Samways and J. Ali, been studying the vertical up-flow of water or a water/kerosene mixture in a circular pipe. The flow system is a recirculating test facility with a storage tank with a volume of about $2\,m^3$. A cylindrical hot-film probe is being used to measure mean and turbulent quantities in both flow cases. The original measurements were carried out without controlling the water temperature and with a relatively coarse bypass filtration unit connected to the water storage tank. As observed in many other water-flow investigations with cylindrical hot-film probes, a substantial change occurred to the probe calibration over a relatively short period of time, which made it difficult to interpret the anemometer signals correctly.

Two integrated modifications to the flow loop have virtually eliminated both the probe-contamination and the temperature-drift problem. Firstly, a bypass chiller unit has been connected to the storage tank and water drawn from the tank is passed through the cooler unit and returned to the tank. Secondly, in the connection pipe between the tank and the chiller, two filtration units ($8\,\mu m$ and $2\,\mu m$) have been installed. The chiller system is switched on throughout the day, and it is currently set to a temperature of $16\,°C$ ($\pm 0.1\,°C$). The chiller system has two beneficial effects. It keeps the temperature of the fluid in the pipe test section virtually constant, and selecting a relatively low temperature setting of $16\,°C$ has minimized the growth of algae in the system.

The system was originally filled with de-ionized water, and the $2\,\mu m$ filter effectively removes all significant contaminations. (That the $2\,\mu m$ filter is essential is demonstrated by the necessity to replace the filter every two to

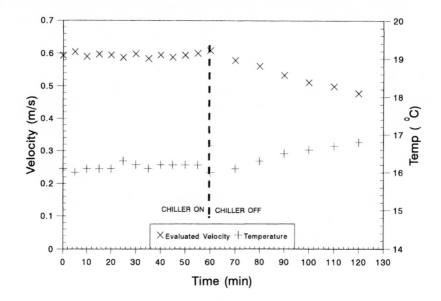

Fig. 4.16. The stability of a hot-film probe in water: (a) with water-temperature control (chiller on), and (b) variation in the water temperature and hot-wire signal when the chiller is turned off.

three months.) The combined effect of cooling and filtration has resulted in the water remaining virtually crystal clear over a period of more than a year. The corresponding improvement in the long-term stability of the hot-film probe is demonstrated in Fig. 4.16. When the chiller is on, the water temperature can be seen to remain virtually constant and, during a period of 60 min with a constant-flow condition, the anemometer output voltage, and consequently the interpreted velocity, did not change. Further studies have demonstrated that there are no significant changes to the probe calibration during a full working day. The effect of temperature contamination is also demonstrated in Fig. 4.16. With the chiller turned off, the water temperature was observed to rise linearly with time at a rate of about $0.75\,^{\circ}\mathrm{C}\,\mathrm{h}^{-1}$. During this period, the velocity interpreted from the hot-film signal was seen to drop continuously at a rate of $20\text{--}25\%\,\mathrm{h}^{-1}$. It is therefore essential to control the water temperature in order to obtain reliable hot-film results, and filtration is also necessary to avoid probe contamination.

Frequency response Water has a much higher heat-transfer coefficient than air. The convective heat transfer is therefore large compared with the heat-transfer process between the film element and the substrate (see Section 2.3.3.2). For a cylindrical hot-film probe the dynamic substrate effect is so small that it can be ignored for most practical applications. A static calibration can therefore be used, and the signal analysis will be similar to that of an *SN* hot-wire probe.

Due to their massive substrates, wedge and cone probes have larger dynamic substrate effects. As shown in Section 2.3.3.2, wedge probes have the most complex amplitude-attenuation curve. However, except at low velocities, the maximum attenuation in water is less than about 5%, making an uncorrected signal analysis acceptable for many practical applications. Although the amplitude attenuation for conical probes is simpler than for wedge probes it is also larger, reaching more than 20% at low velocities. An attenuation of this magnitude should be accounted for in the signal analysis. This problem has been considered in the investigations by Warschauer *et al.* (1974) and by Nelson and Borgos (1983).

Calibration relationships In principle, any of the response equations discussed in Section 4.4 for a hot-wire probe can also be applied to a (cylindrical, cone, or wedge) hot-film probe in water. Normally, the polynomial relationship (see eqn (4.12))

$$U = A + BE + CE^2 + DE^3 + FE^4$$

or the simple power law (eqn 4.8)

$$E^2 = A + BU^n$$

are used. The results of Wu and Bose (1993) indicate that an improved curve fit can be obtained at low velocities when an extended power law of the form

$$E^2 = A + BU^n + CU^{2n} \qquad (4.27)$$

is applied to the calibration data. The calibration constants in all three equations are usually determined by a least-squares curve-fitting method applied to a set of calibration points (E_R, U_R) spread evenly over the velocity range investigated. Calibration results for cylindrical film probes have been presented by Richardson and McQuivey (1968), Tan-atichat *et al.* (1973), and Zabat *et al.* (1992), for conical probes by Resch (1970), Morrow and Kline (1971), Saunders and Lawrence (1972), Bonis and van Thinh (1973), and Warschauer *et al.* (1974), and for a wedge probe by Morrow and Kline (1971). The value quoted for n in the power-law relationship was found to depend on the probe type and to be sensitive to the details of the calibration method selected and the velocity range investigated. For water velocities of less than $20\,\mathrm{cm\,s^{-1}}$ the values of n evaluated were usually in the range 0.25–0.3, while in the velocity range 0.5–$4\,\mathrm{m\,s^{-1}}$ values of 0.4–0.45 have often been quoted. These last values are consistent with the exponent values obtained by the author and his co-workers in the velocity range 0.2–$1\,\mathrm{m\,s^{-1}}$.

4.10.2.1 Calibration techniques

Differential-pressure methods The velocities in most water-test facilities are within the range 0.01–$5\,\mathrm{m\,s^{-1}}$. Since the density of water is much higher than air, it is possible to calibrate hot-film probes using a pressure-difference device such as a pitot tube or a calibration nozzle. Jones and Zuber (1978) calibrated their probes in the exit plane of a TSI 1150D nozzle calibrator immersed in a bath of water. The differential pressure between the bath and the stagnation pressure in the nozzle settling chamber was for the lowest velocities measured with an E. Vernon Hill well type of micro-manometer, type C, 15 cm with a calibrated repeatability of $50\,\mu\mathrm{m}$. For the higher range of differential pressures, a 1 m water manometer was used in conjunction with a 1 m mercury manometer. Johnston *et al.* (1983) also used a submerged jet and measured the differential head, H, by using the water surfaces in the two chambers as reflecting elements. An image of a vertical illuminated scale was sighted by a theodolite telescope, as described by Whittington and Clapp (1958). The velocity calibration range was 0.1–$1.0\,\mathrm{m\,s^{-1}}$.

A.L. Samways has developed a calibration technique, for a vertical pipe containing an upward water flow, based on a pitot-tube placed at the centre line at specified axial-measurement positions. The difference between the pitot-pressure, P_{o}, and the corresponding wall static pressure, P, was used to determine the centre-line velocity, $U_{\mathbb{C}}$, from the Bernoulli equation

$$P_{\mathrm{o}} - P = \tfrac{1}{2}\rho U_{\mathbb{C}}^{2}.$$

For each axial position, an inital calibration was carried out to relate the centre-line velocity $U_{\mathbb{C}}$ (obtained from the pitot-tube and the Bernoulli equation), and the water flow rate, $\dot{Q}$, measured with a flow meter installed in the water supply pipe. Least-squares curve-fitting was used to evaluate the corresponding calibration relationship

$$U_{\mathbb{C}} = f(\dot{Q}) \tag{4.28}$$

which was found for all practical purposes to be a linear relation. To account for flow developments in the pipe, it was necessary to determine the calibration, by eqn (4.28), for each axial position. In the subsequent hot-film-probe calibrations, the centre-line velocity was evaluated from eqn (4.28) using direct measurements of $\dot{Q}$.

Placing a hot-film probe at the probe centre-line, and knowing the form of eqn (4.28), it was then possible to carry out a calibration of the anemometer output voltage, E, in terms of the flow rate, $\dot{Q}$, with a subsequent conversion into a conventional $(E, U_{\mathbb{C}})$ calibration. Figure 4.17 shows two corresponding sets of calibration data obtained with the same probe using the same overheat ratio on two different days. Using a complete computer-based technique, each calibration takes less than 10 min, and the

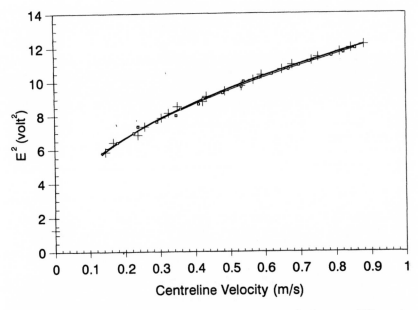

Fig. 4.17. Calibration curves for a hot-film probe in water obtained on two different days, demonstrating long-term calibration stability: (□) $n_1 = 0.52$, and (+) $n_2 = 0.45$.

two sets of data demonstrate the long-term calibration stability for the cooling/filtration system described previously. The value of the optimum exponent, n, in the power-law relationship of eqn (4.8) was found to be about 0.45–0.5. It is worth noting that before the cooling/filtration system was installed a hot-film probe placed in the water became heavily contaminated. The anemometer signal displayed a continuous slow drift, and the optimum value of the exponent, n, was about 0.3–0.4, giving a much poorer sensitivity than for the calibration results presented in Fig. 4.17.

Moving-probe calibration A hot-film probe is only sensitive to the relative velocity between the sensor and the liquid. Calibration facilities with stationary water and moving probes are therefore used frequently. The most common test facility is a towing tank, in which the probe is moved with a constant, but variable, velocity through a long water tank (see, for example, Raichlen 1967; Saunders and Lawrence 1972; Shaukatullah and Gebhart 1977; Persen and Saetran 1983; Wu and Bose 1993). A similar calibration procedure in an oil channel has been reported by Kreplin and Eckelmann (1979a). Anhalt (1973) and Bertrand and Couderc (1978) have described a calibration facility consisting of a circular tank with stationary water and a rotating probe holder.

A recent development of moving-probe calibration is the use of a single sweep with a variable probe velocity. This method has been developed by Zabat *et al.* (1992) for a hot-film probe on a swinging arm. The basic principle of their technique is similar to the swinging-arm calibration method for a hot-wire probe in air (Haw and Foss 1990), as discussed in Section 4.7.

Velocity-profile method A calibration can be carried out in a pipe or jet flow, provided the velocity profile, $U(r)$, can be measured as part of the calibration procedure. This procedure, which requires several iterations, has been applied to a free-jet flow in a recirculating test facility (Pluister and Nagib 1975).

Rotating tanks Seed and Wood (1969) and Pichon (1970) have described calibration experiments in which the fluid in a tank is first rotated until a (nearly) solid-body motion, $U = \omega r$, is obtained in the tank. A probe calibration over the required velocity range can be carried out by varying the probe position, r, and the angular velocity, ω. Due to the disturbance caused by the probe, it was retracted from the liquid except for the brief periods of the individual measurements. The disadvantages of this method include the parabolic configuration of the free-water surface, and secondary flows caused by the centrifugal forces acting on the boundary layer on the bottom of the tank. The rotating-tank technique has also been used by Hirano *et al.* (1989) for calibration at low velocities ($U < 10 \, \text{cm s}^{-1}$).

Free-jet flows Calibration experiments have been carried out in the free jet issuing from a sharp-edged orifice in a large tank (Rubatto 1970; Pichon 1970; Franke and Preuss 1970). The jet velocity, U, was calculated from the height, H, of the water in the tank above the orifice

$$U = C_v (2gH)^{1/2}$$

where C_v is a velocity coefficient, which may be obtained by measuring mass-flow rates by weighing. This method is quoted as only giving reliable results for $U > 1 \, \text{m s}^{-1}$.

4.10.3 Measurements in other liquids

Most hot-film probe studies have been reported for water flows, and detailed turbulent boundary-layer studies have also been reported in an oil-channel flow (Eckelmann and Reichardt 1971; Eckelmann 1972, 1974). *HWA* has also been used in a variety of other fluids, including polymer solutions, glycerine, blood, and mercury. The application of *HWA* in these liquid flows is often very specialized, and the reader is referred to the book by Lomas (1986) for a discussion of this topic.

4.11 MEASUREMENTS WITH AN *SN*-PROBE

By following the step-by-step procedure outlined in Chapter 3 and the information contained in Sections 4.1 to 4.10, the reader will be able to select an *SN* hot-wire/film probe, a *CT* anemometer system, and to perform a calibration to determine the calibration constants in the response equation selected. The criteria for evaluating a suitable digital measurement system, comprising a low-pass filter unit, a signal-conditioning unit, an *A/D* converter, and a computer have also been discussed (Section 3.5.1), and such a system is assumed to be available. Before data acquisition can commence it is necessary to identify the mean-flow direction.

4.11.1 Identification of the mean-flow direction

A fixed *SN*-probe can only measure one velocity component. It is therefore essential that the wire element is placed perpendicularly to the mean-flow direction and the probe-stem orientation selected, usually the orientation shown in Fig. 4.1. For flow over, for example, a flat plate, the mean-flow direction is known, but in many flow situations it must be determined experimentally. Provided that the mean-flow field is known to be two-dimensional, then an *SN*-probe with a perpendicular stem-orientation can be used to identify the mean-flow direction by rotating the probe around its axis and recording the anemometer signal, E, as a function of the angle of rotation (also called the roll angle), γ (Bond and Porter 1967; Bissonnette and Mellor 1974). The evaluation of two angle positions (60–90° apart) corresponding to a selected value of E and then bisecting these two angles will identify the mean-flow direction. A similar procedure using an *SN*-probe with a parallel-stem orientation was used by Müller (1982a). This method is easy to implement in a low turbulence intensity flow. In a flow with a moderate/high turbulence intensity, the time-averaged signal, $\bar{E}$, must be used. An alternative method using the midpoint between the two minima of the $E(\gamma)$ curve was used by Vagt and Fernholz (1979).

4.11.2 Data acquisition and storage

Before data acquisition can commence, the parameters of the digital-measurement system must be set. The purpose of the signal-conditioning unit (see Fig. 3.2) is considered first. An n-bit *A/D* converter operates in discrete voltage steps, ΔE, with 2^n different output values. A twelve-bit converter can therefore only select 4096 different output values, and if the voltage range of the *A/D* converter is 0–10 V (or ±5 V) then $\Delta E = 2.5$ mV. Consequently, if a nonlinear voltage signal with small fluctuations is fed directly to the *A/D* converter, then significant resolution errors may occur in the fluctuating part of the acquired signal.

A practical way of minimizing this problem is to match the fluctuating part of the hot-wire signal to the *A/D* converter's voltage range, by using a signal-conditioning unit. This procedure is illustrated in Fig. 4.18. The signal-conditioning steps are as follows. First, an offset voltage, E_{off},

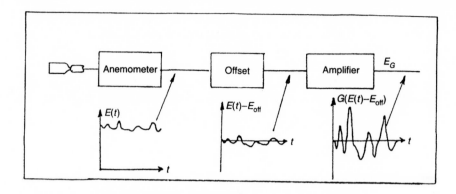

Fig. 4.18. The principle of a signal-conditioning unit.

which is approximately equal to the time-mean value, $\bar{E}$, is subtracted from the direct hot-wire signal, $E(t)$. Provided that the value of E_{off} is known, then any reasonable value of E_{off} can be used. The signal, $E(t) - E_{off}$, is then amplified, with a gain G, so that the fluctuating output signal, $E_G(t)$, from the conditioning unit,

$$E_G(t) = G(E(t) - E_{off}), \qquad (4.29)$$

nearly covers the complete A/D converter voltage range (for example, ± 5 V). Having set the offset, E_{off}, and the gain, G, of the signal-conditioning unit and set the optional low-pass filter (see Section 3.5), the only remaining step is to specify the sampling rate, SR, and the sampling time, T (or the number of samples N), as described in Chapter 12. Data acquisition can then be initiated, and on completion a digital time-series record, $E_G(m)$, will be available. These values may be stored in the computer using an array format, or in the form of address locations. The latter format greatly facilitates the data handling, and, in particular, it enables data conversion to be implemented by a look-up table method as described in Section 4.12.2.2.

4.12   DATA ANALYSIS FOR AN *SN*-PROBE

In principle two different methods can be used. In the first, the signal analysis is carried out on the direct nonlinear voltage signal, $E(t)$, providing statistical values such as $\bar{E}$ and $\overline{e^2}$. The related velocity quantities $\bar{U}$ and $\overline{u^2}$ can be evaluated from the calibration equations, eqns 4.24 and 4.25. This was the standard procedure for analog-signal analysis, and it is also used when the sensitivity, dE/dU, is determined by a dynamic calibration of the *SN*-probe. In the second method, the nonlinear voltage signal, $E(t)$, is first converted into the corresponding velocity time-history, $U(t)$, by inverting the selected hot-wire relationship. The required statistical velocity information can then be obtained by a time-series analysis of $U(t)$ (see Chapter 12). This latter method is well suited for implementation on a computer system and it is recommended by the author; but it should be stressed that, in terms of the accuracy of the results, both procedures should in principle give similar results for both $\bar{U}$ and $\overline{u^2}$.

4.12.1 Voltage-signal analysis

This approach can be implemented using either an analytical form for the calibration relationship, or the sensitivity, dE/dU, in eqn (4.25) can be determined by a dynamic calibration. Since static and dynamic calibration normally give similar results, the analytical approach in the form of a simple power law, will be used. For low turbulence intensity flow, and assuming $\bar{V} = \bar{W} = 0$, the value of $\bar{U}$ can be evaluated from eqn (4.24). The corresponding equation for the fluctuating quantities is

$$\overline{e^2} = \left(\frac{dE}{dU}\right)^2 \overline{u^2},$$

with dE/dU given by eqn (4.15).

For flows with moderate/high turbulence intensities it is necessary to retain higher-order terms in the series expansions derived in numerous publications, including those by Hinze (1959), Rose (1962), and Bruun (1972). If second-order terms in the equation for $\bar{U}$ and third-order terms in the equation for $\overline{u^2}$ are retained, the relationships between the true values of $\bar{U}$ and $\overline{u^2}$ and the measured quantities $\bar{U}_m$ and $\overline{u_m^2}$ are

$$\bar{U}_m = \bar{U}\left(1 + \frac{n-1}{2}\frac{\overline{u^2}}{\bar{U}^2} + \frac{1}{2}h^2\frac{\overline{w^2}}{\bar{U}^2}\right) \tag{4.30}$$

and

$$\overline{u_m^2} = \overline{u^2}\left(1 + (n-1)SuTu + h^2R_{uw^2}Tu\right), \tag{4.31}$$

where

$$Su = \frac{\overline{u^3}}{(\overline{u^2})^{3/2}}, \qquad R_{uw^2} = \frac{\overline{uw^2}}{(\overline{u^2})^{3/2}}; \qquad Tu = \frac{(\overline{u^2})^{1/2}}{\bar{U}}.$$

In this series-expansion analysis, which is based on eqns (3.1) and (3.2), terms containing k^2 have been omitted since $k^2 \ll h^2$ (see Chapter 5). The magnitude of the higher-order terms in eqns (4.30) and (4.31) will be compared in Section 4.13 with similar equations for the velocity-analysis method.

4.12.2 Velocity-analysis method

4.12.2.1 Data conversion

The data acquired by the computer is a time-series record of the voltage, $E_G(m)$. The corresponding anemometer voltage values, $E(m)$, can be obtained by inverting eqn (4.29) on the computer giving

$$E(m) = \frac{1}{G} E_G(m) + E_{off}. \tag{4.32}$$

To evaluate the required velocity information it is necessary to apply a calibration equation specified either as $E = F_1(U)$ or as $U = F_2(E)$ (see Section 4.4). For measurement purposes, it is necessary in both equations to replace U by an effective velocity V_e.

The most common function for the $E = F(U)$ calibration is the simple power law of eqn (4.8). By performing the inversion process

$$V_e(m) = \left[\frac{E(m)^2 - A}{B}\right]^{1/n} \tag{4.33}$$

on the computer for all the $E(m)$ values, a digital time-history record, $V_e(m)$, is obtained. As proposed by George et al. (1981), V_e may also be evaluated from a $U = F(E)$ relationship in the form (eqn (4.12))

$$V_e = A + BE + CE^2 + DE^3 + FE^4$$

In this method, it is necessary for each value of E to calculate $E^n (n = 1\text{-}4)$ and to insert these values into the equation above to obtain the digital time-history record, $V_e(m)$. The related complete conversion time for one data point was found to be similar for a power-law inversion, for a fourth-order polynomial in E, and for a third-order polynomial in E^2 (Bruun et al. 1988). However, whether the single-power law or the polynomial fits are used, the problem of the conversion time becomes important if thousands or millions of conversions are to be carried out. One time-saving solution to this problem is to combine digital data acquisition with a look-up table method as described in the next section.

4.12.2.2 Look-up table method

The most direct look-up table method utilizes the n-bit nature of an A/D converter. Consider the following two A/D principles. Firstly, an n-bit A/D converter can only identify 2^n different output values over the complete input voltage range, for example, ± 5 V, and for a typical twelve-bit converter there are only 4096 different output values. Secondly, the direct output from the A/D converter is an integer value, I. For a twelve-bit, ± 5 V converter, the relationship between I and the A/D output voltage, E_{out}, is

$$E_{out} = -5 + I \frac{10}{4095} \quad (0 \leqslant I \leqslant 4095). \tag{4.34}$$

For each input voltage, E_G to the A/D converter, the nearest value of I which will minimize $E_G - (-5 + 10I/4095)$ is selected, and at the end of the data acquisition an integer time-history record, $I(m)$, will be available. Setting $E_{out}(m)$ equal to the A/D input voltage record, $E_G(m)$, we have

$$E_G(m) = -5 + I(m) \frac{10}{4095}. \tag{4.35}$$

In the conventional data-conversion method (outlined in Section 4.12.2.1) it was necessary first to evaluate the corresponding anemometer voltage record, $E(m)$, using eqn (4.32), and then to carry out an inversion process, for example, by eqn (4.33), to obtain $V_e(m)$. These steps can be combined in a look-up-table method. Inserting eqn (4.32) into eqn (4.33) the following relationship between $V_e(m)$ and the acquired voltage, $E_G(m)$, can be obtained:

$$V_e(m) = \left[\frac{(E_G(m)/G + E_{off})^2 - A}{B} \right]^{1/n}. \tag{4.36}$$

Finally, by introducing the digital relationship between $E_G(m)$ and $I(m)$ of eqn (4.35), the expression for $V_e(m)$ becomes

$$V_e(m) = \left[\frac{((-5 + 10I(m)/4095)/G + E_{off})^2 - A}{B} \right]^{1/n}. \tag{4.37}$$

In eqn (4.37) A, B, and n are known calibration constants determined from the calibration of the *SN*-probe, and G and E_{off} are known preset values for a set of measurement points. The only variable in eqn (4.37) is the integer variable I, and for the twelve-bit A/D converter considered I can only have 4096 different values. Therefore, if a large number of data conversions are required, it is worthwhile considering the use of a look-up table method due to the savings in the data-conversion time.

The look-up table method may be implemented in the following way:

1. The probe is first calibrated and the signal-conditioning unit set. A look-up table, $V_e(J)$, $0 \leqslant J \leqslant 4095$, is then created using eqn (4.37) (or any similar relationship).

2. For each data acquisition, the input voltage record, $E_G(m)$, is stored in its equivalent A/D *integer* format, $I(m)$; see eqn (4.35).

3. Data conversion is achieved by poking each value of $I(m)$ into the look-up table and retrieving and storing the related $V_e(I(m)) = V_e(m)$.

This method, which is much faster than the procedure outlined in Section 4.12.2.1, has been implemented by Farrar (1988).

4.12.2.3 Analysis of $V_e(m)$-records

The data conversion described in the above sections will provide a digital time-history record of the effective velocity, $V_e(m)$. For the parallel probe-stem orientation shown in Fig. 4.1, V_e may be expressed as

$$V_e^2 = U^2 + k^2V^2 + h^2W^2. \tag{4.38}$$

In a low turbulence intensity flow, with $\bar{V} = \bar{W} = 0$, the following equation can be obtained by a first-order series expansion of eqn (4.38):

$$V_e(m) = U(m) = \bar{U} + u(m). \tag{4.39}$$

A time-series analysis can be applied to $U(m)$ to obtain the velocity information required, as described in Chapter 12. For moderate/high turbulence intensity flows, additional terms must be included in the series expansion. Retaining the assumptions that $\bar{V} = \bar{W} = 0$ and neglecting terms containing k^2, since $k^2 \ll h^2$, it follows, to second order, that

$$\bar{U}_m = \bar{U}\left(1 + \frac{1}{2}h^2\frac{\overline{w^2}}{\bar{U}^2}\right) \tag{4.40}$$

and to third order that

$$\overline{u_m^2} = \overline{u^2}(1 + h^2R_{uw^2}Tu) \tag{4.41}$$

where

$$Tu = \frac{(\overline{u^2})^{1/2}}{\bar{U}}, \qquad R_{uw^2} = \frac{\overline{uw^2}}{(\overline{u^2})^{3/2}}.$$

It has been proposed, for high turbulence intensity flows, that evaluations in terms of V_e^2 would reduce the error in the signal analysis since no series expansion is required in this method. Retaining the assumptions that $\bar{V} = \bar{W} = 0$, and using the $k^2 \ll h^2$ simplification, the expression for $\overline{V_e^2}$ becomes

$$\overline{V_e^2} = \bar{U}^2 + \overline{u^2} + h^2\overline{w^2}. \tag{4.42}$$

This equation contains both mean-velocity and fluctuating terms, and to separate these terms the mean velocity is usually evaluated from the conventional series expansion of eqn (4.40) (Acrivlellis and Felsch 1979; Acrivlellis 1982). The measured fluctuating term $\overline{v_e^2} = \overline{u_m^2}$ is related to $\overline{V_e^2}$ and $\bar{V}_e^2$ by

$$\overline{v_e^2} = \overline{V_e^2} - \bar{V}_e^2, \tag{4.43}$$

and the equation for $\overline{u_m^2}$ can therefore be expressed as

$$\overline{u_m^2} = \overline{u^2}[1 + O(4)], \tag{4.44}$$

where $O(4)$ represents fourth-order terms in the series expansion. Comparing eqns (4.41) and (4.44) it should be noted that the first equation contains third-order terms, while the latter only contains fourth-order terms. It has therefore been proposed (Rodi 1975; Acrivlellis and Felsch 1979; Acrivlellis 1982) that the V_e^2 method will provide accurate results for the Reynolds stresses, including $\overline{u^2}$. However, as shown in Section 4.13.2, this method has other severe limitations.

4.13 UNCERTAINTY ANALYSIS, *SN*-PROBES

To determine the signal-analysis errors which occur in moderate- and high-turbulence intensity flows, it is necessary to measure/estimate the higher-order terms in the series-expansion relationships presented.

4.13.1 $V_e(t)$- and $E(t)$-methods

Mean velocity, $\bar{U}$ The magnitude of the truncation errors introduced in the nonlinear signal analysis (eqn (4.30)) and in the $V_e(t)$ method-(eqn (4.40)) can be evaluated if it is assumed that $\overline{u^2} = \overline{v^2} = \overline{w^2}$ (Bruun 1976b). The related errors are shown in Table 4.4 as a function of $Tu = (\overline{u^2})^{1/2}/\bar{U}$, for values of Tu up to 30%. For both methods, the corrections are fairly small and the nonlinear method gives a more accurate result due to the (negative) term $\frac{1}{2}(n-1)\,\overline{u^2}/\bar{U}^2$ in eqn (4.30). A related numerical error analysis for $V_e(t)$ has been carried out by Swaminathan *et al.* (1986b). In their analysis they assumed that the true hot-wire relationship was given by the V_e^2-relationship of eqn (4.38) and that the hot-wire was exposed to a random turbulent flow field $(\bar{U} + u, v, w)$ with a joint-Gaussian-probability distribution. By inserting this flow field into eqn (4.38) they calculated the true values of $\bar{V}_e$ (and $\overline{v_e^2}$) as functions of the turbulence intensity, Tu. The corresponding 'measured' value, $\bar{U}_m$, was obtained from the first-order series expansion in eqn (4.39). Knowing the true value of $\bar{U}$, the error in the measured value $\bar{U}_m$ was identified as a function of Tu, as shown in

TABLE 4.4 *The turbulence corrections for mean-velocity evaluations, $\bar{U}_m/\bar{U}$, by the $E(t)$- and $V_e(t)$-methods*

Tu (%)	$E(t)$-method	$V_e(t)$-method
10	1.003	1.006
20	1.01	1.02
30	1.03	1.05

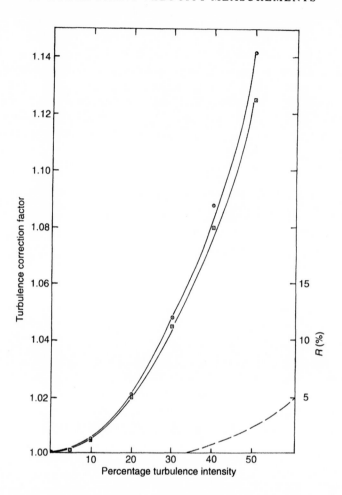

Fig. 4.19. The variation with turbulence intensity of the turbulence correction factor for the mean velocity, $\bar{U}$: ($\odot$) exact correction, ($\square$) second-order correction, ($-\!-\!-$) R = frequency of occurrence of the reverse flow condition ($U < 0$) (From Swaminathan *et al.* 1986*b* and Kawall *et al.* 1983. Reproduced with kind permission of Cambridge University Press from the *Journal of Fluid Mechanics*.)

Fig. 4.19. This error evaluation is consistent with the results in Table 4.4, which were based on the inclusion of second-order terms only. The results by Swaminathan *et al.* (1986*b*) demonstrate that the magnitude of the third- and higher-order terms will be small for values of $Tu < {\sim}40\%$. Figure 4.19 also contains (from Kawall *et al.* 1983) the frequency of occurrence, R, of the reverse-flow condition, $U < 0$, which defines rectification. It is observed that rectification effects occur when $Tu > {\sim}35\%$, and that the occurrence, R, is less than 2% when $Tu \leqslant 50\%$.

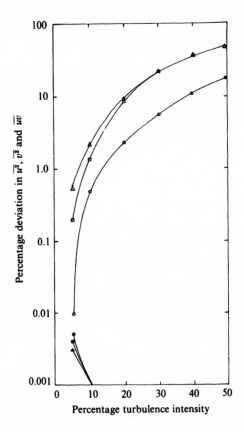

Fig. 4.20. The effect of the turbulence intensity on the evaluated Reynolds stresses: for method 1, ($\odot$) $\overline{u^2}$, ($\square$) $\overline{v^2}$, ($\triangle$) $\overline{uv}$ and for method 2, ($\bullet$) $\overline{u^2}$, ($\blacksquare$) $\overline{v^2}$, ($\blacktriangle$) $\overline{uv}$. From Swaminathan *et al.* 1986*b*. Reproduced with kind permission of Cambridge University Press from the *Journal of Fluid Mechanics*.)

Normal stress, $\overline{u^2}$ To evaluate the error in $\overline{u^2}$ caused by truncation of the series expansion for either the nonlinear voltage (eqn (4.31)) or $V_e(t)$ (eqn (4.41)) methods requires measurement or estimation of third- and possibly higher-order terms (see, for example, Bruun 1976*b*). The nonlinear voltage method contains an additional term involving the skewness factor, $Su = \overline{u^3}/(\overline{u^2})^{3/2}$, and the relative accuracy of the two procedures will therefore depend on the sign of Su. The numerical error analysis by Swaminathan *et al.* (1986*b*) also predicted the error in $\overline{u^2}$ as a function of Tu, and these results (method 1) are shown in Fig. 4.20. The error in $\overline{u^2}$ increases with Tu, reaching 10% for $Tu \approx 40\%$. In general, the relative error for $\overline{u^2}$ due to series-expansion truncation is similar to the corresponding error in $\bar{U}$. The errors in the values of $\bar{U}$ and $\overline{u^2}$ measured with an *SN*-probe

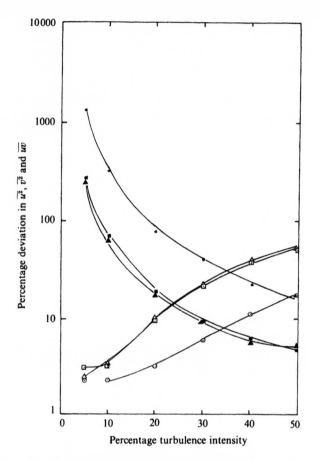

Fig. 4.21. The effect of turbulence intensity and calibration uncertainty (1%) on evaluated Reynolds stresses. For method 1, (○) $\overline{u^2}$, (□) $\overline{v^2}$, (△) $\overline{uv}$, and for method 2, (●) $\overline{u^2}$, (■) $\overline{v^2}$, (▲) $\overline{uv}$. (From Swaminathan *et al.* 1986*b*. Reproduced with kind permission of Cambridge University Press from the *Journal of Fluid Mechanics*.)

will therefore be relatively small, provided that $Tu < {\sim}30\text{--}40\%$. However, note that for time-series analysis of a finite record, the corresponding error in $\bar{U}$ is much smaller than the error in $\overline{u^2}$ (see Chapter 12). Figure 4.21 contains results of a similar error evaluation, but with an additional 1% uncertainty in the calibration parameters. As can be seen, no significant increase in the total error is observed except at low turbulence intensities.

4.13.2   The $\overline{V_e^2}$-method

The form of eqn (4.44) suggests that this method may provide an accurate evaluation of $\overline{u^2}$. However, eqn (4.44) was derived from eqn (4.43) and in low- and moderate turbulence intensity flows $\overline{v_e^2}$ is a small difference between two large quantities. In practical HWA, such a procedure can lead to significant errors in $\overline{v_e^2}$ and hence in $\overline{u^2}$. This is also the conclusion reached for $\overline{u^2}$ based on the error analysis for $\overline{V_e^2}$ (method 2) by Swaminathan *et al.* (1986*b*), and the error results for $\overline{V_e^2}$ are included in Figs 4.20 and 4.21. The $\overline{V_e^2}$-technique does not involve any truncation errors and in this respect it can be seen (Fig. 4.20) to be superior to the $V_e(t)$-analysis (method 1). However, as shown in Fig. 4.21, its sensitivity to calibration uncertainties makes it useless at low and moderate turbulence intensities. The value of Tu must be increased to ~50% before the errors in the $V_e(t)$- and $\overline{V_e^2}$-methods are of a similar magnitude (~20%). Consequently, there appears to be no advantage or justification for the use of the $\overline{V_e^2}$-technique for an *SN*-probe.

5

TWO-COMPONENT VELOCITY MEASUREMENTS

This chapter describes hot-wire/film-probe methods used for measuring two components of the velocity vector at a point. It will be assumed in this chapter that the selected plane of the two velocity components contains the mean-flow direction.

5.1 HOT-WIRE PROBE TYPES

Two-component velocity measurements are carried out either with a probe with a single wire inclined relative to the mean-flow direction or with a probe with two wires placed in an X-configuration. The related hot-film probes are discussed in Section 5.6. An X-probe enables simultaneous measurements of two velocity components to be made, whilst a single inclined wire can only provide statistical information about the velocity field. The geometry of a typical single yawed (SY) hot-wire probe with a parallel-stem orientation is shown in Fig. 5.1(a–b), unplated and plated, respectively. To minimize the probe-disturbance effects on the hot-wire signal, the plated configuration (Fig. 5.1(b)) is recommended. In certain flow situations it may be necessary to place the probe-stem perpendicular to the mean-flow direction, and with this support orientation a single normal (SN) probe can also be used for two-component velocity measurements by rotating the

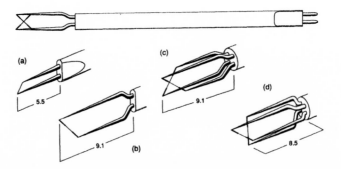

Fig. 5.1. SY and X-hot-wire probes: (a) an unplated SY probe, (b) a plated SY probe, (c) an X-probe with sensor plane parallel to the probe-stem, (d) an X-probe wth the sensor plane perpendicular to the probe-stem. The dimensions are in millimetres. (Reprinted with the permission of DANTEC Measurement Technology.)

probe around its axis so that the wire element becomes inclined relative to the mean-flow direction. Similarly, for an X-probe, the wire and the prong configuration depend on the probe-stem orientation. For an X-probe with a parallel probe-stem, the plane of the two sensors is parallel to the probe axis (Fig. 5.1(c)), whilst the sensor plane is perpendicular to the probe axis for the probe shown in Fig. 5.1(d).

5.2   HOT-WIRE PROBE DESIGN AND INTERFERENCE EFFECTS

5.2.1   Single yawed hot-wire probes

The design of the SY-probes shown in Fig. 5.1(a–b) are very similar to the related SN-probes shown in Fig. 4.2(a–b). Merati and Adrian (1984) have extended the slender-body theory for SN-probes (Adrian *et al*. 1984) to SY-probes. Their results demonstrated that if the prong spacing, s, is kept constant, as for plated DANTEC 55P01 (SN) and 55P02 (SY) probes ($s = 3$ mm), then the disturbance effects for the two probe types will be nearly the same. However, if the total wire length, L, is the same for the two probe types, as is the case for the TSI SN and SY film probes investigated by Merati and Adrian (1984), then the disturbance effects will be greater for the SY-probe.

5.2.2 X-hot-wire probes

An X-hot-wire probe consists of two inclined wires placed close together to form an X. For signal-analysis purposes it is usually assumed that the two wires are contained in the same plane. For this reason, some of the early commercial probes (for example, DISA 55-A-32) were manufactured with a very small distance of about 0.16 mm between the two wires. Subsequent investigations have, however, shown that this arrangement can lead to significant problems.

5.2.2.1   Thermal-wake interference

The problem of the thermal wake from one wire affecting the output from the other wire was reported by Guitton and Patel (1969) and Jerome *et al*. (1971). Studying the pitch, β, behaviour of a DISA 55-A-32 probe, they observed different output voltage values from the downstream wires, when the upstream wire was cold and hot. The largest effect was observed at low velocities, and it was explained by a narrowing of the wake width with increasing values of the Reynolds number (Jerome *et al*. 1971). When a modified probe with a wire spacing, h, of about 1 mm was used, the hot-wake effect was found to be insignificant provided that β was less than $\sim 25°$. Many commercial X-hot-wire probes are therefore now manufactured with a wire spacing of about 1 mm. Strohl and Comte-Bellot (1973) have investigated the wake effect in turbulent flows by varying the wire spacing ratio, h/ℓ, in the range 0.3–1.5 for an X-probe with an active wire length of $\ell = 1$ mm. Measurements of $\overline{u^2}$, $\overline{v^2}$, and $\overline{uv}$ were carried out in the centre of a turbulent round jet at an axial position $x/D = 14$ where the value of $(\overline{w^2})^{1/2}/\bar{U}$ was about 20%. For the case of a zero-degree mean pitch angle ($\beta = 0°, \overline{W} = 0$), they did not observe any systematic wake effect on the values of $\overline{u^2}$, $\overline{v^2}$, and $\overline{uv}$. They concluded that the fluctuating w-component resulted in a flapping motion of the hot wakes from both wires, and the value of h/ℓ had no significant effect on the measurements.

5.2.2.2 Prong-wake problems

An additional prong-wake problem has been pointed out by Wygnanski and Ho (1978). They carried out yaw-calibration experiments for X-probes with both parallel- and perpendicular-stem orientations. For X-probes with a parallel-stem orientation (Fig. 5.1(c)) they observed, as expected, a near-cosine dependence of the evaluated effective velocity. The results for an *unplated* X-probe with a perpendicular-stem orientation (Fig. 5.1(d)) gave a very different calibration curve for the wire attached to the shorter prongs. A near cosine dependence was observed when the yaw angle, α, was increased from $\alpha = 0°$ (with the short wire normal to the velocity) to $\alpha \simeq 45°$. In the region $45° < \alpha < 90°$, a substantial irregular deviation from a cosine law was noticed. This was caused by the wake from one of the longer prongs intersecting this wire, resulting in a velocity deficit for that part of the wire. This explanation was confirmed by related rms measurements of the fluctuating signal. An *unplated* X-hot-wire probe with a perpendicular-stem orientation is therefore not suitable for turbulent flow measurements. However, the perpendicular-support orientation can be used, provided a *plated* X-probe is used. It follows from a simple geometry evaluation that a plated X-probe with an active wire ratio ℓ/L of about 1/3 will have no significant prong-wake interference in the yaw-angle range $\alpha = \bar{\alpha} \pm 20/25°$. When larger flow-angle variations are expected to occur, then it is recommended that only X-probes with a parallel stem orientation are used.

5.2.2.3 Aerodynamic-disturbance effects of X-probes

The most detailed investigation of aerodynamic-disturbance effects for X-probes was carried out by Strohl and Comte-Bellot (1973). The procedure adopted was similar to the SN-probe investigation by Comte-Bellot *et al.* (1971) (see Section 4.2.2). A reference X-probe was created by using two inclined SN-probes, and these probes were inserted from either side of a calibration jet flow to create the reference X-probe. The calibration of the reference X-probe was carried out without the stem and prongs investigated. These authors investigated the disturbance of the reference X-probe signal caused by a variety of stem and prong configurations. Both the stem and the prongs introduced perturbations in u and v, and they also changed the sensitivity coefficients of the two wires. They concluded that the level of the errors can be kept within about ±2% by using the following probe dimensions:

Stem diameter	4 mm	Prong spacing	3 mm
Prong length	20 mm	Wire spacing	1 mm
Prong diameters	0.4 mm (base)	Wire length	1 mm
	0.2 mm (tip)		

Further investigations by Vagt (1979) have shown that the prong length may be reduced to 10–12 mm with only a small increase in the uncertainty. In practical HWA most of these uncertainty problems can be minimized by the calibration procedure. Provided the unperturbed velocity at the probe position is used as the calibration velocity, then (most of) the aerodynamic-disturbance effects caused by the prong and stem configuration will be compensated for by the calibration technique.

5.3 SINGLE INCLINED HOT-WIRE PROBES

For two-component measurements, it is necessary to determine the response of the inclined SN or SY probe to the velocity components U_N and U_T in the (x', y') plane of the wire-fixed coordinate system. Figure 5.2 illustrates both the wire-fixed and the mean-velocity aligned probe-stem coordinate system. In this chapter, the same yaw notation will be used for single inclined SN and SY probes.

During a probe calibration, the component, U_B, perpendicular to the (x', y')-plane is set equal to zero and the response of the hot-wire probe is usually determined in terms of the magnitude of the flow vector, $\tilde{V}$, and the yaw angle, α; that is,

$$E = F(\tilde{V}, \alpha). \tag{5.1}$$

Equation (5.1) is often assumed to be the power law of eqn (3.1),

$$E^2 = A + BV_e^n,$$

where the effective velocity, V_e, can be expressed as

$$V_e = \tilde{V}f(\alpha), \tag{5.2}$$

where $f(\alpha)$ is a yaw function as defined in the next section.

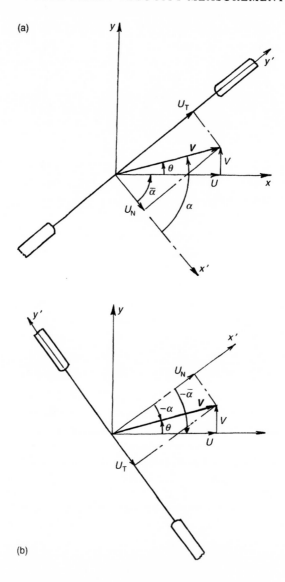

Fig. 5.2. The notation for the wire-fixed and probe-stem coordinate systems for inclined *SN* or *SY* probes: (a) mean yaw position, $\bar{\alpha}$, (b) mean yaw position, $-\bar{\alpha}$.

5.3.1 The calibration of SN and SY probes

To obtain an accurate description of $F(\tilde{V}, \alpha)$, a complete $(\tilde{V}, \alpha)$ matrix calibration should be carried out. The advantages and complexity of this approach are discussed in Section 5.4.3.2. However, both the calibration procedure and the related signal analysis are often simplified by the introduction of the effective-velocity concept. Several expressions have been proposed for the yaw function, $f(\alpha)$, in eqn (5.2). The most common method, proposed by Hinze (1959), has several signal-analysis advantages, since V_e can be expressed in terms of either $(\tilde{V}, \alpha)$ or the velocity components (U_N, U_T) in the wire-fixed coordinate system (x', y') (see Fig. 5.2(a)):

$$V_e = \tilde{V}f(\alpha) = \tilde{V}(\cos^2\alpha + k^2 \sin^2\alpha)^{1/2} \tag{5.3a}$$

$$= (U_N^2 + k^2 U_T^2)^{1/2}. \tag{5.3b}$$

Furthermore, as suggested by Jørgensen (1971a), a similar extended relationship, $V_e^2 = U_N^2 + k^2 U_T^2 + h^2 U_B^2$, can be applied in a three-dimensional flow. For these reasons, the yaw calibration will initially be described in terms of this method.

5.3.1.1 The calibration of inclined SN-probes

The early yaw-response investigations were carried out with SN-probes. The velocity calibration described in Chapter 4 for SN-probes is usually carried out with the stem parallel or perpendicular to the mean-flow direction. Both of these stem orientations have also been used in investigations of inclined SN-probes, and different yaw responses were obtained for the two probe orientations. Yaw investigations of SN-probes have been reported by Hinze (1959), Webster (1962), Champagne *et al.* (1967), Davies and Bruun (1968), Fujita and Kovasznay (1968), Friehe and Schwarz (1968), Jørgensen (1971a), Bruun (1971, 1972), Bruun and Tropea (1980, 1985), Bremhorst (1981), Fulachier *et al.* (1982), Samet and Einav (1985), Mobarak *et al.* (1986), and Wagner and Kent (1988a).

The normal position of a SN-probe corresponds to $\alpha = 0°$, and the velocity-calibration part of the above yaw investigations were therefore carried out at $\alpha = 0°$. At this position $V_e = U$, and a velocity calibration, as described in Section 4.4.1, will determine the calibration constants A, B, and n in the simple power law of eqn (4.8). The yaw calibration consists of measuring $E(\alpha)$ at a constant velocity, $\tilde{V}$, and the value of k as a function of α can be evaluated from

$$k = \frac{1}{\sin \alpha} \left[\left(\frac{E^2(\alpha) - A}{E^2(0) - A} \right)^{2/n} - \cos^2 \alpha \right]^{1/2}. \tag{5.4}$$

The yaw results in Fig. 5.3 obtained by Bruun and Tropea (1980, 1985)

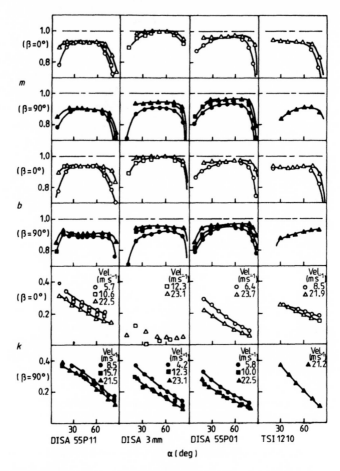

Fig. 5.3. The variations in the yaw coefficients k, b, and m with velocity and yaw for four SN hot-wire probe types.

demonstrate most of the features observed individually in other yaw investigations.

Perpendicular-stem orientation ($\beta = 90°$) The results in Fig. 5.3 for typical SN-probes (DANTEC 55P11, 55P01, and TSI 1210) show that the value of k decreases with α and $\tilde{V}$. Removing the wire from the prongs by a plating technique has a beneficial effect, as the value of $k(45°) \simeq 0.3$ for the unplated DISA 55P11 is reduced to $k(45°) \simeq 0.2$ for the plated DISA 55P01 probe. Furthermore, increasing the active wire length from 1.25 mm to about 3 mm (unplated DISA 3 mm) results in a small reduction in the value of $k(45°)$ from ~0.3 to ~0.2. Related observations have been made by Webster (1962), Fulachier *et al.* (1982), and Wagner and Kent (1988*a*).

Parallel-stem orientation ($\beta = 0°$) A different ℓ/d dependence is observed when a 'parallel' probe-stem is used. In this case, the value of $k(45°)$ is reduced to 0.2–0.25 for the unplated DISA 55P11 and to 0.15–0.2 for the plated DISA 55P01 probe. Also k has a significant ℓ/d dependence, with the value of k being about 0–0.05 when $\ell/d \simeq 600$, as observed previously by Champagne *et al.* (1967) and Bruun (1972). A $\cos \alpha$ yaw dependence therefore applies for very long wires, when a parallel probe-stem orientation is used.

Other yaw functions Several functions have been proposed for the yaw response $f(\alpha)$ in eqn (5.2). The most common expressions for $f(\alpha)$ are

$$f(\alpha) = \begin{cases} (\cos^2 \alpha + k^2 \sin^2 \alpha)^{1/2} & (5.5a) \\ \cos^m \alpha & (5.5b) \\ [1 - b(1 - \cos^{1/2} \alpha)]^2 & (5.5c) \\ \cos \alpha_e & (5.5d) \\ \cos \alpha + \varepsilon(\cos \alpha - \cos 2\alpha), & (5.5e) \end{cases}$$

and the related yaw coefficients are k, m, b, $\bar{\alpha}_e$ ($= \bar{\alpha} + \Delta \alpha_e$), and ε. Equation (5.5b) was proposed by Davies and Bruun (1968) eqn (5.5c) by Friehe and Schwarz (1968), the effective-angle, α_e, method of eqn (5.5d) by Bradshaw (1971), and eqn (5.5e) by Fujita and Kovasznay (1968). Equations (5.5b) and (5.5c) were also used in the comparative yaw investigations of *SN*-probes by Bruun and Tropea (1980, 1985) and the resulting values of b and m are included in Fig. 5.3.

5.3.1.2 The calibration of an SY-probe

Velocity calibration at $\alpha = 0°$ Bruun (1972) has evaluated the yaw response of an *SY*-probe by applying the approach adopted for inclined *SN*-probes. Consider an *SY*-probe with a mean yaw angle, $\bar{\alpha}$ (see Fig. 5.2(a)). To carry out the velocity calibration, the probe was rotated so that the wire was placed at $\alpha = 0°$. This was followed by a yaw calibration at a constant velocity using eqn (5.4). The $k(\alpha)$ and $m(\alpha)$ values measured for plated *SY*-probes were found to be similar to the results for an inclined *SN*-probe with a parallel-stem orientation.

Velocity calibration at $\bar{\alpha}$ A velocity calibration at $\alpha = 0°$ permits the direct evaluation of all three calibration constants A, B, and n in the power-law relationship of eqn (4.8). However, the yaw angle ($\alpha = 0°$) used during the velocity calibration differs from the measurement position $\bar{\alpha}$ (usually $\simeq$ 45°). A second type of calibration has therefore been developed, in which the velocity calibration is carried out at $\bar{\alpha}$. This procedure was originally proposed by Bradshaw (1971) for the effective-angle, α_e, function in eqn (5.5d)) and is applied in the following to the four yaw functions in eqn (5.5a–d).

Introducing $V_e = \tilde{V}f(\alpha)$ (eqn (5.2)) into the power-law relationship of eqn (3.1) gives

$$E^2 = A + B[f(\alpha)\tilde{V}]^n = A + \hat{B}(\alpha)\tilde{V}^n, \qquad (5.6)$$

with $\hat{B}(\alpha)$ defined by

$$\hat{B}(\alpha) = Bf(\alpha)^n. \qquad (5.7)$$

It follows from eqn (5.6) that a velocity calibration at $\bar{\alpha}$ will determine A, $\hat{B}(\bar{\alpha})$, and n. However, eqn (5.7) contains two unknowns, B and the yaw coefficient (k, m, etc.) in the selected yaw function, $f(\alpha)$. If the yaw calibration at a constant velocity is restricted to small variations, $\theta = \alpha - \bar{\alpha}$, in the yaw angle (see Fig. 5.2), then it can be assumed that B, and the yaw coefficient being investigated, is independent of α and therefore of θ. In this case, the following ratio relationship applies

$$E_\theta = \left[\frac{E_\alpha^2 - A}{E_{\bar{\alpha}}^2 - A}\right]^{1/n} = \frac{f(\alpha)}{f(\bar{\alpha})}, \qquad (5.8)$$

which only contains the yaw dependence and not B. As originally demonstrated by Bradshaw (1971) for $f(\alpha) = \cos\alpha_e$, eqn (5.8) may be rearranged and plotted in the form $Y = aX$, with the gradient, a, determining the yaw parameter, $\bar{\alpha}_e$, in the form $\tan\bar{\alpha}_e$. A similar approach has been developed by Khan (1991) for the other three yaw functions in eqn (5.5a–c), and the related expressions for X, Y, and a are given in Table 5.1. Having obtained the relevant yaw coefficient from the yaw calibration the corresponding value of B can be determined from eqn (5.7) by setting $\alpha = \bar{\alpha}$. A detailed yaw-calibration study of inclined SN and SY hot-wire probes has been reported by Bruun *et al.* (1990b). Using the $\bar{\alpha}$ reference method, it was found that, for plated SY DANTEC probes, the value of $k(\bar{\alpha})$ was about 0.12–0.15. Similar values have also been reported by Bremhorst (1981). These results supersede some of the results and conclusions made by Bruun and Tropea (1980, 1985). In these investigations A and n were correctly identified, in the power-law relationship, as being functions of the yaw

TABLE 5.1   *The $Y = aX$ curve-fitting relationship for the yaw functions of eqns (5.5a–d)), $\theta = \alpha - \bar{\alpha}$, E_θ is defined by eqn (5.8)*

Yaw equation	Y	X	a
(5.5a)	$E_\theta^2 - 1$	$E_\theta^2 \sin^2\bar{\alpha} - \sin^2\alpha$	$1 - k^2$
(5.5b)	$\log E_\theta$	$\log(\cos\alpha / \cos\bar{\alpha})$	m
(5.5c)	$E_\theta^{1/2} - 1$	$E_\theta^{1/2}(1 - \cos^{1/2}\bar{\alpha}) - (1 - \cos^{1/2}\alpha)$	b
(5.5d)	$\cos\theta - E_\theta$	$\sin\theta$	$\tan\bar{\alpha}_e$

angle $\alpha(= \bar{\alpha} + \theta)$ (see Section 5.4.1). It was concluded that this yaw dependence of A and n had a significant effect on the value of the calculated yaw coefficient (for example, k), and that under certain conditions negative values could be obtained for k^2. Negative values for k^2 have subsequently been reported by Swaminathan *et al.* (1984a), and by Samet and Einav (1985). However, the subsequent detailed yaw calibration study by Bruun *et al.* (1990b) has predicted positive values for k^2 in all cases. Negative values of k^2 can be obtained if approximate evaluation methods are used, or if a significant uncertainty exists in the value of the yaw angle, α. As shown in Section 5.9 an error in the mean yaw angle, $\bar{\alpha}$, has a substantial effect on the value of the yaw coefficient, k. In this respect the effective-angle method is superior, since it does not require knowledge of the value of the mean yaw angle, $\bar{\alpha}$.

5.3.1.3 Combined velocity and yaw calibration of an SY-probe

If eqn (5.3a) is inserted into eqn (3.1) the hot-wire response equation becomes

$$E^2 = A + B[\tilde{V}(\cos^2\alpha + k^2\sin^2\alpha)^{1/2}]^n. \tag{5.9}$$

By measuring the output voltage, E, at discrete points throughout the applicable $(\tilde{V}, \alpha)$ calibration range, and applying a least-squares curve-fitting technique to eqn (5.9), the optimum values of A, B, n, and k for the selected velocity and yaw-angle range can be obtained. Such a method is described by Al-Kayiem and Bruun (1991).

5.3.1.4 The dynamic calibration of an SY or X-probe

To avoid the problems related to a static calibration procedure and the selection of a suitable response equation, Morrison *et al.* (1972) have extended the dynamic calibration method to SY and X-probes. The probe was placed in a uniform flow and oscillated by a shaker mechanism in either the U- or V-direction, as shown in Fig. 5.4. This enabled the evaluation of the velocity sensitivities $S_u = \partial E/\partial U$ and $S_v = \partial E/\partial V$ in the following relationship between the fluctuating voltage signal, e, and the fluctuating velocity components, u and v:

$$e = \frac{\partial E}{\partial U} u + \frac{\partial E}{\partial V} v. \tag{5.10}$$

Note that for the wire orientation shown in Fig. 5.4 (corresponding to Fig. 5.2(a)) the value of S_v will be negative. For comparison purposes, Morrison *et al.* (1972) also carried out a static calibration, in which it was assumed that the exponent, n, in the power-law relationship was 0.5 and

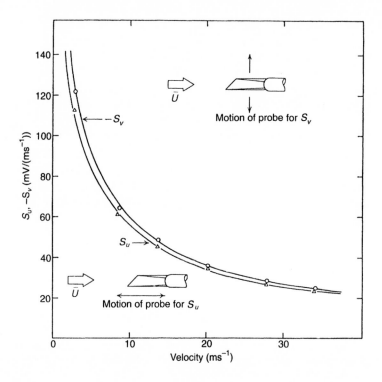

Fig. 5.4. The dynamic calibration of an inclined hot-wire. (From Morrison *et al.* 1972. Reproduced with kind permission of Cambridge University Press from the *Journal of Fluid Mechanics*.)

that the mean yaw angle, $\bar{\alpha}$ was 45°. However, a value of $n = 0.5$ results in a poor velocity curve fit, and the value of $\bar{\alpha}$ often deviates by several degrees from 45°. Consequently, no valid conclusion can be made from the differences between their static- and dynamic-calibration results. Any consistent difference between static and dynamic calibration is caused by a significant dynamic-prong effect (see Section 4.9). A comparative calibration investigation of X-probes has been reported by Mulhearn and Finnigan (1978). The dynamic calibration involved oscillating the X-probe in a closed ellipsoidal loop. No significant difference was observed between their dynamic-calibration measurements and the results of a careful static calibration. Kühn and Dressler (1985) have carried out a detailed study of the dynamic-prong effect by spinning an inclined SN-probe around its stem in a nozzle-calibration flow. From an elaborate signal-analysis procedure, they found that the dynamic-prong effect depended primarily on the active-wire's length-to-diameter ratio, ℓ/d, as shown in Fig. 5.5. They concluded

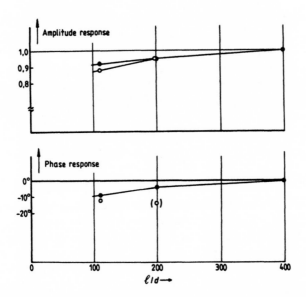

Fig. 5.5. The influence of the length-to-diameter ratio on the step in the frequency response at low frequencies: ($\bullet$) $\bar{V} \simeq 30 \, \mathrm{m \, s^{-1}}$, (O) $\bar{V} \simeq 9 \, \mathrm{m \, s^{-1}}$. (From Kühn and Dressler 1985.)

that the dynamic-prong effect should be taken into account for short wires with $\ell/d < 200$. For probes with $\ell/d > \sim 300$ the effect will be so small that it can he ignored in most hot-wire anemometry (HWA) applications. For practical purposes this criterion is satisfied by most TSI and DANTEC hot-wire probes.

5.3.2 Measurements with an inclined SN or SY probe

Most of the steps in the measurement procedure for an inclined SN or SY probe are identical to those described in Chapter 4 for SN-probes. To implement measurements, the SY-probe must first have been calibrated, as described in Section 5.3.1, and a suitable digital measurement system must have been selected (as discussed in Section 4.11). Finally, the probe must be aligned with the direction of the mean flow, which must be known or determined (using, for example, the procedure described in Section 4.11.1).

5.3.2.1 Data acquisition and storage

Before measurements can commence, the parameters of the digital-measurement system must be selected. As described in Section 4.11.2 for SN-probes, it is good data-acquisition practice to feed the direct hot-wire signal, E, through a signal-conditioning unit (see Fig. 4.18) containing an offset, E_{off}, and a gain, G, to match the fluctuating signal component, $E_G = G(E(t) - E_{off})$ (eqn (4.29)) to the voltage range of the analog/digital (A/D) unit used. Having set E_{off} and G, set the low-pass filter, and specified the data-acquisition parameters (N, number of samples; sampling rate, SR (samples s^{-1}), or T, the total sampling time—see Chapter 12), data acquisition can be initiated, and a digital time-series record $E_G(m)$, is available on completion for subsequent data analysis.

5.3.3 Data analysis for inclined SN or SY probes

Statistical analysis can be carried out on either: (i) the voltage signal, with subsequent conversion to the related velocity quantities; or (ii) the voltage record (which is first converted into an effective-velocity record, $V_e(m)$, and then a time-series analysis of $V_e(m)$ is made).

5.3.3.1 Voltage-signal analysis

This method can be implemented using either an analytical relationship (Davies and Bruun 1968; Papavergos and Hedley 1979) or it can be based on eqn (5.10), where the sensitivities $\partial E/\partial U$ and $\partial E/\partial V$ can be obtained from a dynamic calibration (Morrison *et al.* 1972). The time-mean value, $\bar{E}$, is evaluated from the digital time record, $E(m)$, and the corresponding time-mean velocity, $\bar{U}$, can, for a low turbulence intensity flow, be determined from

$$\bar{E}^2 = A + B\bar{V}_e^n, \tag{5.11a}$$

with

$$\bar{V}_e = \bar{U} \left(\cos^2 \bar{\alpha} + k^2 \sin^2 \bar{\alpha} \right)^{1/2}. \tag{5.11b}$$

If the wire element of the inclined SN or SY probe is placed sequentially in the (x, y)-plane at mean yaw angles, $\bar{\alpha}$ and $-\bar{\alpha}$, relative to the mean-flow direction (see Fig. 5.2), then the response equations for the fluctuating quantities for the two positions will be for $\bar{\alpha}$

$$e_1 = S_u u + S_v v, \tag{5.12a}$$

and for $-\bar{\alpha}$

$$e_2 = S_u u - S_v v, \tag{5.12b}$$

where S_u and S_v are the values of $\partial E/\partial U$ and $\partial E/\partial V$ for the probe placed at a mean yaw angle of $\bar{\alpha}$ (see Fig. 5.2(a)). The analysis in Section 5.3.1.4 demonstrates that the value of S_v is negative for the $\bar{\alpha}$ position. Since the two measurements are carried out sequentially, only statistical quantities can be obtained. For a statistically stationary flow the following relationships apply

$$\overline{e_1^2} = S_u^2 \overline{u^2} + S_v^2 \overline{v^2} + 2S_u S_v \overline{uv}, \tag{5.13a}$$

$$\overline{e_2^2} = S_u^2 \overline{u^2} + S_v^2 \overline{v^2} - 2S_u S_v \overline{uv}, \tag{5.13b}$$

and the value of $\overline{uv}$ can be calculated by subtraction from

$$\overline{e_1^2} - \overline{e_2^2} = 4S_u S_v \overline{uv}. \tag{5.14}$$

Placing the same (or a different) probe at $\alpha = 0°$ will provide the value of $\overline{u^2}$, and the related value of $\overline{v^2}$ can then be calculated from eqn (5.13a or b).

5.3.3.2 Velocity-analysis method

The acquired digital-voltage record, $E_G(m)$ must first be converted into the corresponding effective-velocity record, $V_e(m)$.

Data conversion In principle, this data conversion involves two steps. The anemometer voltage record $E(m)$ is calculated from the acquired $E_G(m)$-values using eqn (4.32)

$$E(m) = \frac{1}{G} E_G(m) + E_{\text{off}}.$$

The corresponding $V_e(m)$-values are then obtained using, for example, the inverted calibration relationship of eqn (4.33)

$$V_e(m) = \left[\frac{E(m)^2 - A}{B} \right]^{1/n}.$$

These two conversion steps can be implemented directly on a computer, or they may be combined into a look-up table method, using the same procedures as for a *SN*-probe; the reader should refer to Section 4.12.2.2 for the details.

Analysis of the $V_e(m)$-record The effective velocity is often expressed in the form given by eqn (3.2)

$$V_e^2 = U_N^2 + k^2 U_T^2 + h^2 U_B^2,$$

where U_N, U_T and U_B are the instantaneous velocity components in the wire-fixed coordinate system, as shown in Fig. 3.1. The following transformations from the wire-fixed to the probe-stem coordinate system apply to the $\bar{\alpha}$ orientation (Fig. 5.2(a))

$$U_N = U \cos \bar{\alpha} - V \sin \bar{\alpha}, \tag{5.15a}$$

$$U_T = U \sin \bar{\alpha} + V \cos \bar{\alpha}, \tag{5.15b}$$

$$U_B = W, \tag{5.15c}$$

where (U, V, W) are the velocity components in the (x, y, z) probe-stem coordinate system. Inserting eqns (5.15a–c) into eqn (3.2), the expression for V_e^2 becomes for $\bar{\alpha}$

$$
\begin{aligned}
V_{e1}^2 &= (U\cos\bar{\alpha} - V\sin\bar{\alpha})^2 + k^2(U\sin\bar{\alpha} + V\cos\bar{\alpha})^2 + h^2W^2 \\
&= U^2(\cos^2\bar{\alpha} + k^2\sin^2\bar{\alpha}) + V^2(\sin^2\bar{\alpha} + k^2\cos^2\bar{\alpha}) \\
&\quad - UV(1 - k^2)\sin 2\bar{\alpha} + h^2W^2
\end{aligned}
\tag{5.16a}
$$

The corresponding equation for the $-\bar{\alpha}$ orientation (Fig. 5.2(b)) is

$$
\begin{aligned}
V_{e2}^2 &= U^2(\cos^2\bar{\alpha} + k^2\sin^2\bar{\alpha}) + V^2(\sin^2\bar{\alpha} + k^2\cos^2\bar{\alpha}) \\
&\quad + UV(1 - k^2)\sin 2\bar{\alpha} + h^2W^2
\end{aligned}
\tag{5.16b}
$$

Equations (5.16a–b) are expressed in terms of V_e^2, and it has been proposed that an analysis in terms of $\overline{V_e^2}$ will give accurate results because a series expansion is not required. This method is considered in Section 5.8. However, most signal analyses are related to V_e rather than to $\overline{V_e^2}$, and from eqn (5.16a) the equation for V_{e1} becomes

$$
\begin{aligned}
V_{e1} &= [U^2(\cos^2\bar{\alpha} + k^2\sin^2\bar{\alpha}) + V^2(\sin^2\bar{\alpha} + k^2\cos^2\bar{\alpha}) \\
&\quad - UV(1 - k^2)\sin 2\bar{\alpha} + h^2W^2]^{1/2}
\end{aligned}
\tag{5.17a}
$$

and the equation for V_{e2} is

$$
\begin{aligned}
V_{e2} &= [U^2(\cos^2\bar{\alpha} + k^2\sin^2\bar{\alpha}) + V^2(\sin^2\bar{\alpha} + k^2\sin^2\bar{\alpha}) \\
&\quad + UV(1 - k^2)\sin 2\bar{\alpha} + h^2W^2]^{1/2}
\end{aligned}
\tag{5.17b}
$$

To obtain a simple expression for V_e, a series expansion must be applied to eqn (5.17a–b). It will be assumed that the probe-stem is aligned with the mean-flow direction, $\bar{U}$ (that is, $\bar{V} = \bar{W} = 0$), and if the fluctuating components u, v, and w are small compared with $\bar{U}$, then it is justified to retain only first-order terms in the series expansion, giving for $\bar{\alpha}$

$$
V_{e1} = f(\bar{\alpha})[\bar{U} + u - g(\bar{\alpha})v]
\tag{5.18a}
$$

and for $-\bar{\alpha}$:

$$
V_{e2} = f(\bar{\alpha})[\bar{U} + u + g(\bar{\alpha})v]
\tag{5.18b}
$$

where

$$
f(\bar{\alpha}) = (\cos^2\bar{\alpha} + k^2\sin^2\bar{\alpha})^{1/2}
\tag{5.19a}
$$

$$
g(\bar{\alpha}) = \frac{(1 - k^2)\cos^2\bar{\alpha}}{\cos^2\bar{\alpha} + k^2\sin^2\bar{\alpha}}\tan\bar{\alpha}
\tag{5.19b}
$$

Equations (5.18a–b) can be applied to any of the yaw functions, $f(\alpha)$, in eqns (5.5a–e). The value of $g(\bar{\alpha})$ may be obtained either from a series expansion method or by noting (Fujita and Kovasznay 1968) that $\mathrm{d}(f(\alpha))/$

TABLE 5.2 *The yaw functions $f(\bar{\alpha})$ and $g(\bar{\alpha})$ corresponding to eqns (5.5a–d)*

Equation number	$f(\bar{\alpha})$	$g(\bar{\alpha})$
(5.5a)	$(\cos^2\bar{\alpha} + k^2\sin^2\bar{\alpha})^{1/2}$	$\dfrac{\cos^2\bar{\alpha}(1 - k^2)}{\cos^2\bar{\alpha} + k^2\sin^2\bar{\alpha}}\tan\bar{\alpha}$
(5.5b)	$\cos^m\bar{\alpha}$	$m\tan\bar{\alpha}$
(5.5c)	$[1 - b(1 - \cos^{1/2}\bar{\alpha})]^2$	$\dfrac{[b(1 - b)\cos^{1/2}\bar{\alpha} + b^2\cos\bar{\alpha}]}{[1 - b(1 - \cos^{1/2}\bar{\alpha})]^2}\tan\bar{\alpha}$
(5.5d)	$\cos\bar{\alpha}_e$	$\tan\bar{\alpha}_e$

$d\alpha = -f(\alpha)g(\alpha)$. The expressions for $f(\bar{\alpha})$ and $g(\bar{\alpha})$ corresponding to eqns (5.5a–d) are listed in Table 5.2.

By placing the inclined probe at mean yaw angles of $\bar{\alpha}$ and $-\bar{\alpha}$ (positions 1 and 2) and the same probe (or a different probe) at $\alpha = 0°$, and evaluating $\bar{V}_{e0}$, $\overline{v_{e0}^2}$, $\overline{v_{e1}^2}$, and $\overline{v_{e2}^2}$ from the measured voltage records, it follows from the above analysis (Fujita and Kovasznay 1968; and Bruun 1972) that

$$\bar{V}_{e0} = \bar{U}, \tag{5.20a}$$

$$\overline{v_{e0}^2} = \overline{u^2}, \tag{5.20b}$$

$$\overline{v_{e1}^2} = f(\bar{\alpha})^2[\overline{u^2} + g(\bar{\alpha})^2\,\overline{v^2} - 2g(\bar{\alpha})\overline{uv}], \tag{5.20c}$$

$$\overline{v_{e2}^2} = f(\bar{\alpha})^2[\overline{u^2} + g(\bar{\alpha})^2\,\overline{v^2} + 2g(\bar{\alpha})\overline{uv}]. \tag{5.20d}$$

From these four equations, $\bar{U}$, $\overline{u^2}$, $\overline{v^2}$, and $\overline{uv}$ can be evaluated. To improve the accuracy of the results, Fujita and Kovasznay (1968) proposed that more than three angle positions should be used in conjunction with a least-squares curve-fitting method.

5.4 X-HOT-WIRE PROBES

For calibration and evaluation purposes, it is usually assumed that the two wires of the X-probe are placed in the same plane. The notation used is shown in Fig. 5.6. When a parallel probe-stem orientation (Fig. 5.1(c)) is used, then the probe-stem is often aligned with the x-axis of the space-fixed coordinate system selected. The greatest advantage of an X-probe, compared with an SY-probe is that it allows the simultaneous measurement of two velocity components. Provided the velocity component W perpendicular to the X-probe is small compared with the two in-plane components

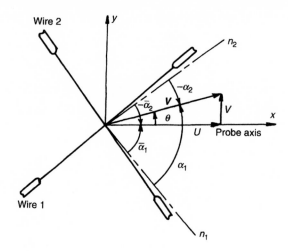

Fig. 5.6. The velocity-component and angle notation for X-hot-wire probes. n_1 and n_2 are the normals to wires 1 and 2.

U and V, then the simultaneous response equations for the two wires can be expressed as

$$E_1 = F_1(U, V), \tag{5.21a}$$

$$E_2 = F_2(U, V). \tag{5.21b}$$

Alternatively, the calibration and signal analysis may be performed in terms of the magnitude of $\tilde{V}$ and the flow angle, θ, of the velocity vector, V (see Fig. 5.6):

$$E_1 = F_3(\tilde{V}, \theta), \tag{5.22a}$$

$$E_2 = F_4(\tilde{V}, -\theta), \tag{5.22b}$$

which use the $\pm\theta$ relationship for wire, 1 and 2. The flow angle, θ, and the yaw angles, α_1 and $-\alpha_2$, for the two wires (see Fig. 5.6) are related for wire 1 by

$$\alpha_1 = \bar{\alpha}_1 + \theta, \tag{5.23a}$$

and for wire 2 by

$$-\alpha_2 = -\bar{\alpha}_2 + \theta. \tag{5.23b}$$

In the effective-angle method and in some of the more recent, computer-based look-up matrix methods, the calibration is carried out directly in terms of $\tilde{V}$ and θ and no knowledge of $\bar{\alpha}_1$ and $-\bar{\alpha}_2$ is required. The corresponding values of U and V can be evaluated from $(\tilde{V}, \theta)$ as

$$U = \tilde{V}\cos\theta, \tag{5.24a}$$

$$V = \tilde{V}\sin\theta. \tag{5.24b}$$

When the yaw relationships in eqns (5.5a–d) are used, the response equations must be expressed in terms of the yaw angles by

$$E_1 = F_3(\tilde{V}, \alpha_1, \bar{\alpha}_1), \tag{5.25a}$$

$$E_2 = F_4(\tilde{V}, -\alpha_2, -\bar{\alpha}_2). \tag{5.25b}$$

This is the most common calibration relationship, and it requires measurements of the mean yaw angles, $\bar{\alpha}_1$ and $-\bar{\alpha}_2$.

5.4.1 The calibration of an X-probe

To apply eqns (5.21a–b) to eqns (5.25a–b) for calibration and subsequent signal analysis, it is necessary to demonstrate that there is a unique relationship between (E_1, E_2) and $(\tilde{V}, \theta)$, or (U, V).

Limiting V_e-lines The yaw dependence for a hot-wire probe deviates from a $\cos\alpha$ yaw variation. Consequently, if the outputs from an X-probe are plotted on a (V_{e1}, V_{e2})-map, then all points corresponding to $-45° \leqslant \theta \leqslant 45°$ will fall within two limiting lines contained in the first quadrant of this map. To demonstrate these limiting V_e-lines, consider an X-probe with two identical wires set at $\bar{\alpha}_1 = \bar{\alpha}_2 = 45°$. Since the two wires are perpendicular to each other, the same wire-fixed coordinate system can be applied to both wires. Using a power-law relationship, the response equations for the two wires are

$$E_1^2 = A + BV_{e1}^n, \tag{5.26a}$$

$$E_2^2 = A + BV_{e2}^n, \tag{5.26b}$$

and assuming $U_B = 0$ then in the wire-fixed coordinate system corresponding to wire 1

$$V_{e1}^2 = U_N^2 + k^2 U_T^2, \tag{5.27a}$$

$$V_{e2}^2 = U_T^2 + k^2 U_N^2. \tag{5.27b}$$

To achieve a unique signal interpretation, the velocity vector must remain within the approach quadrant of the X-probe and the two related limits are:

1. Flow parallel to wire 1. This corresponds to $U_N = 0$, and eqns (5.27a–b) become $V_{e1} = kU_T$ and $V_{e2} = U_T$, giving

$$V_{e1} = kV_{e2}. \tag{5.28a}$$

2. Flow parallel to wire 2, that is, $U_T = 0$. Consequently, $V_{e1} = U_N$, $V_{e2} = kU_N$ and

$$V_{e2} = kV_{e1}. \tag{5.28b}$$

The principle of these limiting lines on a (V_{e1}, V_{e2}) calibration map was first demonstrated by Tutu and Chevray (1975).

X-probe calibration map To use an X-probe a unique relationship must exist between (E_1, E_2) and $(\tilde{V}, \theta)$ (or (U, V)) throughout the velocity and angle range investigated. If the X-probe has mean flow angles of $\bar{\alpha}_1$ and $-\bar{\alpha}_2$ then the applicable flow angle range will be $-90° + \bar{\alpha}_2 \leqslant \theta \leqslant 90° - \bar{\alpha}_1$. When $\bar{\alpha}_1 = \bar{\alpha}_2 = 45°$, the flow-angle range is $-45° \leqslant \theta \leqslant 45°$. The unique mapping principle has been demonstrated by Bruun *et al.* (1990c) for a standard plated X-probe (DANTEC 55P51) for the flow-angle range $-45° \leqslant \theta \leqslant 45°$ and for the velocity range 5–50 m s^{-1}. It was shown that over the complete velocity and flow-angle range the X-probe calibration relationship could be expressed as

$$E_1^2 = A_1(\theta) + B_1^*(\theta)\tilde{V}^{n_1(\theta)}, \tag{5.29a}$$

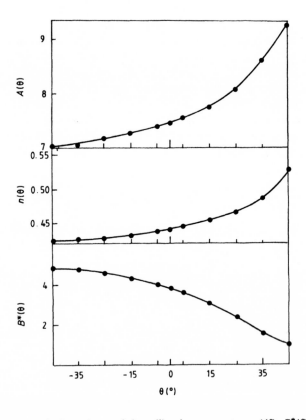

Fig. 5.7. The flow-angle dependence of the calibration parameters, $A(\theta)$, $B^*(\theta)$, and $n(\theta)$; see eqn (5.29).

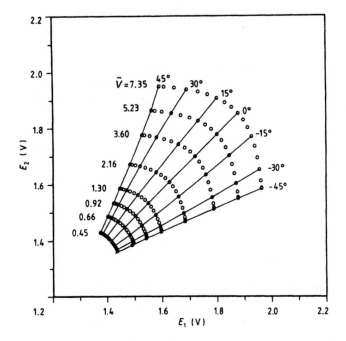

Fig. 5.8. Unique (E_1, E_2) calibration map for a DISA 55P51 *X*-hot-wire probe. The velocity, $\tilde{V}$, is in m s^{-1}. (From Abdel-Rahman *et al.* 1989.)

$$E_2^2 = A_2(-\theta) + B_2^*(-\theta)\,\tilde{V}^{n_2(-\theta)}. \qquad (5.29b)$$

Typical variations in the calibration parameters $A(\theta)$, $B^*(\theta)$, and $n(\theta)$ are shown in Fig. 5.7. The calibration relationship of eqn (5.29a–b), in conjunction with a monotonic variation in $A(\theta)$, $B^*(\theta)$, and $n(\theta)$ in Fig. 5.7, demonstrates that a unique mapping relationship exists between (E_1, E_2) and $(\tilde{V}, \theta)$. A related unique (E_1, E_2) calibration map (Fig. 5.8) has been presented by Abdel-Rahman *et al.* (1989) for the same type of *X*-hot-wire probe. The uniqueness domain of *X*-probes has also been considered by Beuther *et al.* (1987), by LeBoeuf and George (1990), and by Pompeo and Thomann (1993).

A non-unique mapping relationship has been reported by Johnson and Eckelmann (1983) for an *X*-hot-film probe for values of $|\theta| > 30°$. These angle limitations were most probably caused by the use of an unusual prong configuration.

5.4.1.1 V_e-calibration methods

In V_e-calibration methods, each wire in the *X*-probe is considered independently, and the calibration procedure for the single inclined probe is used

(Section 5.3.1). The velocity calibration of the two wires can be carried out simultaneously if the probe is initially placed at its normal $(\bar{\alpha}_1, -\bar{\alpha}_2)$ orientation. Least-squares curve-fitting applied to the velocity calibration data will determine the calibration constants A_1, $\hat{B}_1(\bar{\alpha}_1)$, n_1, and A_2, $\hat{B}_2(-\bar{\alpha}_2)$, n_2 in the power-law relationship of eqn (5.6). For the subsequent yaw calibration, the mean yaw angles $\bar{\alpha}_1$ and $-\bar{\alpha}_2$ must normally be measured, since they are required in yaw evaluations based on eqns (5.5a–c). Due to the $\pm\theta$ relationship for wires 1 and 2 (see eqns (5.23a–b)) the yaw calibration around $\bar{\alpha}$ can be carried out at the same time for both wires. The values of the selected yaw coefficient $(k, m, b, \text{ or } \bar{\alpha}_e)$ in eqns (5.5a–d) can be evaluated by a yaw curve-fitting procedure related to eqn (5.8). The author and his students have carried out a substantial number of calibrations using plated X-probes with: (a) a parallel-stem orientation (DANTEC 55P51), (b) a perpendicular-stem orientation (DANTEC 55P52) and unplated X-probes with a parallel-stem orientation (DANTEC 55P61). The calculated values of $k(\bar{\alpha})$ were similar to those for the corresponding types of single inclined probes (Section 5.3.1). Similar observations have been reported by Müller (1982a) and Chew and Ha (1988) for an unplated X-probe (DISA 55P61).

Pitch dependence The results of Müller (1982b) and Chew and Ha (1988) for an unplated (DANTEC 55P61) X-probe indicate that the larger number of prongs will increase the value of the pitch coefficient, h, to about 1.2.

5.4.1.2 Multi-angle calibration methods

Calibration and analysis procedures based on the effective velocity concept involve a number of simplifications, which can lead to significant errors. More accurate methods have therefore been introduced (as described in Section 5.4.3.2) which utilize a direct relationship between (E_1, E_2) and $(\check{V}, \theta)$ or (U, V). These methods normally require a complete $(\check{V}, \theta)$ calibration within the specified velocity and flow-angle ranges.

5.4.2 Measurements with X-probes

Most of the steps in the digital-measurement procedure for an X-probe are similar to those for an SN or SY probe. However, a two-channel digital-measurement system is required, since an X-probe provides two output signals. The two signals must be acquired simultaneously by the A/D unit, and this is often achieved by placing a sample-and-hold device on the input side of each channel of the A/D unit. Alternatively, a computer system with two separate A/D units may be used.

5.4.2.1 Data acquisition

The parameters E_{off} and G of the signal-conditioning unit connected to each wire signal are adjusted separately, as described for the SN-probe (Section 4.11.2). The *same* setting is selected for the two low-pass filters, and the *same* data-acquisition parameters (N, number of samples, and SR, sampling rate, or T, total sampling time) are specified for both hot-wire signals (see Chapter 12). Data acquisition can then be initiated, and on completion there are two simultaneous digital time-series records, $E_{G1}(m)$ and $E_{G2}(m)$.

5.4.3 Signal analysis for X-probes

The signal analysis is usually carried out either in terms of the effective velocities, V_{e1} and V_{e2}, for the two wires, or by using a look-up method which relates the measured (E_1, E_2) values to the corresponding $(\tilde{V}, \theta)$ or (U, V) values.

5.4.3.1 The V_e-analysis method (sum and difference)

Data conversion The first step is to convert the two digital records $E_{G1}(m)$ and $E_{G2}(m)$ into the corresponding effective velocity records $V_{e1}(m)$ and $V_{e2}(m)$. The conversion for each voltage record is identical to the procedure described for the SY-probe (Section 5.3.3.2). If large amounts of data are to be converted, then it is worthwhile applying the look-up table method described in Section 4.12.2.2. The application of this procedure to X-probes is discussed by Westphal and Mehta (1984), by Khan and Bruun (1990), by Khan (1991), and by Hooper and Westphal (1991).

Analysis of $V_e(m)$-records The response equations presented in Section 5.3.3.2 for the single inclined probe can also be applied to the two wires in an X-probe. If the turbulence intensity is low, $\bar{V} = \bar{W} = 0$, and the mean yaw angles are $\bar{\alpha}_1$ and $-\bar{\alpha}_2$, then from eqns (5.18a-b)

$$V_{e1} = f_1(\bar{\alpha}_1)[U - g_1(\bar{\alpha}_1)v], \qquad (5.30a)$$

$$V_{e2} = f_2(\bar{\alpha}_2)[U + g_2(\bar{\alpha}_2)v]. \qquad (5.30b)$$

For an X-probe, which can be rotated in the plane of the wires, the mean-flow direction corresponding to $\bar{V} = 0$ can be determined from eqns (5.30a-b) as the angular position for which $\bar{V}_{e1}/f_1(\bar{\alpha}_1) = \bar{V}_{e2}/f_2(\bar{\alpha}_2)$. In practice, the alignment criterion, $\bar{V} = 0$, can be relaxed. If $\bar{W} = 0$ and $w/\bar{U} \ll 1$, an exact relationship exists between (V_{e1}, V_{e2}) and (U, V) for the $\cos \bar{\alpha}_e$ yaw dependence (eqn (5.5d)):

$$V_{e1} = U \cos \bar{\alpha}_{e1} - V \sin \bar{\alpha}_{e1}, \qquad (5.31a)$$

$$V_{e2} = U \cos \bar{\alpha}_{e2} + V \sin \bar{\alpha}_{e2}. \qquad (5.31b)$$

In this method the only additional restriction is that the instantaneous velocity vector must remain within the approach quadrant to avoid any reverse-flow ambiguity. However, in practice, the value of $\bar{V}/\bar{U}$ should be fairly small to avoid a reduction in the applicable turbulence-intensity range. As shown by Bruun (1975b), removing the $\bar{V} = 0$ restriction for the other three yaw functions of eqns (5.5a–c) will only result in a small error in the angle range $-35° \leqslant \theta \leqslant 35°$. It is therefore justifiable in many applications to replace the fluctuation component v in eqns (5.30a–b) with the related instantaneous component $V(=\bar{V} + v)$. In many HWA applications, this enables the X-probe to be aligned with the space-fixed coordinate system instead of the mean-flow direction.

Sum-and-difference procedure In the early X-probe investigations, variations between wires 1 and 2 were usually ignored, and it was assumed that $f_1(\bar{\alpha}_1) = f_2(\bar{\alpha}_2) = f(\bar{\alpha})$ and $g_1(\bar{\alpha}_1) = g_2(\bar{\alpha}_2) = g(\bar{\alpha})$. In this case the values of U and V can be obtained by applying a simple sum-and-difference method to eqns (5.30a–b) giving

$$U = \frac{V_{e1} + V_{e2}}{2f(\bar{\alpha})}, \tag{5.32a}$$

$$V = \frac{V_{e2} - V_{e1}}{2f(\bar{\alpha})g(\bar{\alpha})}. \tag{5.32b}$$

However, the values of the yaw functions $f(\bar{\alpha})$ and $g(\bar{\alpha})$ (defined in Table 5.2) normally vary between the two wires; this is primarily because $\bar{\alpha}_1$ often differs from $\bar{\alpha}_2$ by several degrees and secondly because the value of the yaw parameter for each wire will be slightly different. Therefore, to obtain accurate results, the actual values of $f(\bar{\alpha})$ and $g(\bar{\alpha})$ for wires 1 and wire 2 must be used, and the related *modified* sum-and-difference equations become

$$U = \frac{[V_{e1}/f_1(\bar{\alpha}_1)]g_2(\bar{\alpha}_2) + [V_{e2}/f_2(\bar{\alpha}_2)]g_1(\bar{\alpha}_1)}{g_1(\bar{\alpha}_1) + g_2(\bar{\alpha}_2)}, \tag{5.33a}$$

$$V = \frac{[V_{e2}/f_2(\bar{\alpha}_2)] - [V_{e1}/f_1(\bar{\alpha}_1)]}{g_1(\bar{\alpha}_1) + g_2(\bar{\alpha}_2)}. \tag{5.33b}$$

Equations (5.33a–b) apply to any of the yaw expressions for $f(\bar{\alpha})$ and $g(\bar{\alpha})$ listed in Table 5.2. Bruun et al. (1990b) have studied the variation in the results for U and V for the four yaw functions of eqn (5.5a–d). Using the same calibration data from a plated X-probe (DANTEC 55P51), the values of k, m, b, and $\bar{\alpha}_e$ were first evaluated by the calibration method related to eqn (5.8). This method basically curve fits the calibration data to the function $f(\bar{\alpha})$ around $\bar{\alpha}$. Table 5.3 contains the yaw-coefficient results for a hot-wire element with a mean yaw angle, $\bar{\alpha}$, of 45°. The values of $f(\bar{\alpha})$ and $g(\bar{\alpha})$ for the four yaw functions are also given in this table. The 1%

TABLE 5.3 *The values of the yaw coefficients and the yaw functions* $f(\bar{\alpha})$ *and* $g(\bar{\alpha})$ *(Table 5.2) for the four yaw functions of eqns (5.5a–d)* $(\bar{\alpha} = 45°)$ *measured with a Dantec X-probe (55P51)*

	Equation number			
	(5.5a)	(5.5b)	(5.5c)	(5.5d)
Yaw coefficient	$k^2 = 0.014$	$m = 0.944$	$b = 0.961$	$\bar{\alpha}_e = 44.13°$
$f(\bar{\alpha})$	0.712	0.721	0.717	0.718
$g(\bar{\alpha})$	0.973	0.944	0.954	0.970

variation in $f(\bar{\alpha})$ between the four methods probably reflects the accuracy obtainable by such a curve fit. The $g(\bar{\alpha})$-values are related to the slope of the four functions for $f(\bar{\alpha})$ at $\bar{\alpha}$, and the 3% variation corresponds to slightly different slopes of the four $f(\bar{\alpha})$ functions at $\bar{\alpha}$. These differences are not important in many practical *HWA* applications.

Analysis procedures for V_e have been developed in terms of $\tilde{V}$ and θ to avoid the series-expansion limitations related to eqns (5.30a–b). For an X-probe with two wires which have the same value of the yaw parameter m in the yaw function $\cos^m\alpha$, the response equations can be expressed as

$$V_{e1} = \tilde{V}\cos^m(\bar{\alpha}_1 + \theta), \tag{5.34a}$$

$$V_{e2} = \tilde{V}\cos^m(\bar{\alpha}_2 - \theta). \tag{5.34b}$$

Bruun and Davies (1972) have shown that these simultaneous equations can be rearranged and solved for θ and then for $\tilde{V}$. The corresponding values of U and V can then obtained from eqns (5.24a–b).

The corresponding equations for the k-method using eqn (5.5a) are

$$V_{e1} = \tilde{V}[\cos^2(\bar{\alpha}_1 + \theta) + k_1^2\sin^2(\bar{\alpha}_1 + \theta)]^{1/2}, \tag{5.35a}$$

$$V_{e2} = \tilde{V}[\cos^2(\bar{\alpha}_2 - \theta) + k_2^2\sin^2(\bar{\alpha}_2 - \theta)]^{1/2}, \tag{5.35b}$$

and similarly for the b-method using eqn (5.5c)

$$V_{e1} = \tilde{V}[1 - b_1(1 - \cos^{1/2}(\bar{\alpha}_1 + \theta))]^2, \tag{5.36a}$$

$$V_{e2} = \tilde{V}[1 - b_2(1 - \cos^{1/2}(\bar{\alpha}_2 - \theta))]^2. \tag{5.36b}$$

Direct-solution procedures do not exist either for eqns (5.35a–b) or for eqns (5.36a–b). Andreopouls (1981) used an iterative method to obtain $\tilde{V}$ and θ from eqns (5.36a–b). A similar simultaneous solution using eqns (5.35a–b) has been reported by O'Brian and Capp (1989).

However, taking into account the various sources of uncertainty introduced by the effective-velocity method, it is unlikely that either of these methods will give significantly more accurate results than the modified sum-

and-difference method. Due to their computational complexity, there appears little justification for their use.

5.4.3.2   The $(\tilde{V}, \theta)$ analysis method

The direct relationship between (E_1, E_2) and $(\tilde{V}, \theta)$ illustrated in Fig. 5.8 can be utilized in the evaluation of U and V.

Polynomial curve fits for $\tilde{V}$ and θ   Oster and Wygnanski (1982) have proposed, for a restricted velocity $(6.16-13.5 \text{ m s}^{-1})$ and flow-angle $(-27° \leqslant \theta \leqslant 27°)$ range, that $\tilde{V}$ and θ can be evaluated using third-order polynomials in E_1 and E_2. They plotted the calibration data corresponding to six velocities and eleven flow angles on an (E_1, E_2) map and identified curves for constant values of $\tilde{V}$ and θ. The equations for $\tilde{V}$ and θ were assumed to be of the form

$$\tilde{V} = a_1 E_1^3 + a_2 E_1^2 E_2 + a_3 E_1 E_2^2 + \ldots + a_8 E_1 + a_9 E_2 + a_{10}, \qquad (5.37a)$$

$$\theta = b_1 E_1^3 + b_2 E_1^2 E_2 + b_3 E_1 E_2^2 + \ldots + b_8 E_1 + b_9 E_2 + b_{10}. \qquad (5.37b)$$

A third-order surface fitting was applied to obtain the twenty calibration constants. The calibration was only accepted if the errors in U and V were less than 1% and 2%, respectively. The calibration constants were stored and used for the conversion of each set of measured voltage pairs into the corresponding $(\tilde{V}, \theta)$ values. The related velocity components U and V were then obtained from eqns (5.24a–b).

5.4.3.3   Look-up matrix methods

It was shown in Section 4.12.2.2 that a look-up table method can be used to convert the digital signal from an *SN*-probe. This section describes related look-up matrix methods for *X*-probes. A direct extension of the look-up table method was proposed by Cheesewright (1972). In his method (personal communication) a $(\tilde{V}, \theta)$ calibration was carried out using eight different velocities and fifteen angular positions. The calibration data were curve-fitted using separate fifth-order polynomials in E_1, and E_2 for U and V. The equations for U and V were used to create a look-up matrix containing values corresponding to all possible digital values for E_1 and E_2. This approach is the logical extension of the look-up-table method.

Cheesewright used a ten-bit converter, and in his method it was necessary to create a look-up matrix containing $1024 \times 1024 \simeq 10^6$ different address locations. A matrix of this size is too large to be stored in the central memory of most computers. To overcome the memory problem, the full look-up matrix was stored on a backup disk. By applying velocity-vector tracking to the measured data, Cheesewright could identify the applicable matrix range for a given flow situation, and this sub-part of the full matrix was then transferred to the central memory of the computer. This was a rather cumbersome procedure, and later methods have used a matrix of much smaller size linked to an interpolation scheme.

The first of these methods was introduced by Willmarth and Bogar (1977), who applied it to a very small X-hot-wire probe. Other look-up matrix methods have subsequently been presented by Zilberman (1981), Johnson and Eckelmann (1984), Lueptow et al. (1988), Browne et al. (1989a), and Schewe and Ronneberger (1990). All of these methods utilize two similar procedures:

1. The creation of a reference look-up matrix, which contains (in discrete form) the relationship between the X-probe voltage values $(E_1(i), E_2(j))$ and the corresponding velocity field $(U(i,j), V(i,j))$ or $(\tilde{V}(i,j), \theta(i,j))$. (Alternatively the reference matrix may be specified in terms of $(\tilde{V}(i), \theta(j))$ and the related voltage values $(E_1(i,j), E_2(i,j))$.)

2. An interpolation scheme to interpret data points (E_1, E_2), which do not coincide with points in the reference look-up matrix. The interpolation is usually carried out in two stages. First, the nearest lower $(E_1(i), E_2(j))$ values in the reference look-up matrix is identified and an interpolation method is then applied using the matrix values for $E_1(i)$, $E_1(i+1)$, $E_2(j)$, and $E_2(j+1)$.

A number of different methods have been used to evaluate the reference look-up matrix, and several different interpolation schemes have been proposed. During the X-probe calibration, Willmarth and Bogar (1977) recorded the voltage pair as a function of the flow angle, θ, which was varied periodically while the free-stream velocity, $\tilde{V}$, was slowly decreased. For the probe used, they found that unambiguous measurements could be obtained in the flow-angle range $-51° < \theta < 38°$. Within the unambiguous θ-range, they constructed a relatively coarse (20×20) look-up matrix, corresponding to equidistant values in E_1 and E_2. A bilinear interpolation scheme was used, and their method required the storage of the values of U, V and the corresponding four partial derivatives: $\partial U/\partial E_1$, $\partial U/\partial E_2$, $\partial V/\partial E_1$, and $\partial V/\partial E_2$ for each point in the look-up matrix.

In the method proposed by Johnson and Eckelmann (1984), the look-up matrix was created directly from the original calibration data using little or no smoothing. They measured the voltage pairs $(E_1(i,j), E_2(i,j))$ corresponding to $i = 7$ different values of the velocity $\tilde{V}(i)$ and to $j = 19$ different flow angles $\theta(j)$ ($5°$ increments). To improve the accuracy of the data conversion, they introduced an interpolation method based on a second-order Taylor-series expansion. This necessitated the evaluation and storage of twelve first- and second-order partial derivatives for each point in the look-up matrix. The authors claim that a very high accuracy was obtained by their method, but the interpolation method is cumbersome and time-consuming to implement. To improve the accuracy of the look-up matrix, Schewe and Ronneberger (1990) used linearized anemometer signals as inputs to the A/D converter and they included a weighting function in the least-squares curve-fit for their calibration data. Also, to enable data evaluation to be made in real time, they applied a linear interpolation scheme. Lueptow *et al.* (1988) have proposed (based on a method by Zilberman 1981) a relatively simple interpolation procedure and a larger look-up matrix. The initial calibration involved the measurements of the voltage pair (E_1, E_2) as functions of $\tilde{V}(i)$ and $\theta(j)$, as in the investigation by Johnson and Eckelmann (1984). They then introduced a relatively fine equidistant (E_1, E_2) grid, and applied a multistage cubic spline or polynomial regression fits to identify the values of $\tilde{V}$ and θ, and thereby U and V, at each of the grid points. The values for U and V, given arbitrary values of E_1 and E_2, were found by a bilinear interpolation between the nearest points in the look-up matrix. This interpolation method required the evaluation and storage of six calibration coefficients for each point in the look-up matrix. Applying a 40×40 matrix to a calibration range of $1.2\,\mathrm{m\,s}^{-1} \leqslant \tilde{V} \leqslant 20.2\,\mathrm{m\,s}^{-1}$ and $-30° \leqslant \theta \leqslant 30°$, they estimated that the velocity-component error bands were: $0.3\% < U < 1.7\%$ and $0.6\% < V < 1.9\%$, with the largest errors occurring at low velocities.

To evaluate the accuracy of the various look-up matrix methods, an accurate reference inversion method must be applied. Such a method has been developed by Bruun (1987) and Bruun *et al.* (1990c). A related method is also described by Browne *et al.* (1989a). Chew and Ha (1990) have also used the reference inversion method by Browne *et al.* (1989a) to determine the error of the conventional X-probe evaluation methods, including the assumption of a constant yaw coefficient, k, in eqn (3.2).

5.5 OTHER HOT-WIRE PROBE TYPES

The size of standard hot-wire probes prevents them from being placed close to a surface. For measurement of the $\overline{uv}$ shear stress, the plane of the X must be perpendicular to the surface, and the centre of standard plated and unplated X-probes cannot be placed closer than $1.5\,\mathrm{mm}$ and $0.6\,\mathrm{mm}$, respectively, to the surface. One solution is to use subminiature X-probes (Willmarth and Bogar 1977; Ligrani *et al.* 1989c). This enables measurements to be made as close as $0.25\,\mathrm{mm}$ from the wall.

5.5.1 Hot-wake sensor probes

The principle of these probes is based on the detection of the hot wake from a heated sensor. McConachie and Bullock (1976) have presented a theory for the temperature wake of a heated cylinder, based on the assumption that the near wake was controlled by molecular diffusion, and the heated

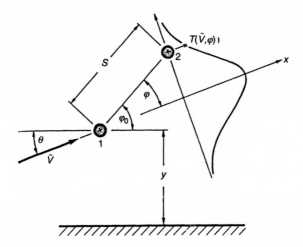

Fig. 5.9. The principle of (and notation for) hot-wake sensors: (1) a heated hot-wire, and (2) a temperature detecting wire. (From McConachie and Bullock 1976.)

upstream wire was a line source. The theory of hot-wake sensors assumes that the laminar heated wake is being waved to and fro by the instantaneous velocity vector $(\tilde{V}, \theta)$ and that the temperature distribution, $T(\tilde{V}, \varphi)$, in the wake will be symmetrically centred around the velocity vector. This principle is illustrated in Fig. 5.9, where φ_0 is an offset angle for the temperature detecting resistance-wire (2). The wake temperature $T(\tilde{V}, \varphi)$ will produce a voltage

$$E = KT(\tilde{V}, \varphi) \tag{5.38}$$

from the resistance-wire (2). The value of K and the scaling parameters for $T(\tilde{V}, \varphi)$ are determined by calibration. During measurements, $\tilde{V}$ is obtained from the heated hot-wire (1), and the corresponding instantaneous wake angle, φ, can be determined from eqn (5.38). It is therefore a fundamental requirement of the wake-sensing method that the two wires are separated by less than the smallest longitudinal scale of turbulence. In the investigation by Walker and Bullock (1972) the separation distance was 0.08 mm.

Wake-sensing probes with one or two temperature detection wires have been investigated. When one detection wire is used (Fig. 5.9), it is necessary to introduce an offset angle, φ_0, for the detection wire (2) to avoid ambiguity in the wake-angle evaluation. This probe type has been investigated by Walker and Bullock (1972) and McConachie and Bullock (1976). They observed major problems related to fine-scale turbulence, wake buoyancy, the wake dynamic response, and the wire spacing.

Hot-wake sensors with two temperature-detection wires, Fig. 5.10, have been studied by Rey (1973), Beguier *et al.* (1973), Rey and Beguier (1977),

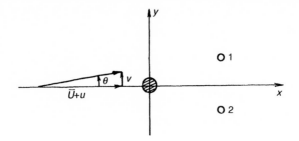

Fig. 5.10. A three-parallel-wire hot-wake-sensor probe.

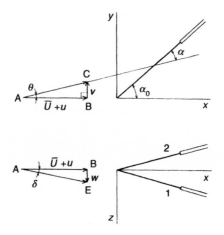

Fig. 5.11. A symmetrically bent V-shaped hot-wire probe. Coordinate system and velocity components with respect to the sensing element. (From Hishida and Nagano 1988b.)

and Durst (1977). This type of probe is usually aligned with the mean-flow component, $\bar{U}$, and the corresponding instantaneous value $U(=\bar{U} + u)$ was measured by the heated upstream wire. Theoretical derivations and experimental results have shown that the difference $E_D = E_1 - E_2$ between the output from the two temperature detection wires 1 and 2 is a linear function of θ over a substantial flow-angle range on either side of $\theta = 0°$

$$E_D = E_1 - E_2 = K\theta \cong K\frac{v}{\bar{U}}. \tag{5.39}$$

It was found (Rey and Beguier 1977) that the calibration constant K in eqn (5.39) varies significantly with the local turbulence intensity, $(\overline{u^2})^{1/2}/\bar{U}$. It is therefore necessary to determine the value of K for any given flow condition $(\bar{U}, \overline{u^2})$ by rotating the probe around $\theta = 0°$ $(K = \Delta E_D/\Delta\theta)$. Another problem with this probe type is that the probe geometry must be matched to the flow velocity occurring. Otherwise a nonlinear distortion can be introduced into the (E_D, θ)-curve around $\theta = 0°$ (Rey and Beguier 1977; Durst 1977).

5.5.2 V-shaped hot-wire probes

The prongs of a conventional hot-wire probe prevent measurements close to a wall and they may also distort the flow field. These problems can be reduced by using a symmetrically bent V-shaped hot-wire as proposed by Hishida and Nagano (1988a,b) and Tsuji and Nagano (1989). The probe geometry and coordinate system used for turbulence measurement is shown in Fig. 5.11, in which θ is the instantaneous flow angle, and $\alpha = \alpha_0 - \theta$ denotes the instantaneous (two-dimensional) yaw angle; α_0 is the mean yaw angle. Their theory predicts that the instantaneous effective velocity, V_e, can be expressed as

$$V_e = \tilde{V}F, \tag{5.40}$$

with

$$F = \tfrac{1}{2}\left[(1 - K_1 \cos^2 \gamma_1)^{0.5} + (1 - K_2 \cos^2 \gamma_2)^{0.5} \right]. \tag{5.41}$$

In these equations, $\tilde{V}$ is the magnitude of the instantaneous fluid velocity, $K_i = 1 - k_i^2$ $(i = 1, 2)$, where k_i denotes the tangential sensitivity coefficient, and γ_i is the instantaneous angle between $\tilde{V}$ and the axis of the hot-wire. The subscripts 1 and 2 correspond to each side of the V-shaped wire shown in Fig. 5.11. Denoting the vertex angle of the V-shaped hot-wire by 2ψ, Hishida and Nagano derived the following relationship between V_e and the instantaneous velocity field $(\bar{U} + u, v, w)$

$$
V_e = \tfrac{1}{2}\Bigg\{ \left[(\bar{U} + u)^2 + v^2 + w^2 - K_1 \cos^2 \psi \cos^2 \alpha_0 \left(\bar{U} + u + v \tan \alpha_0 \right. \right.
$$
$$
\left. \left. + \frac{w \tan \psi}{\cos \alpha_0} \right)^2 \right]^{0.5} + \left[(\bar{U} + u)^2 + v^2 + w^2 - K_2 \cos^2 \psi \cos^2 \alpha_0 \right.
$$
$$
\left. \left(\bar{U} + u + v \tan \alpha_0 - \frac{w \tan \psi}{\cos \alpha_0} \right)^2 \right]^{0.5} \Bigg\}. \tag{5.42}
$$

Applying a series expansion to eqn (5.42) and measuring V_e for different values of the yaw angle, α_0, the values of $\bar{U}$, $\overline{u^2}$, $\overline{v^2}$ and $\overline{uv}$ can be obtained. This analysis procedure is similar to the multiposition yaw-angle method proposed by Fujita and Kovasznay (1968). Hishida and Nagano concluded that the response of a V-shaped probe to velocity and yaw is similar to that for an inclined SN hot-wire probe. The main difference is that the V-shaped probe is less sensitive to w-fluctuations than the SN-probe.

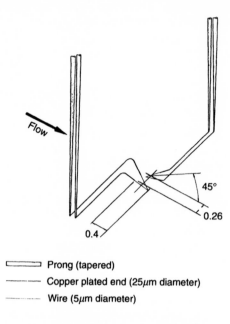

Flow

45°

0.26

0.4

☐▭ Prong (tapered)
——— Copper plated end (25μm diameter)
-------- Wire (5μm diameter)

Fig. 5.12. An X-hot-wire probe consisting of two symmetrically bent V-shaped hot-wires. All dimensions are in millimetres. (From Hishida and Nagano 1988b.)

An array with two V-shaped wires placed at α_0 and $-\alpha_0$ in an X-arrangement is shown in Fig. 5.12. It can be used to measure simultaneously the instantaneous values $\bar{U} + u$ and v by a sum-and-difference technique based on a first-order series expansion of eqn (5.42) and the corresponding equation for the V-probe placed at $-\alpha_0$.

5.6 INCLINED HOT-FILM PROBES

Measurements of two velocity components at points in a flow field can also be carried out with hot-film probes. Cylindrical SN, SY, and X hot-film probe types are commercially available, and the sensor element of the SY and X probe types is usually similar to the element of the corresponding SN-probe. An alternative probe configuration is the V-shaped wedge probe shown in Fig. 5.13. The inclined hot-film probes can be used in both air and water, subject to similar principles and restrictions as for SN hot-film probes (Section 4.10).

5.6.1 Calibration

The velocity and yaw calibration is normally carried out by procedures similar to those described for inclined hot-wire probes. In most investiga-

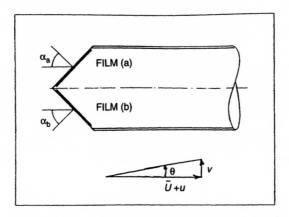

Fig. 5.13. A schematic representation of the double-wedge hot-film probe. (From Resch 1973.)

tions, a power-law relationship eqn (3.1)) has been used in conjunction with the effective-velocity concept (eqn (5.3)) for both calibration and subsequent measurement and data analysis. In general, the velocity-calibration results for inclined hot-film probes in both air and water are similar to those obtained for *SN* hot-film probes. The related yaw-response results in air and water flows are described below.

Air flows Yaw calibration of cylindrical inclined *SN* hot-film probes has been reported by Friehe and Schwarz (1968), Jørgensen (1971*a*), and Andreas (1979*a*). For a typical hot-film probe with an ℓ/d ratio of 18–20, the value of the yaw coefficient, k, in eqn (5.5a) was about 0.2–0.3, which is similar to that of a typical 5 μm tungsten hot-wire probe with an ℓ/d ratio of about 250.

Water flows Only a few studies have been reported for the yaw response of inclined hot-film probes in water flows: inclined cylindrical *SN*-probes have been reported by Bertrand and Couderc (1978), inclined *SN* conical probes by Bertrand and Couderc (1978), inclined *SN* wedge probes by Mollenkopf (1972), and *V*-shaped double-wedge probes by Resch (1973) and by Giovanangeli (1980).

For the *V*-shaped wedge probe, consistent low values for $k \simeq 0.02$–0.035 were obtained by Resch (1973). The results of Giovanangeli (1980) also gave relatively low values for k. However, larger values of about 0.25–0.3 have been reported by Ezraty (1970), Ezraty and Coantic (1970), and Wu and Bose (1993). The published values for k for inclined cylindrical hot-film probes are, in general, relatively high. Gourdon *et al.* (1981) obtained values of $k \simeq 0.17$–0.28 when calibrating a triple-film probe, whilst Bertrand and Couderc (1978) obtained a value of $k \simeq 0.4$. Detailed *X*-probe calibration by the author and co-workers have produced values for k of 0.15–0.25.

Due to the apparent uncertainty of the correct value of k for inclined hot-film probes, the value of k has been assumed to be zero in a number of investigations.

5.6.2 Measurements with inclined hot-film probes

Having calibrated the hot-film probe, the measurements and signal analysis can be carried out using the methods described for the inclined SN, SY, and X hot-wire probes. However, additional considerations apply to hot-film anemometry. The related advantages and disadvantages for air- and water-flow measurements are discussed in Section 4.10. Examples of measurements with hot-film probes in liquids include: inclined SN-probe techniques in water (McQuivey and Richardson 1969), X-probes in water (Hatano and Hotta 1983; Hotta 1986), X-probes in a water/air two-phase flow (Serizawa et al. 1983), X-probes in oil (Johnson and Eckelmann 1984; Schewe and Ronneberger 1990) and a V-probe in oil (Kreplin and Eckelmann 1979b).

5.7 SPLIT-FILM PROBES

Details of the operational principle of the split-film anemometer sensor, developed by TSI, were first reported by Olin and Kiland (1970). The sensor, shown in Fig. 5.14(a), has the same physical characteristics as the corresponding $150\,\mu$m diameter cylindrical hot-film probe, except that the thin platinum film ($1000\,\text{Å}$ thick) is split longitudinally into two separate sensor elements. The probe is assumed to respond to the velocity components $(\bar{U} + u, v)$ contained in a plane perpendicular to the sensor, and to have an insignificant response to the tangential velocity component along the sensor element. This makes the applicability of the probe similar to that of an X-probe. The main advantage in shear flows is its much smaller size in the direction of the mean shear, which minimizes the spatial averaging in regions of severe velocity gradients and permits measurements closer to a surface than with an X-probe. The geometry of the probe also allows, in principle, a larger flow-acceptance angle than for an X-probe.

Investigations of the probe's (static) response characteristics have been reported by Olin and Kiland (1970), Spencer and Jones (1971), Blinco and Sandborn (1973), and TSI (Technical Bulletin No. 20). The theory of operation for the split-film probe is based on the nonuniform heat-transfer distribution around a heated cylinder in a cross-flow, with maximum heat transfer occurring in the region of the upstream stagnation point (see Fig. 5.14(b). The heat-transfer distribution is assumed to remain fixed relative to the direction of the velocity vector $(\bar{U} + u, v)$, and to respond instantaneously to changes in the flow angle, θ and the magnitude, $\tilde{V}$, of the velocity vector. When the flow direction deviates from the plane of the

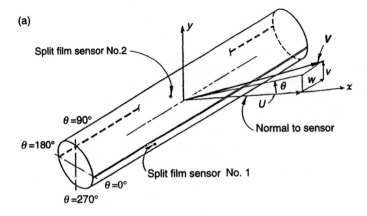

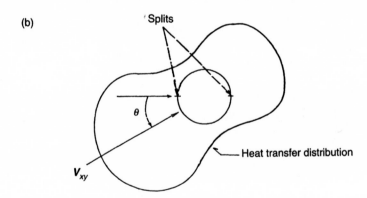

Fig. 5.14. The split-film cylindrical sensor: (a) a schematic of the split-film sensor, (b) the heat-transfer distribution around a circular cylinder.

split, an unequal heat transfer will take place from the two film sensors. For an ideal split-film probe (with a perfect frequency response), consisting of two identical films heated to the same temperature, the response equations for the anemometer output voltages from the elements 1 and 2 (see Fig. 5.14(a)) can be expressed as

$$E_1^2 = (A + B\tilde{V}^n) f(\theta), \qquad (5.43a)$$

$$E_2^2 = (A + B\tilde{V}^n) f(-\theta). \qquad (5.43b)$$

A number of expressions have been proposed for $f(\theta)$, including

$$f(\theta) = \begin{cases} 1 + a \sin \theta & \text{(5.44a)} \\ 1 + b\theta. & \text{(5.44b)} \end{cases}$$

Spencer and Jones (1971) found that eqn (5.44a) gave a good agreement with their experimental data in the flow-angle range $-35° \leqslant \theta \leqslant 35°$, while Blinco and Sandborn (1973) demonstrated that eqn (5.44b) gave a satisfactory representation of their calibration data in the angle range $-25° \leqslant \theta \leqslant 25°$. Provided that the calibration constants A, B, n, and the flow-angle coefficients a or b are independent of both $\tilde{V}$ and θ, then the two-dimensional velocity vector $V = (U, V)$, corresponding to a set of (E_1, E_2)-values can be obtained by applying a sum-and-difference method. Using, for example, eqn (5.4a) for $f(\theta)$ we get

$$E_1^2 + E_2^2 = 2(A + B\tilde{V}^n) \tag{5.45a}$$

$$E_1^2 - E_2^2 = 2(A + B\tilde{V}^n)a \sin \theta. \tag{5.45b}$$

Equation (5.45a), which is independent of θ, determines the value of $\tilde{V}$, and knowing $\tilde{V}$ the value of $\sin \theta$, and therefore the value of θ, can be evaluated from eqn (5.45b). The corresponding instantaneous velocity components U and V can be obtained from eqns (5.24a–b).

More recently Stock *et al.* (1977), Stock and Jaballa (1985), Shook *et al.* (1990) and Ra *et al.* (1990*a*) have proposed the use of normalized response equations, in order to extend the applicable flow-angle range. Richter (1985) used a look-up matrix method to extend the flow-angle range to $-70° < \theta < 50°$.

5.7.1 Practical split-film anemometry

Measurements with split-film probes have been reported for air flows (Spencer and Jones 1971; Kiya and Sasaki 1983; Stock and Jaballa 1985; Duncan and Hartmann 1985; Hartmann and Dengel 1989), in water (Blinco and Sandborn 1973), and in water/air bubbly flow mixtures (Boerner and Leutheusser (1984). These studies have shown that the operation and dynamic response of a split-film anemometer deviate from the simple theory presented, and that the following additional aspects must be considered.

Frequency response The greatest uncertainty of the split-film probe is its complex thermal response, which is caused by two separate film elements being deposited on a substantial substrate. This configuration may lead to thermal crosstalk between the elements, as studied by Ho (1982) using an electric-perturbation technique.

It is instructive to compare the response to a sine-wave test in air of a hot-wire probe (Fig. 2.16), a standard $50 \mu m$ diameter hot-film probe

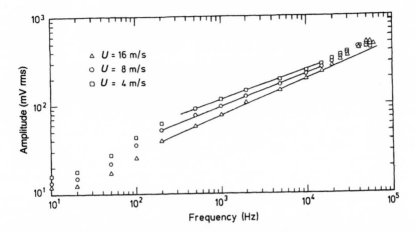

Fig. 5.15. A sine-wave test on a TSI 150 μm diameter split-film probe, using a TSI-IFA100 anemometer. (From Hartmann and Dengel 1989. Reprinted with kind permission of Springer-Verlag, © Springer-Verlag.)

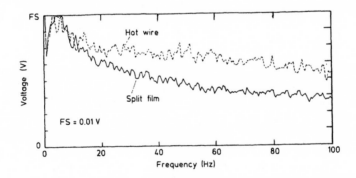

Fig. 5.16. A comparison of the spectra measured with a hot-wire and with a split-film probe in a turbulent air-flow boundary layer. (From Hartmann and Dengel 1989. Reprinted with kind permission of Springer-Verlag, © Springer-Verlag.)

(Fig. 2.18) and a 150 μm diameter split-film probe (Fig. 5.15) all operated in the *CT* mode. Using a log–log plot, the response for the hot-wire probe was found to be flat up to $f = 1/(2\pi M)$, where M is the time constant of the uncompensated wire element, followed by an f^1 response up to a frequency of about 10^5 Hz. The initial frequency response for the hot-film probe was similar, but the f^1-dependence was found, Fig. 2.18, only to extend to about 300 Hz. Above a frequency of 300 Hz attenuation occurred due to heat retention in the substrate during part of the heating cycle, resulting in an $f^{1/2}$-dependence. For the larger split-film probe ($d = 150\,\mu$m) it should be noted from Fig. 5.15 that the f^1-region has disappeared and frequency attenuation takes place when $f > \sim 10\text{--}20\,\text{Hz}$. This frequency attenuation for the split-film probe in air is demonstrated in Fig. 5.16 by comparative spectral measurements with a hot-wire and a split-film probe. Similar observations were made by Young (1976).

Blinco and Sandborn (1973) reported measurements in a turbulent boundary layer of an open-water-channel flow, in which the highest expected frequency of the turbulence was about $150\,\text{Hz}$. Due to the limited frequency range, they observed good agreement between results obtained with a split-film probe and with a miniature hot-film probe.

Equivalization of the two sensor temperatures The two film elements should be operated at the same temperature, T_f, to avoid heat transfer between them. By extending the general hot-wire relationship of eqn (2.36a) to the response of a split-film element, then for the voltage, E_f, across the film element

$$\frac{E_f^2}{R_f} = (A + B\tilde{V}^n) f(\theta) (T_f - T_a), \qquad (5.46)$$

and the resistance, R_f, of the heated hot-film element is related to the corresponding operational temperature, T_f, by

$$R_f = R_a[1 + \alpha(T_f - T_a)]. \qquad (5.47)$$

In practice, the two film elements will not be identical due to geometrical and physical differences, and the same overheat ratio cannot, therefore, be used for the two hot-film elements. Measuring the cold resistances R_{a1} and R_{a2} for the two elements, and knowing the values of the temperature coefficients α_1 and α_2, the operational resistances, R_{f1} and R_{f2}, corresponding to a specified temperature, T_f, can be evaluated from eqn (5.47). Provided that the calibration constants A, B, and n in eqn (5.45) have the same values for the two film elements, then the corresponding summation equation can (Olin and Kiland 1970; Blinco and Sandborn 1973) be expressed as

$$\frac{R_{a1}\alpha_1}{R_{f1}(R_{f1} - R_{a1})} E_{f1}^2 + \frac{R_{a2}\alpha_2}{R_{f2}(R_{f2} - R_{a2})} E_{f2}^2 = A + B\tilde{V}^n. \qquad (5.48a)$$

Provided that the setting for sensor 1 remains unaltered, eqn (5.48a) can be expressed as

$$E_{f1}^2 + K^2 E_{f2}^2 = A + B\tilde{V}^n, \qquad (5.48b)$$

with

$$K^2 = \frac{\alpha_2}{\alpha_1} \frac{R_{a2}}{R_{a1}} \frac{R_{f1}}{R_{f2}} \frac{(R_{f1} - R_{a1})}{(R_{f2} - R_{a2})}. \qquad (5.49)$$

The value of K may be determined using eqn (5.49), or the overheat ratio of one of the sensors may be adjusted until the ratio E_{f1}/E_{f2} is virtually independent of the velocity (TSI Technical Bulletin No. 20).

Angular response The plane of the splits between the two films of commercial probes is normally aligned optically to within $\pm 2°$ of a probe reference direction. This angular uncertainty and the differences between the two film strips may result in a different angular dependence for $f(\theta)$ when $\theta > 0°$ and $\theta < 0°$, as reported by Blinco and Sandborn (1973). To minimize this problem, average values are normally used.

Near-wall measurements One of the main advantages of split-film probes is their small size, which enables them to be placed very close to a wall. However, a number of investigators (Spencer and Jones 1971; Blinco and Sandborn 1973) have reported discrepancies between split-film-probe and hot-wire-probe results when the split-film is placed very close to the wall. It was concluded that this was due to aerodynamic interference between the probe and the wall. Depending on the flow condition, the split-film should not be used closer than about five diameters (~ 0.75 mm) from a wall.

5.8 UNCERTAINTY ANALYSIS, SY-PROBES

5.8.1 V_e-method

To obtain accurate results with the SY-probe method presented in Section 5.3 it is necessary that: (i) the probe is aligned with the mean-flow direction $\bar{V} = \bar{W} = 0$); (ii) the calibration coefficients A, B, n, and k in the corresponding hot-wire response equations, eqn (3.1) and (3.2) are independent of the velocity and the yaw angle (to avoid repetition, the effect of the flow-angle dependence of A, B, and n will be discussed in Section 5.9 for X-probes); and (iii) the turbulence intensity is low/moderate.

Constant yaw-coefficient assumption The yaw-calibration results in Fig. 5.3 demonstrate that k, m, and b are functions of α. Jørgensen (1971a) has (for several DANTEC probes) evaluated the error in the calculated effected velocity, with k assumed to be constant. A similar error evaluation for all four yaw coefficients (k, m, b, $\bar{\alpha}_e$) is shown in Fig. 5.17 for a plated SY-probe (DANTEC 55P02), using a typical value of $k(\bar{\alpha}) = 0.12$ (Bruun *et al*. 1990b). Within the angle range $0° \leqslant \alpha \leqslant 70°$, the k, m, and b-methods are seen to predict the effective velocity to within 1%, but the α_e-method gave a slightly higher uncertainty of 2% at $\alpha = 70°$. As α is increased beyond 70° the approximations obtained by the m, b, and α_e-method rapidly become very poor, but the error in the k-method is much smaller, only reaching 15% at $\alpha = 90°$.

High turbulence intensity flows In moderate/high turbulence intensity flows errors will be introduced due to: (i) the omission of higher-order terms in the series expansion for V_e in eqn (5.17); and (ii) signal ambiguity, when the velocity vector falls outside the approach-acceptance angle. The corresponding errors in $\overline{v^2}$ and $\overline{uv}$ were also evaluated in the numerical

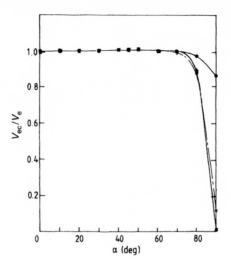

Fig. 5.17. Analytical approximations of V_e using four yaw functions: ($\bullet$) eqn (5.5a) k, ($\blacktriangle$) eqn (5.5b) m, ($\blacksquare$) eqn (5.5c) b, (∇) eqn (5.5d) $\bar{\alpha}_e$.

simulation experiment by Swaminathan *et al.* (1986*b*), using the technique described for the *SN*-probe (Section 4.13.1). The results, which are included in Fig. 4.20, demonstrate the increase in the error with the turbulence intensity, reaching 10% for $\overline{v^2}$ and $\overline{uv}$ at $Tu \cong 20\%$. A similar uncertainty in $\overline{u^2}$ (measured with an *SN*-probe) is first observed when $Tu \cong 35\%$. The reduction in the applicable turbulence-intensity level for inclined probes (*SY* or *X*) is linked to a smaller unambiguous yaw-angle range of only $\pm 45°$ compared with a $\pm 90°$ range for an *SN*-probe. Müller (1992) has shown that the series-expansion errors can be reduced significantly by including and measuring triple-order terms in the calculation of the second-order terms.

5.8.2 $\overline{V_e^2}$-method

It has been proposed, in particular for methods involving multipositioning of an *SY*-probe (Rodi 1975; Acrivlellis and Felsch 1979; Acrivlellis 1982) that an analysis in terms of $\overline{V_e^2}$ may be more accurate than the V_e-method described above. The uncertainty of the $\overline{V_e^2}$-method was also studied in the comparative investigation by Swaminathan *et al.* (1986*b*). If eqn (5.16a–b) for positions $\bar{\alpha}$ and $-\bar{\alpha}$ and a similar equation for $\alpha = 0°$ are applied to a velocity vector ($\bar{U} + u, v, w$) then the following exact relationships apply, for $\alpha = 0$

$$\overline{V_{e_0}^2} = \bar{U}^2 + \overline{u^2} + k^2\overline{v^2} + h^2\overline{w^2}, \tag{5.50a}$$

for $\bar{\alpha}$

$$\overline{V_{e1}^2} = A_1(\bar{U}^2 + \overline{u^2}) + A_2\overline{v^2} - A_3\overline{uv} + h^2\overline{w^2}, \qquad (5.50b)$$

for $-\bar{\alpha}$

$$\overline{V_{e2}^2} = A_1(\bar{U}^2 + \overline{u^2}) + A_2\overline{v^2} + A_3\overline{uv} + h^2\overline{w^2}, \qquad (5.50c)$$

where $A_1 = \cos^2\bar{\alpha} + k^2\sin^2\bar{\alpha}$, $A_2 = \sin^2\bar{\alpha} + k^2\cos^2\bar{\alpha}$, and $A_3 = (1 - k^2)$ $\sin^2\bar{\alpha}$.

Related $\pm\bar{\alpha}$ measurements in the (x, z)-plane will give two additional sets of equations. The resulting five equations can be solved without simplifications or assumptions to give $\bar{U}^2 + \overline{u^2}$, $\overline{v^2}$, $\overline{uv}$, $\overline{w^2}$, and $\overline{uw}$. In this method it is not possible to separate $\bar{U}^2$ and $\overline{u^2}$, and $\bar{U}$ is usually obtained from a series-expansion method. The uncertainty in the values of $\overline{u^2}$, $\overline{v^2}$, and $\overline{uv}$ as a function of the turbulence intensity are also included (method 2) in Fig. 4.20. The $\overline{V_e^2}$-procedure does not involve any truncation errors, and in this respect it is superior to the V_e-method. However, the results in Fig. 4.21 show that the $\overline{V_e^2}$-method is very sensitive to uncertainties in the calibration constants at low and moderate turbulence intensities ($Tu < 20\%$). A similar observation has been made by Müller (1987). In the $\overline{V_e^2}$-method, the uncertainty in $\overline{v^2}$ and $\overline{uv}$ falls below 10% when Tu is increased to ~30%. However, the velocity vector will, at high turbulence intensities, regularly move outside the approach quadrant of the two positions ($\pm\bar{\alpha}$) of the SY-probe. This may result in the wire element being placed in the wake of a prong and/or the stem, and these significant effects have not been included in the analysis by Swaminathan *et al.* (1986*b*). Furthermore, Kawall *et al.* (1983) has also pointed out that large errors may occur in the $\overline{V_e^2}$-method for $\overline{uv}$, even for small values of $\bar{V}$.

5.9 UNCERTAINTY ANALYSIS, X-PROBES

5.9.1 Sum-and-difference method

The most common signal-analysis procedure for X-probes is the sum-and-difference method based on a first-order series expansion for V_e eqn (5.17). A number of factors contribute to the uncertainty in the results obtained by this method.

Uncertainty in the mean yaw angle The yaw functions, eqn (5.5a–c), used in many sum-and-difference methods, require measurements of the mean yaw angle, $\bar{\alpha}$. In many investigations, large variations have been reported in the value of the yaw coefficient k, eqn (5.5a), even for nominally similar probes. However, this variation is, to a large extent, due to uncertainties in the values of $\bar{\alpha}$. Using the yaw-calibration method outlined in Section 5.3.1.2, Al-Kayiem (1989) and Nabhani (1989) have studied the effect of an assumed yaw-angle error, $\Delta\bar{\alpha}$, on the value of the calculated yaw coeffi-

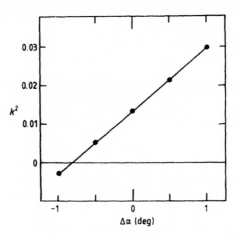

Fig. 5.18. The variation in k^2 with the uncertainty, $\Delta\alpha$, in the value of $\bar{\alpha}$.

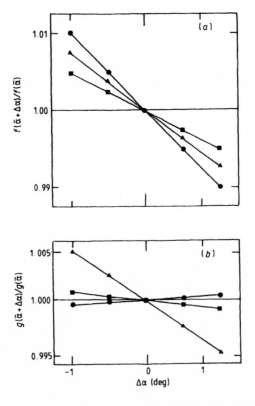

Fig. 5.19. The variation in (a) $f(\bar{\alpha})$ and (b) $g(\bar{\alpha})$ with $\Delta\alpha$, for: (●) eqn (5.5a) k, (▲) eqn (5.5b) m, (■) eqn (5.5c) b.

cient k^2. A typical $k^2(\Delta\bar{\alpha})$ variation is shown in Fig. 5.18, which demonstrates that an error in $\bar{\alpha}$ of only $-1°$ can produce negative values for k^2. However, the corresponding error in U and V, based on the sum-and-difference method of eqns (5.33a–b), depends only on the sensitivity of $f(\bar{\alpha})$ and $g(\bar{\alpha})$ to variations in $\bar{\alpha}$. In these calculations, corresponding values of $\bar{\alpha} + \Delta\bar{\alpha}$ and the yaw coefficients (for example, $k^2(\Delta\bar{\alpha})$) must be used. Fig. 5.19 contains a typical plot of $f(\bar{\alpha} + \Delta\bar{\alpha})/f(\bar{\alpha})$ and $g(\bar{\alpha} + \Delta\bar{\alpha})/g(\bar{\alpha})$ for the four yaw functions of eqns (5.5a–d). Within the angle range $-1° \leqslant \Delta\bar{\alpha} \leqslant 1°$ the variation in $f(\bar{\alpha})$ is about $\pm 1\%$ and the similar variation in $g(\bar{\alpha})$ is less than $\pm 0.5\%$ for all four yaw functions. Exact knowledge of $\bar{\alpha}$ is, therefore, of less importance than has been previously claimed. This result also applies to the $\pm\bar{\alpha}$ SY-probe method since eqn (5.20c–d) only contains $f(\bar{\alpha})$ and $g(\bar{\alpha})$. An in-depth study of the effect of angle errors on the uncertainty of signal analysis for SN, SY, and X-probes has been presented by Yoshino *et al.* (1989).

The constant calibration-coefficient approximation The sum-and-difference method is based on the effective velocity, which is normally evaluated from the simple power law of eqn (3.1) with constant values of A, B, and n. However, it was shown in Section 5.4.1 that, if an accurate calibration relationship is to be obtained over a large angle range, the flow-angle dependence of A, B, and n must be taken into account (see eqns (5.29a–b)). The errors in U and V caused by the assumptions of constant A, B, and n have been evaluated by Bruun *et al.* (1990c) using a reference inversion method. These errors, $\Delta U(\%)$ and $\Delta V(\%)$, are shown in Fig. 5.20 as functions of $\tilde{V}$ and θ. Since eqn (3.1) with constant values for A, B, and n is a poorer curve-fit to the calibration data than eqn (5.29), errors will occur even for small values of θ, in particular for the cross-velocity component. Within the angle range $-25° \leqslant \theta \leqslant 25°$, the errors in U and V can be seen to be nearly independent of θ and to only vary slowly with the magnitude, $\tilde{V}$, of the velocity vector.

Errors in high-intensity turbulent flow Tutu and Chevray (1975) have carried out an analysis of the errors introduced in high-intensity turbulent flow for an X-probe with two identical wires placed at $\bar{\alpha} = \pm 45°$. The true response equations for V_{e1} and V_{e2} were eqns (5.17a–b), with $\overline{W} = 0$, and retaining w. A first-order series expansion and a sum-and-difference method were used to evaluate the 'measured' u_m- and v_m-values. A simulated turbulence field with a joint Gaussian probability distribution for (u, v, w) and a single mean-velocity component, $\bar{U}$, was applied to these response equations to identify the analysis errors. This investigation highlighted the effects of the w velocity component and of the rectification errors. In particular, the errors for the shear stress $\overline{uv}$ might reach 28% when $Tu = 35\%$. The corresponding effect of an out-of-plane mean-velocity component $\overline{W}$ has been investigated by Francis *et al.* (1978), Andreas (1979b), Clausen and Wood (1989), and Brüun *et al.* (1991).

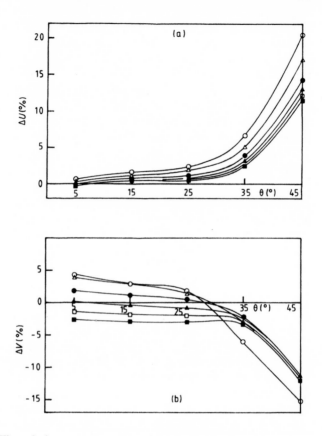

Fig. 5.20. The velocity component errors in the sum-and-difference method, due to the assumption of constant values for A, B, n, and k: (a) ΔU, (b) ΔV. (○) 5 m s^{-1}, (△) 10 m s^{-1}, (●) 20 m s^{-1}, (▲) 30 m s^{-1}, (□) 40 m s^{-1}, (■) 50 m s^{-1}.

A related error analysis based on an acceptance cone for the velocity vector and a Gaussian probability density function (pdf) has been presented by Castro and Cheun (1982) Castro (1986), and Browne *et al.* (1989*b*). However, as discussed by Müller (1987), many turbulent-flow measurements do not show the long tail of a Gaussian distribution, and the probability of a vector lying outside the acceptance cone is reduced considerably. Also, as found by Dengel *et al.* (1981) and by Jaroch (1985), pdf corrections might even worsen the accuracy of the turbulence measurements in cases where the pdf is not distributed normally.

A detailed error study has been carried out by Kawall *et al.* (1983). Using a numerical simulation procedure they investigated the effects on an X-probe of: (i) rectification; (ii) the spanwise component, *w*; (iii) axial cooling; (iv) calibration-parameter uncertainties, and (v) spanwise wire separa-

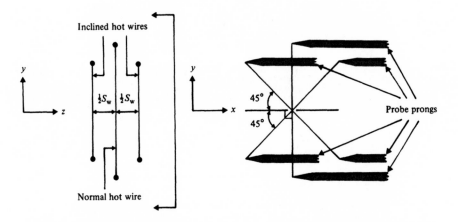

Fig. 5.21. A schematic diagram of the three-wire probe. (From Kawall *et al.* 1983. Reproduced with kind permission of Cambridge University Press from the *Journal of Fluid Mechanics*.)

tion. Tagawa *et al.* (1992) have also studied the error caused by the spanwise separation by assuming a lateral correlation coefficient, $g(r_2)$, as defined in Fig. 2.26, and a simulated Gaussian velocity field. To extend the applicable turbulence-intensity range Kawall *et al.* (1983) proposed the use of a special three-wire probe, in which all three wires are contained in the same plane, as shown in Fig. 5.21. In this analysis the lateral velocity component w is taken into account, but, as pointed out by Müller (1987), the sign of w cannot be resolved by this probe. The analysis of Kawall *et al.* shows that the applicable yaw-angle range for this three-wire probe is increased to $\pm 90°$ (that is, it is similar to that of an SN-probe) compared with $\pm 45°$ for an X-probe. The use of a similar three-wire probe has also been described by Legg *et al.* (1984), who reported difficulties in measuring $\overline{w^2}$. The proposed three-wire probe is much more difficult to calibrate and operate, and the response equations developed are very complex, which substantially increases both the experimental and computational time, when compared with the use of an X-probe.

5.9.2 Look-up-matrix method

This procedure removes several of the sources of errors present in the sum-and-difference method. In particular all restrictions associated with analytical approximations are avoided, removing errors caused by uncertainties in the mean yaw angle, the assumption of constant calibration coefficients, and the first-order series-expansion errors. The look-up matrix method is, however, affected by the out-of-plane velocity-component and rectification errors. An additional uncertainty is the interpolation errors caused by the finite number of points in the look-up matrix.

6

THREE-COMPONENT VELOCITY
MEASUREMENTS

If only the mean-velocity components and the Reynolds stresses are required, multiposition techniques involving single normal (*SN*), single yawed (*SY*), or *X* probes can be applied. Related techniques and solution procedures for one-component, two-component and three-component mean-velocity flow fields are described in Sections 6.1 and 6.2.

For more detailed flow studies, simultaneous measurements of three velocity components of the velocity vector at a point can be carried out with a hot-wire probe containing three (or four) sensors. The techniques discussed in section 6.3 can be applied provided that the velocity vector remains within the approach octant defined by the three sensors.

6.1 MULTIPOSITION SINGLE-SENSOR TECHNIQUES

These techniques are implemented by placing either (i) an *SN*-probe, (ii) an *SY*-probe, or both an *SN* and an *SY* probe, at a number of spatial orientations at each measurement point. Implementing these techniques with an *SY*-probe requires the use of a traversing mechanism which can rotate the probe around its stem, as described by, for example Johnston (1970), Müller (1982*a*), Shayesteh and Bradshaw (1987), Löfdahl (1988), Fardad (1989), Fardad and Bruun (1991), and Lau *et al*. (1993). The velocity vector, *V*, can be specified either by its magnitude, $\tilde{V}$, and two spatial angles or by the instantaneous velocity components (*U*, *V*, *W*) in a space-fixed coordinate system (*x*, *y*, *z*). A number of different procedures have been proposed for the first method. Hoffmeister (1972) based his technique on the direct nonlinear voltage signal, while Hirsch and Kool (1977), Okiishi and Schmidt (1978), De Grande and Kool (1981), Kuroumaru *et al*. (1982), Sherif and Pletcher (1983, 1986, 1987), Shin and Hu (1986), and Wagner and Kent (1988*a*) introduced different expressions for the effective velocity, V_e. A common format for both methods can be obtained if the Jørgensen equation for the effective velocity, V_e, is used

$$V_e^2 = U_N^2 + k^2 U_T^2 + h^2 U_B^2,$$

where (U_N, U_T, U_B) are the velocity components in the wire-fixed coordinate system, as shown in Fig. 6.1. The signal analysis is simplified if the probe-stem is aligned with one of the coordinate axes of the selected space-fixed coordinate system, and the stem orientation will be assumed to be

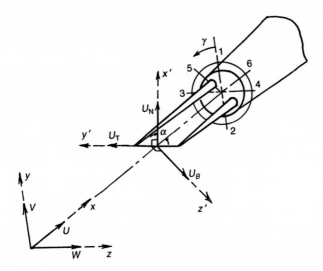

Fig. 6.1. Wire-fixed and space-fixed coordinate systems for the single-element multiposition method. The probe-stem is aligned with the x-axis.

parallel with the predominant mean-flow component unless stated otherwise. The effective velocity, V_e, is usually obtained from the measured anemometer voltage, E, using the power-law relationship of eqn (3.1). The calibration of SN and SY probes for velocity, yaw and pitch is described in Sections 4.2.1, 4.3 and 5.3.1, and in the following derivations for V_e the values of k and h are assumed to be known. Buresti and Talamelli (1992) describe a method which optimizes for A, B, n, k, and h during a complete ($\tilde{V}$, α, β)-calibration.

6.1.1 General response equations for a single-sensor probe

The sensing element of the SY-probe is assumed to have a yaw angle, α, as defined in Fig. 6.1, and the probe is free to rotate around the probe-stem with a roll angle of γ. The reference angle $\gamma = 0°$ corresponds to the probe being placed in the (x, y)-plane (position 1). It has been shown by for example, Acrivlellis (1978) that the wire-fixed velocity components (U_N, U_T, U_B) are related to the space-fixed velocity components (U, V, W) by

$$U_N = U \cos \alpha + [V \cos \gamma - W \sin \gamma] \sin \alpha, \tag{6.1a}$$

$$U_T = - U \sin \alpha + [V \cos \gamma - W \sin \gamma] \cos \alpha, \tag{6.1b}$$

$$U_B = V \sin \gamma + W \cos \gamma. \tag{6.1c}$$

Inserting eqns (6.1a–c) for U_N, U_T, and U_B into eqn (3.2) gives

$$V_e^2 = A_1 U^2 + A_2 V^2 + A_3 W^2 + A_4 UV + A_5 UW + A_6 VW, \quad (6.2)$$

with

$$A_1 = \cos^2 \alpha + k^2 \sin^2 \alpha, \tag{6.3a}$$

$$A_2 = (\sin^2 \alpha + k^2 \cos^2 \alpha) \cos^2 \gamma + h^2 \sin^2 \gamma, \tag{6.3b}$$

$$A_3 = (\sin^2 \alpha + k^2 \cos^2 \alpha) \sin^2 \gamma + h^2 \cos^2 \gamma, \tag{6.3c}$$

$$A_4 = (1 - k^2) \cos \gamma \sin 2\alpha, \tag{6.3d}$$

$$A_5 = - (1 - k^2) \sin \gamma \sin 2\alpha, \tag{6.3e}$$

$$A_6 = - (\sin^2 \alpha + k^2 \cos^2 \alpha - h^2) \sin 2\gamma. \tag{6.3f}$$

By separating each velocity component into mean and fluctuating components, a signal-analysis procedure can be developed for V_e.

If it is assumed that the turbulence intensity is low; that is,

$$\frac{(\overline{u^2})^{1/2}}{\bar{U}} \ll 1, \qquad \frac{(\overline{v^2})^{1/2}}{\bar{U}} \ll 1, \qquad \frac{(\overline{w^2})^{1/2}}{\bar{U}} \ll 1,$$

then to second order the mean value of the effective velocity, $\bar{V}_e$, can be written as (Buresti and Di Cocco 1987)

$$\bar{V}_e = B_1 \bar{U} + B_2 \bar{V} + B_3 \bar{W} + B_4 \left(\frac{\bar{V}^2 + \overline{v^2}}{\bar{U}} \right) + B_5 \left(\frac{\bar{W}^2 + \overline{w^2}}{\bar{U}} \right)$$
$$+ B_6 \left(\frac{\bar{V}\bar{W} + \overline{vw}}{\bar{U}} \right), \tag{6.4}$$

and similarly if first-order mean-velocity terms are included in the equation for the fluctuating effective velocity, v_e,

$$v_e = B_1 u + \left(B_2 + 2B_4 \frac{\bar{V}}{\bar{U}} + B_6 \frac{\bar{W}}{\bar{U}} \right) v + \left(B_3 + B_6 \frac{\bar{V}}{\bar{U}} + 2B_5 \frac{\bar{W}}{\bar{U}} \right) w. \tag{6.5}$$

The corresponding response equation for $\overline{v_e^2}$ can be expressed as

$$\overline{v_e^2} = C_1 \overline{u^2} + \left(C_2 + 2C_{10} \frac{\bar{V}}{\bar{U}} + C_{12} \frac{\bar{W}}{\bar{U}} \right) \overline{v^2} + \left(C_3 + C_{13} \frac{\bar{V}}{\bar{U}} + 2C_{15} \frac{\bar{W}}{\bar{U}} \right) \overline{w^2}$$
$$+ \left(C_4 + 2C_7 \frac{\bar{V}}{\bar{U}} + C_8 \frac{\bar{W}}{\bar{U}} \right) \overline{uv} + \left(C_5 + C_8 \frac{\bar{V}}{\bar{U}} + 2C_9 \frac{\bar{W}}{\bar{U}} \right) \overline{uw}$$
$$+ \left(C_6 + C_{11} \frac{\bar{V}}{\bar{U}} + C_{14} \frac{\bar{W}}{\bar{U}} \right) \overline{vw}. \tag{6.6}$$

The coefficients B_1–B_6 and C_1–C_{15} are given in Table 6.1. By including first-order terms in $\bar{V}/\bar{U}$ and $\bar{W}/\bar{U}$ in eqns (6.4) to (6.6) this technique can

TABLE 6.1   *The coefficients B_1–B_6 and C_1–C_{15} in eqns (6.4) to (6.6)*

$B_1 = (A_1)^{1/2}$	$C_6 = 2B_2B_3$
$B_2 = A_4/2B_1$	$C_7 = 2B_1B_4$
$B_3 = A_5/2B_1$	$C_8 = 2B_1B_6$
$B_4 = (A_2 - B_2^2)/2B_1$	$C_9 = 2B_1B_5$
$B_5 = (A_3 - B_3^2)/2B_1$	$C_{10} = 2B_2B_4$
$B_6 = (A_6 - 2B_2B_3)/2B_1$	$C_{11} = 2B_2B_6 + 4B_3B_4$
$C_1 = A_1$	$C_{12} = 2B_2B_6$
$C_2 = B_2^2$	$C_{13} = 2B_3B_6$
$C_3 = B_3^2$	$C_{14} = 2B_3B_6 + 4B_2B_5$
$C_4 = A_4$	$C_{15} = 2B_3B_5$
$C_5 = A_5$	

be applied to low/moderate turbulence intensity flows ($Tu < {\sim}15\text{--}20\%$) with cross mean-velocity components with magnitudes of up to about $0.2\bar{U}$. To maximize the applicable turbulence-intensity range, the probe should be approximately aligned with the mean-flow direction so that $\bar{V}/\bar{U}$ and $\bar{W}/\bar{U}$ are small.

In swirling-flow investigations (for example, in the investigation by Bissonnette and Mellor 1974; Bank and Gauvin 1977; Nabhani 1989), it is often necessary to insert the probe radially, that is perpendicularly to the mean-flow direction. When this probe orientation is used, the roll-angle position must be selected carefully to avoid prong-interference effects on the hot-wire signal (Bruun *et al.* 1993). The selection of a radial probe orientation may also be due to accessibility constraints (Russ and Simon 1991). This situation is illustrated in Fig. 6.2. If the change in the space-fixed velocity components between Figs. 6.1 and 6.2 is taken into account then eqn (6.2) for V_e^2 can also be applied to the swirling-flow situation provided that the coefficients A_1–A_6 are redefined as

$$A_1 = (\sin^2\alpha + k^2\cos^2\alpha)\sin^2\gamma + h^2\cos^2\gamma,$$

$$A_2 = \cos^2\alpha + k^2\sin^2\alpha,$$

$$A_3 = (\sin^2\alpha + k^2\cos^2\alpha)\cos^2\gamma + h^2\sin^2\gamma,$$

$$A_4 = -(1 - k^2)\sin\gamma\sin 2\alpha, \tag{6.7}$$

$$A_5 = -(\sin^2\alpha + k^2\cos^2\alpha - h^2)\sin 2\gamma,$$

$$A_6 = (1 - k^2)\cos\gamma\sin 2\alpha$$

When the predominant mean-velocity component is $\bar{U}$ (axial) then the series-expansion method described by eqns (6.4) and (6.6) can be applied using the coefficients A_1–A_6 defined by eqn (6.7). If the tangential mean-

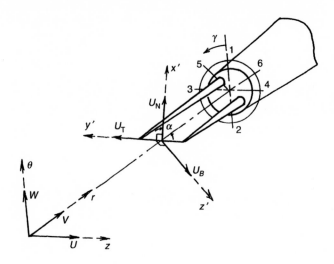

Fig. 6.2. The wire-fixed and space-fixed coordinate systems for the single-element multiposition method. The probe-stem is aligned with the radial direction, r.

velocity component, $\overline{W}$, is the largest component, then the series expansion must be carried out in terms of $\overline{W}$. The resultant modified equations are described by Nabhani (1989), Nabhani and Bruun (1990), and Bruun *et al.* (1993).

Using eqns (6.4) and (6.6), solutions may be obtained in either of the space-fixed coordinate systems shown in Figs 6.1 and 6.2. Alternatively, the mean-flow direction may first be determined, and the hot-wire probe then aligned with this direction. Such a technique is described by Bissonnette and Mellor (1974).

6.1.2 V_e-solution procedures

Equations (6.4) and (6.6) for $\overline{V}_e$ and $\overline{v_e^2}$ are coupled with the mean-velocity components and the Reynolds stresses occurring in both equations. Different solution procedures have been proposed depending on whether the mean-velocity vector has one, two, or three components in the coordinate system selected for the signal analysis.

6.1.2.1   One-component mean velocity

A simple solution procedure exists when the probe-stem is aligned with the mean-flow direction, $\overline{U}$. In this coordinate system, $\overline{V}$ and $\overline{W}$ are zero, and eqn (6.6) for $\overline{v_e^2}$ will not contain any mean-velocity terms. The solution of the Reynolds stresses is therefore decoupled from the mean-velocity evaluation. De Grande and Kool (1981) have presented an SY-probe method, in which the probe was placed at twenty one different positions at each point to increase the accuracy of the measured Reynolds stresses. However, to simplify the signal analysis, most procedures have involved both an SN and an SY probe. For an SN-probe ($\alpha = 0°$) placed in the (x, y)-plane ($\gamma = 0°$), the equations for $\bar{V}_e$ and $\overline{v_e^2}$ are to second order (Section 4.12.2.3)

$$\bar{V}_{e0} = \bar{U}\left(1 + \frac{1}{2} h^2 \frac{\overline{w^2}}{\bar{U}^2}\right) \tag{6.8}$$

and

$$\overline{v_{e0}^2} = \overline{u^2}. \tag{6.9}$$

Equation (6.6) for an SY-probe contains contributions from all six Reynolds stresses for most values of γ. However, equations with only two of the velocity fluctuations can be obtained if the SY-probe is located in either the (x, y)- or (x, z)-plane. On placing the SY-probe at positions 1 ($\gamma = 0°$), 2 ($\gamma = 180°$) in the (x, y)-plane, and 3 ($\gamma = 90°$) and 4 ($\gamma = 270°$) in the (x, z)-plane (see Fig. 6.1), the corresponding equations for $\overline{v_e^2}$ will be

$$\overline{v_{e1}^2} = D_1\overline{u^2} + D_4\overline{v^2} + D_3\overline{uv}, \tag{6.10a}$$

$$\overline{v_{e2}^2} = D_1\overline{u^2} + D_4\overline{v^2} - D_3\overline{uv}, \tag{6.10b}$$

$$\overline{v_{e3}^2} = D_1\overline{u^2} + D_4\overline{w^2} - D_3\overline{uw}, \tag{6.10c}$$

$$\overline{v_{e4}^2} = D_1\overline{u^2} + D_4\overline{w^2} + D_3\overline{uw}, \tag{6.10d}$$

where $D_1 = A_1 = \cos^2\alpha + k^2\sin^2\alpha$, $D_3 = (1 - k^2)\sin 2\alpha$, and $D_4 = D_3^2/(4D_1)$, as listed in Table 6.2.

Taking the difference between eqns (6.10a–b) and similarly for eqns (6.10c–d) the shear stresses $\overline{uv}$ and $\overline{uw}$ can be evaluated from

$$\overline{uv} = \frac{1}{2D_3}\left(\overline{v_{e1}^2} - \overline{v_{e2}^2}\right), \tag{6.11a}$$

$$\overline{uw} = \frac{1}{2D_3}\left(\overline{v_{e4}^2} - \overline{v_{e3}^2}\right), \tag{6.11b}$$

TABLE 6.2 *The coefficients D_1–D_6*

$D_1 = A_1 = \cos^2\alpha + k^2\sin^2\alpha$
$D_2 = \sin^2\alpha + k^2\cos^2\alpha$
$D_3 = (1 - k^2)\sin 2\alpha$
$D_4 = D_3^2/(4D_1)$
$D_5 = (D_2 - D_4)/(4D_1)$
$D_6 = h^2/(4D_1)$

and a summation of eqns (6.10a–b) and (6.10c–d) gives two linear equations in $\overline{u^2}$, $\overline{v^2}$, and $\overline{w^2}$.

$$\overline{v_{e1}^2} + \overline{v_{e2}^2} = 2D_1\overline{u^2} + 2D_4\overline{v^2}, \qquad (6.12a)$$

$$\overline{v_{e3}^2} + \overline{v_{e4}^2} = 2D_1\overline{u^2} + 2D_4\overline{w^2}. \qquad (6.12b)$$

By measuring $\overline{u^2}$ with an *SN*-probe (eqn (6.9)), the corresponding values of $\overline{v^2}$ and $\overline{w^2}$ can be calculated from eqns (6.12a–b). Knowing the value of $\overline{w^2}$ the value of $\bar{U}$ can be determined from eqn (6.8).

The value of the shear stress $\overline{vw}$ can be calculated from two additional measurements at, for example, position 5 ($\gamma = 45°$) and 6 ($\gamma = -45°$); see Fig. 6.1:

$$\overline{v_{e5}^2} = D_1\overline{u^2} + \frac{1}{2}D_4\overline{v^2} + \frac{1}{2}D_4\overline{w^2} + \frac{\sqrt{2}}{2}D_3\overline{uv} - \frac{\sqrt{2}}{2}D_3\overline{uw} - D_4\overline{vw}, \quad (6.13a)$$

$$\overline{v_{e6}^2} = D_1\overline{u^2} + \frac{1}{2}D_4\overline{v^2} + \frac{1}{2}D_4\overline{w^2} + \frac{\sqrt{2}}{2}D_3\overline{uv} + \frac{\sqrt{2}}{2}D_3\overline{uw} + D_4\overline{vw}, \quad (6.13b)$$

By taking the difference between eqns (6.13a–b), $\overline{vw}$ can be calculated from

$$\overline{vw} = \frac{1}{2D_4}\left[(\overline{v_{e6}^2} - \overline{v_{e5}^2}) - \sqrt{2}D_3\overline{uw}\right], \qquad (6.14)$$

with $\overline{uw}$ being determined from eqn (6.11b). Vagt and Fernholz (1979) have described a special *SY*-probe and a related signal-analysis technique for evaluating $\overline{vw}$. It should be noted that the six equations in eqns (6.10a–d) and (6.13a–b) for the six Reynolds stresses are not independent, due to the symmetry of the coefficients. An additional, independent equation is needed; this is usually eqn (6.9) for the *SN*-probe. To overcome the two-probe problem, Hooper (1980) and Sampath *et al.* (1982) modified an *X*-probe into a probe containing an *SN* and an *SY* element, which reduced the experimental time significantly.

6.1.2.2 Two-component mean velocity

In many flow situations the mean-velocity vector will have two significant components in the space-fixed coordinate system selected. Bissonnette and Mellor (1974) investigated a swirling flow, which had significant axial, $\bar{U}$, and tangential, $\bar{W}$, mean-velocity components (see the notation in Fig. 6.2). They used both an *SN* and an *SY* hot-wire probe, which were inserted radially, that is perpendicularly to the mean flow field. At each point the magnitude and direction of the mean-velocity vector was first identified by means of the *SN*-probe. Rotating the probe around its stem, they recorded the voltage output as a function of the roll angle, γ. As described in Section 4.11.1, this type of measurement can determine the magnitude, $\tilde{V}$, and direction, $\bar{\gamma}$ (tan $\bar{\gamma} = \overline{W}/\overline{U}$), of the mean-velocity vector. Having identified this direction at each measuring point, they then used both the *SN* and the *SY* probe to measure the Reynolds stresses corresponding to the velocity fluctuations (u^*, v^*, w^*) in a coordinate system (x^*, y^*, z^*) aligned with the mean-flow direction. In this coordinate system, $\overline{V}^*$ and $\overline{W}^*$ are zero and the related $\overline{v_c^2}$-equation (eqn (6.6)) therefore contains only Reynolds stresses. The functional forms of the response equations for $\overline{v_c^2}$ for both the *SN* and *SY* probes were determined by Bissonnette and Mellor (1974) from curve-fitting procedures applied to the measured $\overline{V}_e(\gamma)$ curves. The response equations can also be obtained using the theory presented in Section 6.1.1 with A_1–A_6 defined by eqn (6.7). For the *SN* probe ($\alpha = 0°$), eqn (6.6) for $\overline{v_c^2}$ becomes

$$\overline{v_{e0}^2} = A_1\overline{u^{*2}} + \frac{A_5^2}{4A_1}\overline{w^{*2}} + A_5\overline{u^*w^*},\tag{6.15}$$

with

$$A_1 = k^2\sin^2\gamma + h^2\cos^2\gamma,$$

$$A_5 = (h^2 - k^2)\sin 2\gamma.$$

Using eqn (6.15), the values of $\overline{u^{*2}}$, $\overline{w^{*2}}$, and $\overline{u^*w^*}$ can be obtained from measurements with the *SN*-probe placed at $\gamma = 0°$, $45°$, and $-45°$. This method is similar to the single inclined probe method described in Section 5.3.3.2. In principle, all Reynolds stresses can be obtained from a multiposition *SY*-probe method, but Bissonette and Mellor found that such a procedure gives unreliable results. This was most likely to be due to prong-interference effects, as reported by Bruun *et al.* (1993). Instead, the *SY*-probe was only used to evaluate $\overline{v^{*2}}$, $\overline{u^*v^*}$, and $\overline{v^*w^*}$, with the values of and $\overline{u^{*2}}$, $\overline{w^{*2}}$, and $\overline{u^*w^*}$ treated as known quantities obtained by the *SN*-probe. A similar procedure was used by Yowakim and Kind (1988).

The velocity results obtained in the coordinate system (x^*, y^*, z^*) aligned with the mean-flow direction must be converted to the related components in the space-fixed coordinate system (x, y, z). Using the transformation equations of eqn (3.8a–b) and replacing (U_s, V_s, ψ) by (u^*, w^*, γ) gives:

$$\overline{u^2} = \overline{u^{*2}}\cos^2\bar{\gamma} - 2\overline{u^*w^*}\sin\bar{\gamma}\cos\bar{\gamma} + \overline{w^{*2}}\sin^2\bar{\gamma},\tag{6.16a}$$

$$\overline{v^2} = \overline{v^{*2}},\tag{6.16b}$$

$$\overline{w^2} = \overline{u^{*2}}\sin^2\bar{\gamma} + 2\overline{u^*w^*}\sin\bar{\gamma}\cos\bar{\gamma} + \overline{w^{*2}}\cos^2\bar{\gamma},\tag{6.16c}$$

$$\overline{uv} = \overline{u^*v^*}\cos\bar{\gamma} - \overline{v^*w^*}\sin\bar{\gamma},\tag{6.16d}$$

$$\overline{vw} = \overline{u^*v^*}\sin\bar{\gamma} + \overline{v^*w^*}\cos\bar{\gamma},\tag{6.16e}$$

$$\overline{uw} = (\overline{u^{*2}} - \overline{w^{*2}})\sin\bar{\gamma}\cos\bar{\gamma} + \overline{u^*w^*}(\cos^2\bar{\gamma} - \sin^2\bar{\gamma}).\tag{6.16f}$$

A similar coordinate transformation is given in Johnston (1970).

An alternative approach was developed by Barrett (1987, 1989). He placed an SN hot-wire probe perpendicularly to the plane of a two-dimensional mean-velocity field. The probe was placed sequentially at four roll angles; $\gamma = 0°$, $45°$, $-45°$, and $90°$. Using combinations of pairs of readings from the four positions, Barrett developed a procedure for the evaluation of the mean flow angle in the range $\pm 90°$. However, the related equations for the corresponding Reynolds stresses ($\overline{u^2}$, $\overline{v^2}$, and $\overline{uv}$) were restricted to much smaller flow angles, particularly for $\overline{v^2}$ and $\overline{uv}$ ($|\gamma| < \sim 25°$).

6.1.2.3 Three-component mean velocity

If the mean velocity vector has significant components in all three directions of the selected space-fixed coordinate system then it may first be necessary to determine the mean-flow direction by a separate hot-wire or pressure-probe method. By aligning the probe-stem with this direction, the Reynolds stresses can then be evaluated by the procedures outlined in the preceding two sections.

However, if the flow has a predominant mean-flow component, say $\bar{U}$, then solution procedures based on eqns (6.4) and (6.6) for $\bar{V}_e$ and $\overline{v_e^2}$ can be applied. Since $\bar{V}$ and $\bar{W}$ have non-zero values, the two sets of equations will be coupled, but it is possible to apply a procedure which first evaluates the mean-velocity components. If measurements are taken with an SY-probe at position 1 ($\gamma = 0°$), position 2 ($\gamma = 180°$), position 3 ($\gamma = 90°$), and position 4 ($\gamma = 270°$) in Fig. 6.1, then it follows from eqn (6.4) that $\bar{V}$ and $\bar{W}$ can be determined from

$$\bar{V} = \frac{\bar{V}_{e1} - \bar{V}_{e2}}{(D_3/B_1)},$$ (6.17a)

$$\bar{W} = \frac{\bar{V}_{e4} - \bar{V}_{e3}}{(D_3/B_1)}.$$ (6.17b)

The corresponding equation for $\bar{U}$ can be expressed as

$$\bar{U} = \frac{1}{2B_1}(\bar{V}_{e1} + \bar{V}_{e2}) - D_5\left(\frac{\bar{V}^2 + \overline{v^2}}{\bar{U}}\right) - D_6\left(\frac{\bar{W}^2 + \overline{w^2}}{\bar{U}}\right),$$ (6.18)

where $B_1 = D_1^{1/2}$ and D_1–D_6 are defined in Table 6.2. Related series-expansion derivations have been presented by Bank and Gauvin (1977) and Phillips (1985), using the magnitude $\tilde{V}$ ($= (\bar{U}^2 + \bar{V}^2 + \bar{W}^2)^{1/2}$) of the complete mean-velocity vector as a reference velocity. In their methods, $\tilde{V}$ was determined by a separate procedure. The corresponding solution procedure for a probe aligned with an (r, θ, z)-coordinate system is described by Bruun et al. (1993).

The mean cross-velocity components $\bar{V}$ and $\bar{W}$ can be calculated directly

from measurements of $\bar{V}_e$ (eqns (6.17a–b)). The evaluation of $\bar{U}$ is slightly
more involved. Ignoring the turbulence terms initially, eqn (6.18) can be
solved for $\bar{U}$ by rearranging it in the form $\bar{U}^2 + B\bar{U} + C = 0$, taking $\bar{V}$
and $\bar{W}$ as known quantities determined from eqns (6.17a–b). Subsequently,
when $\overline{v^2}$ and $\overline{w^2}$ have been evaluated from eqn (6.6), the value of $\bar{U}$ can
be corrected using eqn (6.18). In the study by Gessner and Arterberry (1982)
the magnitude and the direction of the mean-velocity vector was also determined from four measurements with an SY-probe. Their theory was based
on a series expansion of the hot-wire response equation, assuming a low
turbulence intensity, and retaining terms up to the fourth order in $\bar{V}/\bar{U}$ and
$\bar{W}/\bar{U}$. By time averaging the results and combining expressions for the four
positions they obtained two coupled fourth-order, nonlinear, algebraic
equations for $\bar{V}/\bar{U}$ and $\bar{W}/\bar{U}$. These equations were solved by the Newton–
Raphson method.

Having calculated the values of $\bar{U}$, $\bar{V}$, and $\bar{W}$, the terms $\bar{V}/\bar{U}$ and $\bar{W}/\bar{U}$
in eqn (6.6) can now be treated as known quantities in the n equations for
$\overline{v_e^2}$; this enables the calculation of the six Reynolds stresses to be decoupled from the mean-velocity evaluation. The simplest procedure for obtaining the six Reynolds stresses is to place the SY-probe at six different roll
angles, and then to apply a matrix-inversion solution to the six equations
for $\overline{v_e^2}$. However, as pointed out by Müller (1987), to obtain accurate
results with this method attention has to be paid to the conditioning of the
coefficient matrix of the response equations. To improve the accuracy of
the calculated Reynolds stresses, the SY-probe can be placed at a large
number of roll angles and the corresponding set of equations can be solved
by a least-squares curve-fitting method (De Grande and Kool 1981 used
twenty-one positions and Kuroumaru *et al.* 1982 used twelve orientations).
Löfdahl and Larsson (1984) employed both an SN and an SY probe for
the study of turbulence near the stern of a ship model. Placing the two
probes sequentially at each measuring position, they applied a least-squares
curve-fitting procedure to multiposition readings from both probes. Alternatively, more elaborate procedures can be used. Gessner and Arterberry
(1982) combined an SN-probe placed in one position with an SY-probe
placed in certain 'optimum' symmetric positions relative to the mean-
velocity vector. In these optimum positions, the SY-probe was only sensitive
to two of the velocity fluctuations. This principle is similar to the selected
positions for the one-dimensional mean-flow case in Section 6.1.2.1.

6.1.3 $\overline{V_e^2}$-analysis methods

If each velocity component is expressed in terms of its time-mean value and
its fluctuating component then an exact equation for $\overline{V_e^2}$ can be derived
from eqn (6.2).

$$\overline{V_e^2} = A_1(\overline{U}^2 + \overline{u^2}) + A_2(\overline{V}^2 + \overline{v^2}) + A_3(\overline{W}^2 + \overline{w^2})$$
$$+ A_4(\overline{U}\overline{V} + \overline{uv}) + A_5(\overline{U}\overline{W} + \overline{uw}) + A_6(\overline{V}\overline{W} + \overline{vw}), \quad (6.19)$$

with the coefficients A_1–A_6 defined by eqns (6.3a–f). However, this equation does not allow the mean-velocity components and the related Reynolds stresses to be separated. Therefore, to use the $\overline{V_e^2}$-method, the direction and magnitude of the mean-velocity vector must be determined by an independent procedure.

6.1.3.1 One-component mean velocity

If the mean-flow field is one-dimensional, and the probe-stem is aligned with the direction of $\overline{U}$, then eqn (6.19) can be simplified to

$$\overline{V_e^2} = A_1(\overline{U}^2 + \overline{u^2}) + A_2\overline{v^2} + A_3\overline{w^2} + A_4\overline{uv} + A_5\overline{uw} + A_6\overline{vw}. \quad (6.20)$$

This is the flow situation investigated by Acrivlellis and Felsch (1979), by Acrivlellis (1982), and by Swaminathan *et al.* (1986b). Equation (6.20) becomes, for an SN-probe ($\alpha = 0°$) placed in the (x, y) plane ($\gamma = 0°$),

$$\overline{V_{e0}^2} = \overline{U}^2 + \overline{u^2} + k^2\overline{v^2} + h^2\overline{w^2}. \quad (6.21)$$

For measurements taken with an SY-probe at position 1 ($\gamma = 0°$), position 2 ($\gamma = 180°$) in the (x, y)-plane, and at position 3 ($\gamma = 90°$) and 4 ($\gamma = 270°$) in the (x, z)-plane (see Fig. 6.1), the following relationships apply:

$$\overline{V_{e1}^2} + \overline{V_{e2}^2} = 2D_1(\overline{U}^2 + \overline{u^2}) + 2D_2\overline{v^2} + 2h^2\overline{w^2}, \quad (6.22a)$$

$$\overline{V_{e3}^2} + \overline{V_{e4}^2} = 2D_1(\overline{U}^2 + \overline{u^2}) + 2D_2\overline{w^2} + 2h^2\overline{v^2}. \quad (6.22b)$$

The coefficients D_1–D_3 are defined in Table 6.2. Solving these three linear equations, eqns (6.21) and (6.22a–b), gives $\overline{U}^2 + \overline{u^2}$, $\overline{v^2}$, and $\overline{w^2}$. To separate $\overline{U}^2 + \overline{u^2}$, a series-expansion method is usually applied to obtain $\overline{U}$, as described in Section 5.8. The shear stresses $\overline{uv}$ and $\overline{uw}$ can be evaluated from the following difference equations:

$$\overline{uv} = \frac{1}{2D_3}(\overline{V_{e1}^2} - \overline{V_{e2}^2}), \quad (6.23a)$$

$$\overline{uw} = \frac{1}{2D_3}(\overline{V_{e4}^2} - \overline{V_{e3}^2}). \quad (6.23b)$$

The final shear stress, $\overline{vw}$, can be obtained from SY-probe measurements at $\gamma = \pm 45°$. The advantage of this type of signal-analysis procedure is that it does not involve any series-expansion error. However, as shown by Swaminathan *et al.* (1986b), the $\overline{V_e^2}$-method cannot be applied to low/moderate turbulence intensity flow due to large uncertainties in the evaluated Reynolds stresses (see Section 5.8). Also, if the Reynolds stresses are obtained by matrix inversion, then, as Tsiolakis *et al.* (1983) have shown, the coefficient matrix is ill-conditioned. Furthermore, if $\bar{V}$ and $\bar{W}$ are not zero due to probe misalignment, then eqns (6.23a) and (6.23b) become

$$\bar{U}\bar{V} + \overline{uv} = \frac{1}{2D_3}(\overline{V_{e1}^2} - \overline{V_{e2}^2}), \tag{6.24a}$$

$$\bar{U}\bar{W} + \overline{uw} = \frac{1}{2D_3}(\overline{V_{e4}^2} - \overline{V_{e3}^2}). \tag{6.24b}$$

and significant errors can be introduced if eqns (6.24a–b) are interpreted only in terms of $\overline{uv}$ and $\overline{uw}$. In earlier work by Acrivlellis (1978) a method was proposed which included equations for both $\overline{V_e^2}$ and $\overline{v_e^2}$. However, the coefficients in his equation for $\overline{v_e^2}$ were incorrectly identified as being the same as those for $\overline{V_e^2}$ (see eqn (6.20); also see the comments by Bartenwerfer 1979). The same mistaken procedure has also been used in the paper by Tuckey *et al.* (1984).

6.1.3.2 Two-component mean velocity

Rodi (1975) has applied a $\overline{V_e^2}$-method to measurements in a turbulent jet with a two-dimensional mean velocity $(\bar{U}, \bar{V})$. In this investigation, an SN-probe was first used to identify the direction of the mean-velocity vector. A one-dimensional mean-velocity technique was then applied, which involved an elaborate multiposition and curve-fitting procedure. The Reynolds stresses evaluated were transformed into the corresponding space-fixed components using transformation equations similar to eqns (6.16a–f).

6.1.4 V_e^2-analysis method

An unusual signal-analysis procedure has been proposed by Dvorak and Syred (1972) for the single-wire multiposition method. The signal-analysis method was developed for a rotating SN-probe, with the probe-stem aligned with the predominant mean-velocity component. The probe was placed sequentially at three different roll-angle positions, $\gamma = -45°, 0°$, and $45°$. For each position, response equations were developed for the square of the instantaneous effective velocity, V_e^2, using eqn (3.2). This part of the procedure is similar to the evaluation of eqn (6.2). However, in general, the subsequent signal-analysis assumption is not correct. Although the three signals were measured sequentially, the quantities in the three response equations were assumed to have been measured simultaneously, in order to derive equations for the three instantaneous velocity components (U, V, W) in terms of the three instantaneous effective velocities. This assumption can only be applied to a flow containing highly correlated regular fluctuations (see the phase-locked averaging technique in Section 11.1). Series expansion and time-averaging procedures were then applied to obtain response equations for the mean-velocity components and Reynolds stresses. Related six-position methods for an *SN* probe have since been proposed by King (1979), Janjua *et al.* (1983), and Jackson and Lilley (1986). In low turbulence intensity flows, where the difference between instantaneous and time-mean values are small, such a method can probably give reasonably accurate values for the mean-velocity components. However, the amplitude and phase relationships between the fluctuating components u, v, and w vary with time in a turbulent flow. Therefore, even if empirical correlation factors are introduced, it is unlikely that this type of signal-analysis approach can be developed into an accurate, general method for Reynolds-stress evaluations.

6.2 MULTIPOSITION X-PROBE TECHNIQUES

An X-probe can also be used for multiposition measurements, provided it can be rotated around its stem. If the probe-stem is aligned with the mean-flow direction and the turbulence intensity is low, then simultaneous measurements of two velocity components can be obtained at any roll-angle position. Placing the probe sequentially in the (x, y)- and (x, z)-plane will provide measurements of $(\bar{U} + u, v)$ and $(\bar{U} + u, w)$, from which the values of $\bar{U}$, $\overline{u^2}$, $\overline{v^2}$, $\overline{w^2}$, $\overline{uv}$, and $\overline{uw}$ can be calculated. The shear stress $\overline{vw}$ can be evaluated from measurements and response equations corresponding to roll angles of $\gamma = \pm 45°$ (Bradshaw and Terrell 1969; Johnston 1970; Mojola 1974; Cutler and Bradshaw 1991).

However, it is necessary to consider the response of each wire in the X-probe to a three-dimensional flow field if it is not possible to approximately align the probe stem with the mean flow direction (that is, if the values of $\bar{V}/\bar{U}$ and $\bar{W}/\bar{U}$ cannot be assumed to be small) or if the turbulence intensity is high. In both cases the simple two-dimensional techniques for X-probes presented in Section 5.4 cannot be applied; instead a multiposition X-probe method with response equations for $\bar{V}_{e1}$, $\overline{v_{e1}^2}$ and $\bar{V}_{e2}$, $\overline{v_{e2}^2}$ for the two wires should be used. Since the fluctuating signals from the two wires can be measured simultaneously, response equations can also be derived for $\overline{(v_{e1} + v_{e2})^2}$ and $\overline{(v_{e1} - v_{e2})^2}$ (Elsenaar and Boelsma 1974; Müller and Krause 1979). An X-probe can therefore provide four different response equations for the Reynolds stresses for each value of γ. A detailed study of the accuracy of turbulence measurements with this method has been reported by Müller (1982*b*).

6.3 TRIPLE-SENSOR (3*W*) PROBES

A probe containing three wire or film elements can provide information about the complete instantaneous velocity vector at a point. A large number of different probe configurations have been developed, and Lakshimina-rayana (1982*a*) has given a review of existing triple-sensor probes and their related data-reduction techniques. Of the commercially available probes, two types have been designed for minimum aerodynamic-disturbance effects; namely the TSI cylindrical hot-film probe 1299-20-18 and the 'claw-type' DANTEC 55P91 illustrated in Figs 6.3 and 6.4, respectively. For both probes the three sensor elements are nominally mutually orthogonal.

6.3.1 Calibration of a 3*W*-probe

Most 3*W*-probe techniques are based on the effective-velocity concept. The effective velocity for each wire V_{ei} ($i = 1$–3) is obtained from the measured anemometer voltages, E_i, using one of the calibration equations $E_i = F(V_{ei})$ described in Section 4.4. The power-law relationship of eqn (3.1),

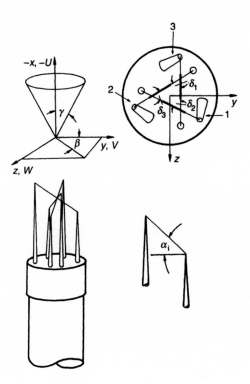

Fig. 6.3. The TSI triple-sensor probe geometry and coordinate systems. (From Lekakis *et al.* 1989.)

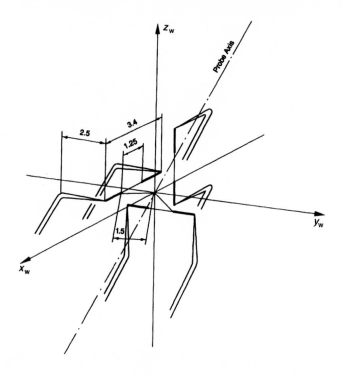

Fig. 6.4. The wire and prong configuration in the DISA 55P91 claw-type $3W$-probe. (From Gaulier 1977.)

$$E^2 = A + BV_e^n,$$

is selected in many applications. The effective velocity for each wire is normally expressed in terms of the Jørgensen equation

$$V_e^2 = U_N^2 + k^2 U_T^2 + h^2 U_B^2,$$

where (U_T, U_B, U_N) are the velocity components in a coordinate system aligned with the wire under consideration. The objective of the calibration procedure is to determine the values of A, B, n, k, and h for each wire. Complex calibration procedures for triple-wire probes in air have been presented by Huffman (1980), Skinner and Rae (1984), Lakshminarayana and Davino (1980), and Gieseke and Guezennec (1993), but the values of k and h are often determined using techniques related to eqns (4.2) and (5.4) for the calibration of a single wire. These and most other calibration techniques require accurate measurements of the yaw, α, and pitch, β, angles (see Fig. 3.1) for each wire. Detailed calibrations of the DANTEC 55P91 triple wire probe in air have been carried out by Jørgensen (1971*b*), who quoted typical values of $k = 0.15$ and $h = 1.02$. These results are in close agreement with the values $k = 0.20$. and $h = 1.01$. obtained by Chew and Ha (1988). Also, the results of both these investigations are very similar to the values obtained for k and h for plated single-wire probes, as described in Chapters 4 and 5. The values of k and h have been observed to be very sensitive to errors in the yaw and pitch angles. Uncertainties in these angles may explain the significantly different values for k and h quoted in some investigations; for example, Andreopoulos (1983*a*) found a similar value for k of 0.15, but his calculated value for h was as high as 1.12. If probes with unplated wires or cylindrical hot-film elements are used, then the values of k and h may deviate significantly from the values quoted above. For the TSI cylindrical hot-film probe 1299-20-18, the typical values in air are quoted as $k = 0.2$. and $h = 1.08$. (Russ and Simon 1990). Calibration experiments have also been performed in water with a DANTEC 55R91 triple cylindrical hot-film probe by Gourdon *et al.* (1981), who quoted values for k in the range 0.17–0.28.

6.3.2 Data-acquisition criteria and procedures for a 3*W*-probe

To be able to measure the velocity vector at a point, the signals from the three anemometers connected to the 3*W*-probe must be acquired simultaneously by the A/D unit of the digital-measurement system used. This is often achieved by placing a sample-and-hold device on the input side of the three channels of the A/D unit.

In an investigation by Moffat *et al.* (1978) the sampling rate, SR, was determined by the stringent criteria of preserving the waveform and relative phase between the three velocity components up to 10 kHz. This requires a sampling rate of 150 kHz for each channel. However, for most turbulent flow studies it is not necessary to apply such a demanding sampling-rate criterion, particularly if the primary requirements are mean-velocity components and Reynolds stresses. Even for spectral measurements, it is usually not necessary to sample much faster than twice the highest frequency of interest in conjunction with the use of an anti-aliasing filter, as described in Chapter 12. Consequently, a three channel A/D unit with a maximum sampling rate of about 50 kHz per channel would be adequate for most turbulent flow studies, except for high-speed compressible flows.

6.3.3 Signal-analysis methods for 3*W*-probes

Before digital computers with fast multi-channel data-acquisition facilities became available, it was necessary to evaluate the time-averaged voltage values, $\bar{E}_i$, $\overline{e_i e_j}$ ($i = 1\text{-}3$, $j = 1\text{-}3$), from the direct wire voltages E_i. Gorton and Lakshminarayana (1976) have presented a technique for converting these measurements into the corresponding mean velocities and Reynolds stresses. This technique required the probe axis to be closely aligned with the mean-velocity vector. A similar method has been described by Gaulier (1977); and Chew and Simpson (1988) have developed an evaluation procedure, which relates $\bar{V}_{ei}$, $\overline{v_{ei}v_{ej}}$ to the corresponding mean-velocity components and Reynolds stresses.

In investigations by Yavuzkurt *et al.* (1977), Moffat *et al.* (1978), and Frota and Moffat (1983), the data analysis was carried out by an analog system, due to their stringent waveform criteria. However, modern A/D systems are now capable of dealing with any realistic data-acquisition criteria, and the following analysis will be based on digital-data acquisition and analysis methods. Two different $3W$-probes and related signal-analysis techniques will be discussed.

6.3.3.1   The TSI triple-hot-film probe 1299–20–18

The probe configuration used for the TSI 1299–20–18 probe is a development of the compact triple-hot-wire probe used by Müller (1983, 1992). This TSI probe contains three nominally mutually orthogonal, hot-film sensors, each with an active length $\ell = 1$ mm and a diameter $d = 50\,\mu m$. Parallel offset prongs are used to minimize the aerodynamic interference resulting in an overall diameter of the probe head of 4.6 mm. The sensing elements are contained within a 2.6 mm diameter sphere, which is smaller than many other triple-sensor probes. The geometry of this triple-sensor probe is defined in Fig. 6.3 in terms of the angles, δ_i ($i = 1$–3), formed by the projections of the sensors on a plane normal to the probe axis, and in terms of the yaw angles, α_i. Signal-analysis procedures for this probe have been presented by Buddhavarapu and Meinen (1988), by Lekakis *et al.* (1989), and by Russ and Simon (1990). A related signal-analysis procedure has also been presented by Chang *et al.* (1983) for a $3W$-probe containing an X-wire probe and a third sensor inclined at 45° with respect to the X-wire plane, The TSI probe has been designed to have orthogonal sensors, but the signal analysis procedure by Lekakis *et al.* (1989), outlined below, allows for a nonorthogonal probe configuration caused by misalignment during probe manufacture. The selected space-fixed coordinate system is attached to the probe with the y-axis normal to the plane of sensor 3 and its supporting prongs, and the x-axis parallel to the probe axis. Thus, $\beta = 30°$ or $210°$, $150°$ or $330°$, and $270°$ or $90°$ correspond to the planes containing sensors 1, 2, and 3, respectively, for a probe geometry with $\delta_1 = \delta_2 = \delta_3 = 60°$.

The effective velocity for each wire V_{ei} ($i = 1$–3) can be expressed, in terms of the space-fixed components (U, V, W) of the instantaneous velocity vector, V, and the probe geometry as:

$$V_e^2 = [U\cos\alpha_1 + (W\cos\delta_2 + V\sin\delta_2)\sin\alpha_1]^2$$
$$+ k_1^2[U\sin\alpha_1 - (W\cos\delta_2 + V\sin\delta_2)\cos\alpha_1]^2$$
$$+ h_1^2(V\cos\delta_2 - W\cos\delta_2)^2 \tag{6.25a}$$

$$V_{e2}^2 = [U\cos\alpha_2 + (W\cos\delta_1 - V\sin\delta_1)\sin\alpha_2]^2$$
$$+ k_2^2[U\sin\alpha_2 - (W\cos\delta_1 - V\sin\delta_1)\cos\alpha_2]^2$$
$$+ h_2^2(V\cos\delta_1 + W\sin\delta_1)^2, \tag{6.25b}$$

$$V_{e3}^2 = (U\cos\alpha_3 - W\sin\alpha_3)^2 + k_3^2(U\sin\alpha_3 + W\cos\alpha_3)^2$$
$$+ h_3^2 V^2. \tag{6.25c}$$

This formulation allows for variations in the yaw and pitch coefficients k_i and h_i for the three sensors. The (U, V, W) components of the instantaneous velocity vector, V, can be expressed in terms of its magnitude, $\tilde{V}$, and the polar and azimutal angles, β and γ, as defined in Fig. 6.3:

$$U = \tilde{V}\cos\gamma, \tag{6.26a}$$

$$V = \tilde{V}\sin\gamma\cos\beta, \tag{6.26b}$$

$$W = \tilde{V}\sin\gamma\sin\beta. \tag{6.26c}$$

By substituting eqns (6.26a–c) into eqns (6.25a–c), the relationship between the measured effective velocities and the instantaneous velocity vector (identified in terms of $\tilde{V}$, β, and γ) becomes

$$V_{e1}^2 = \tilde{V}^2\{\sin^2\gamma[G_{11}\sin^2(\beta+\delta_2)+G_{21}] + G_{31}\sin(2\gamma)\sin(\beta+\delta_2)+G_{41}\}, \tag{6.27a}$$

$$V_{e2}^2 = \tilde{V}^2\{\sin^2\gamma[G_{12}\sin^2(\beta-\delta_1)+G_{22}] + G_{32}\sin(2\gamma)\sin(\beta-\delta_1)+G_{42}\}, \tag{6.27b}$$

$$V_{e3}^2 = \tilde{V}^2[\sin^2\gamma(G_{13}\sin^2\beta+G_{23}) - G_{33}\sin(2\gamma)\sin\beta+G_{43}], \tag{6.27c}$$

where

$$G_{1i} = k_i^2 - h_i^2 + (1 - k_i^2)\sin^2\alpha_i,$$
$$G_{2i} = h_i^2 - 1 + (1 - k_i^2)\sin^2\alpha_i,$$
$$G_{3i} = 0.5(1 - k_i^2)\sin(2\alpha_i),$$
$$G_{4i} = 1 - (1 - k_i^2)\sin^2\alpha_i,$$

where $i = 1$–3.

Equations (6.27a–c) are linear in $\tilde{V}$ and nonlinear with respect to γ and β. It is therefore possible to separate the direction from the magnitude of the velocity vector leaving only two unknowns, the angles β and γ, to be resolved at each instant in time. Equations (6.27a–c) can be rewritten in the form

$$A \equiv \frac{V_{e1}^2}{V_{e3}^2} = f_1(\beta,\gamma),$$

$$B \equiv \frac{V_{e2}^2}{V_{e3}^2} = f_2(\beta,\gamma),$$

and the solution of these two equations yields β and γ. The magnitude, $\tilde{V}$, can then be obtained from eqn (6.27c). By further trigonometric manipulation, a fourth-order polynomial equation — in terms of a single variable $\omega = \tan \beta$ — can be obtained. The Laguerre method of finding the zeros of a polynomial can be used to find all the possible solutions for $\tan \beta$. The corresponding γ values can be evaluated from a related equation for $\tan \gamma$. The sets of angles (β, γ) angles thus obtained specify the directions of all the instantaneous velocity vectors which would produce the observed effective velocities. The correct set of (β, γ) angles can be identified from a (β, γ) 'uniqueness domain'. It can be demonstrated, for each value of A and B, that only one (β, γ) solution falls inside this uniqueness domain. This signal-analysis procedure is rather laborious, but it has the advantage that it can be applied to nonorthogonal probes.

For probes with orthogonal or nearly orthogonal sensors, a simpler solution procedure can be developed if the pitch coefficients h_1, h_2, and h_3 are taken to equal unity, as shown by Russ and Simon (1990) for the TSI probe. A general analysis procedure for orthogonal sensors will be demonstrated for the DANTEC 55P91 triple-wire probe.

6.3.3.2 The DANTEC 55P91 triple-wire probe

Dantec 55P91 triple-wire probes uses gold-plated tungsten wires which are 3.2 mm long with a 1.2 mm sensitive length. The prongs, which are in a claw configuration, are attached perpendicularly to the wire elements, as shown in Fig. 6.4, to minimize the aerodynamic-interference effect. The three wires are nominally mutually orthogonal and they are inclined at a yaw angle, α, of 54.7° relative to the probe axis. The three wire elements lie inside a sphere of 3 mm diameter, and the overall diameter of the probe head is about 6 mm, which restricts the use of this probe type close to walls.

The signal from the probe is expressed in terms of the Jørgensen equation for the effective velocity; and it is assumed that the three wires form an orthogonal coordinate system, so that the tangential velocity component for one wire is normal for another wire and binormal for the third (Fingerson 1968; Jørgensen 1971b; Gaulier 1977; Yavuzkurt et al. 1977; Moffat et al. 1978; Frota and Moffat 1983a; Andreopoulos 1983a). The following notation has been adopted: E_1, E_2, and E_3 are the instantaneous voltages from the anemometers connected to wires 1, 2, and 3; and the corresponding instantaneous effective velocities are V_{e1}, V_{e2}, and V_{e3}. A wire-fixed coordinate system (x_w, y_w, z_w) is aligned with wires 1, 2, and 3, as shown in Fig. 6.5, and the corresponding instantaneous wire-fixed velocity components are denoted (U_w, V_w, W_w). In this coordinate system the following three simultaneous equations apply

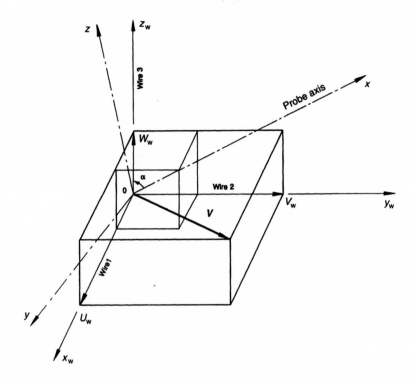

Fig. 6.5. The wire-fixed and space-fixed coordinate axis for the DISA 55P91 triple wire probe. (From Gaulier 1977.)

$$\left(\frac{E_1^2 - A_1}{B_1}\right)^{2/n_1} = V_{e1}^2 = k_1^2 U_w^2 + V_w^2 + h_1^2 W_w^2, \qquad (6.28a)$$

$$\left(\frac{E_2^2 - A_2}{B_2}\right)^{2/n_2} = V_{e2}^2 = h_2^2 U_w^2 + k_2^2 V_w^2 + W_w^2, \qquad (6.28b)$$

$$\left(\frac{E_3^2 - A_3}{B_3}\right)^{2/n_3} = V_{e3}^2 = U_w^2 + h_3^2 V_w^2 + k_3^2 W_w^2, \qquad (6.28c)$$

This formulation allows for different values of k and h for the three wires. Equations (6.28a–c) can be written in matrix form as

$$\begin{Bmatrix} V_{e1}^2 \\ V_{e2}^2 \\ V_{e3}^2 \end{Bmatrix} = D \begin{Bmatrix} U_w^2 \\ V_w^2 \\ W_w^2 \end{Bmatrix} \qquad (6.29)$$

where

$$D = \begin{Bmatrix} k_1^2 & 1 & h_1^2 \\ h_2^2 & k_2^2 & 1 \\ 1 & h_3^2 & k_3^2 \end{Bmatrix} \tag{6.30}$$

Equations (6.28a–c) form a system of three linear equations, and the unknown instantaneous velocity components in the wire-fixed coordinate system (x_w, y_w, z_w) can be obtained by a matrix-inversion method

$$\begin{Bmatrix} U_w^2 \\ V_w^2 \\ W_w^2 \end{Bmatrix} = D^{-1} \begin{Bmatrix} V_{e1}^2 \\ V_{e2}^2 \\ V_{e3}^2 \end{Bmatrix} \tag{6.31}$$

The general form of the matrix D^{-1} has been given by Frota and Moffat (1983). However, if $k_i = k$ and $h_i = h$ ($i = 1$–3) then the matrix D^{-1} can be expressed as

$$D^{-1} = \frac{1}{\Delta} \begin{Bmatrix} k^4 - h^2 & h^4 - k^2 & 1 - k^2h^2 \\ 1 - k^2h^2 & k^4 - h^2 & h^4 - k^2 \\ h^4 - k^2 & 1 - k^2h^2 & k^4 - h^2 \end{Bmatrix} \tag{6.32}$$

where Δ is the determinant of the matrix D given by

$$\Delta = k^6 + h^6 - 3k^2h^2 + 1.$$

A unique solution of eqn (6.31) exists when the velocity vector, V, stays inside the approach octant of the (x_w, y_w, z_w) coordinate system, since the velocity components U_w, V_w, and W_w will all have positive values.

The velocity components (U_w, V_w, W_w) evaluated in the wire-fixed coordinate system (x_w, y_w, z_w) can be transformed to any other orthogonal components, if the direction cosines of the angles between the original and the transformed axis are known. For the space-fixed coordinate system (x, y, z) in Fig. 6.5, in which the x-axis is aligned with the probe stem and wire 3 (z_w-axis) is lying in the (x, z)-plane, the transformation relationship for the velocity components U, V, and W are

$$\begin{Bmatrix} U \\ V \\ W \end{Bmatrix} = N \begin{Bmatrix} U_w \\ V_w \\ W_w \end{Bmatrix} \tag{6.33}$$

with

$$N = \begin{Bmatrix} \cos 45 \cos 35.3 & \cos 45 \cos 35.3 & \cos 54.7 \\ -\cos 45 & \cos 45 & 0 \\ -\cos 45 \sin 35.3 & -\cos 45 \sin 35.3 & \cos 35.3 \end{Bmatrix} = \begin{Bmatrix} \dfrac{\sqrt{3}}{3} & \dfrac{\sqrt{3}}{3} & \dfrac{\sqrt{3}}{3} \\ -\dfrac{\sqrt{2}}{2} & \dfrac{\sqrt{2}}{2} & 0 \\ -\dfrac{\sqrt{6}}{6} & -\dfrac{\sqrt{6}}{6} & \dfrac{\sqrt{6}}{3} \end{Bmatrix}.$$

(6.34)

Consequently, provided the velocity vector remains within the approach octant of the (x_w, y_w, z_w) coordinate system, eqns (6.28a–c) and (6.31) to (6.34) can be used to obtain the values of (U, V, W) corresponding to any measured set of values of (E_1, E_2, E_3).

In some applications it may be necessary to position the $3W$-probe with a pitch angle, β, in, for example, the (x, y)-plane and rotate it by a roll angle, γ, around its stem, as shown in Fig. 6.6. In this case, an additional transformation is required (Frota and Moffat 1983)

$$\begin{Bmatrix} U \\ V \\ W \end{Bmatrix} = CN \begin{Bmatrix} U_w \\ V_w \\ W_w \end{Bmatrix},$$

(6.35)

where

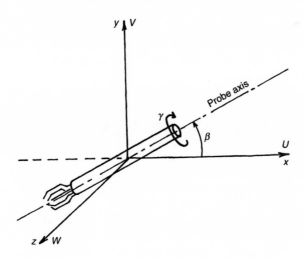

Fig. 6.6. The roll, γ, and pitch, β, angles for $3W$-probes.

$$C = \left\{ \begin{pmatrix} \cos\beta & -\sin\beta & 0 \\ \sin\beta & \cos\beta & 0 \\ 0 & 0 & 1 \end{pmatrix} \begin{pmatrix} 1 & 0 & 0 \\ 0 & \cos\gamma & \sin\gamma \\ 0 & -\sin\gamma & \cos\gamma \end{pmatrix} \right\}. \qquad (6.36)$$

The reference angles for both the pitch and roll angles are $0°$.

The above analysis assumes that the three wires are mutually orthogonal and that the yaw and pitch coefficients k and h are constants which are independent of the direction of the velocity vector. In many investigations it has been found that the yaw coefficient is a function of the yaw angle, $k(\alpha)$, and Andreopoulos (1983a) also found that his pitch parameter, h, was a function of the pitch angle. His values for h were unusually large, and to correct for this variation he applied an iterative method to the above procedure. However, provided the more common value for h of $h \simeq 1.02$ (Jørgensen 1971b; Chew and Ha 1988) is used, such an iteration should not be necessary.

Errors can also be introduced if the three wires are not mutually orthogonal. The angles between the three wires can be determined in the following way. First the yaw angles, α_i, between the three wires and the probe-stem direction are measured optically. The real angles between the three wires can then be calculated (using the notation in Fig. 6.5, Andreopoulos 1983a) from

$$\cos \widehat{X_w O Y_w} = \cos\alpha_1 \cos\alpha_2 - \tfrac{1}{2}\sin\alpha_1 \sin\alpha_2, \qquad (6.37a)$$

$$\cos \widehat{Y_w O Z_w} = \cos\alpha_2 \cos\alpha_3 - \tfrac{1}{2}\sin\alpha_2 \sin\alpha_3, \qquad (6.37b)$$

$$\cos \widehat{X_w O Z_w} = \cos\alpha_1 \cos\alpha_3 - \tfrac{1}{2}\sin\alpha_1 \sin\alpha_3. \qquad (6.37c)$$

For an ideal probe, $\alpha_1 = \alpha_2 = \alpha_3 = 54.74°$ and $\widehat{X_w O Y_w} = \widehat{Y_w O Z_w} = \widehat{X_w O Z_w} = 90°$. The manufacturers usually guarantee that the three angles are $90°$ within a tolerance of $1°$ or $2°$. However, if a probe is repaired substantially, larger deviations may occur. Andreopoulos (1983a) quotes an example of one wire forming an angle of $82°$ with the other two wires. When this non-orthogonality was not taken into account, errors of 3% for $\overline{u^2}$ and 5% for $\overline{uv}$ were reported.

It is well-known that a $3W$-probe cannot interpret the hot-wire signals uniquely in a reversing flow. This problem was resolved by Müller (1983) by adding a thermal-wake detector to the $3W$-probe. The wake detector was used to eliminate the ambiguity in the signal interpretation caused by rectification of the anemometer output, by conditionally sampling only that part of the signal that fell within the approach octant of the probe. By rotating the probe, all relevant octants could be covered and the mean-velocity components and Reynolds stresses evaluated.

6.3.4 Other 3W-probe types and signal-analysis procedures

As shown in Section 6.1.1 it is possible, for a single-wire element, to directly relate (see eqn (6.2) the instantaneous effective velocity and the space-fixed velocity components (U, V, W). Applying this relationship to the three wires in a 3W-probe, the equations for V_{ei}^2 ($i = 1$–3) can be expressed as

$$V_{ei}^2 = A_{1i}U^2 + A_{2i}V^2 + A_{3i}W^2 + A_{4i}UV + A_{5i}UW + A_{6i}VW, \quad (6.38)$$

with the coefficients A_{1i}–A_{6i} ($i = 1$–3) depending on the yaw, α_i, and roll angle, γ_i, angles for each sensor. This is the approach adapted by Paulsen (1983) for a probe containing one SN-wire ($\alpha = 0°$) and two SY-wires ($\alpha = 45°$). A Newton–Raphson iterative technique was used to obtain the instantaneous values of (U, V, W) for any evaluated set of (V_{e1}, V_{e2}, V_{e3}) values. A similar method has also been applied by Bergström and Högström (1987) to a triple-hot-film probe for meteorological-turbulence measurements.

Equation (6.38) is expressed in terms of V_e^2, and Acrivlellis (1980) has developed a related solution procedure for a probe with three SN-wires in the same plane, as shown in Fig. 6.7. In this way the prong interference is considerably reduced, since the prongs are all placed downstream. Acrivlellis showed that direct solutions can be obtained for U^2, V^2, and W^2 for each measured set of $(V_{e1}^2, V_{e2}^2, V_{e3}^2)$, which enable velocity-*magnitude* evaluations to be made inside the four octants of a half sphere. However, as pointed

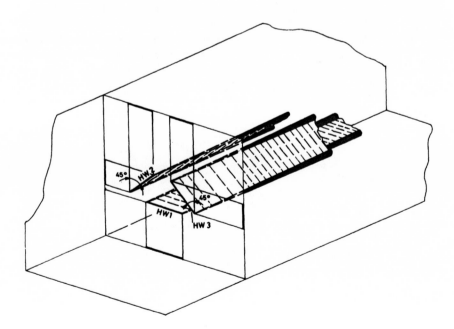

Fig. 6.7. The arrangement of the SN hot-wires in the triple-wire probe. (From Acrivlellis 1980.)

out by Müller (1982c), this technique does not allow determinations of the sign of V and W to be made. To obtain the correct sign of the cross-components V and W, it is necessary to introduce additional selection procedures, as described by Acrivlellis (1989).

Mathioudakis and Breugelmans (1985) developed a $3W$-probe technique for flow fields with large variations in the flow angle. The technique was applied in a stalled axial compressor, and in order to have a useful angle range greater than 90°, two adjacent octants were used for taking the measurements. Their technique was based on initial solutions for U^2, V^2, and W^2 in terms of $(V_{e1}^2, V_{e2}^2, V_{e3}^2)$ for a probe containing one SN-wire ($\alpha = 0°$) and two SY-wires ($\alpha = 45°$) forming a V in a plane perpendicular to the SN wire. They introduced three parameters, Y (yaw), P (pitch), and Q (velocity magnitude), and they determined (from a detailed probe calibration) the variation in Y, P, and Q with the yaw and pitch angles. The probe was calibrated for yaw angles up to $\pm 70°$ and for pitch angles up to 35°. The calculated (initial) values for U^2, V^2, and W^2 were used as input values to the three parameters Y, P, and Q, and the response equations had two yaw solutions, one for each of the two selected octants. If the mean flow is known to lie within one octant only, the probe is placed accordingly. For large flow-angle variations, when the flow vector occurs in both octants, it is necessary to carry out measurements with two different probe positions in order to identify the correct octant. Since the two measurements were carried out sequentially, a phase-averaging technique was employed to evaluate the mean-flow quantities.

Another commonly used triple-wire configuration is shown in Fig. 6.8. This probe contains three mutually perpendicular sensors placed symmetrically with respect to the probe axis. The nominal yaw angle, α, between each wire and the probe axis is 54.7°. The disadvantage of this probe is that

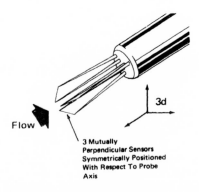

Fig. 6.8. A triple-wire probe with three central prongs. (Published with the permission of TSI Inc.)

it contains three central prongs, which results in significant aerodynamic prong interference. For this reason, related probes with spatially offset sensors have been developed to minimize the aerodynamic interference (DANTEC 55P91 and TSI 1299–20–18, see Figs 6.3 and 6.4).

Some analysis techniques evaluate the velocity vector in the space-fixed coordinate system in terms of its magnitude and two or more spatial angles. In the method of Lekakis et al. (1989), described in Section 6.3.3.1, the effective velocity is related to the velocity vector in terms of its magnitude and its polar and azimutal angles (see Fig. 6.3). Butler and Wagner (1982, 1983), using a probe of the type shown in Fig. 6.8, have presented a solution procedure for V in terms of equations for the magnitude, $\tilde{V}$, and the flow angles, θ_i ($i = 1$–3), which the instantaneous velocity vector makes with the three sensors. Based on the effective velocities, V_{ei}, from the three wires, they introduced three parameters for, respectively, the velocity magnitude, the pitch angle, and the yaw angle. The instantaneous velocity vector was determined from quadratic relationships for $\tilde{V}$ and θ_i ($i = 1$–3), in terms of these three parameters. The coefficients in these equations were obtained by least-squares curve-fitting of the calibration data.

Sammler and Kitzing (1990) have also used a probe of the type shown in Fig. 6.8. They related the velocity vector, specified by its magnitude, $\tilde{V}$, and the yaw, α, and pitch, β angles for the probe, directly to the measured anemometer voltage in the form

$$E_i^2 = A_i + B_i(\alpha, \beta)\tilde{V}^n. \tag{6.39}$$

The functions $B_i(\alpha, \beta)/B_i(0\,^\circ, 0\,^\circ)$ ($i = 1$–3) were determined by direct calibration. For signal-analysis purposes, the calibration results were combined as B_1/B_2 and B_1/B_3, and a search procedure was applied to identify the correct values of (α, β). The corresponding value of $\tilde{V}$ was then determined from eqn (6.39).

6.4 FOUR-WIRE PROBES

It has been suggested in a few investigations that a four-wire (4W) probe can increase the uniqueness domain of the velocity-vector solution in comparison with the domain of a 3W-probe. The four-wire probe used by Pailhas and Cousteix (1986) is similar to the modified Kovasznay-type vorticity probe used by Vukoslavčević and Wallace (1981) shown in Fig. 10.4. This probe has eight prongs placed in pairs at the corners of a small rectangle. The hot-wire response equation for each wire was assumed to be a simple power law of the form,

$$E^2 = E_0^2 + BV_e^n,$$

where E_0 is the voltage at zero velocity. The Jørgensen equation for the effective velocity, eqn (3.2), was applied to each wire assuming $k = 0$ and

$h = 1$. Provided each wire is inclined the same angle, α, relative to the probe axis and that the velocity components (U, V, W) are uniform across the probe volume, then it follows from eqn (10.10) that the response equation for the four wires can be expressed as

$$V_{e1}^2 = (U \cos \alpha - W \sin \alpha)^2 + V^2, \tag{6.40a}$$

$$V_{e3}^2 = (U \cos \alpha + W \sin \alpha)^2 + V^2, \tag{6.40b}$$

$$V_{e2}^2 = (U \cos \alpha + V \sin \alpha)^2 + W^2, \tag{6.40c}$$

$$V_{e4}^2 = (U \cos \alpha - V \sin \alpha)^2 + W^2. \tag{6.40d}$$

Solution procedures to eqns (6.40a–d) have been presented by Pailhas and Cousteix (1986) and by Müller (1992) in terms of the magnitude and two spatial angles of the velocity vector.

Samet and Einav (1987) constructed a probe with four slanted wires to form two V-configurations 1, 3 and 2, 4, as shown in Fig. 6.9(a). They carried out a combined velocity, yaw. and pitch calibration using seven velocities in the range 2.18–53.55 m s^{-1} and fifteen yaw, α, and fifteen pitch,

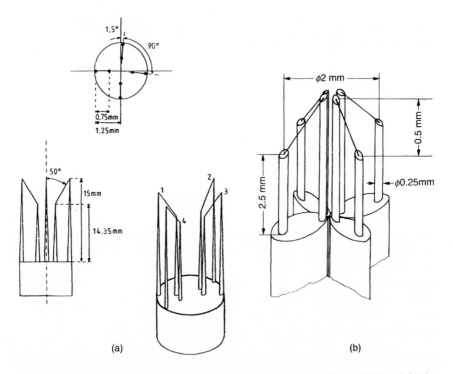

Fig. 6.9. Four-wire probes: (a) the configuration used by Samet and Einav (1987) and (b) the probe developed by Döbbeling *et al.* (1990*b*.)

β, angles with each calibration angle equally spaced in the range $-37.8°$ to $37.8°$. The yaw plane of the probe was defined by the plane of wire 1 and wire 3 and the pitch plane by wire 2 and wire 4. The calibration data was subdivided into fifteen pitch orientations for wires 1 and 3 and into fifteen yaw orientations for wires 2 and 4, and analysed in the form

$$\alpha|_{\beta\,=\,\text{const}} = \alpha(E_3, E_1), \tag{6.41a}$$

$$\beta|_{\alpha\,=\,\text{const}} = \beta(E_4, E_2). \tag{6.41b}$$

Similar relationships were developed for the magnitude, $\tilde{V}$, of the velocity vector. Each equation corresponds to a surface. Samet and Einav found unique relationships in all cases, and they applied surface curve fits to all thirty equations. The velocity vector, V, specified in terms of $\tilde{V}$, α, and β corresponding to a set of measurements (E_1, E_2, E_3, E_4) was obtained by applying a look-up matrix method (see Section 5.4.3.3) to all the (E_1, E_3) and (E_2, E_4) surface relationships in eqns (6.41a-b). The two sets of surface equations provided discrete possible solutions in the form $\alpha = \alpha_1(\beta)$ and $\beta = \beta_2(\alpha)$. By curve fitting these relationships, the correct values of α and β were obtained as the intersection point of these two curves.

Döbbeling *et al.* (1990a, b, 1992) used the 4W-probe shown in Fig. 6.9(b), and developed a simplified calibration and analysis procedure. When the magnitude of the velocity vector $\tilde{V}$ was higher than $5\,\text{m s}^{-1}$, they found that the yaw and pitch dependence of the probe was independent of the velocity. The effective velocities, V_{ei} ($i = 1$-4), from the four wires could therefore be expressed in the form

$$V_{ei}^2 = \tilde{V}^2 g_i(\alpha, \beta). \tag{6.42}$$

The yaw and pitch directional function $g_i(\alpha, \beta)$ for each wire was determined by calibration, and the effective velocities were calculated from the measured voltages using a ninth-order polynomial. Using an initial guess for the values of $\tilde{V}$, α, and β, the correct values corresponding to a set of E_i-values ($i = 1$-4) were obtained by an iterative procedure which minimized the error squared, ε,

$$\varepsilon = \sum_{i=1}^{4} \left[\frac{1}{\tilde{V}^2} V_{ei}^2 - g_i(\alpha, \beta) \right]^2. \tag{6.43}$$

Comparative evaluations showed that the 4W-probe gave a better angular resolution than a 3W-probe over an increased yaw and pitch range.

To obtain a good spatial resolution, Pompeo and Thomann (1993) developed the miniature probe shown in Fig. 6.10 consisting essentially of one X-configuration (wires 1 and 3) and a V-configuration (wires 2 and 4). For each sensor configuration the look-up matrix method described in Section 5.4.3.3 for the two-dimensional response of an X-probe was extended to three dimensions by obtaining (E_1, E_2) calibration maps at different

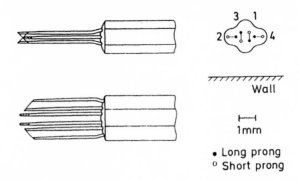

Fig. 6.10. A miniature four-wire probe. (From Pompeo and Thomann 1993.)

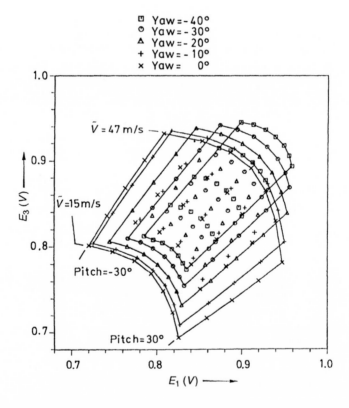

Fig. 6.11. The calibration fields for the X-array configuration of the probe in Fig. 6.10. (From Pompeo and Thomann 1993.)

yaw angles as shown in Fig. 6.11. Similar calibration maps apply to the
V-configuration, and an extended look-up table data reduction method was
applied for measurement with this probe.

The disadvantages of 4W-probes are their complex geometry, and the
necessity for accurate laborious calibration procedures and computer-
intensive solution procedures.

6.5 UNCERTAINTY ANALYSIS, THREE-VELOCITY-COMPONENT EVALUATIONS

The multiposition single-wire method In the comparative investigation by
Pierce and Ezekwe (1976) and by Ezekwe *et al.* (1978), the results obtained
by the multiposition SY-probe method were in general agreement with data
from an X-probe, except for occasional erratic values. The error analysis
by Yavuzkurt (1984) indicates that the error in the Reynolds stresses for
the multiposition method is about twice that of an X-probe or of a 3W
probe. Two aspects can contribute significantly to the uncertainty in the
evaluated turbulent quantities. Firstly, the method is based on sequential
measurements at a number of roll-angle positions, and the evaluated quantities are therefore very sensitive to changes in the flow conditions or in the
hot-wire calibration during the measurements. Secondly, for measurements
in three-dimensional mean-flow fields, if only six roll positions are used
then it is necessary to pay attention to the coefficient matrix of the response
equations (Müller 1987). One solution to the latter problem is to use a larger
number of roll angles in conjunction with a least-squares curve-fitting procedure (De Grande and Kool 1981; Kuroumaru *et al.* 1982; Russ and Simon
1991). An alternative method is first to align the probe with the mean-
flow direction (see Section 6.1.2.1) and to combine the multiposition
SY-probe technique with a single SN-probe measurement. In this case,
separate solution procedures exist for $(\overline{u^2}, \overline{v^2}, \overline{uv})$ and $(\overline{u^2}, \overline{w^2}, \overline{uw})$,
which increases the accuracy of the measured Reynolds stresses.

The multiposition X-probe technique This method reflects most of
the features for the single-wire method above. Firstly, individual response
equations are derived for the two wires in the X-probe. Secondly, since
signals can be obtained simultaneously from the two wires, response equa-
tions can also be developed for $\overline{(v_{e1} + v_{e2})^2}$ and $\overline{(v_{e1} - v_{e2})^2}$, thus providing
four response equations for each roll-angle position of the X-probe. The
solution procedures and the related uncertainty analysis have been
described by Müller (1982b) and Cutler and Bradshaw (1991).

3W-probes Andreopoulos (1983b) has evaluated statistical errors in measured Reynolds stresses associated with the probe geometry and the turbulence intensity, Tu. The yaw angle, α, between each wire and the probe axis
was selected as a variable parameter in the range $35° \leqslant \alpha \leqslant 90°$. The

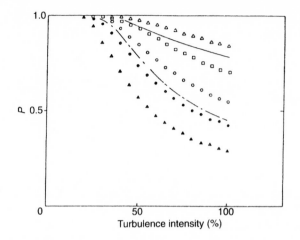

Fig. 6.12. Probability distributions, P, for various wire angles of a $3W$-probe. (▲) $\alpha = 35°$, (●) $\alpha = 45°$, (○) $\alpha = 55°$, (□), $\alpha = 70°$, (△) $\alpha = 90°$ (From Andreopoulos 1983b); (——) a pulsed wire with $\alpha = 70°$, (— — —) a pulsed wire with $\alpha = 45°$. From (Castro (1986.)

analysis evaluated the magnitude of the errors resulting from high angular excursions of the velocity vector. By assuming a joint Gaussian probability density function for the fluctuating velocity field, Andreopoulus calculated the probability, P, of the velocity vector remaining within the approach cone of the $3W$-probe. The variation in P with the turbulence intensity, Tu, and the wire angle, α, is shown in Fig. 6.12 for probes aligned with the mean-flow direction. For a $3W$-probe with orthogonal wires ($\alpha = 54.7°$), the probability, P, is observed to be over 99 per cent for turbulence intensities up to 20 per cent, after which it starts to decrease. Andreopoulos (1983b) concludes that even in the case where the mean-flow direction is not known beforehand, mean-velocity components and turbulence intensities can be measured quite accurately for turbulence intensities up to 30 per cent. This analysis ignores the flow disturbance caused by the probe. Frota *et al.* (1983) have studied the flow disturbance caused by a DANTEC 55P91 probe placed in a uniform flow. For an aligned probe the flow deflection was zero, and the distortion in the flow passing the probe was only ~2° when the probe's pitch angle (see Fig. 6.6) was 20°. This observation was consistent with measured (spurious) errors in V and W.

In the uncertainty analysis above the fluctuating velocity field was also assumed to be spatially uniform over the volume containing the three sensing elements. For measurements close to a wall this criterion cannot be satisfied. The uncertainties in measurements obtained with the claw-type $3W$-probe (DANTEC 55P91) have been studied in detail by qualification tests in a two-dimensional flow channel (Yavuzkurt *et al.* 1977; Moffat *et al.*

1978; Frota and Moffat 1983). The probe was mounted in a two-axis probe holder so that it could be rotated around the axis (roll angle γ) and also tilted (pitch angle, β) against the approach flow (see Fig. 6.6), and it could be traversed to several different distances from the wall. The performance of the $3W$-probe system was evaluated by a complete boundary-layer traverse, including the zero-shear region (at the centre of the channel) and the high-shear region (near the wall). The results from the $3W$-probe were compared with results from an SN-probe, a pitot-static tube, and the linear shear stress distribution calculated from the pressure gradient along the channel. An important operational feature is the probe's ability to operate in fluctuating three-dimensional turbulent flows of unknown flow direction. The boundary-layer qualification tests showed that the accuracy of a $3W$-probe is a function of both the roll and the pitch angle, and the probe does not respond equally to positive and negative pitch. In three-dimensional flows where the probe axis makes large angles with the unknown flow direction, the turbulence quantities may have large errors, particularly in high velocity gradient regions. However, provided the probe axis is approximately aligned with the mean-flow direction, the errors can be reduced considerably. Selecting $\gamma = 90°$ as the best overall choice for turbulent measurements, it was found that $\bar{U}$ and $\overline{uv}$ could be measured within 2 per cent and 4 per cent, respectively, provided the velocity vector remained within a 20° half-apex angle around the probe axis.

7

TEMPERATURE EFFECTS: CORRECTION METHODS FOR DRIFT IN THE FLUID TEMPERATURE; MEASUREMENTS OF FLUID TEMPERATURE FLUCTUATIONS

7.1 INTRODUCTION

In many fluid flows the temperature of the fluid, T_a, may vary with time. This change may take the form of either a slow drift caused by, for example, electrical heat from the fan motor in a closed-loop wind-tunnel or as temperature fluctuations in a nonisothermal flow. In both cases the output voltage from a hot-wire placed in the flow will be modified by the variation in T_a. This chapter deals with (i) correction procedures for CT anemometer signals caused by drift in the ambient-fluid temperature, and (ii) measurements of the fluctuating ambient-fluid temperature, $\theta \, (= T_a - \bar{T}_a)$. To implement the first objective, the velocity and temperature dependence of the anemometer signal must first be established.

7.2 THE VELOCITY AND TEMPERATURE DEPENDENCE OF CT ANEMOMETER SIGNALS

Most of the proposed velocity and temperature response equations for CT hot-wire probes can be classified into three categories (Freymuth 1970):

1. In the simplest method, the heat transfer from the probe is assumed to be proportional to a product of the temperature difference $T_w - T_a$ and a function of the velocity, where T_w is the temperature of the heated wire. Expressed in terms of the anemometer output voltage, E,

$$E^2 = f(U)(T_w - T_a). \tag{7.1}$$

2. The convective heat transfer is expressed in a nondimensional form involving a relationship between the Nusselt number, Nu, the Reynolds number, Re, and the Prandtl number, Pr.

3. Direct calibration of the variation in the anemometer voltage, E, with the velocity, U, and the fluid temperature, T_a, for a given hot-resistance setting, R_w.

For practical *HWA*, method 1 involves the least amount of calibration and it is relatively easy to use for data analysis. Method 3 is the most complex to implement, but the calibration data obtained by this method can reveal to what extent method 1 (or 2) can be used to evaluate the temperature and velocity sensitivity of a hot-wire probe operated in the *CT* mode.

7.2.1 Direct velocity and temperature calibration of a *CT* hot-wire probe

The most accurate way of establishing the velocity and temperature sensitivity of a *CT* hot-wire probe, operated at a fixed hot resistance, R_w, is to measure the anemometer output voltage, E, as a function of the velocity, U, and fluid temperature, T_a. This type of calibration is often carried out by performing a velocity calibration at a number of different fluid temperatures. Such a set of calibration data are shown in Fig. 7.1.

Calibration plots in the form $E = f(U)_{T_a = \text{const}}$ (Fig. 7.1) or $E = f(T_a)_{U = \text{const}}$ have been presented by: Pessoni and Chao (1974) for $U = 6\text{--}30\,\text{m s}^{-1}$,

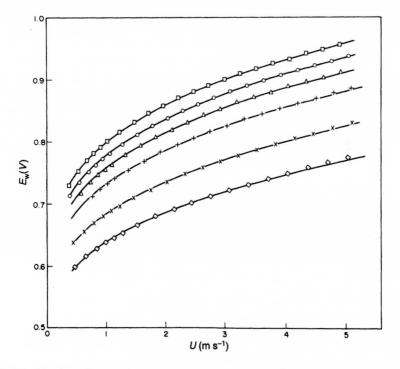

Fig. 7.1. The dependence of the hot-wire voltage, E_w, on the air velocity, U, for the following ambient fluid temperatures, T_a: ($\square$) 283 K, ($\bigcirc$) 293 K, ($\triangle$) 303 K, ($+$) 313 K, ($\times$) 333 K, ($\diamondsuit$) 353 K. $T_w = 473$ K. (From Koppius and Trines 1976. Reprinted with kind permission of Elsevier Science Ltd from the *International Journal of Heat and Mass Transfer*.)

$T_a = 22\text{--}60\,°C$; Koppius and Trines (1976) for $U = 0.5\text{--}5\,\mathrm{m\,s^{-1}}$, $T_a = 10\text{--}80\,°C$; Fiedler (1978) for $U = 2\text{--}20\,\mathrm{m\,s^{-1}}$, $T_a = 21\text{--}50\,°C$; Dekeyser and Launder (1983) for $U = 7.6\text{--}15.4\,\mathrm{m\,s^{-1}}$, $T_a = 25\text{--}35\,°C$; Lemieux and Oosthuizen (1984) for $U = 0.5\text{--}4\,\mathrm{m\,s^{-1}}$, $T_a = 25\text{--}55\,°C$; and Bremhorst (1985) for $U = 1.5\text{--}35\,\mathrm{m\,s^{-1}}$, $T_a = 20\text{--}80\,°C$.

The functional form of the calibration data may be evaluated as

$$E = F(U, T_a)_{T_w = \text{const}}, \tag{7.2}$$

but more commonly the calibration data have been interpreted in the form

$$E = F(U, T_w - T_a)_{T_w = \text{const}}. \tag{7.3}$$

To evaluate T_w it is necessary to measure the hot resistance, R_w, and this requires a knowledge of the probe and cable resistance, R_L (see Section 2.2.1.2). The mean wire temperature, T_w, corresponding to R_w is normally determined by using eqn (2.27), and it may be necessary to determine the temperature coefficient, α_0, for each probe used. To obtain a very accurate value of T_w at high overheat ratios, it may be necessary to apply eqn (2.24).

For SN-probes it was shown in Section 4.5 that a simple power law will be a very good approximation to the velocity-calibration data, obtained at a constant value of T_a, provided the calibration constants are determined by a least-squares curve-fit. This procedure has been applied by Koppius and Trines (1976) to their velocity calibration data using the wire-voltage relationship

$$\frac{E_w^2}{R_w(R_w - R_a)} = A + BU^n, \tag{7.4}$$

which is obtained by combining eqns (2.35) and (2.36). This curve-fitting procedure gave the most accurate results, but A, B, and n were found to be functions of T_a. When a constant value of n ($= 0.45$) was selected A and B also became constants, and the increase in the uncertainty is insignificant for most practical HWA applications.

A similar approach was adopted by Lemieux and Oosthuizen (1984). They expressed their calibration relationship in the form

$$E^2 = A^* + B^*U^n, \tag{7.5}$$

and for each value of T_a they determined the values of A^*, B^*, and n by a least-squares curve-fitting procedure. In their subsequent signal analysis the optimum value for n was selected as being the average value from their four calibration curves. The corresponding calibration coefficients A^* and B^* (see Fig. 7.2(a,b)) were found to vary linearly with T_a:

$$A^* = A_1 + A_2 T_a, \tag{7.6a}$$

$$B^* = B_1 + B_2 T_a. \tag{7.6b}$$

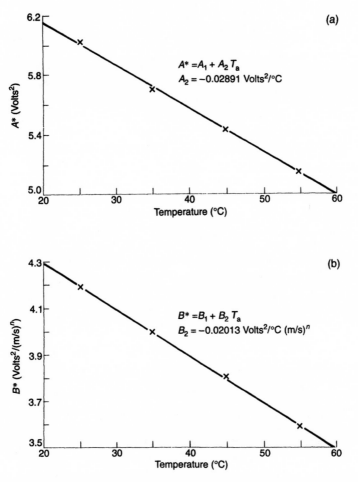

Fig. 7.2. The variation of the calibration parameters (a) A^* and (b) B^* in eqn (7.6a-b) with the fluid temperature, T_a. (From Lemieux and Oosthuizen 1984.)

A linear variation of E^2 with T_a has also been reported by Fulachier (1978), by Champagne (1978), by Dekeyser and Launder (1983), and by Bremhorst (1985). In a comparative study, Bowers *et al.* (1988) found that the method proposed by Bremhorst (1985) was the most accurate.

The data of Lemieux and Oosthuizen (1984) were obtained with a $5\,\mu m$ diameter tungsten wire probe (DISA 55P11) using an overheat ratio R_w/R_a of 1.8, which corresponds to a temperature difference $T_w - T_a$ of about $220\,°C$. Using this value of T_w the data in Fig. 7.2(a,b) can be interpreted as

$$A^* = A(T_w - T_a), \tag{7.7a}$$

$$B^* = B(T_w - T_a), \tag{7.7b}$$

where A and B are calibration constants. Temperature-independent coefficients A and B in eqn (7.4) were also obtained by Koppius and Trines (1976) and by Keffer et al. (1978). Frota and Moffat (1982) have shown that the same principle can be applied to inclined wires in a triple-wire probe, provided that U is replaced by the effective velocity, V_e.

In conclusion, most experimental investigations of this type, covering small or moderate variations in T_a, have demonstrated that the signal from a hot-wire probe operated in the CT mode is directly proportional to a product of the temperature difference $T_w - T_a$ and a function of the velocity. Consequently, provided the hot resistance, R_w, is kept constant, then the simple relationship of eqn (7.1) can be applied. This method does not require a lengthy temperature calibration; it only requires the measurement of the fluid temperature, T_a, and the evaluation of T_w from the measured hot resistance, R_w. Equation (7.1), with $f(U) = A + BU^n$, is also simple to use for signal-analysis purposes. Mayle and Anderson (1991) have extended the principle of the polynomial curve fitting $U = \Sigma A_n E^n$ to include temperature variations, and they have expressed the heat-transfer relationship in terms of a fourth-order polynomial; that is.

$$U = B_0 + B_1 \left(\frac{E}{\Delta T}\right)^2 + B_2 \left(\frac{E}{\Delta T}\right)^3 + B_3 \left(\frac{E}{\Delta T}\right)^4,$$

where $\Delta T = T_w - T_a$.

7.2.2 Nondimensional heat-transfer relationships

The identification of a generally applicable nondimensional heat-transfer relationship for hot-wire probes has been the subject of many studies. However, the correct interpretation of nondimensional relationships is complicated by a number of factors:

1. The equations derived usually apply to very long wires, and corrections for conductive end losses are necessary.

2. The values of the fluid properties ρ, μ, and k vary with the temperature, and they must be evaluated at a suitable reference temperature. This reference temperature is not well-defined, although most investigators have used the film temperature $T_f = \frac{1}{2}(T_w + T_a)$.

3. The heat-transfer relationship corresponds to the voltage drop across the wire, E_w. Details of the Wheatstone-bridge configuration and the resistance, R_L, of the probe and cable are required in order to relate E_w to the anemometer output voltage, E.

4. The wire temperature, T_w, is normally evaluated from R_w using

eqn (2.27), and to obtain accurate results it may be necessary to measure the temperature coefficient, α_0, for each hot-wire probe used.

All these factors have resulted in a considerable variation between the proposed nondimensional heat-transfer relationships.

A commonly used nondimensional heat-transfer equation for a specific fluid (with a constant Prandtl number, Pr) is (eqn (2.1))

$$Nu = A + BRe^{1/2},$$

where $Re = \rho U d / \mu$ and

$$Nu = \frac{E_w^2}{R_w \pi \ell k (T_w - T_a)}. \tag{7.8}$$

Inserting the expressions for Nu and Re into eqn (2.1) and rearranging gives

$$\frac{E_w^2}{R_w(T_w - T_a)} = Ak + k\left(\frac{\rho}{\mu}\right)^{1/2} BU^{0.5}, \tag{7.9}$$

where π, ℓ, and d have been included in the constant coefficients A and B.

Many investigators have used eqn (7.9), with the fluid properties evaluated at the film temperature $T_f = \frac{1}{2}(T_w + T_a)$ (King 1914; Kramers 1946; Grant and Kronauer 1962; Cimbala and Park 1990; Cardell 1993). However, detailed experimental study by Kostka and Vasanta Ram (1992) has shown that this formulation will not collapse data for different conditions onto a single curve. In their detailed hot-wire study, Collis and Williams (1959) found, using the film reference temperature, that it was necessary to include a temperature-loading factor in the Nusselt number equation, eqn (2.3),

$$Nu \left(\frac{T_f}{T_a}\right)^{-0.17} = A + BRe^{0.45}.$$

It has been claimed in several investigations (for example, Bradbury and Castro 1972, Abdel-Rahman et al. 1987) that eqn (2.3) gives a good representation of the calibration data for large variations in both T_w and T_a.

Extending the principle of the polynomial curve fit $U = \Sigma A_n E^n$ of eqn (4.12) to a polynomial heat-transfer law, George et al. (1989a) have proposed the following relationship

$$Re = \sum_{n=0}^{N} C_n Nu^{n/2}.$$

In this relationship, the Reynolds number is evaluated at the ambient fluid temperature and the Nusselt number is evaluated at the film temperature. Using a fourth-order polynomial ($n = 4$) the authors claimed that the data collapsed on a single curve in the temperature range $T_a = 24-47\,°C$.

The temperature dependence of the fluid properties ρ, μ, and k can be expressed (see for example, Collis and Williams 1959; Grant and Kronauer 1962; Morrison 1974; Koppius and Trines 1976) as

$$\frac{k}{k_r} = \left(\frac{T}{T_r}\right)^a, \quad \frac{\mu}{\mu_r} = \left(\frac{T}{T_r}\right)^b, \quad \frac{\rho}{\rho_r} = \frac{T_r}{T}, \qquad (7.10)$$

where T_r is some reference temperature and variables with a subscript r denote values measured at this reference temperature, $a = 0.8\text{--}0.86$, $b = 0.76\text{--}0.9$, and T is measured in K.

It follows from these relationships that the temperature dependence of the term $k(\rho/\mu)^{1/2}$ in eqn (7.9) is very small and it can usually be ignored. The term Ak, however, has a stronger temperature dependence. Introducing the expressions for Nu and Re into eqn (2.3), applying eqn (7.10) with $a = 0.86$, and assuming that $k(\rho/\mu)^n$ is constant, eqns (7.9) and (2.3) become

$$\frac{E_w^2}{R_w(T_w - T_a)} = \left[\left(\frac{T_w + T_a}{T_w + T_{a,r}}\right)^{0.86} A + BU^{0.5}\right]. \qquad (7.11)$$

and

$$\frac{E_w^2}{R_w(T_w - T_a)} = \left(\frac{T_w + T_a}{2T_a}\right)^{0.17} \left[\left(\frac{T_w + T_a}{T_w + T_{a,r}}\right)^{0.86} A + BU^{0.45}\right]. \qquad (7.12)$$

These equations demonstrate the sensitivity of nondimensional heat-transfer relationships to the reference temperature and the temperature loading. Lienhard and Helland (1989) and Boman (1992) have proposed the use of a modified form of eqn (7.11). To summarize, there is currently no generally accepted nondimensional heat-transfer relationship for hot-wire probes, although the results of a number of investigations have indicated that eqn (2.3) can be used with reasonable accuracy for parameter-dependence evaluations in *HWA* applications.

Finally, a number of investigators have also presented nondimensional heat-transfer relationships for different types of hot-film probes operated in water (for example, Resch 1970; Bonis and van Thinh 1973; Giovanangeli 1980; Prahl and Win 1987).

7.2.3 The recommended heat-transfer relationship for *CT* hot-wire probes

For a given application, the most reliable results are obtained by a combined velocity and temperature calibration of the hot-wire probe over the anticipated velocity and ambient-fluid-temperature ranges. The results discussed in Section 7.2.1 indicate that, for moderate variations in the fluid temperature, T_a (10–80 °C), the functional dependence of the hot-wire signal, E_w^2, can be expressed (eqn (2.36a)) as

$$\frac{E_w^2}{R_w(T_w - T_a)} = A + BU^n,$$

where A, B, and n are calibration constants. This form of the heat-transfer relationship will be assumed to apply in the following, unless stated otherwise.

It follows from eqn (2.36a) that a change in the ambient temperature, from a reference condition $T_{a,r}$ to a different value T_a, will introduce a change in the wire voltage from the reference value $E_{w,r}$ to E_w. For constant-velocity conditions the change in the wire voltage will be

$$E_w = E_{w,r} \left(\frac{T_w - T_a}{T_w - T_{a,r}}\right)^{1/2}.$$

This is the approach adopted by Kanevče and Oka (1973). However, for computer-based experiments, it is more appropriate to measure the actual ambient temperature, T_a, and to insert it directly into eqn (2.36a) as discussed below.

For a CT anemometer, using the notation in Fig. 2.12 where R_L is the probe and cable resistance, the anemometer output voltage, E, will be related (for a balanced bridge) to the wire voltage, E_w, by

$$E = \frac{R_1 + R_L + R_w}{R_w} E_w. \tag{7.13}$$

Inserting eqn (7.13) into eqn (2.36a), the hot-wire response equation can be expressed directly in terms of the anemometer voltage, E, as

$$E = \frac{(R_1 + R_L + R_w)^2}{R_w} (T_w - T_a)(A + BU^n). \tag{7.14}$$

7.3 CORRECTION METHODS FOR AMBIENT FLUID TEMPERATURE DRIFT

When the ambient fluid temperature, T_a, varies between the calibration and the measurements, a correction procedure should be applied. The correction may be implemented in three different ways:

1. *Automatic compensation* A temperature sensor incorporated into the Wheatstone bridge can compensate for the change in T_a, making a separate measurement of T_a unneccessary.

To implement the other two methods the ambient fluid temperature must be measured independently, often by a thermocouple or resistor probe.

2. *Manual adjustment* Manual adjustments can be made by changing the value of the hot resistance, R_w, to compensate for changes in T_a.

3. *Analytical correction* Analytical corrections can be made by inserting the measured value of T_a into the selected hot-wire heat-transfer relationship. In this method, the hot-wire probe is operated at a fixed hot resistance, R_w.

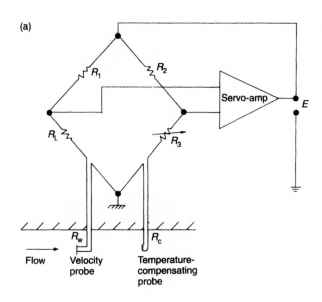

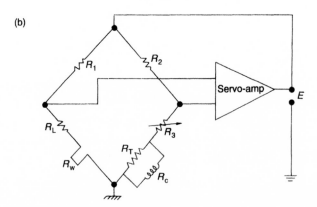

Fig. 7.3. A *SN* hot-wire probe operating in the *CT* mode compensated for variations in the ambient fluid temperature: (a) from Takagi 1986, and (b) a compensation resistor, R_c, in a combined parallel and series arrangement.

7.3.1 Automatic compensation

This technique is implemented by placing a temperature-sensitive element in the Wheatstone bridge, as shown in Fig. 7.3. The bridge configuration in Fig. 7.3(a) contains a variable resistance, R_3, in series with the resistance of the temperature-compensating sensor, R_c, while Fig. 7.3(b) shows an arrangement with a combination of resistors in series and parallel. The compensation element should be operated as a resistance-wire, and it should only be sensitive to the ambient-fluid temperature and not to the velocity of the flow. It must be sufficiently large in size so that, the current through it will not cause any significant heating. For a 1:1 bridge this will give a rather large time-constant, (often several minutes in air). If a 1:5 bridge ratio is used, it may be possible to compensate for temperature fluctuations up to the frequency limit of the hot-wire sensor itself (about 100 Hz). The resistor R_c is often in the form of a large coil element, and this principle is illustrated in the temperature-compensated hot-wire probe shown in Fig. 7.4.

The objective of the temperature compensation is to keep the anemometer voltage, E, independent of changes in T_a. Perfect compensation based on eqn (7.14) is achieved if

$$F = \frac{(R_1 + R_L + R_w)^2}{R_w} (T_w - T_a) \tag{7.15}$$

does not vary with changes in T_a. For the bridge arrangement shown in Fig. 7.3(a) the following resistance equation applies

$$\frac{R_w + R_L}{R_1} = \frac{R_3 + R_c}{R_2},$$

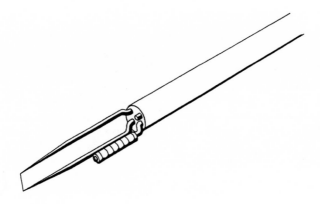

Fig. 7.4. A temperature-compensated hot-wire probe with a coil of wire used as the compensating element. (Reprinted with the permission of Dantec Measurement Technology.)

and for the hot-wire and compensating elements we have

$$R_w = R_{w0}[1 + \alpha_{w0}(T_w - T_0)],$$

$$R_a = R_{w0}[1 + \alpha_{w0}(T_a - T_0)],$$

$$R_c = R_{c0}[1 + \alpha_{c0}(T_a - T_0)],$$

where α_{w0} and α_{c0} are the temperature coefficients of the hot-wire and the compensating elements, respectively. For the bridge arrangement shown in Fig. 7.3(a) the variables are R_3, R_{c0}, and α_{c0}; they must be selected so that F in eqn (7.15), remains independent of the value of T_a. A related procedure has been described by Takagi (1986). The bridge arrangement in Fig. 7.3(b) has additional adjustable variables. Compensation circuits are often set up by hot-wire-probe manufacturers.

7.3.2 Manual adjustments of the hot resistance

A typical CT anemometer was shown in Fig. 2.12 and, provided the probe and cable resistance R_L is included in the analysis, eqn (7.14) will be valid for this standard CT bridge arrangement, For manual adjustment, the objective is to keep the anemometer voltage, E, in eqn (7.14), independent of the change in T_a by changing the hot-resistance, R_w. Several different resistance-setting criteria have been used. As shown by Drubka et al. (1977), a constant resistance difference will lead to undercompensation of the hot-wire system, whereas maintaining a constant overheat ratio will result in a slight overcompensation. Correct manual adjustment is therefore difficult to implement, and with the development of computer-controlled experiments it is rarely used nowadays.

7.3.3 Analytical-compensation techniques

In Section 7.2.1 it was concluded that eqn (2.36a) is an accurate representation of the velocity and temperature dependence of a hot-wire probe operated in the CT mode and exposed to moderate variations in T_a. This is the formulation used for the compensation technique proposed by Bearman (1971). For hot-wire measurements it is convenient to express eqn (2.36a) in terms of the anemometer output voltage, E, as in eqn (7.14). The temperature dependence only occurs via T_a, and if the fluid temperature, T_a, is measured then

$$E^* = E^2 \frac{R_w}{(R_1 + R_L + R_w)^2(T_w - T_a)} = A + BU^n \qquad (7.16)$$

will only be a function of the flow velocity. For a CT anemometer, the ratio $R_w/(R_1 + R_L + R_w)^2$ in eqn (7.16) will be constant, and it can be included in the calibration coefficients A and B. This technique is very simple to implement on a digital computer, and it can also be used to correct for variations in the instantaneous value of T_a.

7.3.4 Temperature effects for hot-film probes in water

When measuring with a hot-film probe in a water flow, large changes can occur in the anemometer voltage, E, even for small changes in T_a, since the temperature difference, $T_w - T_a$, is typically about 20 °C. A number of investigations have proposed various nondimensional heat-transfer relationships and corresponding compensation techniques (Resch 1970, 1973; Morrow and Kline 1971; Burchill and Jones 1971; Bonis and van Thinh 1973; Tan-atichat *et al.* 1973; Giovanangeli 1980; Sherif and Pletcher 1986). In general, most compensation techniques have assumed that eqn (2.36a) also applies to hot-film probes in liquids. If necessary, this relationship can be extended to temperature-dependent coefficients A and B (Sherif and Pletcher 1986). Measurements in water are complicated by the additional problem of changes in the heat-transfer relationship due to accumulation of deposits on the sensing element of the probe. This is particularly noticeable for cylindrical and wedge-shaped probes, and to a lesser extent for conical probes, as discussed in Section 4.10. Therefore to correct for both temperature and contamination effects, frequent velocity calibrations may be more appropriate.

7.4 MEASUREMENTS OF THE FLUCTUATING FLUID TEMPERATURE

In nonisothermal flows the velocity fluctuations are accompanied by fluctuations in the fluid temperature. The following sections describe the methods used to measure the ambient fluid temperature fluctuation, $\theta \ (=T_a - \bar{T}_a)$, and its correlation with the fluctuating-velocity field.

7.4.1 The multiple overheat ratio method

The multiple-overheat ratio method utilizes a single CT hot-wire probe operated sequentially at different overheat ratios, as proposed by Corrsin (1947, 1949) for low-velocity flows and by Kovasznay (1950a, 1953) for supersonic flows. The response equation for a SN hot-wire probe will be assumed to be given by eqn (2.36a), and the corresponding relationship between the fluctuating voltage, velocity, and temperature is (eqn (2.38b))

$$e = S_u u + S_\theta \theta,$$

where $S_u = \partial E/\partial U$ and $S_\theta = \partial E/\partial \theta$. As shown in Section 2.1.2.4, the ratio S_u/S_θ is a function of the overheat ratio, R_w/R_a.

Evaluating $\overline{e^2}$ from eqn (2.38b) gives

$$\frac{\overline{e^2}}{S_\theta^2} = \left(\frac{S_u}{S_\theta}\right)^2 \overline{u^2} + 2\left(\frac{S_u}{S_\theta}\right)\overline{u\theta} + \overline{\theta^2}. \tag{7.17}$$

Equation (7.17) represents a parabolic relationship between $\overline{e^2}/S_\theta^2$ and S_u/S_θ, with $\overline{u^2}$, $\overline{u\theta}$, and $\overline{\theta^2}$ being constant parameters. In principle, the values of these parameters can be determined from three different overheat ratios, but more accurate results are obtained by using a larger number of overheat ratios in conjunction with a least-squares curve-fitting method (Kovasznay 1953; Arya and Plate 1969; Fulachier and Dumas 1976). In addition, Fulachier and Dumas evaluated velocity and temperature spectra by a narrow-band filter technique.

The multiple overheat-ratio method is relatively simple to use, and it provides second-order velocity and temperature statistics such as $\overline{u^2}$, $\overline{u\theta}$, and $\overline{\theta^2}$. However, it does not permit simultaneous measurements of u and θ, so higher moments and probability density functions cannot be obtained. Also the evaluated results are very sensitive to small changes in S_u and S_θ during the measurement period and, in general, the temperature quantities are subject to considerable uncertainty. This method is therefore not recommended.

7.4.2 Dual CT hot-wire probes

A parallel-wire probe, operated with the two wires having different velocity and temperature sensitivities, was first proposed by Corrsin (1949). The two hot-wire sensors are placed close together so that they are exposed to the same velocity and temperature field. Subsequent investigations of the parallel-wire probe operated in the CT mode have been reported by Sakao (1973), Blair and Bennett (1987), and Lienhard and Helland (1989).

If the two sensors are identical and operated at two different overheat ratios then their response equations can be expressed as

$$\frac{E_{w1}^2}{R_{w1}} = (A + BU^n)(T_{w1} - T_a), \tag{7.18a}$$

$$\frac{E_{w2}^2}{R_{w2}} = (A + BU^n)(T_{w2} - T_a). \tag{7.18b}$$

Taking the difference between eqns (7.18a) and (7.18b) gives

$$\frac{E_{w1}^2}{R_{w1}} - \frac{E_{w2}^2}{R_{w2}} = (A + BU^n)(T_{w1} - T_{w2}), \tag{7.19}$$

which is independent of the ambient fluid temperature, T_a. This technique was described by Sakao (1973), who built an electronic circuit to implement eqn (7.19). In practice it is not possible to manufacture two identical hot-wire sensors, and the values of A and B in eqns (7.18a–b) will be different for the two hot-wires. A relatively simple solution procedure can be developed, if it is assumed that the exponent, n, is the same for the two hot-wires. This is not a serious restriction since the results in Section 4.4.1 have demonstrated that the optimum value of n is about 0.41–0.45. Introducing these conditions, eqns (7.18a–b) can be written as

$$E_{w1}^2 = R_{w1}(A_1 + B_1 U^n)(T_{w1} - T_a),\qquad(7.20a)$$

$$E_{w2}^2 = R_{w2}(A_2 + B_2 U^n)(T_{w2} - T_a),\qquad(7.20b)$$

which are two simultaneous nonlinear equations in U and T_a. Expressing these equations in the form

$$E_{w1}^2 = a_1 + a_2 T_a + (a_3 + a_4 T_a)U^n,\qquad(7.21a)$$

$$E_{w2}^2 = b_1 + b_2 T_a + (b_3 + b_4 T_a)U^n,\qquad(7.21b)$$

where

$$a_1 = R_{w1}A_1 T_{w1}, \qquad b_1 = R_{w2}A_2 T_{w2},$$
$$a_2 = -R_{w1}A_1, \qquad b_2 = -R_{w2}A_2,$$
$$a_3 = R_{w1}B_1 T_{w1}, \qquad b_3 = R_{w2}B_2 T_{w2},$$
$$a_4 = -R_{w1}B_1, \qquad b_4 = -R_{w2}B_2,$$

and eliminating U^n, gives a second-order equation in T_a

$$c_3 T_a^2 + c_2 T_a + c_1 = 0,\qquad(7.22)$$

with

$$c_1 = (E_{w2}^2 - b_1)a_3 - (E_{w1}^2 - a_1)b_3,$$
$$c_2 = (E_{w2}^2 - b_1)a_4 - (E_{w1}^2 - a_1)b_4 + a_2 b_3 - b_2 a_3,$$
$$c_3 = a_2 b_4 - b_2 a_4.$$

For measurement purposes, the wire voltages E_{w1} and E_{w2} can be expressed in terms of the direct anemometer output voltages E_1 and E_2 using eqn (7.13)

$$E_1 = \frac{R_1 + R_{L1} + R_{w1}}{R_{w1}} E_{w1},\qquad(7.23a)$$

$$E_2 = \frac{R_1 + R_{L2} + R_{w2}}{R_{w2}} E_{w2}.\qquad(7.23b)$$

Solving eqn (7.22) for each set of measured $E_1(t)$- and $E_2(t)$-values will determine $T_a(t)$. Then, knowing $T_a(t)$, the corresponding value of $U(t)$ can be evaluated from eqn (7.20a) or eqn (7.20b). The calibration and measurement technique for such a probe is simple to implement on a digital computer, and the probe appears to have the advantages of durability, high band width, and it can be driven by commercially available CT equipment.

A signal-analysis procedure similar to (7.22) was proposed by Lienhard and Helland (1989). They used a slightly more complex heat-transfer relationship

$$E^2 = (AT_f^{0.84} + BU^{0.45})(T_w - T_a), \qquad (7.24)$$

where $T_f = (T_w + T_a)/2$ is the film temperature (in K). Eliminating $U^{0.45}$, they obtained a third-order polynomial equation for T_a instead of the simpler second-order equation, of eqn (7.22). Comparative measurements were carried out with a resistance-wire system. They concluded that the parallel-wire probe worked well in low turbulence intensity flows containing temperature signals, which were not too small; and spectral measurements demonstrated that the parallel CT probe had a better frequency response than a resistance-wire probe. However, the parallel-wire probe was shown, by spectral measurements, to suffer from severe noise problems when the temperature fluctuations were small. For these conditions the authors recommended the use of the resistance-wire method.

The parallel-wire method was extended to X-hot-wire anemometry by Blair and Bennett (1987), who mounted a third similar wire equidistant between the wires of the X-configuration and parallel to one of the wires in the X-configuration. In selecting the wire separation in their multi-wire probe, they considered the trade-off between prong- and sensor-interference effects and the loss of both spectral and spatial resolution. Sandborn (1972), Fulachier (1978), and Beguier *et al.* (1978*b*) have shown that the

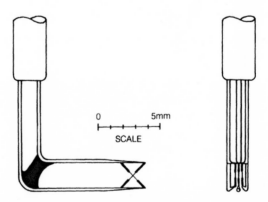

Fig. 7.5. A diagram of X and resistance-wire configuration for a CT mode probe. Nominal dimensions: active length, 0.50 mm, diameter, 2.5 μm, transverse sensor separation, 0.35 μm. (From Blair and Bennett 1987.)

transverse correlation coefficient, $g(r)$, falls rapidly with increasing sensor separation, but at small separation the sensor- and prong-interference effects become large (Guitton and Patel 1969). Blair and Bennett (1987) concluded that a transverse spacing-to-length ratio of $h/\ell = 0.75$ corresponded to a near optimum condition for their X-hot-wire probe, which is shown in Fig. 7.5.

7.4.3 The resistance-wire method

The most common method for measuring fluctuating fluid temperatures is to operate a very thin wire at a low overheat ratio in the CC mode. A Wollaston wire, consisting of a thin platinum or platinum-alloy wire covered by a thick sheet of silver, is normally used as the resistance-wire, as it is available with Platinum diameters as small as $0.25\,\mu m$. Due to the frailty of the sensor element, most resistance-wire probes are made by the user, by first soldering the Wollaston wire to the prongs and then etching it to expose the thin Platinum wire. To achieve accurate 'cold-wire' probe results a number of criteria must be satisfied.

Time constant M The time constant for a hot-wire element of length ℓ and diameter d (see eqn (2.47)),

$$M = \frac{\rho_w c_w (\pi/4) d^2 \ell}{\alpha_0 R_0 (A + BU^n - I^2)},$$

can be expressed in terms of the Nusselt number, Nu, eqn (7.8) as

$$M = \frac{\rho_w c_w d^2}{4k\mathrm{Nu}} \frac{R_w}{R_a} \cong \frac{\rho_w c_w d^2}{4k\mathrm{Nu}}, \tag{7.25}$$

since $R_w/R_a \cong 1$ at a low overheat ratio. For very fine wires the effect of the mean free path, λ, of the gas molecules should be taken into account by including the Knudsen number, $\mathrm{Kn} = \lambda/d$, in the heat-transfer relationship. Collis and Williams (1959) proposed the following relationship

$$Nu = (1.18 + 2\mathrm{Kn} - 1.10 \log_{10} \mathrm{Re})^{-1} \tag{7.26}$$

in the Reynolds number range $0.02 < \mathrm{Re} < 0.5$.

To demonstrate the variation in M with the wire diameter, d, calculations were carried out for platinum wires in the diameter range 0.25–$2.5\,\mu m$ for a flow velocity of $10\,\mathrm{m\,s^{-1}}$. In the air, corresponding Reynolds and Nusselt number ranges are 0.16–1.6 and 0.4–1.0, respectively. A comparison of theoretical (see eqn (7.25)) and measured values of the cut-off frequency, $f_c = 1/(2\pi M)$, at $U = 10\,\mathrm{m\,s^{-1}}$ is shown in Fig. 7.6. The figure demonstrates the increase in the value of f_c when the diameter decreases. When $d \geq 1\,\mu m$ the measured and calculated values are in close agreement, but for the finest wires the measured values for f_c are substantially lower than the theoretical values (LaRue *et al.* 1975; Højstrup *et al.* 1976; Antonia *et al.* 1981a). This difference is particularly noticeable when $d = 0.25 \mu m$.

The time constants and transfer functions of resistance-wires have been studied as a function of the running time (see, for example, LaRue *et al.* 1975; Smits *et al.* 1978; Paranthoen *et al.* 1982a). As indicated in Fig. 7.6, for the 'old' wire results, dust contamination can significantly increase the time constant of a resistance-wire. Washing of the wire in Javel water will often remove most of the deposited dust and (nearly) restore the original value of the time constant (Paranthoen *et al.* 1982a).

To compensate for the wire's thermal inertia, it is necessary to determine the time constant, M, or the complete amplitude transfer function, $H_\theta(\omega)$ for selected velocities covering a specified velocity range. The response characteristics of resistance-wires can be determined by a number of experimental techniques.

Resistance-wire response The response of a wire element can be measured by placing the wire in a constant velocity, constant temperature flow and varying the electrical heating. Two different types of tests have been used. (i) Sinusoidal variations in the heating current at various frequencies. This will determine the amplitude transfer function, $H_\theta(f)$, of the wire element itself (LaRue *et al.* 1975). For this method to give correct results Bremhorst and Krebs (1976) and Bremhorst and Graham (1990) have recommended the use of a balanced bridge and cable compensation. (ii) A step increase (square-wave) in the heating current (Bradbury and Castro 1972; LaRue *et al.* 1975; Hishida and Nagano 1978). As shown by eqn (2.50), the response is an exponential decay $(1 - e^{-t/M})$, and the time constant, M, is equal to the time, t, at which the signal has decayed by 63%. Both of the internal-heating methods will only heat the sensor element and not the prongs, and the measured response is therefore that of the wire element itself. Typical transfer functions, $H_\theta(f)$, for 0.25 and 0.63 μm diameter platinum resistance-wire probes are shown in Fig. 7.7, demonstrating the increase in the cut-off frequency, f_c, with increasing velocity and decreasing wire diameter.

A number of external-heating techniques have been used to measure the response of resistance-wires, but Fiedler (1978) has shown that different results can be obtained depending on whether only the wire element or also the prongs are heated. External heating by either a chopped laser beam (Fiedler 1978; Weeks *et al.* 1988) or a radiant heat flux (Smits *et al.* 1978) may not expose the prongs to the same temperature fluctuations as the wire element. These methods are therefore most reliable if used to heat only the wire element of long resistance wires ($\ell/d \geqslant 1000$). A related method is the pulsed-wire technique, in which a pulsed-wire is placed perpendicularly to, and upstream of, the resistance-wire (Antonia *et al.* 1981a; Shah and Antonia 1986).

Resistance-probe response To obtain the total probe response, including

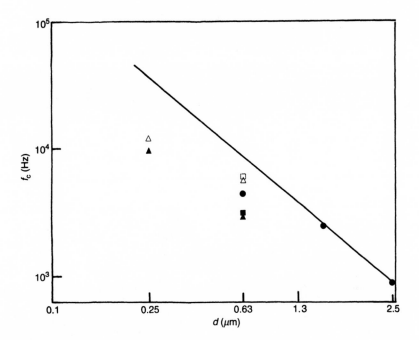

Fig. 7.6. Measured and calculated cut-off frequencies, f_c (-3dB), for resistance-wires at 10 m s^{-1}: (—) eqn (7.25); measured data: (●) Antonia *et al.* (1981*a*); (△) new, (▲) old, LaRue *et al.* (1975); (□) new, (■) old; Paranthoen *et al.* (1982*a*).

the dynamic wire–prong interaction (see Section 2.3.2.3) it is necessary to use an external-heating source. Paranthoen *et al.* (1982*b*) and Petit *et al.* (1985) placed the resistance-wire parallel to and in the wake of an upstream wire, which was heated by a sinusoidal current at a frequency of f; this exposed both the wire and the prongs to temperature fluactuations at a frequency of $2f$. By varying the frequency f, the transfer function $H_\theta(f)$ of the complete probe was measured (as illustrated in Fig. 2.10). The transfer function, $H_\theta(f)$, has also been measured by Højstrup *et al.* (1977) by exposing the probe to temperature fluctuations caused by a strong sound field. It was shown in Section 2.3.2.3 that the low-frequency attenuation can be estimated from eqn (2.53)

$$H_p = 1 - 2\ell_c/\ell,$$

and for a resistance-wire with $R_w/R_a \cong 1$ the cold length, ℓ_c, from eqn (2.15) can be expressed as

$$\ell_c = \frac{d}{2}\left(\frac{k_w}{k}\frac{1}{\text{Nu}}\right)^{1/2}. \tag{7.27}$$

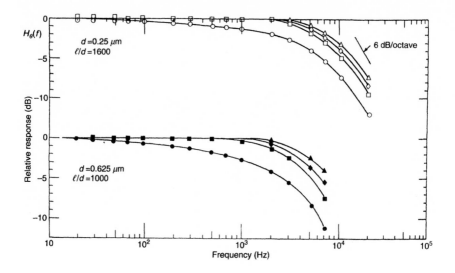

Fig. 7.7. Measured transfer functions, $H_\theta(f)$, for 0.25 and 0.6 μm diameter platinum resistance-wires: (○, ●) $U = 0$, (□, ■) $U = 2\,\mathrm{m\,s^{-1}}$, (◇, ◆), $4\,\mathrm{m\,s^{-1}}$, (△, ▲) $8\,\mathrm{m\,s^{-1}}$. (From LaRue *et al.* 1975.)

For an air velocity of $10\,\mathrm{m\,s^{-1}}$ it can be estimated from eqn (7.27) that $\ell_c = 35d$ for a $1\,\mu\mathrm{m}$ diameter wire and it equals $40d$ for a $0.25\,\mu\mathrm{m}$ diameter wire. To reduce the low-frequency attenuation to less than 5%, the ℓ/d ratio should be 1400 for a $1\,\mu\mathrm{m}$ diameter wire ($\ell \cong 1.4$ mm) and it should be 1600 for a $0.25\,\mu\mathrm{m}$ diameter wire ($\ell \cong 0.40$ mm).

Frequency compensation The cut-off frequency, f_c, of a resistance-wire, (see Fig. 7.6) is a strong function of its diameter, d; and the frequency response of an uncompensated wire may not be adequate for fluctuating-temperature measurements unless a very fine wire is used. Whether frequency compensation is necessary depends both on the type of measurements required and on the diameter of the resistance-wire. The main contribution to $\overline{\theta^2}$ is from the large energy-containing eddies, and for an uncompensated resistance-wire the errors caused by signal attenuation at high frequencies are usually small. Fulachier (1978) has shown that the error in $\overline{\theta^2}$ for a $1\,\mu\mathrm{m}$ diameter platinum wire operated at a heating current of $I = 0.15$ mA is only about 4%. However, for temperature dissipation and spectral measurements with an uncompensated wire, large errors will often occur.

If the conductive end losses are ignored (which is justified for large values of ℓ/d), and the forced-convection heat transfer is expressed in terms of the Nusselt number, Nu, using eqn (7.8), then the unsteady heat transfer from the total wire element of length ℓ, can be expressed, using eqns (2.45), and (7.8) as

$$\frac{\rho_w c_w (\pi/4) d^2 \ell}{\alpha_0 R_0} \frac{dR_w}{dt} + \frac{\pi \ell k}{\alpha_0 R_0} (R_w - R_a) \text{Nu} = I^2 R_w. \tag{7.28}$$

Rearranging this equation gives

$$M \frac{dR_w}{dt} + R_w = R_a \frac{\pi \ell k \text{Nu}}{(\pi \ell k \text{Nu} - I^2 \alpha_0 R_0)}, \tag{7.29}$$

where

$$M = \frac{\rho_w c_w (\pi/4) d^2 \ell}{\pi \ell k \text{Nu} - I^2 \alpha_0 R_0} \approx \frac{\rho_w c_w d^2}{4 k \text{Nu}}$$

is the time constant of the resistance-wire. Multiplying eqn (7.29) by the current, I, through the wire, and introducing the measured voltage across the resistance-wire

$$E_m = IR_w, \tag{7.30}$$

the following equation can be obtained

$$M \frac{dE_m}{dt} + E_m = IR_a \frac{\pi \ell k \text{Nu}}{(\pi \ell k \text{Nu} - I^2 \alpha_0 R_0)}. \tag{7.31}$$

The corresponding theoretical voltage E_T for an ideal wire with no thermal inertia ($M = 0$) is

$$E_T = IR_a \frac{\pi \ell k \text{Nu}}{(\pi \ell k \text{Nu} - I^2 \alpha_0 R_0)}. \tag{7.32}$$

Note that E_T is a direct measure of R_a and therefore of the ambient fluid temperature, T_a, and for a low value of the heating current, I, eqn (7.32) can be simplified to

$$E_T = IR_a, \tag{7.33}$$

Substituting eqn (7.32) for E_T in eqn (7.31) gives

$$M \frac{dE_m}{dt} + E_m = E_T, \tag{7.34}$$

which is a first-order differential equation relating the voltage E_T (corresponding to the fluid temperature, T_a) to the measured wire voltage, E_m. Provided that the value of M is known, the instantaneous values of $E_T(t)$ (and therefore of $T_a(t)$) can be obtained from the measured $E_m(t)$ by implementing eqn (7.34) either by an analog device or digitally. For a given resistance-wire, the time constant, M, is a function of the flow velocity, and its value may vary by a factor of four between 0 and 50 m s^{-1} (Bremhorst and Graham 1990). It is therefore necessary to adjust the value of M when a significant variation occurs in the mean velocity, as, for example, during the traverse of a boundary layer flow.

Automatic compensation for the (velocity dependent) time constant of a resistance-wire can be achieved by using a dual-wire probe with one wire operated as a resistance-wire in the CC mode and the second as a velocity sensor operated at a high overheat ratio in the CT mode. The second wire is used to measure the flow velocity, $U(t)$; and, having determined $M(U)$ either theoretically (Hishida and Nagano 1978) or experimentally (Bremhorst and Graham 1990), $E_T(t)$ (and therefore $T_a(t)$) can be obtained by either an analog or digital implementation of eqn (7.34). A related compensation technique was presented by Weeks *et al.* (1988), who fed the output from the velocity sensor to an analog switch, which selected one of eight resistors representing $M(U)$ in a stepwise fashion.

Velocity and temperature sensitivity All heated probes are, in principle, sensitive to velocity and temperature variations. To achieve accurate temperature measurements with a resistance-wire the ratio between the velocity sensitivity, $S_{u,cc}$ (see eqn (2.39a)), and the temperature sensitivity, $S_{\theta,cc}$ (see eqn (2.39b)), should be minimized. In nondimensional form this ratio can be expressed (Wyngaard 1971a) as

$$\frac{S_{u,cc}}{S_{\theta,cc}} = -\frac{I^2 R_w (0.25 \mathrm{Re}^{0.45})}{\pi k \ell \bar{U}(0.24 + 0.56 \mathrm{Re}^{0.45})^2}$$

$$= -\frac{\chi_w \mathrm{Re}^{0.45}}{\pi^2 k \bar{U}(0.24 + 0.56 \mathrm{Re}^{0.45})^2} \frac{I^2}{d^2}. \tag{7.35}$$

For a resistance-wire with a given diameter, d, this ratio can be seen to vary with I^2, and I should therefore be as small as possible to minimize the velocity component of the resistance-wire signal. If the diameter, d, is reduced then the current, I, must also be reduced if a specified value for the ratio is to be maintained. Predictions based on eqn (7.35) are in close agreement with result measured by Mestayer and Chambaud (1979). In practice, the value of I has a lower limit due to signal-to-noise criteria (LaRue *et al.* 1975. Mestayer and Chambaud 1979), and the large amplification required for resistance-wire signals. If (d, I)-values of $(d = 0.25\,\mu\mathrm{m}, I = 0.05\,\mathrm{mA})$, $(d = 0.63\,\mu\mathrm{m}, I = 0.10\,\mathrm{mA})$, and $(d = 1\,\mu\mathrm{m}, I = 0.15\,\mathrm{mA})$ are selected then the contamination of the resistance-wire signal by the fluctuating velocity will produce an error of less than 2% in $\overline{\theta^2}$ and $\overline{\theta u}$. Fulachier (1978) reported difficulties in using a heating current below 0.15 mA, but Fiedler (1982) has shown that this is likely to have been due to an instrumentation problem.

The temperature senstitivity in eqn (2.39b) can, for a resistance-wire with $R_w/R_a \simeq 1$, be expressed as

$$S_{\theta,cc} = \alpha_0 R_0 I \tag{7.36}$$

A platinum resistance-wire with $d = 1\,\mu\text{m}$ and $\ell = 0.4\,\text{mm}$ will have a resistance of $R_a \simeq 50\,\Omega$, and when heated by a current $I = 0.15\,\text{mA}$ the temperature sensitivity will be about $30\,\mu\text{V}\,^\circ\text{C}^{-1}$. Temperature measurements with resistance-wires therefore require both low drift, low noise CC-mode anemometers, and high-quality amplifiers with typical gains of 1000–5000. (Tavoularis 1978; Peattie 1987; Haugdahl and Lienhard 1988). If the resistance-wire is heated by a current $I = 0.15\,\text{mA}$, then the 'hot' resistance R_w will only deviate from R_a by $(R_w - R_a)/R_a \simeq 0.0004$, and the corresponding temperature difference $T_w - T_a$ will be less than $0.1\,^\circ\text{C}$. It is therefore justifiable to set $R_w = R_a$, with

$$R_a = R_0[1 + \alpha_0(T_a - T_0)]. \tag{7.37}$$

For practical applications, it is recommended that a temperature calibration of the resistance-wire is used to determine the calibration constants A and B in the relationship

$$R_a = A + BT_a. \tag{7.38}$$

7.4.3.1 The temperature dissipation rate, ε_θ, and the microscale, λ_θ

Specification of a fluctuating-temperature field requires the evaluation of parameters for both the large-scale energy-containing eddies and the small dissipating eddies. The small-scale temperature fluctuations are often described by the temperature dissipation rate, ε_θ, and by the microscale, λ_θ. This section outlines their definition and measurement, and the effect of the time constant, M, of the resistance-wire on the measurement of these quantities. Based on an energy balance for $\frac{1}{2}\overline{\theta^2}$, and using the notation of Beguier *et al.* (1978*a*), the temperature dissipation rate can be expressed as

$$\varepsilon_\theta = \alpha\left(\overline{\left(\frac{\partial\theta}{\partial x}\right)^2} + \overline{\left(\frac{\partial\theta}{\partial y}\right)^2} + \overline{\left(\frac{\partial\theta}{\partial z}\right)^2}\right) = \alpha(\varepsilon_{\theta,x} + \varepsilon_{\theta,y} + \varepsilon_{\theta,z}), \tag{7.39}$$

where α is the thermal diffusivity. Provided that the small-scale temperature fluctuations are isotropic, eqn (7.39) can be expressed as

$$\varepsilon_\theta = 3\alpha\,\overline{\left(\frac{\partial\theta}{\partial x}\right)^2}. \tag{7.40}$$

Measurements of $\varepsilon_{\theta,x} = \overline{(\partial\theta/\partial x)^2}$ and $\varepsilon_{\theta,y} = \overline{(\partial\theta/\partial y)^2}$ have been carried out with a probe containing two parallel resistance-wires by Antonia *et al.* (1977) and by Antonia and Browne (1983). Sreenivasan *et al.* (1977) also measured $\varepsilon_{\theta,z} = \overline{(\partial\theta/\partial z)^2}$ by using a four-wire probe. Krishnamoorthy and Antonia (1987) used a pair of parallel cold wires to measure all three components of the dissipation rate in an approximately self-preserving turbulent boundary layer. The longitudinal component, $\varepsilon_{\theta,x}$, was evaluated mainly by Taylor's hypothesis, and the two transverse components, $\varepsilon_{\theta,y}$ and $\varepsilon_{\theta,z}$, were estimated mainly from the curvature of spatial temperature autocorrelations. In the outer region it was found that $\varepsilon_{\theta,z} > \varepsilon_{\theta,y} > \varepsilon_{\theta,x}$, the total variation being 20–40%. In this region eqn (7.40) is a good approximation of eqn (7.39). However, in the near-wall region, the anisotropy is large and eqn (7.40) is grossly in error, since the ratios $\varepsilon_{\theta,y}/\varepsilon_{\theta,x}$ and $\varepsilon_{\theta,z}/\varepsilon_{\theta,x}$ can be as high as 13 and 7, respectively. The parallel cold-wire method has also been used to measure all three derivatives of the temperature fluctuation, θ, in a fully developed turbulent channel flow (Zhu and Antonia 1993) and in a turbulent round jet (Antonia and Mi 1993b). Antonia and Mi (1993a) have developed corrections which compensate for spectral attenuation caused by the separation between the cold wires.

For measurement purposes, eqn (7.40) can be rewritten using Taylor's hypotheses as

$$\varepsilon_\theta = 3 \frac{\alpha}{\bar{U}^2} \overline{\left(\frac{\partial \theta}{\partial t}\right)^2}. \tag{7.41}$$

Alternatively, ε_θ can be evaluated from

$$\varepsilon_\theta = 3\alpha \int_0^\infty k_1^2 E_\theta(k_1) \, dk_1, \tag{7.42}$$

where $k_1 = 2\pi f/\bar{U}$ is the wavenumber, and $E_\theta(k_1)$ is the one-dimensional wavenumber spectra defined in such a way that

$$\overline{\theta^2} = \int_0^\infty E_\theta(k_1) \, dk_1, \tag{7.43}$$

A temperature microscale, λ_θ, may be introduced in a way similar to that for λ (see eqn (2.79)) by

$$\varepsilon_\theta = 3\alpha \frac{\overline{\theta^2}}{\lambda_\theta^2}, \tag{7.44}$$

and λ_θ can therefore be evaluated from

$$\lambda_\theta = \left[\frac{\overline{\theta^2}}{\int_0^\infty k_1^2 E_\theta(k_1) \, dk_1}\right]^{1/2}. \tag{7.45}$$

The effect of the wire's time-constant, M, on the evaluation of ε_θ (eqn (7.42)) and λ_θ (eqn (7.45)) can be demonstrated by considering the spectrum $E_\theta(k_1) = (\bar{U}/2\pi)E_\theta(f)$. From eqn (7.34), and assuming that M is constant, it follows that the temperature spectrum $E_\theta^m(f)$, measured by an uncompensated resistance-wire, is related to the true spectrum, $E_\theta(f)$, by

$$E_\theta^m(f) = \frac{E_\theta(f)}{(1 + (2\pi f M)^2)}. \tag{7.46}$$

Consequently, if the value of M is large, then a significant attenuation will occur at high frequencies and evaluations of ε_θ and λ_θ based on $E_\theta^m(f)$ will have a very large error (Fulachier 1987; Shishov et al. 1987). However, if the true spectrum $E_\theta(f)$, is first evaluated using eqn (7.46), then good estimates of ε_θ and λ_θ can be obtained using eqns (7.42) and (7.45) provided that the probe has a satisfactory spatial resolution; this is discussed by Wyngaard (1971b), Larsen and Højstrup (1982), Lecordier et al. (1984), and Browne and Antonia (1987).

7.5 SIMULTANEOUS VELOCITY AND TEMPERATURE MEASUREMENTS

If a resistance-wire operated in the CC mode is placed close to one or more hot-wire sensors operated in the CT mode, simultaneous measurements can be obtained of both the fluctuating temperature and velocity. The following analysis is presented in terms of the instantaneous ambient fluid temperature, T_a. In many applications the time-mean ambient fluid temperature, $\bar{T}_a$, is measured separately by a fine thermocouple or resistor probe, and the resistance-wire is used only for the measurement of the fluctuating component θ ($=T_a - \bar{T}_a$).

7.5.1 Dual-sensor probes

Dual sensor probes consist of a resistance-wire and a hot-wire sensor which are both placed perpendicularly to the mean-flow direction.

Resistance-wire equations Provided that the ℓ/d ratio is large and the heating current I is low then the dynamic prong effect and the velocity sensitivity can be ignored. The theoretical wire voltage, E_T, corresponding to a wire with no thermal inertia ($M = 0$) can be expressed, using eqns (7.33) and (7.38), as

$$E_T = IR_a = I(A + BT_a), \tag{7.47}$$

and E_T is related to the measured wire voltage, E_w, by eqn (7.34),

$$M \frac{dE_m}{dt} + E_m = E_T,$$

where $M = M(U)$ is the velocity-dependent time-constant of the resistance-wire. The functional form of $M(U)$ is either specified theoretically (for example, by eqn (7.25)) or it is measured experimentally.

Hot-wire equation The response equation for a hot-wire probe operated in the CT mode at a high overheat ratio can be expressed in terms of the wire voltage, E_w, by eqn (2.36a), and the wire voltage, E_w, is (for a balanced bridge) related to the anemometer voltage, E, by eqn (7.13). In the CT mode the hot-wire is kept at a fixed hot resistance, R_w, and by combining these two equations the hot-wire response equation can be expressed in terms of the anemometer voltage, E, as

$$E^2 = (A + BU^n)(T_w - T_a). \qquad (7.48)$$

In this equation, all the fixed resistance values have been included in the calibration constants A and B.

Measurement procedure The dual-sensor probe will provide simultaneous measurements of $E_m(t)$ from the CC-mode resistance-wire and of $E(t)$ from the CT-mode hot-wire sensor. The corresponding instantaneous values of $U(t)$ and $E_T(t)$ (and therefore of $T_a(t)$) can be obtained by simultaneously solving eqns (7.34) and (7.48) for a known $M(U)$-function. The solution procedure lends itself to real-time analog implementation, and this method has been implemented, with various degrees of simplification, in a number of applications (Chevray and Tutu 1972; Ali 1975; Hishida and Nagano 1988; Weeks *et al.* 1988; Bremhorst and Graham 1990). If real-time data analysis is not required then a digital implementation can be considered, as described by Bremhorst and Graham (1990).

A number of different wire configurations have been used. Chevray and Tutu (1972) and Ali (1975) placed the resistance-wire perpendicular to, and slightly upstream of, the hot-wire sensor to minimize the effect of the cold wake from the resistance-wire on the hot-wire element. Hishida and Nagano (1978) and Antonia *et al.* 1988b) used two parallel wires placed perpendicular to and separated by a distance Δx in the mean-flow direction. The resistance-wire was placed upstream, and Hishida and Nagano found that the effect of the cold-wire wake on the mean velocity, $\bar{U}$, measured by the CT-mode hot-wire sensor could be ignored when $\Delta x = 150d_c$, where d_c is the diameter of the cold wire, It should, however, be noted, as shown in Fig. 2.27, that the wake effect on the measured value of $\overline{u^2}$ can be detected over much larger axial separations. This problem has been considered by Bremhorst (1988). In other investigations (for example, Yeh and van Atta 1973; Beguier *et al.* 1978b; Antonia *et al.* 1981b) the two wires were separated in the lateral direction to ensure that the resistance-wire was outside the thermal wake of the CT-mode hot-wire sensor. To minimize the lateral separation, the resistance-wire has been placed an additional small distance upstream of the hot-wire sensor in a number of investigations (for example, Antonia *et al.* 1980).

7.5.2 Multi-sensor probes

A number of investigations have also been carried out with probes containing an X-hot-wire configuration and a resistance-wire. The analysis pre-

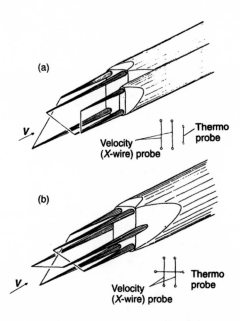

Fig. 7.8. Velocity (*X*-wire) and temperature combination probes. (From Fiedler 1978.)

sented for the $(U(t), T_a(t))$-measurements with a dual-sensor probe can be extended to include the cross-flow component, $V(t)$, by replacing U in eqn (7.48) by the effective velocity, V_e, and applying one of the *X*-probe signal-analysis procedures described in Section 5.4. Graham and Bremhorst (1990, 1991) have described an analog compensation-solution technique for such a probe.

The wire configurations used are in general one of the two types shown in Fig. 7.8(a,b). Bourke and Pulling (1970), Pessoni and Chao (1974), Senda *et al.* (1980), and Subramaniam and Antonia (1981) placed the resistance-wire perpendicular to, and slightly upstream of, the centre of the *X* (Fig. 7.8(a)). In the second configuration, the resistance-wire was placed perpendicular to the mean-flow direction and parallel to the *X*-configuration, either halfway between the two wires of the *X* (Beguier *et al.* 1978*b*; Gibson and Verriopoulos 1984) or next to one of the wires of the *X*-probe (Sreenivasan and Tavoularis 1980; Dekeyser and Launder 1983; and Tureaud *et al.* 1988). Also, in some investigation (for example, Antonia *et al.* 1981*b*), the resistance-wire was, in addition to the lateral separation, also placed slightly upstream of the centre of the *X*-configuration. Finally, measurements with a four-wire probe consisting of a triple hot-wire probe and a resistance-wire have been reported by Fabris (1978, 1983*a,b*) and by Frota and Moffat (1982).

8

HWA TECHNIQUES FOR REVERSING FLOWS AND FOR THE NEAR-WALL REGION

8.1 THE FORWARD-REVERSE AMBIGUITY

Reversing flows occur in many practical engineering applications; they can be caused by, for example, flow over obstacles, sudden expansions in ducts and pipes, flow separation on strongly curved surfaces, and swirling flows (combustion chamber, etc.). These flows cannot be studied with standard *HWA* probes, since a single cylindrical hot-wire probe cannot detect a reversal of the flow direction, due to the rotational symmetry of the sensing element. The related signal ambiguity for a single normal (*SN*) hot-wire probe placed in a one-dimensional flow with periodic flow reversal (negative values of U) is illustrated in Fig. 8.1. The (linearized) *HWA* output voltage, E in Fig. 8.1(b) displays the typical folding or rectification of the anemometer signal (shaded areas), when flow reversal occurs. Due to this rectification, the measured mean value, $\bar{U}_m$, will be too high and $\overline{u_m^2}$ will be too low. Signal ambiguity will also occur for X- and triple-wire ($3W$) probes when the vector falls outside the approach-acceptance angle for the wire array. The following sections describe various sensor/instrumentation solutions to the rectification problems.

8.2 SHIELDED SINGLE-SENSOR PROBES

Neuerburg (1969) developed a directional hot-wire probe that permitted the study of periodically reversing flows. Directional sensitivity was accomplished by means of a simple hood, which partially enclosed the probe wire, in the form of an obliquely cut pipe end. This provided a wind-shadow effect, so that the wire element was only directly exposed to forward motion of the flow. The probe was used to measure the forward and backward flow in the individual inlet pipes of a six-cylinder, four-stroke, Otto engine.

8.3 CYLINDER-WAKE PROBES

The wake created by an unheated circular cylinder, of diameter D, placed in between, and parallel to, two heated wire elements has been used as the flow-detection criterion in two types of probes. Gupta and Srivastava (1979) utilized the regular Kármán vortex sheet shed from the cylinder, and the

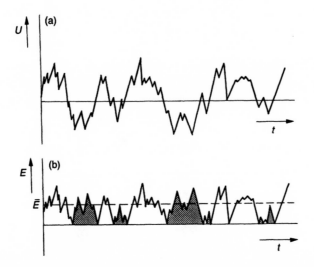

Fig. 8.1. (a) A one-dimensional, reversing-flow field, and (b) the corresponding linearized hot-wire signal.

downstream sensor was identified by the hot-wire signal, which contained the wake of the cylinder superimposed on its normal output signal. Flow reversal will be observed as an interchange in the signals from the two wires. Measurements of the shedding frequency, f, can, in principle, provide the magnitude of the flow velocity, U, as for example Roshko (1954) has shown that the Strouhal number $S = fD/U$, which in general is a unique function of the Reynolds number, is nearly constant ($\simeq 0.2$) at low Reynolds numbers. However, a major disadvantage of this probe is that the hot-wires and the cylinder are all placed in the same plane. Consequently, in turbulent flow, the cylinder wake will often miss or partly miss the downstream wire, making flow reversal difficult to detect.

In a probe developed by Mahler (1982), the wake was detected as a region of reduced mean velocity producing a smaller signal from the downstream sensor. To obtain a nearly one-dimensional flow past the wire arrangement, the cylinder and the two hot-wires were placed in a small tube containing inlet and exit nozzles, as shown in Fig. 8.2. For measurement purposes, the two wires were incorporated on two sides of a Wheatstone bridge with the remaining two sides being composed of fixed resistors. Using a trim pot, the bridge was initially balanced at a zero flow velocity. When the probe was placed in an air flow the bridge became unbalanced; this imbalance will be positive or negative depending upon the flow direction. The magnitude of the imbalance is a measure of the air speed. It is claimed that this probe is capable of measuring flow reversals in the velocity range $0 \pm 150\,\mathrm{cm\,s^{-1}}$.

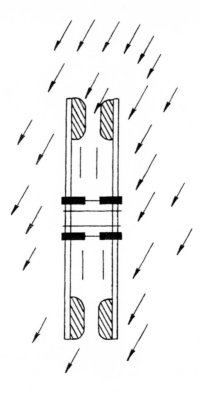

Fig. 8.2. A cylinder-wake hot-wire probe contained in a mounting tube. (From Mahler 1982.)

8.4 CONTINUOUSLY HEATED HOT-WAKE PROBES

In continuously heated hot-wire probes the flow direction is detected by the effect of the heated wake from one sensor on the output from a second sensor. The probes which have been developed contain either two heated sensors operated in the *CT* mode or one heated sensor and two resistance-wires placed on either side of the heated sensor.

A probe containing two hot-wire sensors operated in the *CT* mode was developed by Günkel *et al.* (1971) and used by Cook and Redfearn (1976). The two hot-wires were placed close and parallel to each other inside a shield, which created a one-dimensional flow past the two wires and made the probe insensitive to velocity components that were perpendicular to the shield. A number of tube and disk shields were tested to improve the performance of the probe. The calibration of each wire was performed with the wire in the upstream position, placing the shield normal to the velocity in the calibration unit. The two signals were linearized so that they had the same output characteristics. The turbulence noise caused by the shield was below 1%; this value is tolerable in flows of high turbulence intensity. The direction of the flow was detected by monitoring the decrease in the heat transfer from the downstream wire caused by the hot wake from the upstream wire. At any instant, only the larger of the two signals was passed through the gate of an amplifier, whilst the smaller signal was blocked. The sign of one of the signals was inverted if it passed through the amplifier gate to indicate flow reversal.

Downing (1972) used a probe containing a hot-wire and two sensor wires operated as resistance-wires. The probe configuration, which is shown in Fig. 8.3, is similar to that used by Bradbury and Castro (1971) for their pulsed-wire technique (see Section 8.5). The magnitude of the velocity was obtained from the hot-wire signal, and the flow direction was obtained from the two resistance-thermometer wires. The two sensor wires were operated by *CC*-mode Wheatstone bridges. The outputs from the Wheatstone bridges were fed to a differential amplifier and to a voltage comparator with a logic output, which was used to invert the hot-wire signal when flow reversal took place. This probe is restricted to measurements in pulsating one-dimensional flows.

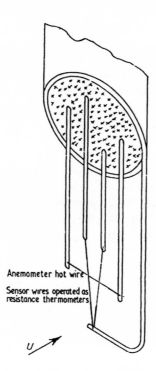

Anemometer hot wire

Sensor wires operated as resistance thermometers

U

Fig. 8.3. A continuously heated hot-wake probe. (From Downing 1972.)

For measurements of flow reversal in blood vessels, Seed (1969), Seed and Wood (1970a,b), and Seed and Thomas (1972) developed probes with surface mounted hot-film probes. Three gold films were deposited, by hand painting, on different shapes of cylindrical Pyrex-glass substrates with diameters of 0.7–1 mm. The central hot-film was used to determine the magnitude of the flow velocity and forward–backward direction sensing was obtained using the two hot-films placed parallel to each other on either side of the central hot-film and connected to two arms of a Wheatstone bridge. The heated wake from the upstream film is convected past the downstream sensor, causing a reduction in the sensitivity of the downstream sensor and a corresponding modification of the output from the Wheatstone bridge.

Horstman and Owen (1974) used two thin platinum film gauges operated at a constant temperature and flush mounted with the surface. The signal-analysis technique was based on cross-correlation of the turbulence characteristics of the output signals from the two films. Rubesin et al. (1975) used buried wire skin-friction gauges consisting of a heated central element with two sensors on either side operated as resistance elements. The difference between the two sensor signals was used to indicate the flow direction. However, there were severe limitations in the dynamic performance of this technique. Regardless of the design of the substrate, significant amounts of thermal conduction and thermal lag were found to occur. The thermal time constant of the substrate could be of the same order of magnitude as the time-scale for flow reversal, and the uncertainty in the zero-velocity position could be as large as the differential signal from the sensors.

Near-wall probes Probes placed just above the surface have been developed to overcome the substrate problems related to flush-mounted, hot-film elements. These probes usually consist of two or three hot-wires placed very close together, parallel to the wall, and perpendicular to the nominal flow direction. These probes are often referred to as thermal tufts. Probes used by Moon (1962) and Ligrani et al. (1983) contained two hot-wires operated in the CT mode. In the study by Moon, both the wires were operated at the same overheat ratio, and the difference in output from the two wires was taken as an indicator of the direction of the flow because the relative magnitude of the two signals depends on which wire is in the wake of the other. In the corresponding probe design developed by Ligrani et al. (1983), the sensitivity of the probe to flow reversal was increased by operating wire 1 at an overheat ratio of 1.6 and wire 2 at 1.1. The probe was used for near-wall measurements on a compressor blade, and the probe was calibrated and the signal interpreted in terms of the variation in the effective velocity ratio, V_{e2}/V_{e1} with velocity magnitude, $\tilde{V}$, yaw angle, α, and pitch angle, β. Flow-phenomena studies indicated that the output from the probe could be used to indicate abrupt changes in flow behaviour; for example, for reverse flow $V_{e2}/V_{e1} > 1$, and for the accelerated boundary-layer situation $V_{e2}/V_{e1} < 0.25$.

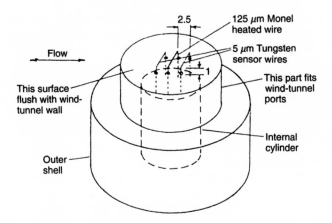

Fig. 8.4. A thermal tuft probe. (From Eaton *et al.* 1979.)

Carr and McCroskey (1979) used a three-wire probe to study the direction and magnitude of the instantaneous velocity on a dynamically stalling airfoil. The central heater wire was operated at an overheat ratio of 1.8, and the outer two sensor wires were used as heat detectors. The outputs from the two heat-detection sensors were electronically conditioned and compared to determine the instantaneous flow direction. Eaton *et al.* (1979) and Eaton and Johnston (1980) have also developed a three-wire wall-flow-direction probe. The probe shown in Fig. 8.4 consists of a 125 μm diameter central wire heated by a direct current (d.c.) of about 1.5 A and of two 5 μm tungsten sensor wires operated as resistance thermometers. The two sensors were operated in a bridge which was balanced when both wires were at the same temperature. The difference in outputs from the two sensors was used electronically to determine the flow direction. The three wires were mounted on six copper-plated sewing needles, which protruded through surface clearance holes. The needles could be extended to about 5 mm from the wall to facilitate wire mounting. Using a specially developed electronic circuitry, the investigators measured the fraction of time the flow was forward or backward at different wall locations. Adams and Johnston (1988) report the use of this technique for the determination of reattachment of flow over a backward-facing step. A probe with two sensors and three heater wires was developed by Shivaprasad and Simpson (1982) to eliminate the insensitivity of earlier near-wall designs to cross-flow. This probe was tested in separated turbulent shear flows produced by adverse pressure gradients.

8.5 THE PULSED-WIRE ANEMOMETER

The Pulsed-Wire Anemometer (*PWA*) technique was first reported by Bauer (1965), who used two parallel wires for measurements in a laminar flow. By pulsing the upstream wire with a voltage pulse of a few micro-seconds, a tracer of heated air was introduced into the air and the time taken for this tracer to reach the downstream wire was measured using the downstream sensor as a resistance wire. However, this parallel-wire arrangement is very sensitive to the flow direction, and in a turbulent flow many of the heat pulses would miss the resistance-wire altogether. Tombach (1969, 1973) and Bradbury (1969) subsequently independently recognized that, by the simple expedient of placing the sensor wire at right angles to the pulsed wire, it is possible to produce an instrument with a wide yaw response, which can be used in highly turbulent flows.

The *PWA* technique was first analysed in detail by Bradbury and Castro (1971); further probe developments were described by Bradbury (1976), by Jaroch (1985), and by Jaroch and Dahm (1988). The first *PWA* shear stress measurements were reported by Bradbury (1978); and Castro and Cheun (1982) have presented a related detailed error analysis. Comprehensive *PWA* reviews have been published by Handford and Bradshaw (1989) and by Castro (1991). The following description of the *PWA* technique is based primarily on these references.

8.5.1 The ideal *PWA* probe response

The Bradbury–Castro type of probe (Fig. 8.5) consists of three wires. The central wire is the pulsed-wire, and at a distance h on either side of this are the two sensor wires, with their axes perpendicular to the pulsed wire

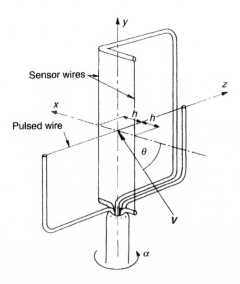

Fig. 8.5. The Bradbury–Castro type of *PWA* probe.

but parallel to each other. The plane of the probe is defined as the (y, z)-plane parallel to the pulsed and the two sensor wires. The angular response of the probe is specified by the yaw angle, α, which corresponds to rotation around the probe stem, and the pitch angle, β, which is related to rotation around the pulsed-wire. To achieve similar acceptance angles for both yaw and pitch, the pulsed and sensor wires are usually of similar length, ℓ.

An ideal PWA probe contains sensors with no thermal inertia. The temperature of the pulsed-wire will, during the period of the pulse heating, jump instantaneously from the ambient temperature, T_a, to a higher temperature, and it will be held there while heated air is convected away with the instantaneous velocity of the fluid passing the probe. Provided that both thermal diffusion and viscous wake effects are small, the time, T, taken for the front of the heated air to reach the sensor wire will be (see Fig. 8.5)

$$T = \frac{h}{\tilde{V}\cos\theta},\qquad(8.1)$$

where $\tilde{V}$ and θ are the magnitude and direction of the instantaneous velocity vector, V. By measuring the time of flight, T, the velocity component $V_x = \tilde{V}\cos\theta$ perpendicular to the (y, z)-plane can be obtained. The use of two sensor wires, one on either side of the pulsed-wire, ensures that the direction of the flow at that instant can be determined unambiguously. If a sufficient number of these time-of-flight measurements are made then it is possible to evaluate both the mean velocity and the Reynolds stress in the x-direction at right angles to the (y, z)-plane of the probe.

However, the real performance of a PWA probe deviates significantly from the above simplified description, due to the thermal inertia of the pulsed and sensor wires, viscous wake effects, and thermal diffusion of the convected heat pulse. To clarify these physical phenomena, the influence of thermal inertia on the response of a PWA probe placed in an ideal fluid without thermal diffusion will be considered first. In this analysis the x-axis of the PWA will be assumed to be aligned with the mean-flow direction.

8.5.2   The response of a PWA probe to a heating pulse

The thermal-energy balance for a heated wire with no appreciable conductive end losses can, as discussed in Chapter 2, be expressed as

$$I^2 R_w = \frac{E_w^2}{R_w} = \rho_w c_w \frac{\pi}{4} d^2 \ell \frac{dT_w}{dt} + \pi k \ell \mathrm{Nu}(T_w - T_a),\qquad(8.2)$$

rate of	rate of heat	rate of convective
electrical	storage in	heat transfer
heat generation	wire	

where T_w and T_a are the wire and flow temperatures, respectively; d is the wire diameter, and ℓ is its length; ρ_w and c_w are the density and specific heat of the wire material; R_w is the wire resistance; I is the current through the wire; E_w is the voltage across the wire; Nu is the Nusselt number; and k is the thermal conductivity of the fluid. Equation (8.2) can, with different simplifications, be applied to both the pulsed and the sensor wires.

The thermal inertia of a wire element is specified by its time constant, M (see eqn (2.47)). Introducing eqn (7.8) for the Nusselt number, Nu, the expression for M becomes

$$M = \frac{\rho_w c_w (\pi/4) d^2 \ell}{\pi \ell k \text{Nu} - I^2 \alpha_0 R_0} \tag{8.3}$$

and, as discussed in Section 2.3.2.1, M varies with $d^{3/2}$ provided that $d > 1\,\mu$m.

Response of the pulsed wire The diameter of the pulsed-wire is typically $5\,\mu$m, and the corresponding time constant, M_p, is about 1 ms at $5\,\text{m s}^{-1}$. This is much larger than the duration, $\hat{t}$ (2–10 μs), of the constant-voltage heating pulse, and during the time $\hat{t}$ the heat convection term in eqn (8.2) can be ignored, giving

$$\frac{E_w^2}{R_w} = \rho_w c_w \frac{\pi}{4} d^2 \ell \frac{dT_w}{dt} \tag{8.4}$$

and the wire resistance, R_w, is usually related to the wire temperature, T_w, by eqn (2.27); that is,

$$R_w = R_0 [1 + \alpha_0 (T_w - T_0)].$$

Equation (8.4) is a first-order differential equation, which can be integrated to give the maximum pulsed-wire temperature, $T_{p,\max}$, at the end of the pulse.

After the almost impulsive rise in temperature, the rate of decay of the pulsed-wire temperature, T_p, due to heat convection is governed by its time constant, M_p,

$$\frac{T_p - T_a}{T_{p,\max} - T_a} = \exp(-t/M_p). \tag{8.5}$$

Consequently, in normal use, the maximum pulsing frequency must be an order of magnitude lower than the reciprocal of the pulsed-wire time constant, M_p, to allow the wire to cool between pulses. The corresponding maximum pulsing frequency for a $5\,\mu$m diameter wire is usually quoted as being about 100 Hz. For measurements down to zero air speed, Handford and Bradshaw (1989) found it necessary to keep the frequency below 70 Hz.

Response of the sensor wire The sensor wires are operated in the *CC* mode as resistance wires at a very low heating current, and the electrical heating term in eqn (8.2) can therefore be neglected. Bradbury and Castro considered the response of a sensor wire with a time constant M_s to a convected heat pulse; they initially neglected the influence of viscous wake and thermal diffusion. The heat pulse was created by heating the pulsed-wire, with a time constant M_p, for a short time period $\hat{t}$. This resulted in a sharp-fronted heat pulse with an exponential decay being convected by a uniform flow field.

When a pulse of heated air arrives at the sensor wire there will be an increase in the wire temperature, T_s, along the wire length, ℓ. The corresponding change in the wire resistance, R_s, and its voltage, E_s, is

$$\Delta E_s = I(R_s - R_a) = IR_0\alpha \frac{1}{\ell} \int_{-\ell/2}^{\ell/2} (T_s - T_a)\, dy. \tag{8.6}$$

Provided the sensor wire is much longer than the width of the heat tracer, the increase in the sensor-wire signal caused by the heat pulse from the pulsed-wire can be expressed as

$$\frac{\int_{-\ell/2}^{\ell/2} (T_s - T_a)\, dy}{(T_{p,\max} - T_a)h} = \frac{\pi Nu_p}{Pe} \frac{M_p}{M_p - M_s} \left[\exp\left(-\frac{t - h/U}{M_p} \right) \right.$$
$$\left. - \exp\left(-\frac{t - h/U}{M_s} \right) \right] H\left(\frac{h}{U} \right), \tag{8.7}$$

where Nu_p is the Nusselt number for the pulsed wire, Pe is the Peclet number, Uh/k, and $H(h/U)$ is a Heaviside step function. This equation shows that the magnitude of the sensor signal does not depend on the distance, h, between the sensor wire and the pulsed wire. There is simply a convective time delay, equal to h/U, between the heat tracer leaving the pulsed-wire and it arriving at the sensor wire a distance h downstream.

The response equation, eqn (8.7) for the sensor wire has a discontinuity in the slope at $t = h/U$ with an initial value in nondimensional form which is given by

$$\frac{h^2}{k} \frac{d}{dt} \left[\frac{\int_{-\ell/2}^{\ell/2} (T_s - T_a)\, dy}{(T_{p,\max} - T_a)h} \right] = \pi Nu_p \frac{h}{UM_s}. \tag{8.8}$$

Consequently, if viscous wake and diffusion processes can be ignored, there are no difficulties in determining the time of flight because of the discontinuity in the slope at $t = h/U$. For signal-analysis purposes, if the signal is differentiated, it will exhibit a step rise at $t = h/U$, which can be used to trigger an electronic comparator to obtain the time of flight.

The maximum amplitude of the signal is given by

$$\left[\frac{\int_{-\ell/2}^{\ell/2} (T_s - T_a)\, dy}{(T_{p,\max} - T_a)h} \right]_{\max} = \frac{\pi Nu_p}{Pe} \left(\frac{M_p}{M_s} \right)^{-M_s/(M_p - M_s)}, \tag{8.9}$$

which occurs at a time

$$\frac{t}{M_s} = \frac{M_p}{M_p - M_s} \log \frac{M_p}{M_s} + \frac{h}{UM_s}. \tag{8.10}$$

The maximum amplitude of the sensor signal is observed to be related to the ratio of the pulsed and sensor-wire time constants, M_p/M_s. As discussed in Chapter 2, the time constant, M, of a wire element (see eqn (8.3)) varies with $d^{3/2}$, and it would therefore be advantageous to use a large-diameter pulsed wire and very thin sensor wires. The practical lower limit for the diameter, d_s, of the sensor wire is 2.5 μm, even using tungsten. As will be discussed below, the sensor wire length, ℓ, must be much larger than h to avoid signal drop-out, and ℓ is therefore an order of magnitude longer than is typical for a hot-wire probe. Consequently, the sensor wires are prone to breakage. Two phenomena restrict the upper value of d_p. The time constant, M_p, controls the cooling of the pulsed wire (eqn (8.5)) and therefore it controls the maximum pulsing frequency. Also, the effect of the viscous wake from the pulsed wire should be kept to a minimum. A practical compromise is to use a pulsed wire of about double the diameter of the sensor wire, that is, of around 5 μm.

For signal-analysis purposes, it is useful to consider the changes in the values of the maximum amplitude (see eqn (8.9)) and the initial slope (see eqn (8.8)) with flow speed. Bradbury and Castro estimated, assuming $Nu_p \sim Re^{0.45}$, that the velocity dependence of the maximum amplitude is $U^{-0.55}$, giving a variation in the signal level of about 5.2 : 1 over a 20 : 1 speed range. In contrast, the initial slope was found to vary much more slowly, varying as $U^{-0.1}$. Consequently, by differentiating the signal and using it to trigger a simple voltage comparator set to a constant threshold level, the time of flight can be determined over a wide speed range. In order to avoid triggering by small spurious signals, it is recommended that the comparator threshold should be set to about one-third of the typical peak values of the differentiated sensor signal (Handford and Bradshaw 1989).

The influence of thermal diffusion and viscous wake on the sensor-wire signal Bradbury and Castro also studied the effect of thermal diffusion in the heated wake of a pulsed-wire. They found that the solution to the two-dimensional energy equation for convection and diffusion of heat in a laminar stream was a strong function of the Peclet number, $Pe = Uh/k$, but they also concluded that when $Pe \gg 1$ longitudinal diffusion is only significant near the front of the pulse. They introduced a diffusion time, T_d, corresponding to the time taken for the diffusive front to pass the sensor wire, and demonstrated that the solution depended on the value of T_d, compared with the value of the time constant, M_s, of the sensor wire.

At a high Peclet number, the diffusion time, T_d, is small compared with M_s, and the main effect of diffusion is to replace the discontinuity in slope

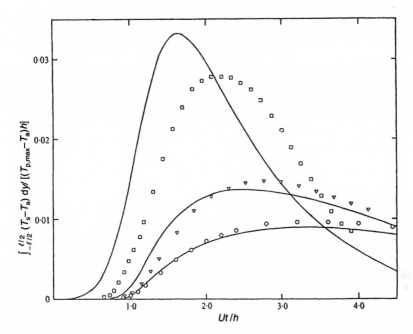

Fig. 8.6. A comparison of (———) theoretical and ($\square$, ∇, $\bigcirc$) experimental sensor-wire signals for Peclet numbers of: ($\square$) 23, (∇) 80, and ($\bigcirc$) 144. (From Bradbury and Castro 1971. Reproduced with kind permission of Cambridge University Press from the *Journal of Fluid Mechanics*.)

(occuring at $t = h/U$ without diffusion) by a transition region. This flow situation is shown for a Peclet number of 144 in Fig. 8.6. Following a short transition region, corresponding to the duration of the diffusion time, the sensor signal attains the nearly linear variation predicted by eqn (8.7). It should be noted that the linear part of the curve can be extrapolated back from the position of maximum slope, and that it intersects the zero-signal axis at $Ut/h = 1$, as suggested initially by Bauer (1965). When the Peclet number is reduced, the longitudinal diffusion effects become significant. This is illustrated in Fig. 8.6 by the solid curve corresponding to a Peclet number of 23. However, the experimental results included in Fig. 8.6 demonstrate that the viscous wake from the pulsed-wire will partly compensate for the longitudinal thermal diffusion. In general, significant diffusion effects will occur when the Peclet number is below 50.

8.5.3 Velocity calibration of a *PWA* probe

It is necessary to perform a velocity calibration for each sensor wire, due to the uncertainty in the exact value of the spacing, h, and due to the effects of thermal diffusion, viscous wake, and wire time constants. The response equation can be expressed in the form

$$U = h/T + f(T),\tag{8.11}$$

where U is the velocity, h is the wire spacing, T is the measured time of flight, and $f(T)$ is some small function introduced to account for the various secondary effects. In practice, h is not known accurately and it is replaced by an empirical constant. Usually, $f(T)$ is taken as being proportional to T^{-n}, so eqn (8.11) becomes

$$U = A/T + B/T^n,\tag{8.12}$$

where A and B are calibration constants determined by a least-squares curve fit to the calibration data. Bradbury (1976) selected $n = 2$ for his calibration work, but most workers have preferred $n = 3$ because the sign of T is included automatically in the data analysis. Handford and Bradshaw (1989) have compared a number of calibration laws, and their results are shown in Fig. 8.7, which demonstrates a similar curve-fitting accuracy for $n = 2$ and 3.

The wire spacing, h, influences the applicable velocity range of a *PWA*. Thermal diffusion has been shown to affect the signal analysis if the Peclet number (Uh/k) is reduced below 50, which in air (for a typical wire spacing of 1.2 mm) corresponds to a velocity of about 1 m s^{-1}. However, provided

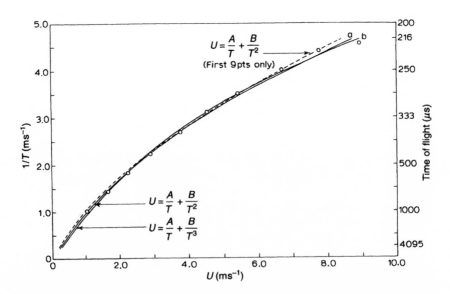

Fig. 8.7. Velocity versus time-of-flight calibration laws for a pulsed-wire anemometer. (From Handford and Bradshaw 1989.)

one can calibrate to much lower velocities, there is no reason why a *PWA* probe should not be used at such velocities (Castro, personal communication). The maximum velocity is restricted by probe-strength and vibration considerations. However, more importantly, the impulsive heating of the pulsed-wire creates an electromagnetic pickup signal in the sensor wire during the pulse time, $\hat{t}$ (2–10 μs). The decay of the induced pulse and subsequent ringing in the sensor wire signal may last for about 50–100 μs. To solve this problem, a commercial system (PELA Flow Instruments Ltd.) has a circuit which suppresses the differentiated signal used to trigger the voltage comparator for a time period, Δt_d, of about 100 μs after the probe is pulsed. The maximum detectable flow velocity is therefore

$$U_{max} = h/\Delta t_d \qquad (8.13)$$

and U_{max} will be about $12 \, \text{m s}^{-1}$ when $h = 1.2 \, \text{mm}$ and $\Delta t_d = 100 \, \mu$s.

8.5.4 Probe geometry and operational considerations

The thermal stresses set up when the pulsed-wire is heated can cause the whole probe to vibrate. If the wires are mounted when stretched the resulting strain-gauging of the sensor wires will result in a contaminated signal. To avoid this problem, the pulsed and sensor wires may be mounted with a small, tensioned, 'kink' in them so that they are straight but not taut nor so prone to vibration (Castro, personal communication). Any change in the wire spacing by the steady deflection will nominally be compensated for by the velocity calibration, but wire vibration and electronic noise will introduce a significant 'artificial' turbulence intensity of about 1–4% (Handford and Bradshaw 1989), which makes the *PWA* technique unsuitable for measurements in low turbulence intensity flows.

The sensor wire is operated as a resistance-wire, and the heating current, I, is a compromise between sensitivity and electronic-noise criteria. To operate the sensor wire as a resistance thermometer, the velocity sensitivity should be minimized. As discussed in Section 7.4.3, when the velocity and temperature field is uniform along the length of the wire, eqn (7.35) applies and $S_{u,cc}/S_{\theta,cc} \approx I^2/d^2$. Under normal operational conditions a 1 μm platinum resistance wire is typically operated with a heating current of 0.15 mA. However, in the *PWA* application the heat pulse usually only covers a small part of the sensor length and signal-to-noise considerations may be more important than the velocity-sensitivity criteria. Handford and Bradshaw (1989) quote a value of 1–2 mA for a 2.5 μm tungsten sensor wire. When the length, ℓ, of the sensor is larger than the width of the heat pulse, the temperature part of the sensor signal remains constant, while the Johnson noise of the sensor element varies with $\ell^{1/2}$ (LaRue *et al.* (1975). Therefore to obtain a large signal-to-noise ratio the sensor wire should be as short as possible.

The ratio of the wire spacing, h, to the sensor length, ℓ, controls the detection-angle range for the heat tracer. From Fig. 8.5 it follows that the maximum yaw acceptance angle at which the sensor wire will detect the heat tracer is

$$\alpha_{\text{max}} = \tan^{-1}\left(\frac{\ell}{2h}\right). \tag{8.14}$$

Bradbury and Castro (1971) demonstrated that an acceptance-cone angle of about 60° is sufficient for accurate measurements of the streamwise mean velocity, $\bar{U}$, and normal stress, $\overline{u^2}$. For general Reynolds-stress measurements (for example, $\overline{v^2}$ and $\overline{uv}$) the probe must be placed at an angle to the mean-flow direction (typically ±45°), and a larger acceptance cone is required to avoid significant errors due to missed heat pulses. For a probe with a wire spacing, h, of about 1.2 mm, a reasonable compromise between signal-to-noise ratio and acceptance-angle criteria is a length, ℓ, of about 10 mm, corresponding to an acceptance angle of about 75°.

Yaw and pitch response The *PWA* theory assumes a perfect cosine angle dependence. However, when the, flow angle is larger than the acceptance-cone angle, the probe will record zero velocity because the heat tracer misses the sensor wire. Bradbury (1976) also observed that the yaw response deviates from a cosine dependence. Jaroch (1985) has investigated, in detail, the yaw and pitch dependence of *PWA* probes with 5 μm diameter tungsten wires. Jaroch demonstrated, as shown in Fig. 8.8, that the standard *PWA*

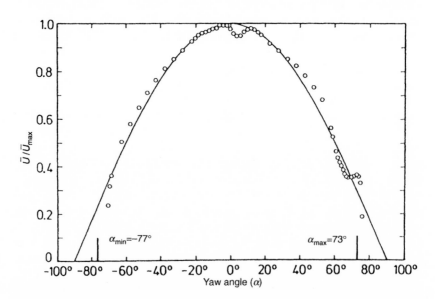

Fig. 8.8. The yaw-response characteristic of the original Bradbury–Castro *PWA*: (———) the cosine function. (From Jaroch 1985.)

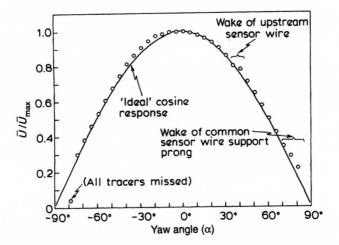

Fig. 8.9. The yaw response of a modified *PWA* at zero pitch. (From Handford and Bradshaw 1989.)

probe had a significant yaw defect at $\alpha = 0°$, caused by the wake velocity deficit from the upstream sensor affecting the downstream sensor. This problem can be reduced by using 2.5 μm wires. Also, by rotating the plane of the sensor wires 30° (or 45°) around the stem axis, the corresponding kink in the yaw response, shown in Fig. 8.9, will be of less consequence because fewer readings will be affected when sampling in a turbulent flow.

Jaroch (1985) and Jaroch and Dahm (1988) have studied the effect of the *PWA* geometry on the measured Reynolds stresses in detail. They found that, within the probe's acceptance cone, prong wakes could significantly affect the measured signals at large flow angles. They proposed and tested various improved probe designs.

8.5.5 Turbulence measurements with a *PWA* probe

A *PWA* probe aligned with the streamwise direction ($\alpha = 0°$) can provide accurate measurements of $\bar{U}$ and $\overline{u^2}$. As first described by Bradbury (1978), by inclining the pulsed wire probe relative to the mean-flow direction additional Reynolds stresses (for example, $\overline{v^2}$ and $\overline{uv}$) can be measured. For a probe placed with a yaw angle, α, and a perfect cosine response, the equation for the measured effective fluctuating velocity can be expressed as

$$\overline{v_{e,\alpha}^2} = \overline{u^2}\cos^2\alpha + \overline{v^2}\sin^2\alpha + \overline{uv}\sin 2\alpha. \tag{8.15}$$

Castro and Cheun (1982) and Jaroch (1985) have shown that, provided the probe geometry is selected to ensure a particularly good yaw and pitch response, measurements at three yaw angles are sufficient to obtain accurate values of $\overline{u^2}$, $\overline{v^2}$, and $\overline{uv}$. For the inclined probe positions, Castro and Cheun (1982) recommend angles of $\pm 45°$ for optimum conditions. The accuracy of the evaluated quantities depends on both the total number of samples, N, and on the sampling rate, SR, as discussed in Chapter 12. For measurements of the mean velocity and Reynolds stresses, using a given number of samples, N, the most accurate results are obtained when the samples are statistically independent, which requires the sampling interval to be at least twice the integral time-scale of the flow (see Chapter 12). From autocorrelation measurements in separated air flows (Castro 1985) the related integral time-scale is typically of the order of 10–50 ms, which is a similar order of magnitude as the maximum sampling rate of about 100 Hz for a *PWA* probe. For measurements in this type of flow, the rather slow pulsing rate of a *PWA* probe is therefore close to the optimum sampling rate for simple statistical turbulent quantities.

The *PWA* technique has been used for Reynolds-stress measurements, mainly in separated flows (for example, Dengel *et al*. 1982; Ruderich and Fernholz 1986; Castro and Haque 1987, 1988; Castro *et al*. 1989; Dianat and Castro 1989; Jaroch and Fernholz 1989; Dengel and Fernholz 1990; and Dianat and Castro 1991). Castro (1991) has presented a review of flow-field studies by the *PWA* technique. The *PWA* technique has been developed further to enable the evaluation of additional turbulent quantities. Dianat and Castro (1991) have shown that third-order turbulence quantities can be evaluated from

$$\overline{v_{e,\alpha}^3} = \overline{u^3}\cos^3\alpha + \overline{v^3}\sin^3\alpha + 3\,\overline{u^2 v}\sin\alpha\cos^2\alpha + 3\,\overline{uv^2}\sin^2\alpha\cos\alpha. \tag{8.16}$$

In principle, the four triple products can be evaluated from measurements at four angle positions. However, to obtain an acceptable level of accuracy Dianat and Castro found that it was necessary to apply a least-squares curve-fitting technique to measurements at 10–12 angles in the range $-60° \leqslant \alpha \leqslant 60°$.

Gaster and Bradbury (1976) and Bradbury (1978) demonstrated how spectral measurements could be made using either a random or an arithmetic sequence of pulses to overcome the (aliasing) difficulty caused by the relatively low maximum sampling rate. This time-domain technique has been developed further by Castro (1985) for both autocorrelation and spectral evaluations. More recently it has been demonstrated by Nasser and Castro (1989) that *PWA* can also be used to obtain spatial velocity correlations (with or without a time delay).

8.5.6 Near-wall *PWA* measurements

A conventional *PWA* cannot be used close to a wall due to its relatively large size, and special *PWA* probes have been developed for the near-wall region. These probes are built around a circular plug designed to fit into an instrument port in a model or wind-tunnel wall, with the top surface of the plug mounted flush with the test surface. The prongs, and in two of the designs the pulsed or sensor wires also, protrude through holes in the top surface of the plug. Using an accurate spring-loaded micrometer assembly, the position of the entire probe head can be varied between about 0.025 and 10 mm from the surface of the plug.

In the probe developed by Westphal *et al.* (1981), the pulsed-wire and the two sensor wires are mounted parallel to each other in a plane parallel to the surface of the plug. This geometry is only appropriate for use where the streamlines are nearly parallel to the wall since a substantial vertical (v) velocity component will cause the heat tracer to miss the sensor wire. For their measurements, Westphal *et al.* (1981) selected $H = 0.2$ mm, and the probe was used primarily for time-dependent skin-friction measurements; this is discussed in Section 8.8.3.

Two probes with a conventional *PWA* wire arrangement have been developed. In the probe used by Castro and Dianat (1987, 1990), the pulsed-wire is parallel to the surface and the two perpendicular sensor wires protrude through surface holes. Devenport *et al.* (1990) mounted the two sensor wires parallel to the surface, and the pulsed-wire protruded through a hole in the surface, as shown in Fig. 8.10. This probe is a development of an initial design by Evans (1973). Detailed measurements with this probe in separated flow regions have been reported by Devenport and Sutton (1991, 1992). Compared with a standard *PWA* probe, the near-wall type has several additional potential sources of error:

(1) Uncertainty in the absolute distance from the wall.

(2) Inadequate spatial resolution of the smaller turbulence scales.

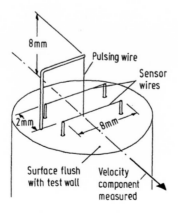

Fig. 8.10. A near-wall pulsed hot-wire probe. (From Devenport *et al.* 1990.)

(3) Transverse thermal diffusion of heat tracers in flow regions with a
    strong mean velocity gradient.

For measurements closer than 1 mm from the wall, the value of y must be
measured very accurately to enable a $U^+(= \bar{U}/U_\tau)$ versus $y^+(= yU_\tau/v)$
comparison, where $U_\tau(= (\tau_w/\rho)^{1/2})$ is the friction velocity and τ_w is the
mean wall shear stress. Procedures for the accurate measurement of near-
wall distances are discussed in Section 8.8.1.

Errors resulting from the inadequate spatial resolution of pulsed-wire
probes have been analysed by Adams *et al.* (1984), who concluded that
the measured turbulence level will be reduced, but not the mean-velocity.
Near the wall of a turbulent boundary layer, the Kolmogorov length scale,
η, may be as small as $100\,\mu m$, and in this region η is about 1.5–2 viscous
lengths (v/U_τ) (Ligrani and Bradshaw 1987*b*). A wire spacing, h, of 1 mm
(Devenport *et al.* 1990) corresponds to about twenty viscous lengths, and
comparative measurements with a *SN* hot-wire probe demonstrated an
underestimation of $(\overline{u^2})^{1/2}/U_\infty$ of about 10% (Fig. 8.11), which is presumably caused mainly by this effect. If the wire spacing is equivalent to forty
viscous lengths then the peak turbulence intensity may be underestimated
by about 30%.

When the probe is exposed to a substantial mean velocity gradient over
its measuring volume the measured mean velocity will be in error due to turbulent lateral diffusion, which spreads the heated tracer in the y-direction to

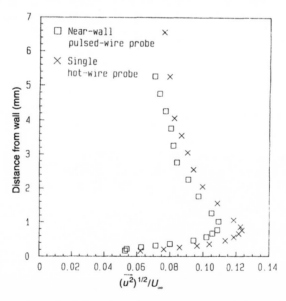

Fig. 8.11. A comparison of turbulence-intensity measurements made in a turbulent boundary
layer by a near-wall *PWA* and an *SN* hot-wire probe. (From Devenport *et al.* 1990.)

a faster moving stream of fluid. As the sensor responds to the fastest part of the pulse front, the probe will record too high a velocity. This effect, which is most noticeable within the linear sublayer, has been studied by Castro and Dianat (1990), who have proposed various correction procedures.

8.6 THE FLYING HOT-WIRE-ANEMOMETER TECHNIQUE

An increasing amount of attention has been devoted to the Flying Hot-wire Anemometer (*FHA*) technique and signal-analysis procedures of moving hot-wire probes. Payne and Lumley (1966) and Sheih *et al*. (1971) measured one-dimensional spectra and the small-scale structure of atmospheric turbulence with an *SN* hot-wire probe attached to an aeroplane. This method was also used by Jacobsen (1977) to study aircraft-wake vortices. Several investigations with a hot-wire probe placed on a rotating arm have been reported. Corsiglia *et al*. (1973) studied the trailing vortices of a rectangular wing tip. Turbomachinery flows have been studied by Gerich (1975), Gorton and Lakshminarayana (1978), Hah and Lakshminarayana (1978), and Kjörk and Löfdahl (1989). More recently the *FHA* method has been used by Walker and Maxey (1985) and by Sirivat (1989) to study the structure of grid turbulence. They tested the applicability of Taylor's frozen-turbulence-pattern hypothesis for transformation of autocorrelation and turbulence statistics (see Section 2.4.2) by comparison between *FHA* and stationary hot-wire-probe data. Related *FHA* studies of Taylors hypothesis have also been reported by Hussein and George (1989*b*) and George *et al*. (1989). Quasi-instantaneous multipoint velocity measurements with a rotating *X*-probe have been described by Wark *et al*. (1990). Finally, Collings *et al*. (1987) and Dinsdale *et al*. (1988) have studied the flow field in a motored internal-combustion engine with a flying hot-wire probe.

8.6.1 The basic principle of the *FHA* system

One of the main restrictions of stationary hot-wire anemometry is the inability to measure and correctly interpret flows containing flow reversals. As the hot-wire probe responds to the velocity relative to the sensors, this problem can be overcome by using a flying hot-wire probe (*FHA*).

The basic principle of the *FHA* can be explained with reference to Fig. 8.12. Consider a surface with a (two-dimensional) separated flow region. A space-fixed coordinate system is introduced, in which the flow velocity vector, V, and its velocity components U and V are to be evaluated. Specified by the geometry of the *FHA* mechanism the sensors (usually an *X*-configuration) of the probe will follow a prescribed curve; for example, curve (a) in Fig. 8.12. At a time t, the probe is assumed to be at a known position (x_p, y_p) and to move with a known probe velocity, V_p.

The moving hot-wire probe is exposed to a relative velocity, V_r, and this

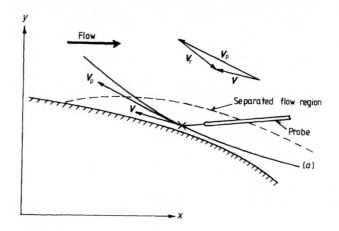

Fig. 8.12. The principle of measurements with an *X*-configuration flying hot-wire probe.

velocity vector is normally measured and evaluated for an *X*-probe in terms of the velocity components (U'_r, V'_r) in a probe-stem aligned coordinate system. Provided the orientation of the probe-stem relative to the space-fixed coordinate system is known, the corresponding space-fixed velocity components (U_r, V_r) can be evaluated. Having measured V_p and V_r the flow vector, V, is obtained from

$$V = V_p + V_r, \tag{8.17}$$

as illustrated in Fig. 8.12. Provided that the magnitude of the probe velocity, $\tilde{V}_p$, is greater than the magnitude of the flow velocity, $\tilde{V}$, then the relative velocity vector, V_r, will remain within the approach quadrant of the *X*-probe, and the hot-wire signals can be interpreted uniquely.

8.6.2 Implementation of an *FHA* system

Any path which cuts through the separated flow region in such a manner that the relative velocity remains within the approach quadrant of the *X*-probe can be used. In practice, three different probe paths and related mechanisms have been used. (i) a circular motion (Cantwell 1976; Wadcock 1978; Coles and Wadcock 1979; Cantwell and Coles 1983; Walker and Maxey 1985; Sirivat 1989; Hussein and George 1989; George *et al.* 1989*b*; Hussein 1990). (ii) A linear motion (Perry 1982; Watmuff *et al.* 1983; Hussein 1990; and Panchapakesan and Lumley 1993). Crouch and Saric (1986) have also reported the development of an oscillatory *SN* hot-wire system. (iii) A curvilinear 'bean-shaped' motion (Thompson 1984, 1987; Thompson and Whitelaw 1984; Jaju 1987; Al-Kayiem 1989; Bruun *et al.* 1990*a*; Al-

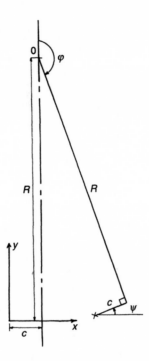

Fig. 8.13. The geometry and notation for a circular probe motion.

Kayiem and Bruun 1991; Badran 1993).

The principle of the mechanical implementation for the circular and bean-shaped curve paths is illustrated in Figs 8.13 and 8.14. These figures also contain the notation used for the evaluation of the position and the velocity of the flying hot-wire probe.

8.6.3 Probe position and probe velocity

In all *FHA* methods it is necessary to specify the positions at which measurements are to be carried out and to evaluate the corresponding probe velocity.

Linear motion For linear motion, the probe motion and hence the probe velocity, V_p, is usually aligned with the space-fixed coordinate system used for the evaluation of the velocity components of the flow velocity, V. Consequently, the components of the probe velocity, V_p, will be $(U_p, 0)$. Provided the probe velocity is nearly constant, U_p can be evaluated from

$$U_p = \frac{\Delta x_p}{\Delta t},$$
(8.18)

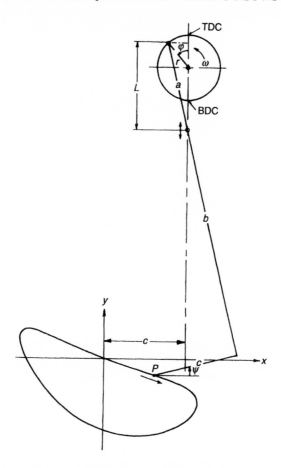

Fig. 8.14. Four-bar flying hot-wire probe mechanism and notation, showing the top dead centre (TDC) and the bottom dead centre (BDC). The geometry is that of the Bradford *FHA* system: $r = 60$ mm, $a = 160$ mm, $b = 410$ mm, and $c = 147$ mm.

by measuring the time Δt corresponding to a specified distance Δx_p.

Circular motion The position (x_p, y_p) of the probe and the related probe velocity, $V_p = (U_p, V_p)$, can be determined from the geometry shown in Fig. 8.13. In this figure, a hot-wire probe, with a support of length c, is mounted at right angles to a rotating arm of length R. The reference angular position $\varphi = 0°$ corresponds to the arm being at the Top Dead Centre (TDC). The origin of the selected space-fixed coordinate system for (x_p, y_p) corresponds to the arm being at the Bottom Dead Centre ($\varphi = 180°$). From geometry considerations it follows that

$$x_p = c + R \sin \varphi + c \cos \varphi, \tag{8.19}$$

$$y_p = R + R \cos \varphi - c \sin \varphi. \tag{8.20}$$

The velocity components U_p and V_p can be obtained by differentiating eqns (8.19) and (8.20) with respect to time, t, and setting $d\varphi/dt = \omega$:

$$U_p = \omega (R \cos \varphi - c \sin \varphi), \tag{8.21}$$

$$V_p = -\omega (R \sin \varphi + c \cos \varphi). \tag{8.22}$$

The angle ψ which the probe and its support makes with the x-axis is required for measurements with an X-hot-wire-probe. For the geometry shown in Fig. 8.13, the following simple angle relationship applies

$$\psi = 180° - \varphi. \tag{8.23}$$

Curvilinear motion Based on derivations given by Thompson and Whitelaw (1984), and using the notation and coordinate system illustrated in Fig. 8.14, it can be shown that the probe position (x_p, y_p) corresponding to the angle φ is given by

$$x_p = r(b/a) \sin \varphi - (c/a)L + c, \tag{8.24}$$

$$y_p = a + b - r + r \cos \varphi - (1 + b/a)L - r(c/a) \sin \varphi. \tag{8.25}$$

For convenience, the distance L, illustrated in Fig. 8.14, has been included in these equations, where

$$L = [a^2 - (r \sin \varphi)^2]^{1/2}. \tag{8.26}$$

The probe velocity components (U_p, V_p) in the space-fixed (x, y) coordinate system are obtained by differentiating eqns (8.24) and (8.25) with respect to time, t, and setting $d\varphi/dt = \omega$, giving

$$U_p = \omega r \left[\frac{b}{a} \cos \varphi + \frac{cr \sin 2\varphi}{2aL} \right], \tag{8.27}$$

$$V_p = \omega r \left[\left(1 + \frac{b}{a} \right) \left(\frac{r \sin 2\varphi}{2L} \right) - \sin \varphi - \frac{c}{a} \cos \varphi \right]. \tag{8.28}$$

The angle, ψ, which the probe-stem makes with the x-axis is required for measurements with an X-hot-wire probe. From the geometry in Fig. 8.14, it can be shown that the angle ψ can be evaluated from

$$\psi = \sin^{-1}[(r \sin \varphi)/a]. \tag{8.29}$$

Equations (8.19) and (8.20) for the circular motion, and eqns (8.24) and (8.25) for the curvilinear motion, show that for a given geometry the probe position (x_p, y_p) can be determined if the angular position φ is known. This information can be obtained by fixing a rotary encoder to the shaft of the driving motor. By monitoring the pulses from the encoder and hence the related angle φ, the position of the probe can be evaluated. Also, in both cases, to evaluate the probe velocity (U_p, V_p) corresponding values of both φ and ω must be known. The value of ω can be determined from

$$\omega = \frac{\Delta\varphi}{\Delta t},\qquad(8.30)$$

where $\Delta\varphi$ (in radians) is the angular rotation between two encoder pulses and Δt is the corresponding time interval. If an encoder with N pulses is used then $\Delta\varphi = 2\pi/N$ (in radians). In the investigation by Al-Kayiem and Bruun (1991), an encoder with 500 pulses and a reference pulse was used. The measurement of the pulse time, Δt, was carried out using a counter board as in the investigation by Thompson and Whitelaw (1984). Using recent electronic development, Δt can now also be measured by time tagging.

8.6.4 Measurement of the relative velocity

Most *FHA* systems utilize one or more X-probes, which are calibrated stationary using one of the calibration methods described in Chapter 5. Al-Kayiem and Bruun (1991) have presented a related *in-situ* static calibration method for a flying X-hot-wire probe. In the following analysis, it will be assumed that the calibration procedure has evaluated the calibration constants in the response equations for the two wires

$$E_1^2 = A_1 + B_1 V_{e1}^{n1},\qquad(8.31a)$$

$$E_2^2 = A_2 + B_2 V_{e2}^{n2},\qquad(8.31b)$$

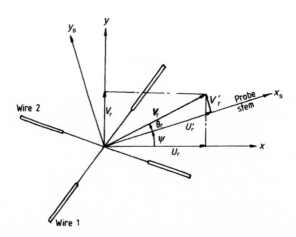

Fig. 8.15. Components of the relative velocity V_r in the probe-stem and space-fixed coordinate systems.

with

$$V_e = \tilde{V}(\cos^2 \alpha + k^2 \sin^2 \alpha)^{1/2},$$

where $\tilde{V}$ is the magnitude of the velocity vector and α is the yaw angle.

An X-hot-wire probe moving in a flow field will respond to the velocity vector seen by the sensor; the measured velocity components are normally evaluated in the probe-stem coordinate system (x_s, y_s) shown in Fig. 8.15. Since the probe is moving with a probe velocity of V_p, the X-probe will not measure the absolute velocity vector, V, but it will respond to the relative velocity vector, V_r, which is related to V by eqn (8.17).

The response equations, eqns (8.31a–b), can therefore also be applied to the moving probe, with the effective velocity, V_e, being related to V_r instead of to V. In this analysis, eqns (8.31a–b) can be applied directly, and the effective velocities V_{e1} and V_{e2} corresponding to each measured voltage pair (E_1, E_2) are obtained by inverting eqns (8.31a–b).

$$V_{e1} = \left(\frac{E_1^2 - A_1}{B_1} \right)^{1/n1}, \tag{8.32a}$$

$$V_{e2} = \left(\frac{E_2^2 - A_2}{B_2} \right)^{1/n2}. \tag{8.32b}$$

In the probe-stem coordinate system (x_s, y_s) of the *FHA* (Fig. 8.15), the magnitude of the velocity vector is $\tilde{V}_r$, the flow angle is θ_r and the corresponding velocity components are denoted U_r' and V_r'. Using the probe-stem equations, (eqns (5.30a–b)), the relationship between (V_{e1}, V_{e2}) and (U_r', V_r') can be expressed to first order as

$$V_{e1} = f(\bar{\alpha}_1)[U_r' - g_1(\bar{\alpha}_1)V_r'], \tag{8.33a}$$

$$V_{e2} = f(\bar{\alpha}_2)[U_r' + g_2(\bar{\alpha}_2)V_r'], \tag{8.33b}$$

where (see Table 5.2)

$$f(\bar{\alpha}) = (\cos^2 \bar{\alpha} + k^2 \sin^2 \bar{\alpha})^{1/2},$$

and

$$g(\bar{\alpha}) = \frac{(1 - k^2)\cos^2 \bar{\alpha}}{\cos^2 \bar{\alpha} + k^2 \sin^2 \bar{\alpha}} \tan \bar{\alpha}.$$

Solving eqns (8.33a–b) for U_r' and V_r' by the modified sum-and-difference method of eqns (5.33a–b) gives

$$U_r' = \frac{[V_{e1}/f_1(\bar{\alpha}_1)]g_2(\bar{\alpha}_2) + [V_{e2}/f_2(\bar{\alpha}_2)]g_1(\bar{\alpha}_1)}{g_1(\bar{\alpha}_1) + g_2(\bar{\alpha}_2)}, \tag{8.34a}$$

$$V_r' = \frac{[V_{e2}/f_2(\bar{\alpha}_2)] - [V_{e1}/f_1(\bar{\alpha}_1)]}{g_1(\bar{\alpha}_1) + g_2(\bar{\alpha}_2)}. \tag{8.34b}$$

Furthermore, for all three probe paths at each specified measuring point the value of the probe-stem angle, ψ, is known. For the linear motion, $\psi = 0°$; for the circular motion, ψ is given by eqn (8.23); and for the curvilinear motion it is given by eqn (8.29). This enables the transformation of (U_r', V_r') into the corresponding (U_r, V_r) velocity components in the space-fixed coordinate system

$$U_r = U_r' \cos \psi - V_r' \sin \psi \qquad (8.35a)$$

$$V_r = U_r' \sin \psi + V_r' \cos \psi \qquad (8.35b)$$

8.6.5 Evaluation of the flow velocity

Having calculated the related values of (U_p, V_p) and (U_r, V_r) at a measuring point, the actual flow-velocity components can be evaluated from eqn (8.17) as

$$U = U_r + U_p \qquad (8.36a)$$

and

$$V = V_r + V_p. \qquad (8.36b)$$

The magnitude of the flow velocity and its direction can be obtained from

$$\tilde{V} = (U^2 + V^2)^{1/2} \qquad (8.37a)$$

and

$$\theta = \tan^{-1}(V/U). \qquad (8.37b)$$

8.6.6 Operational procedure for the Bradford *FHA* system

The control of the *FHA* system and the related data acquisition and analysis is governed by the angular position, φ, of the flywheel, which is attached to the driving motor. The values of and variation in φ were determined from a 500-pulse rotary encoder linked to a position binary counter. The significant angular positions and related pulse counts are shown on Fig. 8.16(a).

The sequence of events for a single sweep were as follows. The flywheel resting position was initially adjusted manually to a value of $\varphi = -74°$. When the power supply to the driving motor is switched on (either as a manual motor start for one cycle or using the timer for continuous operation every 10 s) the flywheel starts to rotate anticlockwise from the REST position. When the flywheel passes the MARK position, the related marker pulse from the encoder causes a position binary counter to be reset to zero. The position binary counter then starts counting the pulses from the encoder, and when the pulse count reaches 256 (the START M position) the position binary counter initiates the data acquisition of the two simul-

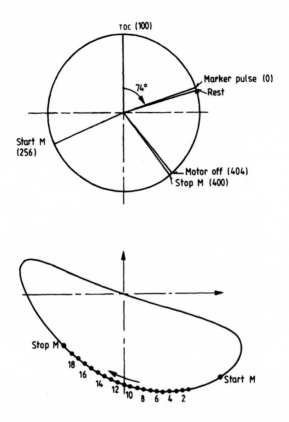

Fig. 8.16. The measurement procedure for the Bradford *FHA* system: (a) the significant position of the flywheel and the related pulse counts, and (b) identification of the measurement region and nineteen selected measurement points.

taneous wire signals from the *X*-probe, E_1 and E_2, and of the time period, Δt, between the two consecutive encoder pulses. As the flywheel continues to rotate, data transfer to the computer takes place for each pulse until a counter reading of 400 is reached (the STOP M position), when the data transfer to the computer is stopped. At count 404 a signal from the binary position counter causes the current to the motor to be cut off. Due to the angular momentum, the flywheel continues to rotate about two further rotations until it is stopped by an arrester mechanism.

The original design of the mechanical and electronic parts of this *FHA* system has been described by Thompson (1984). Jaju (1987), and Al-Kayiem (1989) give details of the modified system used at Bradford University. In our investigations, data were acquired from 144 points in the measurement region during one sweep, and for the study of separated flow 19 evenly spaced points were selected for detailed analysis. The positions of these points are shown in Fig. 8.16(b). To obtain statistical information of the flow field, the sweep was repeated N times and ensemble-averaging techniques were applied. The capabilities of the *FHA* system are illustrated by the measured two-dimensional mean-velocity field behind a backward-facing step (Fig. 3.4(c)). Reversed flow within a large separated region is clearly visible.

8.7 SPLIT-FILM PROBES

Split-film probes have been developed and applied to a number of two-dimensional or three-dimensional reversing-flow situations. The diameter of the fibre on which the films are deposited is large ($d \geqslant 150 \, \mu$m) compared to that of a hot-wire, and the frequency response of these probes in air flows is rather poor, as discussed in Section 5.7.1. It is therefore unlikely that split-film probes can be used for accurate turbulence studies in air flows, but they can provide reasonably accurate mean-velocity results and information related to, for example, the low-frequency unsteadiness of the separated flow behind a backward-facing step (Eaton and Johnston 1983). The performance of a split-film probe in liquid flows is much better, since the turbulence has a much lower frequency content.

8.7.1 Forward flow-fraction measurements using a split-film sensor

Reattachment of a separated flow on a solid surface is of great practical importance due to the large variations in turbulent structures and the heat transfer in the reattachment zone. The thermal-tuft method (Eaton *et al.* 1979), described in Section 8.4, is often used to determine the forward fraction, γ_p, of the flow. An alternative method is to place the split-film probe (see Fig. 5.14) with the plane of the splits perpendicular to the mean flow direction. The use of this technique has been reported by Wentz and Seetharam (1977), by Kiya and Sasaki (1983), and by Ra *et al.* (1990*b*). In this method, the output voltages from the anemometers connected to the two film strips were compared with each other using an analog comparator, and since the output from one anemometer is generally greater than that of the other, the instantaneous flow direction can be determined. For accurate results the overheat ratios of the two films must be closely matched. Figure 8.17 shows a comparison of γ_p data near the wall obtained with thermal-tuft and split-film probes.

8.7.2 Triple-split probes

Jørgensen (1982), DISA (1982) have developed a three-sensor probe, with three film-sensors placed in parallel paths, displaced by 120° on a common 400 μm diameter quartz fibre. The probe, which is shown in Fig. 8.18, can

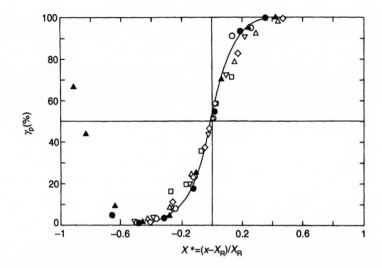

Fig. 8.17. A comparison of near-wall forward-flow-fraction, γ_p, measurements obtained with the split-film and thermal-tuft methods. (●) Ra *et al.* (1990*b*); (▲) Eaton *et al.* (1979); (△, ▽, ○) Eaton and Johnston (1980) for Re_θ = 850, 510, 240, respectively; (□, ◊) Pronchick (1983) thermal tuft and LDA, respectively; and (——) Westphal and Johnston (1984). (From Ra *et al.* 1990*b*.)

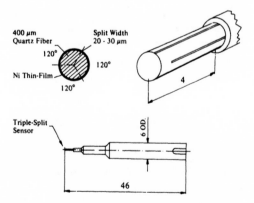

Fig. 8.18. A triple-split film probe for measurements in two-dimensional reversing flows. (Printed with the permission of DANTEC Measurement Technology.)

cover the complete 360° in one plane; it is intended for measurements in two-dimensional reversing flows with the sensor placed perpendicularly to the flow. During calibration the probe is exposed to a number of velocities and rotated 360° around the sensor axis at each velocity setting, while reading the sensor output at selected angular positions. The calibration procedure, which requires 135 combinations of the flow and angle, is performed automatically by computer. A set of 135 calibration constants can be established, which completely describe the relations between the three sensor outputs and the velocity and the angle of the flow. Jørgensen estimates that the velocity vector can be measured with an accuracy of ±3% and the flow angle with a standard deviation of less than 0.5°.

8.7.3 The TSI total-velocity-vector system

The TSI model 1080 total-velocity-vector system was first reported by Olin and Kiland (1970). A feasibility study of the use of this probe for measurements in a simulated tornado flow has been reported by Fattahi and Hsu (1977). The probe consists of three orthogonal split-film sensors, and it can be treated as a $3W$ probe (as described in Section 6.3), with the additional advantage that the axial splits enable the octant of the velocity vector to be determined. However, split-film probes have a rather poor frequency response (Section 5.7.1), which limits the use of this probe for turbulence studies in air flows.

8.8 NEAR-WALL AND SKIN-FRICTION MEASUREMENTS

Detailed information about the flow in the near-wall region is of particular interest for flow and heat-transfer predictions and it is of great importance for turbulence modelling. For a turbulent boundary layer with a zero- or a moderate-pressure gradient, the 'law of the wall' implies that in the inner region the mean velocity and the intensities of the three components of velocity fluctuations are governed by the viscous length, v/U_τ, where v is the kinematic viscosity and the friction velocity, U_τ, equals $(\tau_w/\rho)^{1/2}$ (τ_w is the mean wall shear stress, and ρ is the density of the fluid). This representation requires measurements of τ_w, as described in Section 8.8.2. In nondimensional form, the 'law of the wall' is usually expressed as

$$U^+ = f(y^+) \tag{8.38}$$

where $U^+ = \bar{U}/U_\tau$ and $y^+ = yU_\tau/v$. The turbulent boundary layer can be separated into three regions: a viscous sublayer with a linear velocity distribution, a buffer layer, and a turbulent region with a logarithmic velocity variation. The complete universal profile is given by

$$
\begin{aligned}
y^+ &\leqslant 5 & U^+ &= y^+, \\
5 < y^+ &\leqslant 30 & &\text{buffer layer,} \\
y^+ &> 30 & U^+ &= \frac{1}{\kappa} \ln y^+ + C,
\end{aligned}
\tag{8.39}
$$

where $\kappa = 0.40$ is the universal von Kármán constant and the constant C typically varies between 4.5 and 6.0, depending on the Reynolds number and the flow conditions. As discussed by Alfredsson *et al.* (1988), in the near-wall region the mean-velocity distribution will, for all practical purposes, be the same for channel flows and boundary layers without or with a weak pressure gradient. They also suggested that close to the wall the 'law of the wall' should hold for all one-point velocity moments. For example, the turbulence intensities (u', v', and w' in the x-, y-, and z-directions) may be written as

$$u'/U_\tau = f_1(y^+), \qquad v'/U_\tau = f_2(y^+), \qquad w'/U_\tau = f_3(y^+),$$

where the prime denote the rms values.

By applying Newton's viscosity law within the laminar sublayer, the mean wall shear stress can be expressed as

$$\tau_w = \mu \left.\frac{\partial \bar{U}}{\partial y}\right|_{y=0} = \mu \frac{\bar{U}}{y}, \qquad (8.40)$$

and similarly the corresponding fluctuating components of the wall shear stress (τ_x and τ_z) are related to the u and w velocity fluctuations by

$$\tau_x = \mu \frac{u}{y}$$

and

$$\tau_z = \mu \frac{w}{y} \qquad (8.41)$$

where $\mu = \rho \nu$ is the viscosity of the fluid.

8.8.1 Near-wall measurements

Obtaining accurate near-wall measurements often requires complex signal-analysis methods and/or advanced experimental techniques. For measurements within the viscous sublayer with an *SN* probe, the wire must be straight and parallel to the wall, the surface of which should be flat with variations of less than 0.001 mm (Azad 1983). It is difficult to achieve these conditions, which may explain some of the scatter in results from early near-wall measurements. Careful consideration should also be given to the following experimental aspects.

Probe calibration For measurements in air, the velocity in the near-wall region is usually less than 1 m s^{-1}. For such low velocities, special calibration techniques are required; these methods have been described in Section 4.7.

Aerodynamic interference The presence of the hot-wire probe will disturb the flow field, as described in Section 4.2.2 for an *SN* hot-wire probe placed in an infinite fluid flow, that is, far away from any solid boundary. Near a wall the nonsymmetric probe/wall geometry may cause additional flow disturbances around the wire elements, as discussed by Azad (1983). To minimize this near-wall blockage effect, a plated hot-wire probe should be used. When a probe with a parallel-stem orientation is used, it is recommended that either a boundary-layer probe is used (Fig. 4.2(c)) or that a standard plated probe (Fig. 4.2(b)) is inclined at a small angle ($\sim5°$) relative to the wall to remove the probe stem from the near-wall region. Alternatively, wall probes have been designed with the prongs protruding from the wall, since this configuration eliminates the blockage effect of the probe stems (see, for example, Rogers and Head 1969; Zemskaya *et al.* 1979; Hebbar 1980; Janke 1987).

Spatial resolution Within a boundary layer, the Kolmogorov length scale, η, becomes very small as the wall is approached, and subminiature hot-wire probes may be required to obtain accurate results, as discussed in Section 2.4.2.

Wall distance, y Within the viscous sublayer and buffer region the value of the wall distance, y, is required to great precision. Although methods are used which measure the absolute value of y at all measurement positions, the most common procedures combine one accurate measurement of the absolute distance, y_R, at a reference point R with a fine-scale traversing mechanism without backlash. Measurements with an accuracy of ±0.01 mm can be achieved if the traversing mechanism is aligned with an accuracy better than that mentioned above and if the probe is always traversed perpendicularly to the wall. If the wall has a reflecting surface, then the distance between the wire and its reflection will be twice the wall distance, y, of the wire. This method was used by Wills (1962), who viewed the wire and its reflection in the test wall through a microscope and a 45° mirror, the distance between the two images being measured on a graticule in the eyepiece. Similar methods have also been used by Krishnamoorthy *et al.* (1985) and by Devenport *et al.* (1990). For the case of a transparent wall, Van Thinh (1969) observed the distance of the wire from the wall by means of a microscope situated on the other side of the glass test wall. Orlando (1974) used a wall stop of known size to specify the reference position. The distance of the wire from the wall when the wall stop makes contact was measured by an optical comparitor; and this distance was nominally set to 0.127 mm, with an accuracy of about 0.025 mm. For electrically conducting walls, the distance between the wire and the wall can be measured by means of a third prong or wall stop. When the 'sensing prong' touches a conducting wall an electric circuit is closed and, since the distance between the wire and the end of the sensing prong can be determined very accurately by means of a microscope, the distance between the hot-wire and the wall will be known to the same accuracy. This method has been used by a number of investigators, including Azad and Burhanuddin (1983).

A detailed study of the effect of a vertical wall on the hot-wire signal when the flow velocity is zero has been carried out by Turan *et al.* (1987). Their results clearly demonstrated that, for both conducting and nonconducting wall materials, the anemometer voltage increases when the probe approaches the wall, as shown in Fig. 8.19. The magnitude of the voltage increase, ΔE, can be seen to depend on the value of k_w/k where k_w and k are, respectively, the thermal conductivities of the wall material and of air. This phenomenon has been developed into an accurate method for determining the distance from the wall, as reported by Vagt (1979) and by Vagt and Fernholz (1979). Their method applies to a given probe type, operated at the same overheat ratio and a specific wall material. The method was implemented as follows. With the air flow in the test section switched off, the output voltage was first set to zero, when the hot-wire probe was placed a fixed relatively large distance from the wall. (In principle, this method can also be developed using the direct nonadjusted output voltage from the anemometer.) Using a probe-deformation test, as described in Vagt and Fernholz (1979), it has been demonstrated, as shown in Fig. 8.20, that almost identical ΔE versus y curves can be obtained for probes of a given type manufactured to a tight manufacturing tolerance. Vagt and Fernholz claim that this method can determine the position of the wire relative to the wall with an accuracy of ± 0.010 mm.

A related procedure has been implemented by Bhatia *et al.* (1982) and Durst *et al.* (1987). In their approach the $\Delta E = f(y)$ relationship was

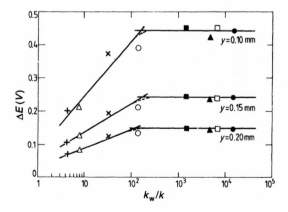

Fig. 8.19. The near-wall effect on an anemometer signal in still air as a function of the relative thermal conductivity: (●) copper, (□) aluminium, (▲) brass, (■) steel, (×) glass. (○) granite, (+) plywood, (△) plexiglass. (From Turan *et al.* 1987.)

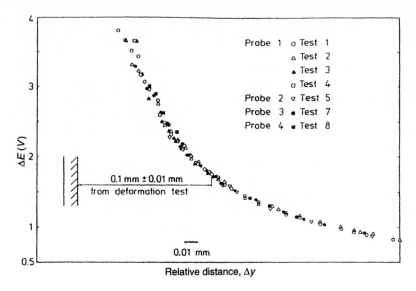

Fig. 8.20. The variation in the hot-wire signal in still air with the wall distance. (From Vagt and Fernholz 1979.)

obtained by performing the calibration in still air outside a wind-tunnel by approaching a plate made of the same material as the test-section plate in the wind-tunnel. The variation in the wire position in this distance calibration was measured with a precision microscope with a traverse accuracy for the scaled objective-lens system of $2\,\mu$m. Using a plate with a reflecting surface, the zero position was determined by observing the wire and its reflected image. The hot-wire probe was then positioned in the wind-tunnel without flow, and position-calibration information similar to that of Fig. 8.20 was used to place the wire at a selected reference position, y_R. To obtain accurate results with this method it is essential that the wire is straight and that it is placed parallel both to the calibration plate and the wind-tunnel test section plate.

In the method used by Ligrani and Bradshaw (1987a) for their subminiature probes, the probe distance, y, was determined by matching time-averaged mean-velocity data at different y-locations to the viscous sublayer equation

$$U^+ = y^+. \tag{8.42}$$

Typical mean-velocity data, U^+, plotted against, y^+, are shown in Fig. 8.21. Results at three different velocities using two different length sensors show a reasonable match to eqn (8.42) for $1.5 < y^+ < 5$. When y^+ is less than 1.5 the data demonstrate increased heat transfer due to the presence of the wall, as discussed in the next paragraphs. Finally, probe-distance methods based on laser lights have been proposed by Takagi (1985) and Janke (1987).

Influence of a wall on the hot-wire signal In air-flow studies, the thermal conductivity, k_w, of the wall is higher than that of the fluid, as illustrated in Table 8.1 for both conducting and nonconducting wall materials. Consequently, in the proximity of a solid wall the heat transfer from the heated element increases due to changes in the velocity and temperature field around the sensor. However, the conductivity of water is much higher, and a similar phenomenon has not been observed in water flows. Van der Hegge Zijnen (1924) appears to have been the first to correct for this additional heat loss of a hot-wire in the vicinity of a solid wall. He measured the increase in the heat loss of the wire in still air at various distances from the wall, and the additional heat loss was then used to correct the output signal of the wire in the moving air. A substantial number of theoretical and experimental studies of the wall-proximity effect have since been carried out. Most of these studies have been reviewed in detail by Azad (1983). Probably the best known is that of Wills (1962) who obtained, for laminar flow, the relationship

$$\mathrm{Nu}\left(\frac{T_w}{T_a}\right)^{-0.17} = A + K_w\left(\frac{2y}{d}\right) + B\mathrm{Re}_d^{0.45}, \qquad (8.43)$$

which was based on the calibration results of Collis and Williams (1959). (However, note that in eqn (2.3) the temperature loading is expressed in terms of the film temperature $T_f = \frac{1}{2}(T_a + T_w)$ and not in terms of T_w.) The function $K_w(2y/d)$ was introduced to account for the wall effect. Equation (8.43) implies that the heat transfer from the wire to the wall is by conduction only, and it can be considered as a modified version of the 'zero-velocity' correction of Van der Hegge Zijnen (1924). For turbulent

TABLE 8.1   *The ratio of the thermal conductivity, k_w, of wall materials to the thermal conductivity of air, k ($=0.02559\,W\,m^{-1}\,K^{-1}$).*

Type of wall material	Material	k_w/k
Conducting	Copper	14 382
	Aluminium	6400
	Brass	4582
	Steel	1400
	Granite	136
Nonconducting	Glass	32
	Plexiglas	8
	Plywood	4.5

From Turan *et al.* 1987.

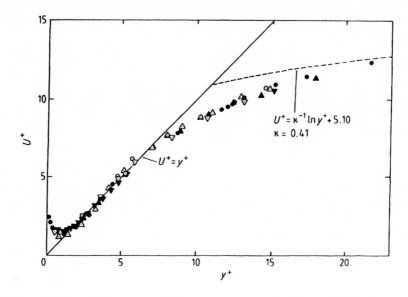

Fig. 8.21. The normalized mean velocity in the viscous sublayer and buffer layer, $d =$ 0.625 μm. For a sensor length of 170 μm: ($\triangledown$) 7.3 ms^{-1}, ($\odot$) 6.5 ms^{-1}, ($\triangle$) 5.7 ms^{-1}. For a sensor length of 430 μm: ($\blacktriangledown$) 7.3 ms^{-1}, ($\bullet$), 6.5 ms^{-1}, ($\blacktriangle$) 5.7 ms^{-1}. (From Ligrani and Bradshaw 1987a.)

flows, Wills (1962) found that eqn (8.43) could not be applied and suggested that half of the value of the laminar correction should be used. Later studies have shown that this method gives unsatisfactory corrections.

A comparison of the results from numerous wall-proximity investigations shows large variations in both the measured quantities and in the predicted results. However, provided that a given wall material, a fixed probe geometry, and a specified overheat ratio are used, then a number of investigators have shown that the data for a turbulent boundary layer fall on a single curve (independent of the values of the free-stream velocity, U_∞, and of the friction velocity, U_τ) when the measured mean velocity, $\bar{U}_m$, is plotted as $U_m^+ = f(y^+)$, where $U_m^+ = \bar{U}_m/U_\tau$ (Oka and Kostic 1972; Zemskaya *et al.* 1979; Polyakov and Shindin 1978; Hebbar 1980; Azad 1983; Janke 1987). A typical set of results for a ceramic wall with a high thermal and low electrical conductivity is shown in Fig. 8.22. Similar results have been obtained for all conducting wall materials (see Table 8.1) and for a hot-wire diameter, d, of 4–5 μm. The measured data, U_m^+, start to deviate from the curve $U^+ = y^+$ at $y^+ \simeq 4$–5. The minimum of the U_m^+-curve occurs at $y^+ \simeq 2$, and the corresponding error $(U_m^+ - U^+)/U^+$ is about 50%. The results for still air (Turan *et al.* 1987) indicate that a smaller but still nonzero effect should be expected for 'nonconducting' walls. Only a few investigators (for example, Polyakov and Shindin

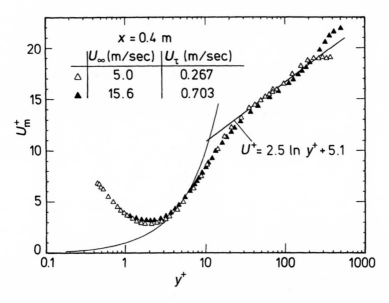

Fig. 8.22. The measured mean-velocity profile for a wall with a high thermal conductivity plotted on a U_m^+ versus y^+ graph. (From Janke 1987. Reproduced with kind permission of Springer-Verlag, © Springer-Verlag.)

1978; Ligrani and Bradshaw (1987a) have presented results for 'nonconducting' walls in the region $y^+ < 2$. These investigations show that deviation from the curve $U^+ = y^+$ occurs at $y^+ \simeq 2\text{-}3$ and the minimum in U_m^+ occurs at $y^+ \simeq 1\text{-}2$. These experimental observations contradict the inferred observation of Bhatia *et al.* (1982) that no correction is required for 'nonconducting' wall materials.

A number of investigators have studied the variation in the excess heat transfer with the wire diameter, the overheat ratio, and the wall material (Polyakov and Shindin 1978; Zemskaya *et al.* 1979; Bhatia *et al.* 1982; Krishnamoorthy *et al.* 1985). These results show a reduction in the near-wall effect when either the diameter or the overheat ratio is reduced and the wall material is thermally nonconducting. Many authors have developed correction procedures for the excess velocity $U_m^+ - U^+$, but no universal correction procedure has been established because of the parameter dependence mentioned above. This is shown in Fig. 8.23, which contains, for a wall with high thermal conductivity, both experimental data and several correction procedures. For the data in Fig. 8.22, which are included in Fig. 8.23, Janke (1987) has evaluated (by a least-squares curve-fitting procedure) that his U_m^+-data can be represented by

$$\frac{\bar{U}_m y}{\nu} = \frac{\bar{U}y}{\nu} - 0.55 \left(\frac{\bar{U}y}{\nu}\right)^{1/2} + 3.2. \tag{8.44}$$

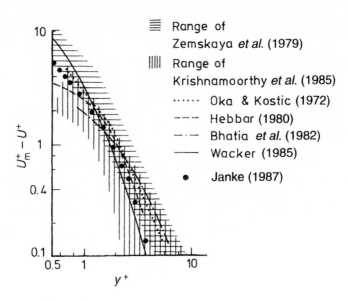

Fig. 8.23. Experimental data and correction procedures for the near-wall excess velocity $U_m^+ - U^+$. (From Janke 1987. Reproduced with kind permission of Springer-Verlag, © Springer-Verlag.)

This formula is valid both in laminar and turbulent flow for a $5\,\mu$m diameter hot-wire probe operated at a high overheat ratio of 1.7–1.8 near a wall with high thermal conductivity (for example, aluminium or a ceramic).

8.8.2 Skin-friction (mean wall shear stress) measurements

The 'law of the wall', eqn (8.39), is expressed in terms of the mean wall shear stress, τ_w, and a substantial research effort has consequently been devoted to measuring τ_w (strictly speaking the notation $\bar{\tau}_w$ should be used, but this distinction will only be introduced for measurements and calibrations which also involve the fluctuating shear stress component, τ_w'). Winter (1977) and Haritonidis (1989) have given in-depth reviews of skin-friction techniques based on force-measurement balances, on the use of the velocity profile, on pressure measurements by surface pitot tubes, on flow about obstacles, and on surface-mounted hot-film/wire sensors. A comparative study of four of these methods is given by Gasser *et al.* (1993). Many of these methods are still used for comparison with *HWA* methods, and for this reason the most common will be briefly outlined.

Clauser plot The inner part of a turbulent boundary layer, with a zero or a moderate pressure gradient, scales on the friction velocity, $U_\tau =$

$(\tau_w/\rho)^{1/2}$, and Clauser (1954) was the first to point out that this scaling enabled the skin-friction coefficient, $c_f = \tau_w/(\frac{1}{2}\rho U_\infty^2)$, to be derived from measurements of the mean velocity profile. The mean wall shear stress, τ_w, can be determined from measurements in the viscous sublayer, which have a linear velocity distribution $U^+ = y^+$ (eqn (8.42)). Solving for U_τ,

$$U_\tau^2 = \frac{\tau_w}{\rho} = v\frac{\bar{U}}{y} \qquad (8.45)$$

This method was used by Andreopoulos *et al.* (1984). However, as pointed out by Alfredsson *et al.* (1988), this method is, in general, difficult to use due to aerodynamical-wall-interference effects and due to the excess heat transfer from the wire in the near-wall region.

Since the viscous sublayer is very thin, the logarithmic region is normally used in the Clauser-plot method. In its original form, the Clauser chart for determining skin friction for an incompressible flow was constructed in the following way. The velocity profile of eqn (8.39) can be expressed as

$$\frac{\bar{U}}{U_\tau} = A \log \frac{yU_\tau}{v} + C, \qquad (8.46)$$

which can be rewritten as

$$\frac{\bar{U}}{U_\infty}\frac{U_\infty}{U_\tau} = A \log \frac{yU_\infty}{v} + A \log \frac{U_\tau}{U_\infty} + C, \qquad (8.47)$$

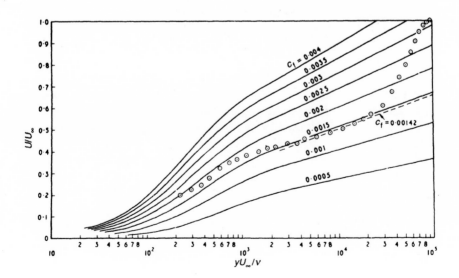

Fig. 8.24. A clauser plot. (From Winter 1977.)

where U_∞ is the free-stream velocity. In this form, a series of curves $\bar{U}/U_\infty = f(yU_\infty/\nu)$ can be drawn with $U_\tau/U_\infty = (c_f/2)^{1/2}$ as a parameter. Such a chart is shown in Fig. 8.24. The value of the skin friction will depend on the values selected for A and C. These vary from study to study (see, for example, Winter 1977; Azad and Burhanuddin 1983). If the skin-friction coefficient is obtained from the slope of the velocity distribution plotted in Clauser format, then the additive constant C has no direct influence on the skin-friction computation (Andreopoulos *et al.* 1984).

Based on an extensive survey of mean-velocity profile measurements, both with and without pressure gradients, Coles (1956) proposed that the velocity profile outside the viscous sublayer could be accurately described by the expression

$$\frac{\bar{U}}{U_\tau} = \frac{1}{\kappa} \ln \frac{yU_\tau}{\nu} + C + \frac{\Pi}{\kappa} w\left(\frac{y}{\delta}\right), \tag{8.48}$$

where δ is the boundary-layer thickness, $\kappa = 0.40$, and $C = 5.1$. Π is a profile parameter, and for flows at constant pressure it has a value of approximately 0.55. The variation in Π with the Reynolds number is discussed by Coles (1962, 1968), by Fernholz and Finley (1980), and by Erm *et al.* (1987). The function $w(y/\delta)$ is termed the 'law of the wake'. The values of U_τ and δ can be determined by applying a least-squares curve-fitting procedure to eqn (8.48). This method has been used by Keith and Bennett (1991).

Preston tube Preston (1954) proposed the measurement of the skin friction in a turbulent boundary layer by a circular Pitot tube resting on the wall. Based on the 'law of the wall' assumption, he proposed that the relationship between the Preston-tube reading and the skin friction can be presented in the following nondimensional form

$$\frac{\tau_w d^2}{4\rho\nu^2} = F\left(\frac{\Delta p d^2}{4\rho\nu^2}\right), \tag{8.49}$$

where Δp is the Preston-tube reading (that is, the difference between the pitot and static pressures) and d is the outer diameter of the Preston tube. Following Preston's work, this method has been widely used, and a considerable amount of related literature has been published.

Patel (1965) has produced what is probably regarded as the definitive calibration relationship, covering a wide range of flow conditions and sizes of Preston tube. His calibration is given in terms of:

$$x^* = \log \frac{\Delta p d^2}{4\rho\nu^2}$$

and

$$y^* = \log \frac{\tau_w d^2}{4\rho\nu^2}$$

$y^* < 1.5$

$$y^* = \tfrac{1}{2}x^* + 0.037,$$

$1.5 < y^* < 3.5$

$$y^* = 0.8287 - 0.1381x^* + 0.1437x^{*2} - 0.0060x^{*3}$$

$3.5 < y^* < 5.3$

$$x^* = y^* + 2\log{(1.95y^* + 4.10)}.$$

$$(8.50)$$

However, as pointed out by Head and Ram (1971), eqns (8.50) do not quite match the changeover, and the expression for the outermost region is inconvenient to use because of its implicit form. Furthermore, $\tau_w d^2/(4\rho v^2)$ varies by more than four orders of magnitude over the full range of Patel's calibration. Head and Ram therefore suggested the use of two alternative forms of calibration, the first, $\Delta p/\tau_w$ versus $\Delta pd^2/(\rho v^2)$, was given in tabulated form, and the second is in effect a Clauser plot for a Preston tube. McAllister *et al.* (1982) have presented further Preston-tube calibration results and Kassab (1993) has produced a simple calibration chart for correlating the Preston-tube reading directly with the friction velocity.

Surface fence The surface fence or sublayer fence was first applied as a means of measuring the magnitude of the skin friction by Konstantinov and Dragnysh (1960), and it was also applied later by Head and Rechenberg (1962) and by Vagt and Fernholz (1973). Basically, this method consists of a projection of a very small height positioned perpendicularly to the wall and the flow, as shown in Fig. 8.25. By means of pressure tappings at the front and rear of the projection, a pressure difference, Δp, is measured;

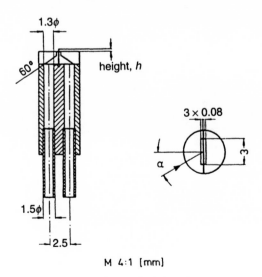

M 4:1 [mm]

Fig. 8.25. A cross section of a surface fence. (From Vagt and Fernholz 1973.)

this pressure difference is linearly related to the wall shear stress, τ_w, provided that the height is sufficiently small

$$\Delta p = k\tau_w \qquad (8.51)$$

The constant k depends on the geometry of the device, and especially on the height of the fence; it has to be determined by calibration using, for example, a Preston tube as a primary reference. Since the response should be the same for forward and reverse flow, the surface fence is particularly well-suited for measurements of the mean and fluctuating shear stress in separating flow regions, as demonstrated by Dengel *et al.* (1987).

Surface-mounted hot-film and hot-wires A considerable amount of research has been carried out into the use of surface-mounted *HWA* sensors for the measurement of the mean wall shear stress (skin friction). Typical configurations are a hot-film sensor mounted on the end of a quartz rod (see Fig. 2.4(d)) or the glue-on probe type, which is particularly useful for curved surfaces. The DANTEC glue-on probe (55R47), shown in Fig. 8.26, consists of a nickel heating film (0.9 mm × 0.7 mm × 0.001 mm) deposited on a polyimide foil (8 mm × 16 mm × 0.05 mm). A thin layer of silicon dioxide is deposited over the film to provide a protective coating. The film is connected to two nickel/gold-plated areas onto which the copper wires (0.7 mm in diameter and 55 mm long) are soldered. The polyimide foil is glued directly onto the wall at the required measuring point. Alternatively, a hot-wire can be used placed: (i) embedded in a plug of low thermal diffusivity (Liepmann and Skinner 1954; Rubesin *et al.* 1975; Murthy and Rose 1978), (ii) resting on the surface of the wall (Sandborn 1979), or (iii) placed a small distance away from the wall, but still within the viscous sublayer (Ajagu *et al.* 1982; Wagner 1991).

Fage and Falkner (1931) derived a relationship between skin friction and the heat convected from a heated platinum strip embedded in a surface. The first practical instrument using the analogy between skin friction and heat transfer was designed by Ludwieg (1949, 1950) who also obtained an analytical solution to the heat-transfer equations governing flush-mounted skin-friction sensors. Subsequently Liepmann and Skinner (1954), Bellhouse and Schultz (1966, 1968), and, more recently, Menendez and Ramaprian

 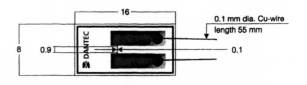

Fig. 8.26. A glue-on hot-film probe. The dimensions are in millimetres. (Printed with the permission of DANTEC Measurement Technology.)

(1985) have analysed the heat transfer from flush-mounted hot-film sensors under the condition that the thermal boundary layer developing above the sensor is entirely contained within the viscous sublayer of the turbulent boundary layer. Steady-state analysis of the thermal boundary layer yields an equation of the form

$$\frac{I^2 R_f^2}{\Delta T} = A\tau_w^{1/3} + B, \qquad (8.52a)$$

where R_f is the resistance of the heated hot-film element, ΔT is the difference in temperature between the heated film, T_f, and the ambient fluid temperature, T_a, and A and B are calibration constants.

The use of a flush-mounted hot-film and a hot-wire placed on the wall surface is shown in Fig. 8.27. The hot-wire extends further into the viscous sublayer than the film gauge, but the thermal layer (which controls the extent of the region over which the gauge samples) is a function of the gauge length in the flow direction, and the film gauge will sample much further out in the sublayer than the hot-wire. For thin boundary layers in air, a film gauge may therefore be too large to sample only in the viscous sublayer. The viscous sublayer in a typical laboratory water flow is substantially thicker than that in air, and the film gauge length is therefore less critical in water-flow studies (Sandborn 1979).

Operating the hot-film element in the CT mode and for steady conditions, eqn (8.52a) may be expressed for measurement purposes as

$$E^2 = A\tau_w^{1/3} + B, \qquad (8.52b)$$

where E is the anemometer voltage and the constant values for R_f and ΔT have been included in the calibration constants A and B.

The heat conducted into the substrate is normally significantly greater than the convective heat transfer to the fluid, and thus only a small proportion of the measured signal is flow-dependent (Mathews and Poll 1985). This relatively large conductive heat loss and the large thermal capacity of

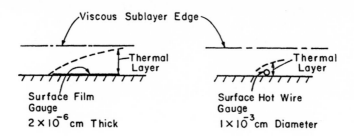

Fig. 8.27. A flush-mounted hot-film and a hot-wire placed on the surface of a wall. (From Sandborn 1979.)

the substrate have a mostly adverse effect on the performance of surface-mounted heated elements.

Using linear, steady, two-dimensional heat-conduction and forced-convection equations, Mathews and Poll (1985) showed that the steady-state calibration equation (8.52b) may more accurately be expressed as

$$E^2 = A\tau_w^{1/n} + B, \tag{8.52c}$$

where n (>3) increases with the substrate-to-fluid thermal conductivity.

It follows from the above discussion that the calibration relationship as expressed by eqn (8.52b) or by eqn (8.52c) will be modified by changes in the fluid temperature, T_a, and/or changes to the temperature distribution in the plug/wall material. The importance of heat losses to the substrate under 'steady-state' conditions has been noticed in a number of studies. Guitton (1969) used a thin film set on the end of a glass rod fitted into a brass plug. He observed that a calibration of his instrument obtained when it was mounted in a brass pipe required corrections when the same instrument was mounted in a Perspex plate. Changes in the calibration when a probe is moved from one test location to another have been reported by Reichert and Azad (1977). They suggested that this was probably due to a small change in the amount of contact between the probe body and the mounting, which can significantly alter the heat conduction through the substrate. Misalignment or protrusion of the surface containing the hot-film element can also significantly change the calibration, particularly in laminar flows. For these reasons a flush-mounted hot-film probe should preferably be calibrated *in situ*.

It has been found for many types of skin-friction gauges that the usual *HWA* practice of calibrating the instrument in a laminar flow and applying the calibration relationship obtained to the sensor output in a turbulent flow will lead to significant errors (Bradshaw and Gregory 1961). The calibration should therefore be carried out in a turbulent flow, and the mean wall shear stress, τ_w, should be determined independently either by, for example, a Preston tube or from the pressure gradient in a fully developed pipe or channel flow (Geremia 1972; Blinco and Simons 1974).

As pointed out by Sandborn (1979), the use of eqn (8.52b) in a turbulent-flow calibration will result in nonlinear averaging errors if the calibration and curve-fitting procedures are based on directly measured $(\bar{E}, \bar{\tau}_w)$-values only. If eqn (8.52b) is solved for τ_w, and setting $A_1 = 1/A$ and $B_1 = -B/A$, then

$$\tau_w = (A_1 E^2 + B_1)^3. \tag{8.53}$$

Applying this equation to the instantaneous values $\tau_w = \bar{\tau}_w + \tau_w'$ and to $E = \bar{E} + e$ gives

$$\bar{\tau}_w + \tau_w' = A_1^3(\bar{E} + e)^6 + 3A_1^2 B_1(\bar{E} + e)^4 + 3A_1 B_1^2(\bar{E} + e)^2 + B_1^3. \tag{8.54}$$

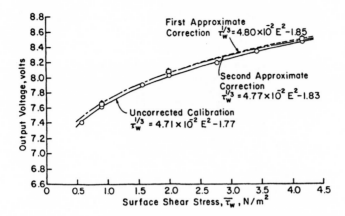

Fig. 8.28. The calibration of the hot-film shear-stress transducer. (From Sandborn 1979.)

Taking the time mean of eqn (8.53) demonstrates that $\bar{\tau}_w$ is a function not only of $\bar{E}$ but also of moments of the fluctuation up to $\overline{e^6}$. The magnitude of the corresponding correction for calibration experiments in a turbulent pipe flow is shown in Fig. 8.28. Equation (8.53) has also been applied by Ramaprian and Tu (1983) to the instantaneous voltage value, E, and evaluation of the time-mean values $\overline{E^2}$, $\overline{E^4}$, and $\overline{E^6}$.

The discussion above relates to the performance of a flush-mounted hot-film probe under steady-state-temperature conditions. In practice, such conditions can never be achieved. The fluid temperature often changes slowly with time and this and variations in the temperature of the air surrounding the complete test facility may modify the temperature distribution in the substrate and therefore the calibration of the hot-film sensor may also alter. To compensate for the variation in the fluid temperature, the placing of a temperature sensor on the wall has been proposed, in order to measure the variation with time of the wall temperature, which is assumed to be identical to the fluid temperature. The wall temperature was measured by Houdeville *et al.* (1983) with a thermocouple embedded in the surface, and Mathews and Poll (1985) have suggested the use of a surface-mounted hot-film element operated as a resistance sensor. However, even when ΔT was measured and compensated for using eqn (8.52a) calibration changes were still observed by Reichert and Azad (1977). They attributed this to a modification in the temperature distribution in the substrate causing a change in the relative magnitude of the conductive heat losses.

Rubesin *et al.* (1975) examined the errors arising from thermal conduction into the substrate. They pointed out that the glass substrate used by Bellhouse and Schultz (1966) had a relatively high thermal conductivity some fourteen times that of the ebonite used by Liepmann and Skinner (1954). To reduce the conductive heat losses they investigated the use of

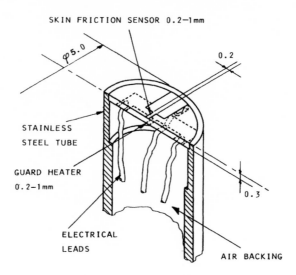

Fig. 8.29. A skin-friction thin-film probe with a guard heater mounted with air backing. (Printed with the permission of DANTEC Measurement Technology.)

a plastic substrate of low thermal conductivity. It has been suggested that the effect of the conductive heat losses to the substrate can be minimized by stabilizing the temperature of the substrate underneath the hot-film sensor. This approach was used by Polyakov *et al.* (1988) for heated boundary-layer studies. They developed a probe containing a heat guard in the form of an identical film buried directly under the flush mounted film on the surface. This design has been modified by DANTEC, who replaced the solid cylindrical substrate with a 0.3 mm glass disc with a sensor placed on either side. The disc is mounted flush on the end of a hollow cylinder (see Fig. 8.29) so that thermal contact to the probe body is reduced to the circumference of the disc.

In studies of surface boundary layers, their state (that is, whether they are laminar or turbulent, separated, or attached) must be determined. Multi-array glue-on probes have been used extensively for the determination of stagnation points, the separation point, laminar/turbulent transition, intermittency factors, etc. The sensors are deposited upon thin 25 or 50μm sheets of polyamide in patterns which are often specially designed for the actual body geometry. The flexibility of the foil allows it to be glued onto curved surfaces, with the total thickness of adhesive (normally a low-viscosity epoxy) and foil so small (50 to 70μm) that it does not affect the flow (Jørgensen, personal communication). Measurements have been reported for turbine cascade blades (Hodson 1984; Sugeng and Fiedler 1986; Capece and Fleeter 1987; Hodson and Addison 1989; Addison and

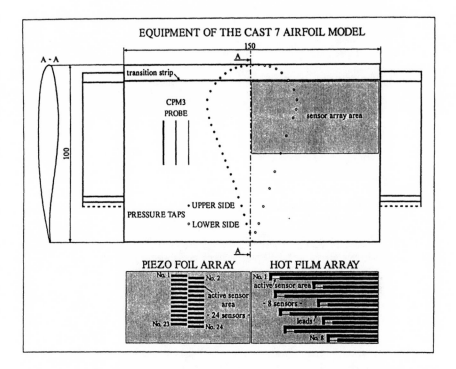

Fig. 8.30. The use of hot-film sensors and piezofoils for the detection of the shock position on a transonic air-foil model (CAST-7). (From Swoboda *et al.* 1991.)

Hodson (1990*a,b*) and for airfoil studies both in wind-tunnels and using flight tests (Holmes *et al.* 1986; Croom *et al.* 1987; Johnson *et al.* 1987; Stack *et al.* 1987, 1988; Feyzi *et al.* 1989; Kornberger and Feyzi 1990; Mirow *et al.* 1990; Carraway 1991; Hall *et al.* 1991; Swoboda *et al.* 1991). Multi-elements are often etched out of a single thin-film coated polyimide sheet. Alternatively, for a 'high fidelity' surface finish, micro-thin film sensors have been deposited directly onto the model surface (Johnson *et al.* 1987; Johnson and Carraway 1989). Glue-on probe designs include V-array probes for determining the flow direction near the wall and multi-array probes with parallel sensors. The application of the latter type in an airfoil study is shown in Fig. 8.30. The array was arranged in an alternate order to minimize mutual thermal interference. An offset sensor arrangement was used in supersonic flat-plate transition detection studies by Johnson and Carraway (1989). Laminar, transitional, and turbulent-flow conditions can be determined from the fluctuating part of the hot-film signals. Tests in a low-speed (Feyzi *et al.* 1989) and supersonic (Johnson and Carraway 1989) flat-plate boundary layer have demonstrated typical characteristics throughout the

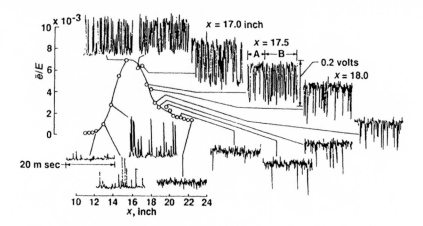

Fig. 8.31. The transition region from a hot-film array and some voltage-versus-time traces for flow over a flat plate at $M_\infty = 1.5$. (From Johnson and Carraway 1989.)

transition region. This is illustrated in Fig. 8.31 for the supersonic-flow case. Corresponding selected energy spectra are shown in Fig. 8.32. In the laminar-flow region the signal had a low energy level, except for a 200 Hz component, which is usually associated with a laminar boundary layer. The transition region begins with positive spikes in the voltage-versus-time

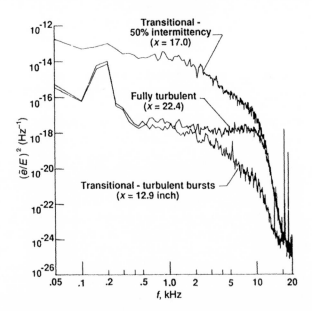

Fig. 8.32. A comparison of transitional and turbulent spectra for flow over a flat p $M_\infty = 2.0$. (From Johnson and Carraway 1989.)

traces, and the corresponding spectra for the film at $x = 12.9$ inches shows a significant increase in energy compared with the film in the laminar region. Moving further into the transitional regions, signals with large-amplitude positive and negative spikes were observed. The corresponding energy spectra show a surge in energy across the spectra, which is associated with the transitional boundary layer. Near the end of the transitional region the voltage-versus-time signal changes again with a reduction in the general level of fluctuations and occasional negative spikes. In the fully turbulent region, the energy spectrum is broad band with a nearly constant amplitude from low frequencies to the cut-off frequency. The amplitude is at a lower level than the spectrum for the 50% intermittency signal, but it is substantially higher than the signal from the laminar region. In the low-speed investigation by Feyzi *et al.* (1989), a value of 500 was quoted for the ratio of the energy levels in the turbulent and laminar boundary layers. Boundary-layer diagnosis requires the use of many hot-film sensors and their associated anemometers. Relatively cheap, reliable, constant, *voltage* anemometers have been developed for this purpose (Mangalam *et al.* 1992; Kuppa *et al.* 1993).

8.8.3 Time-dependent surface-shear-stress measurements

A number of studies have shown that large fluctuations occur in the wall shear stress in turbulent flow, as summarized by Alfredsson *et al.* (1988).

Surface film gauges An in-depth study of the measurement of $\tau_x'/\bar{\tau}_w$ in air, oil, and water with both conventional wall-mounted hot-film gauges and specially constructed sensors has been carried out by Alfredsson *et al.* (1988). Their results showed that the true root-mean-square level of the wall stress fluctuations in the streamwise direction for both channel- and boundary-layer flow is 40% of the mean wall shear stress. They also demonstrated that the wide range of values reported in the literature, from studies using flush-mounted hot-film probes, can be explained mainly by the inadequacy of static-calibration procedures for such probes. Most of the previous results (for example, Eckelmann 1974; Chambers *et al.* 1983 were obtained by methods where the sensor was calibrated against the mean wall shear stress; and, when the flow medium is air or oil, the heat loss to the substrate greatly affects the measured turbulent wall shear stress, resulting in a substantial under-estimation of τ_w'. The investigations by Blinco and Simons (1974) and Alfredsson *et al.* (1988) also showed that, in general, a standard flush-mounted hot-film gauge in a water flow is virtually unaffected by the heat loss to the substrate, due to the high thermal conductivity of water. The results by Madavan *et al.* (1985) also demonstrated that spanwise spatial averaging can reduce the fluctuating output from large film elements.

Near-wall hot-wire probes As discussed in the section above, the heat loss to the substrate greatly affects the dynamic performance of a flush-mounted hot-film probe. One way of reducing this effect is to place the sensor on a plug of low thermal conductivity (Liepmann and Skinner 1954, Rubesin *et al.* 1975). This will reduce the heat loss to the surrounding wall and thus also reduce the influence of the temperature field in the wall on the sensor output. A further reduction in this effect can be achieved by employing an air gap between a flush-mounted film and the surrounding wall (Brown 1967; Bellhouse and Schultz 1968).

A similar approach has been used by Houdeville *et al.* (1983) and by Houdeville and Juillen (1989) who replaced the flush-mounted hot-film with a 70 μm hot-film fibre sensor placed in the surface plane, but with a cavity under it (see Fig. 8.33), so that the heat transfer to the substrate is heavily reduced. The introduction of the cavity modifies the local flow field, and for this reason these authors suggested that the calibration relationship eqn (8.52b) should be modified to $E^2 = A\tau_w^{1/n} + B$, with the value of n determined by a calibration procedure. The ambiguity of the sign of τ_w in separated flows was eliminated by adding two resistance-wires at either side of the heated hot-film fibre sensor to detect the flow direction of the hot wake.

The fluctuating wall shear stress has also been measured using a hot-wire mounted at or a fixed small distance from the wall and calibrated directly in terms of the mean wall shear stress. Alfredsson *et al.* (1988) used a probe with a hot-wire mounted only a few wire diameters above the wall. It was

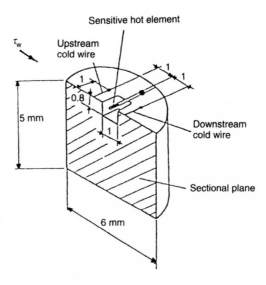

Fig. 8.33. A skin-friction probe with a cylindrical hot-film sensor placed over the cavity. (From Houdeville and Juillen 1989.)

calibrated in a turbulent boundary layer against a Preston tube. However, due to the low velocities very close to the wall, the sensitivity of such a probe is very low at low Reynolds numbers. Wagner (1991) increased the sensitivity of his probe by placing the probe further away from the wall but still within the viscous sublayer. By using the correction obtained by Janke (1987) for the excess heat transfer near the wall, eqn (8.44), Wagner obtained a relationship between the measured mean wall shear stress, τ_m, and the true mean value, τ_w. The calculated value of τ_w was used in the response equation, $E^2 = A\tau_w^{1/n} + B$, to calibrate the hot-wire in a turbulent flow over the full range of skin-friction measurements. The calibration was corrected for nonlinear averaging errors (eqn 8.53). Comparative measurements with a pulsed wall probe (see the next paragraphs) gave values of $\tau_x'/\bar{\tau}_w$ to within $\pm 2\%$ of 0.37, which is consistent with the values quoted by Alfredsson et al. (1988). It has been proposed that the dynamic heat transfer from the wall to a near-wall hot-wire sensor can be minimized by stabilizing the wall temperature. In an investigation by Ajagu et al. (1982) this was implemented by placing a flush-mounted hot-film under a near-wall hot-wire sensor.

The pulsed wall probe Pulsed wall probes are an adaptation of the pulsed-wire anemometer developed by Bradbury and Castro (1971). Their primary purpose is to measure the time-dependent skin friction in low-speed flows and they have been used extensively for the study of separated flows. The principle was investigated by Ginder and Bradbury (1973), who used surface-mounted thin-film elements. However, because of the influence of the substrate, the signal-to-noise ratio was much lower than for the standard pulsed-wire velocity probe. Also, calibration of the probe in a laminar-flow channel led to skin-friction measurements in a turbulent boundary layer that were too high by as much as 20%. More recently, workers at Stanford University (Eaton et al. 1981; Westphal et al. 1981) and at the University of Surrey (Castro and Dianat 1983; Castro et al. 1987) have developed similar probes containing three wires mounted parallel and just above the surface, as shown in Fig. 8.34.

In principle, the probe can be calibrated for velocity, U, against the time of flight, T, as described by eqn (8.12) for the standard PWA probe in Section 8.5.3. Provided that the wire distance, H, from the wall is within the viscous sublayer ($HU_\tau/\nu < 5$) then eqn (8.45) can be applied with $y = H$ giving

$$\tau_w = \mu U/H. \tag{8.55}$$

Measuring the time of flight, T, and using eqn (8.12), the probe can be calibrated directly in terms of the skin friction, τ_w. Eaton et al. (1981) selected

$$\tau_w = A/T + B/T^2; \tag{8.56}$$

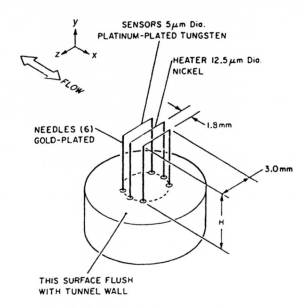

Fig. 8.34. Pulsed wall probe for time-dependent skin-friction measurements. (From Eaton *et al.* 1981.)

while Castro *et al.* (1987) found that

$$\tau_w = A/T + B/T^2 + C/T^3 \qquad (8.57)$$

gave an accurate curve fit over a wide range of wall shear stresses.

In common with other skin-friction techniques the probe should be calibrated in a turbulent flow. Eaton *et al.* (1981) found that a laminar calibration with a linear-profile assumption gave a 5–10% lower mean wall shear stress than was obtained using the direct, turbulent, shear-stress calibration. For accurate measurements, it is not appropriate to fit the average value of T to the corresponding mean wall shear stress, even at moderate rms intensity fluctuations; this is because of nonlinear averaging errors caused by the nonlinear form of eqns (8.56) and (8.57). The correction scheme applied by Eaton *et al.* (1981) provided a 2% correction to the calibration constants A and B in eqn (8.56).

9

TWO-PHASE FLOWS, GAS MIXTURES, AND COMPRESSIBLE FLOWS

This chapter covers measurements in two-phase flow, in gas mixtures, and in flows with significant compressibility effects.

9.1 TWO-PHASE FLOWS

Two-phase flows have been studied extensively, with many investigations devoted to air/water or steam/water flows due to their common occurrence in engineering machinery and plants. However, many of the phenomena observed in gas/liquid flows are also observed in the flow of two immiscible liquids, and the following *HWA* measurement techniques for two-phase flow can be applied to both flow situations.

9.1.1 The probe response to the passing of a bubble

In the study of two-phase flow, a large amount of work has been concentrated on the development of probes and instrumentation for the determination of local fluctuating quantities. Much of this work has been concerned with the measurement of the local void fraction, α, and this has led to the development of optical and resistance probes as notable examples of local void fraction measuring devices (Sheng and Irons 1991; Cartellier 1990; 1992, Cartellier and Achard 1991). Liu and Bankoff (1993b) and Liu (1993) have presented results for the radial profiles of the void fraction, the bubble velocity, and the bubble size using a miniature dual-sensor resistivity probe. Hot-film probes can also provide the local void fraction. Moreover, unlike optical and resistance probes, they have the great advantage of being able to measure the velocity fluctuations in the continuous phase, and they can therefore provide valuable information about the turbulent structure of two-phase flow. Measurements in two-phase flow, using thermal anemometry, have been carried out mainly with cylindrical or conical hot-film probes. Both probe types have been observed to give characteristic signals when placed in a bubbly two-phase flow. However, different signal interpretations have been suggested, and a description of the interaction between the two probe types and a passing bubble, and the related variation in the anemometer signal, is therefore appropriate. For simplicity of presentation it will be assumed that the continuous phase is water.

The theory presented below is based on the assumption that the probe penetrates the bubbles; this is commonly described as a *direct hit*. Investigations by Bremhorst and Gilmore (1976*b*) and by Serizawa *et al*. (1983) have shown that direct hits are likely to occur when the bubble diameter, d, is greater than 2–3 mm. However, for smaller bubbles, $d < \sim 2$ mm, imperfect penetration or *glancing hits* were observed for some of the bubbles. The related bubble-sensor interaction was classified by Serizawa *et al*. (1983) as recoiled, drifting, or crawling interaction. Void-fraction measurements with cylindrical film probes in two-phase flow containing small bubbles must therefore be treated with caution.

It is the experience of Serizawa (personal communication) that bubble penetration may be difficult to achieve with a conical probe, and that the bubble motion is easily deflected by the probe. However, the velocity-field measurement with a conical probe should not be too sensitive to the effect of the bubble escaping the probe.

Cylindrical SN hot-film probes The interaction between a cylindrical single normal (*SN*) *HWA* probe and a bubble and the related signal interpretation has been the subject of several investigations (Hsu *et al*. 1963; Delhaye 1969; Bremhorst and Gilmore 1976*b*; Rehmke 1978; Börner *et al*. 1980; Serizawa *et al*. 1983; Farrar 1988; Bruun and Farrar 1988; Farrar and Bruun 1989; Liu and Bankoff 1990*a, b*). The interaction between a bubble and a cylindrical sensor element placed perpendicular to the flow can be explained with reference to Fig. 9.1 for the condition of a 'direct hit'. Figure 9.1 also contains the corresponding variation in the anemometer signal, E, which has been shown by Farrar and Bruun (1989) to be predominantly related to the change in the surface areas exposed to the two phases.

When a bubble approaches the cylindrical element (A) the anemometer signal, E will increase because the bubble moves with a velocity that is greater than the average water velocity. The signal will, for a direct hit, continue to increase until the bubble front reaches the cylindrical element (B). This conclusion is in general agreement with the flow visualization observations of Jones and Zuber (1978) and Börner *et al*. (1980). A meniscus is then formed, which creeps round the cylindrical element (C) as the undisturbed part of the bubble moves past the probe. This is accompanied by a steep drop in the anemometer signal. At position D, the two menisci will merge behind the cylinder, leaving the cylinder surrounded by nearly still water in the film and the meniscus. The anemometer signal will therefore approach that of still water. A thin film stretching between the prongs of the probe (E) is then formed. Since the shape of the meniscus attached to the sensor will remain virtually constant during this stage, the main probe effect will be a slow evaporation of the water film on the sensor. As a consequence, the slope of the anemometer signal will be much reduced. At some stage the water film between the prongs may break, giving rise to a peak (F). However, in most bubble traverses the film will not break before the back of the bubble arrives. When the back of the bubble arrives (G) the

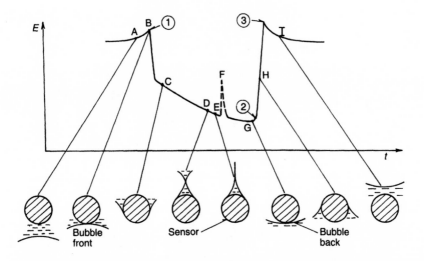

Fig. 9.1. The response of a cylindrical hot-film probe to the passage of a bubble.

continuous-phase liquid (water) will jump around the cylinder forming a meniscus (H). The rapid covering of the sensor with water results in a steep rise in the signal ending in peak 3. The signal part 2–3, described by Bremhorst and Gilmore (1976*b*) as a detachment tail, is caused by a dynamic meniscus effect (Farrar and Bruun 1989) and it should not be included in the continuous-phase velocity evaluations. Following peak 3, the signal drops slowly (I) towards the average velocity of the continuous phase, as the bubble moves away from the probe.

Figure 9.1 demonstrates that the signal from a cylindrical hot-film probe is well-suited for the signal analysis of bubbly two-phase flows. Three events (1, 2, and 3) can be clearly identified. The passage of the undeformed bubble corresponds to the time difference $\Delta t_{12} = t_2 - t_1$, and for the evaluations of the continuous-phase velocity the time interval $\Delta t_{13} = t_3 - t_1$, (corresponding to bubble passage and the dynamic meniscus effect) must be removed from the sample record.

X-hot-film probes The *SN* hot-film probe technique described in the previous section has been extended by Serizawa *et al.* (1983) and Liu and Bankoff (1990*a*) to the signal analysis of *X*-hot-film probes. The change in the anemometer signal for several different types of bubble hits is shown in Fig. 9.2, and it is observed for all cases that the shapes of the curves are similar to Fig. 9.1.

Split-film probes The theory for the use of a split-film probe in a single-phase flow is discussed in Section 5.7. Börner *et al.* (1982) and Börner and Leutheusser (1984) have investigated the application of split-film probes to two-phase flow measurements. When the output signals E_1 and E_2 from

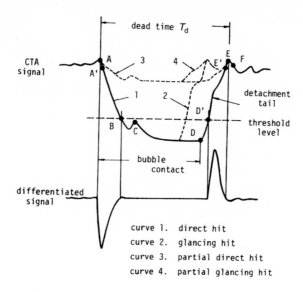

curve 1.   direct hit
curve 2.   glancing hit
curve 3.   partial direct hit
curve 4.   partial glancing hit

Fig. 9.2. The combined threshold level and $\partial E/\partial t$ signal-analysis method for a cylindrical probe. (From Serizawa *et al.* 1983.)

the two elements of the split-film probe are added, its operation is basically that of an *SN* probe. Consequently, the passing of a bubble will produce a signal variation in $E_1^2 + E_2^2$ which is similar to Fig 9.1, and phase discrimination can be carried out with this signal. In the continuous phase, the summation signal $E_1^2 + E_2^2$ will determine the magnitude of the (two-dimensional) velocity vector and the corresponding flow angle can be obtained from the difference signal $E_1^2 - E_2^2$ as discussed in Section 5.7.

Conical probes It is more difficult to interpret correctly the signal from a conical hot-film probe placed in a two-phase flow, since the hot-film element is not positioned at the tip of the cone, but at a small distance, Δx, downstream from the tip. This spatial difference results in the tip and not the hot-film element first making contact with the bubble front and rear. There are consequently several possible ways of interpreting the bubble part of the signal. In the pioneering work by Delhaye (1969), a procedure based on the position of the 'undeformed' bubble relative to the hot-film element was suggested. This method was adopted by Serizawa *et al.* (1975a,b,c) for their turbulence measurements. Wang *et al.* (1986, 1987, 1990) used an identification procedure based on the bubble position relative to the tip, while in the paper by Serizawa *et al.* (1983) the conical hot-film probe signal was analysed in the same manner as for a cylindrical hot-film probe.

The 'direct-hit' interaction between a bubble and a conical probe is illustrated in Fig. 9.3. When a bubble approaches the probe (A) the anemometer

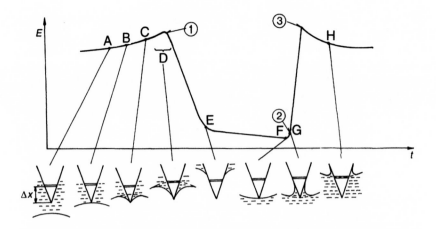

Fig. 9.3. The response of a conical hot-film probe to the passage of a bubble.

signal increases because the water in front of the bubble moves faster than the average water velocity. The signal is still increasing when the front of the bubble reaches the tip of the probe (B). A meniscus is then formed at the tip (C), which slows down the fast-moving fluid in front of the bubble, causing the formation of signal peak 1. At position D, the 'undeformed' bubble front reaches the hot-film element position, placed a distance Δx downstream of the tip. The position of D relative to peak 1 depends strongly on the probe geometry, including the distance Δx, and it may occur just before or just after peak 1. As the undeformed part of the bubble front moves past the probe, the meniscus at the tip will continue to develop until the contact angle of the meniscus reaches the value of the receding contact angle, β_r ($\simeq 0°$). During this period, the velocity in the meniscus surrounding the hot-film element is reduced even further, and this is observed as a relatively steep drop in the probe signal. When the angle of the meniscus reaches the limiting value $\beta_r \simeq 0°$, the meniscus detaches from the tip and moves up the probe leaving a thin stationary water film on the probe (E). The anemometer signal will therefore approximate to a value corresponding to still water. Evaporation of the water film then takes over as the main event, and this period is observed as a much flatter anemometer signal. The back of the bubble reaches the tip of the probe at F, and as the probe is preferentially wetted by water, a meniscus is rapidly formed (G), which may reach the hot-film element immediately or a short time later depending on the value of the distance Δx. This is followed by a dynamic meniscus region resulting in a steep rise in the anemometer signal terminating in peak 3. The voltage then drops towards the average water velocity and the back of the 'undeformed' bubble arrives at the hot-film position at (H).

The conical probe also produces three clearly identifiable events (1, 2, and 3), but none of these events correspond to the position of the undeformed bubble relative to the hot-film element, as observed by flow visualization experiments by Delhaye (1969) who therefore proposed the use of a probability density function method for the evaluation of the void fraction. The application of this method to cylindrical probes has been considered by Herringe and Davis (1974) by Bremhorst and Gilmore (1976*b*), and by Bruun and Farrar (1988). The position of the bubble front and rear relative to the tip of the probe was used as the reference method by Wang *et al.* (1986, 1987). Figure 9.3 shows that this method will usually correctly identify the back of the bubble, points 2, and the related dynamic meniscus region 2–3. However, peak 1 deviates significantly from the arrival (B) of the front of the bubble at the tip of the probe. Consequently, a void-fraction correction must be applied if the tip is used as the reference position (Wang *et al.* 1987, 1990). Alternatively, placing an optical probe next to the conical probe, can be used to identify the bubble passage and the void fraction (Lance and Bataille 1991). For the continuous-phase velocity evaluations, the signal part $\Delta t_{B3} = t_3 - t_B$ should be eliminated. This might be achieved by using an optical probe to provide the trigger signal or, as an approximation only, the well-defined part $\Delta t_{13} = t_3 - t_1$, may be eliminated. However, this procedure will introduce some error if peak 1 is large compared with the velocity fluctuations in the continuous phase.

9.1.2 Signal analysis in two-phase flows

The analysis of the anemometer signal relates primarily to: (i) void-fraction calculations, which require the determination of the times corresponding to the passage of the undeformed bubbles past the probe; and (ii) evaluations of the continuous-phase velocity quantities. For this part of the analysis it is necessary to remove both the bubble and the meniscus-affected portion of the signal.

The following signal analysis applies to a cylindrical hot-film probe, since the discussion related to Fig. 9.1 has shown that the bubble passage and the dynamic meniscus region can be related directly to events 1, 2, and 3 in Fig. 9.1. The first part of the signal-analysis procedure is therefore to identify these three events for each bubble in a sample record, which may typically look like the signal shown in Fig. 9.4. In the early, mainly analog methods, an amplitude threshold was used as a discriminator (Resch and Leutheusser 1972; Resch *et al.* 1974; Abel and Resch 1978; Jones and Zuber 1978; Toral 1981). Any data points in the signal which lie below the voltage threshold are deemed to be associated with the dispersed phase, and those points which are above are assumed to represent the continuous-phase velocity. This method is simple in both concept and application, but it has several disadvantages. The most important is that this method does not cor-

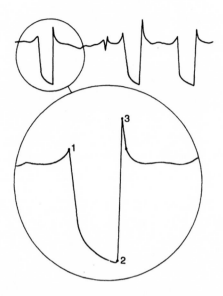

Fig. 9.4. A typical signal from a cylindrical hot-film probe immersed in a bubbly flow. (The insert show an individual bubble signal.)

rectly identify any of the points 1, 2, and 3 in Fig. 9.1, and no distinction is made between points 2 and 3. Serizawa *et al.* (1983) have produced an improved detection technique based on a slope-threshold method, in which the first derivative of the anemometer signal, $\partial E/\partial t$, is compared with one or more threshold levels. However, as shown in Fig. 9.2, to avoid signal misinterpretation when film breakage (C) occurs, a threshold level had to be introduced and, as a consequence, point D will be approximated by D′. To overcome the problems of these methods, Farrar (1988) has developed a digital procedure, which comprises an initial amplitude method for bubble detection only, followed by a digital search method which identifies points 1, 2, and 3 in each of the bubble signals. A further refinement of this method has been presented by Samways and Bruun (1992). A combined threshold and slope method has also been proposed by Liu and Bankoff (1990*a*,*b*), by Wang *et al.* (1990), and by Liu (1993).

9.1.2.1 Void-fraction evaluation

Having evaluated points 1, 2, and 3 for each bubble signal, the void fraction, α, can for conditions when 'direct hits' apply, be determined from

$$\alpha = \sum \Delta t_{12}/T, \tag{9.1}$$

where T is the length of the sample record.

9.1.2.2 Continuous-phase velocity evaluations

To evaluate the statistical properties of the continuous phase, the signal part $\Delta t_{13} = t_3 - t_1$, for each bubble is first eliminated from the complete sample record. The analysis presented applies to measurements with an *SN* cylindrical hot-film probe. Corresponding two-dimensional measurements with a miniature *X*-probe have been described by Serizawa *et al.* (1983), and Wang *et al.* (1986, 1987, 1990) have presented three-component flow measurements obtained with a conical hot-film with the geometry shown in Fig. 9.5. When the hot-film is surrounded by the continuous phase, the signal output *E* from the *SN* probe, can be converted into the longitudinal

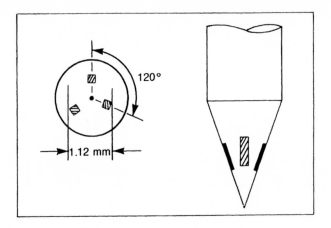

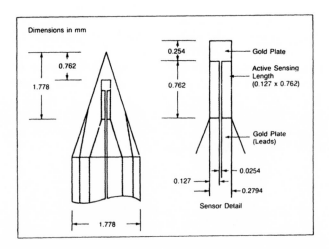

Fig. 9.5. The geometry and dimensions of a three-dimensional conical probe (TSI 1231 AJ-W). (Reproduced with the permission of TSI Inc.)

velocity component U, by using the response equation $E^2 = A + BU^n$, where A, B, and n are calibration constants. The digital procedure for the evaluation of velocity statistics for the continuous phase is as follows.

Mean velocity, $\bar{U}$ The sample record of the converted digitized hot-film signal consists of individual samples, X_n, which only correspond to valid continuous-phase velocity values at times when no bubbles are present at the probe sensor. The mean velocity can therefore be calculated from

$$\bar{U} = \frac{1}{N_c} \sum_{n=1}^{N} X_n \xi_n, \tag{9.2}$$

where $\xi_n = 1$ when the sample data point represents a true continuous-phase velocity, U, and $\xi_n = 0$ otherwise, N_c is the number of data points for which $\xi_n = 1$ (that is, $N_c = \Sigma_{n=1}^{N} \xi_n$), and N is the total number of samples. If the product $X_n \xi_n$ is denoted by U_n, then the original sample record, X_n, can be converted into a U_n sample record by setting to zero all points corresponding to bubble contact and dynamic overshoot, that is, all data samples between points 1 and 3 of each bubble signal (but not including point 1 itself); see Fig. 9.4. If a signal dead-time fraction, γ, is defined in a similar manner to the signal void fraction, α (but using bubble signal points 1 and 3 instead of 1 and 2), as follows

$$1 - \gamma = \frac{N_c}{N} = \frac{1}{N} \sum_{n=1}^{N} \xi_n, \tag{9.3}$$

then the mean velocity, $\bar{U}$, can be evaluated from the total, U_n, sample record as

$$\bar{U} = \frac{1}{(1-\gamma)} \frac{1}{N} \sum_{n=1}^{N} U_n = \frac{\bar{U}_n}{(1-\gamma)}, \tag{9.4}$$

that is, the mean continous-phase velocity is given by the overall mean of all the points in the new sample record, $\bar{U}_n$, divided by $(1 - \gamma)$.

Normal stress, $\overline{u^2}$ Having calculated the mean velocity, its value is subtracted from all data points in the sample record for which $\xi_n = 1$. The new sample record, u_n $(=U_n - \bar{U}_n)$, now consists of (a) continuous-phase velocity fluctuations, u, with a zero mean when $\xi_n = 1$ and (b) zero values when $\xi_n = 0$. The mean square value of the continuous phase-velocity fluctuations is then given by

$$\overline{u^2} = \frac{1}{N_c - 1} \sum_{n=1}^{N} u_n^2, \tag{9.5}$$

while the corresponding value, $\overline{u_n^2}$, for the whole signal is

$$\overline{u_n^2} = \frac{1}{N - 1} \sum_{n=1}^{N} u_n^2. \tag{9.6}$$

For large values of N and N_c the following relationship applies

$$\overline{u^2} = \frac{1}{(1 - \gamma)} \overline{u_n^2}. \tag{9.7}$$

Autocorrelation-coefficient function, $\rho_u(\tau)$ The autocorrelation coefficient function is often evaluated in turbulence studies, and it usually decays to zero relatively quickly. It is this initial region of decay, up to the first or second crossing point on the τ-axis, which is normally of interest. The autocorrelation coefficient function, $\rho_u^n(\tau)$, for the whole signal (including the zeros), u_n, is given by

$$\rho_u^n(\tau) = \frac{1}{\overline{u^2}} \frac{1}{N - r} \sum_{n=1}^{N-r} \xi_L(t, t + \tau)u_n(t)u_n(t + \tau), \tag{9.8}$$

where $\xi_L(t, \ t + \tau) = \xi_n(t) \ \xi_n(t + \tau)$ (hereafter denoted as ξ_L) equals 1 when the signal corresponds to the continuous-phase velocity both at times t and $t + \tau$, and $\xi_L = 0$ otherwise (Lance and Bataille 1983).

The true autocorrelation coefficient function of only the continuous-phase velocity fluctuations is given (Farrar 1988) by

$$\rho_u(\tau) = \frac{1}{\overline{u^2}} \frac{\displaystyle\sum_{n=1}^{N-r} \xi_L(t, t + \tau)u_n(t)u_n(t + \tau)}{\displaystyle\sum_{n=1}^{N-r} \xi_L(t, t + \tau)}, \tag{9.8}$$

and $\rho_u^n(\tau)$ and $\rho_u(\tau)$ are therefore related by

$$\rho_u^n(\tau) = x(\tau)\rho_u(\tau), \tag{9.9}$$

where

$$x(\tau) = \frac{1}{N - r} \sum_{n=1}^{N-r} \xi_L(t, t + \tau). \tag{9.10}$$

At $\tau = 0$ ($r = 0$), ξ_L is equal to ξ_n^2, and therefore

$$x(0) = \frac{1}{N} \sum_{n=1}^{N} \xi_n^2 = 1 - \gamma. \tag{9.11}$$

However, when $\tau \neq 0$, $x(\tau)$ will decrease with increasing values of τ; the asymptotic value of $x(\tau)$ is $(1 - \gamma)^2$ for large values of τ (Farrar 1988).

The one-dimensional spectrum, $E_u(f)$ Spectral analysis can provide valuable information on the flow structure. Because of the discontinuous character of the u_n signal and the $x(\tau)$ function (eqn (9.10)) a definitive method has not yet been adopted for spectral analysis of two-phase flows. Various approaches can be found in the literature: suppression of the (bubble) peaks and patching together of physically meaningful segments of the signal (Lance 1979; Gherson and Lykoudis 1984), linear interpolation inside the bubble or particle peaks (Tsuji and Morikawa 1982; Serizawa *et al*. 1983) and conditional sampling by the characteristic function, ξ_L, of the liquid, which is equivalent to inserting the liquid mean velocity into the data gaps (Resch and Abel 1975; Lance 1979; Lee 1982; Farrar 1988; Wang *et al*. 1990). As discussed by Wang *et al*. (1990), the method of replacing the bubble signal with the known mean velocity of the continuous phase is preferred, since the effect of this operation on the true spectrum can be analysed. However, the characteristic function, ξ_L, is a sequence of rectangular windows which may introduce high frequencies, particularly for gas/liquid flows in which the bubble peaks can be very large compared with the turbulent fluctuations in the continuous phase. To overcome this problem, Lance and Bataille (1991) applied smoothing windows within the bubble part of the u_n-record.

Bubble chord lengths Resch *et al*. (1974), measured the volume fraction and turbulence in a hydraulic jump and they also determined the bubble-diameter distribution using a single conical hot-film probe. Their measuring technique was based on three assumptions: (i) all bubbles had a spherical shape; (ii) each bubble sensed by the probe was intersected along its diameter; and (iii) the bubble velocity, U_b, was the instantaneous velocity of the continuous phase immediately prior to bubble arrival. In practice, most bubbles are oblate spheroids, and the probe may intersect the bubble at any point within a relatively large central area and still give a bubble chord length, which is very close to the length of the minor axis. Provided that assumption (iii) is valid, the cut-chord length, c, can be estimated

$$c = U_b \Delta t_{12}. \tag{9.12}$$

The evaluated individual bubble-passage times and chord lengths can be statistically analysed to give mean values and distributions.

9.1.3 Liquid droplets in a gas flow

Hot-wire anemometry (*HWA*) has been used to measure the impact rate of the droplets and the corresponding concentration flux. A number of liquids have been used to form the aerosol droplets: water and ice crystals (Vonnegut and Neubauer 1952), safflower oil (Goldschmidt and Eskinazi 1966), dibutyl phthalate, sinco prime 70 and safflower oil (Goldschmidt and Householder 1969), dibutyl phthalate (Bragg and Tevaarwerk 1974) and water (Lin and Chang 1989). In experiments by Goldschmidt (1965) and by Goldschmidt and Eskinazi (1966), the size of the droplets was smaller than most of the turbulent eddies. The impact frequency of the droplets will therefore be different from the frequency range of the most energetic eddies of the turbulence in the gas flow, and the impact-signal part was separated from the fluctuations due to turbulence by using a bandpass filter. The geneneration of a square pulse for each droplet strike gave the collision frequency of the droplet and the concentration flux (defined as the number of particles flowing per unit time across a unit area). Due to the inertia effects on the droplets imposed by the presence of the wire, the number of impacts of droplets on the wire will not correspond to a flow area equal to the frontal area of wire. To correct for this effect, Goldschmidt and Eskinazi (1966) introduced an impact coefficient defined as the ratio of the number of particles that collide with the wire to the number in an area equal to the frontal area of the wire.

Attempts have also been made to relate the anemometer output signal to the diameter of the droplets. Based on theoretical derivations and experiments with droplets smaller than $200\,\mu\mathrm{m}$, Goldschmidt and Householder (1969) concluded that the peak voltage output due to impact is linearly proportional to the droplet diameter. However, Bragg and Tevaarwerk (1974) could find no correlation between the droplet size and the hot-wire impact signal. Lin and Chang (1989) studied the impact signals of larger droplets of water (500–$1500\,\mu\mathrm{m}$ diameter) on a conical hot-film probe. Their experiments demonstrated that the diameter of the droplet was related to both the rise time and the peak amplitude of the hot-wire impact signal.

The mean velocity and turbulence intensities in the gas phase of the flow can also be measured by eliminating the droplet impact signals. Hetsroni *et al.* (1969) and Hetsroni and Sokolov (1971) applied an amplitude-discriminator method to obtain such measurements in a low-concentration-mist flow. However, the 'clipper circuit' which was used to reduce the amplitude of the spikes is claimed by Ritsch and Davidson (1992) to produce errors in the flow statistics. They developed a modified technique for gas or liquid flows with much smaller particles.

9.2 CONCENTRATION MEASUREMENTS IN GAS MIXTURES

Analysis by Corrsin (1949) indicated that *HWA* could be used to measure gas-mixture concentrations. Corrsin suggested either operating a single sensor at several different overheat ratios for rms measurements, or using two sensors of different diameters for simultaneous measurements of the velocity and the gas-mixture concentration.

9.2.1 Single-sensor response

The response of an *SN* hot-wire or hot-film probe exposed to concentration variations has been studied by Wasan *et al.* (1968), by Wasan and Baid (1971), by McQuaid and Wright (1973), by Simpson and Wyatt (1973), and by Pitts and McCaffrey (1986).

A method for measuring concentration fluctuations in air/gas mixtures was developed by McQuaid and Wright (1973); they assumed that similar power-law relationships applied to pure air, to the air/gas mixture, and to the pure gas:

$$E_a^2 = A_a + B_a U^n, \tag{9.13a}$$

$$E_m^2 = A_m + B_m U^n, \tag{9.13b}$$

$$E_g^2 = A_g + B_g U^n, \tag{9.13c}$$

where the subscripts a, m, and g denote the air, the mixture, and the pure gas. The variation with the concentration of the calibration coefficients A_m and B_m was represented by polynomials in terms of the concentration, Γ:

$$A_m = A_a + a_1\Gamma + a_2\Gamma^2 + a_3\Gamma^3 + \ldots, \tag{9.14a}$$

$$B_m = B_a + b_1\Gamma + b_2\Gamma^2 + b_3\Gamma^3 + \ldots, \tag{9.14b}$$

For fluctuating measurements, set $E = \bar{E} + e$, $U = \bar{U} + u$, and $\Gamma = \bar{\Gamma} + \gamma$. Provided that the relative intensity of both the velocity and concentration is low, a first-order series expansion for the fluctuating voltage, e, will give

$$e = S_u u + S_c \gamma \tag{9.15}$$

with

$$S_u = \frac{n\bar{B}_m \bar{U}^{n-1}}{2\bar{E}_m} = \frac{n(\bar{E}_m^2 - \bar{A}_m)}{2\bar{E}_m \bar{U}}, \tag{9.16}$$

and

$$S_c = \frac{(a_1 + 2a_2\bar{\Gamma} + \ldots) + (b_1 + 2b_2\bar{\Gamma} + \ldots)\bar{U}^n}{2\bar{E}_m}, \tag{9.17}$$

where $\bar{A}_m$ and $\bar{B}_m$ are time-averaged values of eqns (9.14a–b).

Calibration experiments by McQuaid and Wright (1973) showed that the voltage ratio

$$\frac{\bar{E}_m^2 - \bar{E}_a^2}{\bar{E}_g^2 - \bar{E}_a^2} = \psi(\bar{\Gamma}) \tag{9.18}$$

was virtually independent of the velocity (see Fig. 9.6) and $\psi(\bar{\Gamma})$ varied monotonically with $\bar{\Gamma}$. The calibration data were curve fitted by the function

$$\psi(\bar{\Gamma}) = (1 + c)\bar{\Gamma} - c\bar{\Gamma}^2 \tag{9.19}$$

using a least-squares fit. For air/carbon dioxide $c = 0$, and for air/argon the value of c varied between 0.15 and 0.22 depending on the overheat ratio of the wire. Introducing eqn (9.19), it can be shown that the concentration sensitivity, S_c, can be expressed as

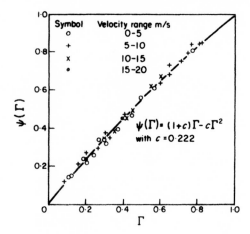

Fig. 9.6. The variation in the voltage ratio, ψ (eqn 9.18), with the concentration, Γ, for an overheat ratio of 1.3. (From McQuaid and Wright 1973. Reproduced with kind permission of Elsevier Science Ltd from the *International Journal of Heat and Mass Transfer.*)

$$S_c = \varphi(\overline{\Gamma}) \frac{\overline{E}_g^2 - \overline{E}_a^2}{2\overline{E}_m}, \qquad (9.20)$$

where

$$\varphi(\overline{\Gamma}) = (1 + c) - 2c\overline{\Gamma}. \qquad (9.21)$$

Equation (9.15) can be used for simultaneous measurements of u and γ by placing two wires with different S_u/S_c ratios next to each other. Alternatively, if only rms values are required, a single wire operated sequentially at different overheat ratios can be used. The practicality of both these methods is discussed by McQuaid and Wright (1973), and the measurement technique presented here has been used by McQuaid and Wright (1974) for turbulence measurements in a nonhomogeneous jet flow.

9.2.2 The interfering-sensor method

Way and Libby (1970) studied the use of a dual-wire interference probe for measurements of the velocity and the concentration in air/helium mixtures. The probe consisted of a thin platinum wire ($d = 2.5\,\mu$m) placed perpendicular to, and slightly upstream of, a cylindrical hot-film ($d = 25\,\mu$m). An initial investigation showed that, to avoid significant interference, the upstream sensor must be operated at a low overheat ratio and it must be placed a relatively large separation distance from the film sensor. In this case, the calibration relationships for the two sensors could be expressed as

$$E_w^2 = A_w(\Gamma) + B_w(\Gamma)U^{1/2}, \qquad (9.22a)$$

$$E_f^2 = A_f(\Gamma) + B_f(\Gamma)U^{1/2}, \qquad (9.22b)$$

where the subscripts w and f refer, respectively, to the wire and the film, and the functions $A_w(\Gamma)$, $B_w(\Gamma)$, $A_f(\Gamma)$, $A_f(\Gamma)$ are known from calibration experiments and related curve-fitting procedures. Eliminating the velocity, U, eqns (9.22a–b) can be rearranged as

$$E_w^2 = A_w[1 - (B_w/B_f)(A_f/A_w)] + (B_w/B_f)E_f^2,$$
$$= a(\Gamma) + b(\Gamma)E_f^2. \qquad (9.23)$$

Consequently, from a pair of measured voltages (E_w, E_f), Γ can be determined, in principle, from eqn (9.23) provided that $a(\Gamma)$ and $b(\Gamma)$, are known. However, the results in Fig. 9.7, corresponding to the noninterference case, show that both $a(\Gamma)$ and $b(\Gamma)$ are relatively weak functions of the concentration; in fact, they are so weak as to make the determination of Γ from eqn (9.23) rather inaccurate. Way and Libby found that the sensitivity to concentration could be greatly enhanced by allowing the thermal field of the film and the wire to interfere. This was achieved by using a small separation ($\Delta x = 0.05$ mm), so that the wire was in the thermal field of the film sensor. This dramatically increased the sensitivity to concentration, as shown in Fig. 9.8. For a velocity range with its lower end at $U = 25$ cm s^{-1} and its upper end dependent on concentration, Way and Libby found that a relationship of the form $E_w^2 \simeq a(\Gamma) + bE_f^2$ could be applied, providing a simple means of obtaining Γ. Further developments

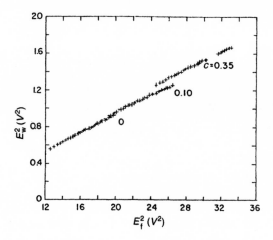

Fig. 9.7. The heat-loss characteristics for a concentration *HWA* probe with a large wire separation. (From Way and Libby 1970. AIAA – Reproduced with the permision of AIAA. © AIAA 1970.)

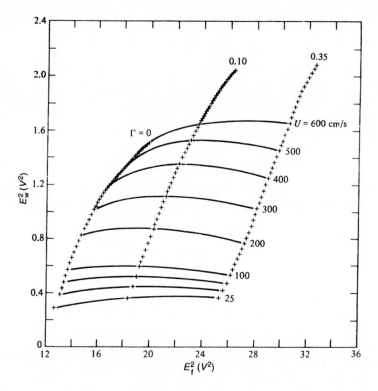

Fig. 9.8. The heat-loss characteristics for an interfering concentration HWA probe. (From Way and Libby 1970. AIAA – Reproduced with the permission of AIAA. © AIAA 1970.)

of this method, including simultaneous measurements of U, V, and Γ, have been described by Way and Libby (1971), by Stanford and Libby (1974), by Libby (1977), and by Libby and LaRue (1978).

9.2.3 Aspirating probes

The earliest reference in the literature to aspirating probes is by Blackshear and Fingerson (1962) who originally designed a water-cooled version to act as a heat-flux probe; they also recognized the ability of this probe to measure the concentration in isothermal binary mixtures. The first researchers to develop and use a choked-nozzle aspirating probe appear to be D'Souza *et al.* (1968) and Leithem *et al.* (1969). Further discussions of the device's design, frequency response, and use in subsonic flows have been presented by Colin and Olivari (1971), Brown and Rebollo (1972), Adler (1972), Brown and Roshko (1974), Adler and Zvirin (1975), Perry (1977), Wilson and Netterville (1981), Ng and Epstein (1983, 1985), Birch *et al.*

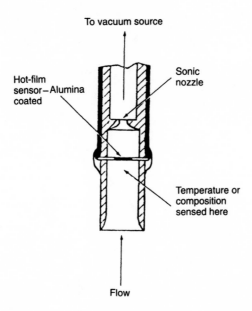

Fig. 9.9. The layout of a TSI aspirating probe. (Published with the permission of TSI Inc.)

(1986), Ahmed and So (1986), and Fedyaev *et al.* (1990). Kotidis and Epstein (1991) have reported concentration measurements in a transonic compressor stage and Kwok *et al.* (1991) and Ninnemann and Ng (1992) have carried out measurements in a supersonic shear flow.

An aspirating probe consists of a sample tube exposed to a gas mixture and linked to a nozzle with a fixed cross-sectional area. A hot-wire or hot-film is placed either upstream or downstream of the nozzle. Figure 9.9 shows the layout of the TSI model 1440 aspirating probe. During operation of the probe, a vacuum is applied so that the pressure ratio across the nozzle exceeds the critical-pressure ratio, giving a choked-nozzle condition.

It can be shown from gas dynamic considerations that the sonic speed in the nozzle depends on the stagnation-flow condition, (P_0, T_0), and on the concentration, Γ, of the gas mixture. The velocity at the wire/film location is a function of the sonic speed in the nozzle, and for constant (P_0, T_0)-conditions this results in a one-to-one relationship between the anemometer output voltage and concentration for any gas mixture. This is illustrated in Fig. 9.10 for the TSI 1296 aspirating probe placed in five different air/gas mixtures.

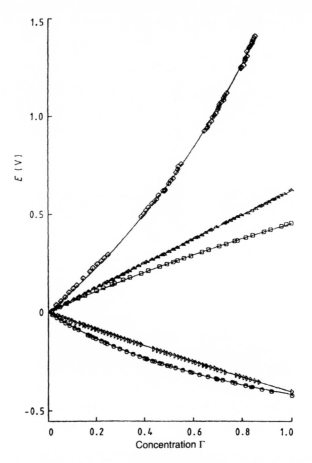

Fig. 9.10. Calibration curves for a TSI 1291 aspirating probe in various gas/air mixtures: (◊) helium, (△) natural gas, (□) ethylene, (▷) argon, (⊙) Freon 12. (From Birch *et al.* 1986.)

9.3 COMPRESSIBLE FLOWS

At high velocities or in low-density flows, the Mach number, M, affects the heat transfer from the *HWA* sensor. For an *SN* hot-wire probe the non-dimensional heat-transfer relationship will be of the form

$$\text{Nu} = f(\text{Re}, \text{Pr}, \text{M}, \tau, \ell/d), \tag{9.24}$$

where Nu is the Nusselt number, ℓ is the active length of the wire, d is the diameter of the wire, Re is the Reynolds number, Pr is the Prandtl number, $\tau = (T_w - T_e)/T_o$ is the temperature loading factor (Kovasznay 1950a), T_w is the wire temperature, T_e is the equilibrium (recovery) temperature, and T_o is the stagnation temperature. If the variation in the Nusselt number with the Prandtl number is small, and compensation is made for the heat conduction to the prongs, then eqn (9.24) reduces to

$$\text{Nu} = f(\text{Re}, \text{M}, \tau). \qquad (9.25)$$

In low-density flows, the flow regime is most conveniently specified by the Knudsen number, Kn, which is related to the Mach number, M, by

$$\text{Kn} = \frac{\lambda}{d} = \left(\frac{\gamma\pi}{2}\right)^{0.5} \frac{\text{M}}{\text{Re}}, \qquad (9.26)$$

where λ is the molecular mean free path, and γ is the ratio of the specific heats of the gas. Three flow regimes are usually defined:

Continuum flow	$\text{Kn} < 0.01$
Slip flow	$0.01 < \text{Kn} < 1$
Free molecular flow	$\text{Kn} > 1$

For a typical 4–5 μm diameter hot-wire sensor in an air flow at atmospheric conditions, $\text{Kn} \simeq 0.02$, which is in the top range of the slip-flow regime. However, it is usually adequate to apply the continuum-flow assumption provided the density changes are small (Fingerson and Freymuth 1983).

Recovery temperature Two temperatures can be defined when a fluid is flowing: the static temperature, T, and the total temperature, T_o, which are related by

$$T_o - T = \frac{U^2}{2c_p}, \qquad (9.27)$$

with T in Kelvin and U in m s^{-1}. For air, $c_p = 1004$ (J kg^{-1} K^{-1}). At low and moderate velocities the difference between T and T_o is insignificant, but the difference can be substantial in high-speed flows. When an unheated *HWA* sensor is placed in a high-speed flow it will be heated to the recovery (or equilibrium) temperature, T_e, which usually differs by a small amount from the total temperature, T_o. The recovery factor

$$\eta = \frac{T_e}{T_o} \qquad (9.28)$$

defines their relationship. The value of η depends primarily on whether the sensor is in the continuum-flow, the slip-flow, or the free-molecular-flow regime. In the continuum-flow regime ($\text{Kn} < 0.01$) the recovery factor varies slowly with the Mach number of the free stream, and η is greater than 0.98 when $\text{M} < 1$ (Baldwin *et al.* 1960; Laurence and Sandborn 1962). Because the temperature measured by the unheated sensor in a high-speed flow is the recovery temperature, T_e, this temperature is normally substituted for the ambient static temperature, T_a, in the heat-transfer relationships selected and in the evaluation and setting of the overheat ratio of the *HWA* sensor.

9.3.1 Transonic flows

HWA can in principle be used to obtain measurements of fluctuating turbulence quantities in transonic flows $(0.9 < M < 1.2)$, but there are few reported investigations for a number of reasons. These include difficulties in determining accurate sensitivity coefficients, as well as wire breakage, vibration, and strain-gauging problems associated with the high dynamic pressures in transonic flow. The principal problem in determining accurate sensitivity coefficients is that the velocity, S_u, and density, S_ρ, sensitivity coefficients vary with Mach number, and in general they are not equal, so that $S_u \neq S_\rho \neq S_{\rho u}$ (Morkovin 1956).

Subsequent to Morkovin's calculations, additional transonic hot-wire-calibration data have been obtained. Behrens (1971) has correlated these data (including the measurements of Laufer and McClellan 1956; Christiansen 1961 and Vrebalovich 1962*a*) in the form of the recovery factor and the Nusselt number as functions of the Mach and Reynolds numbers. Horstman and Rose (1977) have re-evaluated the use of *HWA* for obtaining fluctuating data in transonic flows. They applied the hot-wire response correlations of Behrens (1971) and the conductive end-loss corrections of Dewey (1961) to the detailed hot-wire calibration data of Rose and McDaid (1976) and of Mateer *et al.* (1976). From these data, they evaluated the hot-wire sensitivities S_u and S_ρ and they conducted a parametric study to define a range of Reynolds numbers and temperature loadings, where the velocity-sensitivity coefficient is independent of the Mach number and it is equal to the density-sensitivity coefficient. It was shown in their investigation that when $\tau = (T_w - T_e)/T_o > 0.5$ and the Reynolds number Re > 20, then $S_u = S_\rho = S_{\rho u}$ independent of the Mach number, of the sensor material, and of the ℓ/d ratio. Measurements by Rong *et al.* (1985) are in agreement with these conclusions.

Measurements of velocity and temperature fluctuations in a turbulent transonic flow at low Reynolds number with a *CC* anemometer operated at a low or moderate overheat ratio have been reported by Dupont and Debieve (1990, 1992) and by Barre *et al.* (1992). Under these conditions (as discussed above) the sensitivities to the density, the velocity, and the mass flux are not identical, and it is necessary to determine the aerodynamic and thermal dependence of these sensitivity coefficients. The above authors have carried out an analysis of the parameter dependence of these sensitivity coefficients and developed related calibration procedures for their determination. Measurements have been reported using a modified version of the multiple overheat ratio method (also referred to as the fluctuating-diagram method).

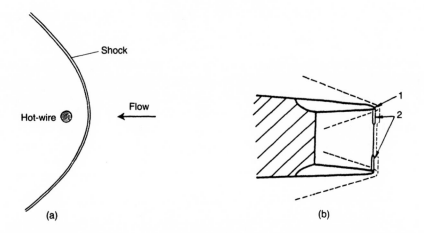

Fig. 9.11. A hot-wire probe in a supersonic flow: (a) the hot-wire and a detached bow shock, and (b) an *SN*-probe design showing the approximate major shocks, (1) soldered joints, (2) copper-plated stubs. (From Smits *et al.* 1983.)

9.3.2 Supersonic flows

Kovasznay (1950a, 1953), Morkovin (1956), and Morkovin and Phinney (1958) pioneered the technique for the interpretation of hot-wire signals in supersonic (1.2 < M < 5) and hypersonic (M > 5) flows. When an *SN* hot-wire probe is placed in a supersonic flow, a detached bow shock forms in front of the wire, which will be positioned in the subsonic flow downstream of a nearly normal shock, as shown in Fig. 9.11. In supersonic flow, Kovasznay (1950a), Morkovin (1956), and Laufer and McClellan (1956) found that the heat transfer from a hot-wire, expressed in nondimensional form by the Nusselt number, was nearly independent of the Mach number. They established that the primary dependence was on the Reynolds number (with the viscosity evaluated at the stagnation temperature) and on the temperature-loading factor, τ ($= (T_w - T_e)/T_o$). As the recovery temperature, T_e, is a function of the Reynolds number in this Mach number range, the hot-wire signal will only be sensitive to fluctuations in the mass-flow rate, ρU, and in the total temperature, T_o. This parameter dependence has been utilized in virtually all subsequent *HWA* studies in supersonic and hypersonic flows, with extensions to the measurement of Reynolds stresses by inclusion of the angular sensitivity of *SY* and *X*-hot-wire probes.

SN hot-wire probes The initial measurements were all carried out using *CC* anemometer techniques. Related measurements with an *SN* probe have been reported by Kistler (1959), Vrebalovich (1962b), Dewey (1965a), Batt and Kubota (1968), Behrens and Ko (1971), Laderman and Demetriades

(1973, 1974) who used a probe developed by Doughman (1972), and Owen *et al.* (1975).

One advantage of the *CC*-mode operation is its ability to achieve a high sensitivity to temperature fluctuation, when operated at a very low overheat ratio in the resistance-wire mode (see Section 2.1.2.4). Following Kovasznay (1950*a*), Morkovin (1956), and Morkovin and Phinney (1958), hot-wire data have been interpreted using the fluctuating-diagram method, in which measurements are repeated for a number of different heating currents. The sensitivity coefficients for the anemometer signal were usually determined from Nusselt-number relationships, which included corrections for the conductive end losses to the prongs (Dewey 1965*b*; Lord 1974; Ko *et al.* 1978).

For a small supersonic wind-tunnel, the 'boundary-layer frequency', U_∞/δ, of the large eddies may typically be about 10 kHz (U_∞ is the free-stream velocity and δ is the boundary-layer thickness). The spectrum of the energy-containing range may therefore easily extend to more than 100 kHz. The frequency response of the anemometer system is therefore of primary importance in supersonic flows. Because the time constant, M, of a hot-wire element is relatively low, it is necessary in the *CC* mode to feed the direct anemometer signal through a frequency-compensation circuit, as described in Section 2.3.2.1. Operation of an anemometer in the *CC* mode is time-consuming and cumbersome, particularly for studies of boundary-layer flows, where large variations occur in the mean-velocity. For each measurement in the flow field the operational condition of the anemometer must be reset to keep the current constant, and adjustments must also be made to the frequency compensation, because the wire's time constant, M, is velocity dependent. Arzoumanian and Debieve (1989) have described a computerized anemometer system, which automatically adjusts and implements these changes. In general, an *SN* hot-wire probe operated in the *CC* mode is not suitable for blow-down supersonic wind-tunnels with typical running times of less than 1 min.

The frequent setting and the adjustment for the thermal lag of the wire can be avoided by the use of a *CT* anemometer, where the wire is maintained at a constant temperature by suitable feedback electronics. However, difficulties with the frequency response of a *CT* anemometer can be experienced at low overheat ratios. The frequency response of an ideal *CT* anemometer without any parasitic circuitry effects may be estimated (Hinze 1959; Owen *et al.* 1975) from

$$M_{\mathrm{CT}} = \frac{M}{1 + 2[(R_{\mathrm{w}} - R_{\mathrm{e}})/R_{\mathrm{e}}]R_{\mathrm{w}}g_{\mathrm{tr}}}, \qquad (9.29)$$

where M_{CT} and M are the time constants of the *CT* anemometer system and the wire, respectively, R_{w} is the hot resistance, R_{e} is the the resistance of the unheated wire, and g_{tr} is the anemometer's transconductance. At low overheat ratios $R_w \simeq R_e$, and the anemometer-system time constant therefore approaches that of the wire element itself. Owen *et al.* (1975) have assessed the performance of CT and CC anemometers by assuming that the anemometers respond to fluctuations in the total temperature at low overheat ratios and to mass-flux fluctuations at high overheat ratios. The comparative results presented in Fig. 9.12 demonstrate the difficulties of operating a CT anemometer at a very low overheat ratio. The CC mode results in Fig. 9.12 were obtained using frequency-compensating amplifiers.

Summaries of CC and CT methods in supersonic flows have been given by Bonnet *et al.* (1986) and by Smits and Dussauge (1989). Bonnet and Alziary de Roquefort (1980) and Bastion *et al.* (1983) have carried out in-depth studies of the performance of both CC and CT anemometers in supersonic flows. They concluded that the two systems are comparable, with regard to the bandwidth and the signal-to-noise ratio provided that the overheat ratio, $(R_w - R_e)/R_e$, is not too small. For their experimental CT set-up they were able to achieve a satisfactory frequency response of up to 250 kHz for overheat ratios down to 0.07. Their results proved that the square-wave test could be used for frequency optimization at high overheat ratios. However, this technique proved unsatisfactory at low overheat ratios, where a sine-wave test in conjunction with a symmetrical bridge, a flat amplifier frequency response, and fine tuning of the offset voltage, e_{off}, (see Fig. 2.12) were needed to achieve the optimum frequency response. Even with the optimum setting, it may not be possible to obtain a

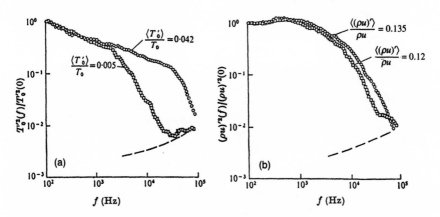

Fig. 9.12. A comparison of normalized power spectra obtained with a frequency-compensated, constant-current and constant-temperature system in a hypersonic boundary-layer flow: (○) constant current, (□) constant temperature. (a) At a low overheat ratio (total-temperature fluctuations); and (b) at a high overheat ratio (mass-flow fluctuations). (From Owen *et al.* 1975. Reproduced with kind permission of Cambridge University Press from the *Journal of Fluid Mechanics*.)

hot-wire frequency response which is sufficiently high to capture all of the turbulent kinetic energy. Walker and Walker (1990) have developed a method for post-detection frequency compensation.

Most of the recent *HWA* studies of supersonic flows have been carried out with the hot-wire operated in the *CT* mode, and related early results with an *SN* probe have been reported by Demin and Zheltukhin (1973), by McLaughlin and McColgan (1974), and by Ko *et al.* (1978). As is now the case for subsonic-flow studies, response equations are usually derived which specify the relationship between the anemometer output, E, and the flow variables ρU and T_0 for *SN* probes and also the flow angle, θ, for *SY* and *X*-probes. Smits *et al.* (1983) assumed a semi-empirical form for the non-dimensional heat-transfer relationship

$$\text{Nu} = A + B\text{Re}^n, \qquad (9.30)$$

where A and B are functions of the temperature-loading factor $\tau = (T_w - T_e)/T_0$. For a perfectly balanced anemometer bridge, eqn (9.30) can be expressed in terms of the anemometer output voltage, E, as

$$\frac{E^2 R_w}{\pi k \ell (R_1 + R_L + R_w)^2 (T_w - T_e)} = F(\tau) + G(\tau)\text{Re}^n, \qquad (9.31)$$

where R_w is the wire resistance at the operating temperature, T_w, T_e is the recovery temperature, k is the heat conductivity of the fluid, and ℓ is the wire length, R_L is the probe and cable resistance, and R_1 is the anemometer bridge resistance, as defined in Fig. 2.12. All the fluid properties are evaluated at the stagnation temperature, T_0, and the recovery factor $\eta = T_e/T_0$ is assumed to be constant. The functions $F(\tau)$ and $G(\tau)$ are found experimentally. Equation (9.31) indicates that E is sensitive to changes in the stagnation temperature and the mass-flux.

For calibration purposes, eqn (9.31) is usually written in the form

$$E^2 = L + M(\rho U)^n, \qquad (9.32)$$

where ρU is the instantaneous mass flux along the axis of the probe, L and M are constants for a particular wire at a given overheat ratio and stagnation temperature, and the exponent n is found from the line of best fit to the data. A mass flux exponent of 0.55 was found to provide a good fit to the calibration data.

The anemometer response is commonly expressed in terms of small perturbations in T_0 and ρU. Due to the variations in both ρ and U, *the fluctuating quantities will be denoted by primes in this section.* Setting $E = \bar{E} + e'$, $\rho U = \overline{\rho U} + (\rho u)'$, and $T_0 = \bar{T}_0 + T_0'$, the response equation for the fluctuating voltage signal can be expressed as

$$e' = \left.\frac{\partial E}{\partial \rho U}\right|_{T_0} (\rho u)' + \left.\frac{\partial E}{\partial T_0}\right|_{\rho U} T_0' \qquad (9.33a)$$

or

$$e' = S_m(\rho u)' + S_{T_0} T'_0, \tag{9.33b}$$

where S_m is the mass-flow sensitivity and S_{T_0} is the total-temperature sensitivity. Alternatively, eqn (9.33b) has often been expressed as

$$\frac{e'}{\bar{E}} = F\frac{(\rho u)'}{\overline{\rho U}} + G\frac{T'_0}{\bar{T}_0}, \tag{9.34a}$$

with the corresponding sensitivity coefficients defined as

$$F = \frac{\partial(\ln E)}{\partial(\ln \rho U)}, \tag{9.34b}$$

$$G = \frac{\partial(\ln E)}{\partial(\ln T_0)}.$$

Smits *et al.* (1983) have shown that the sensitivities S_m and S_{T_0} can be expressed, using eqns (9.31) and (9.32), as

$$S_m = \frac{nM}{2\bar{E}}\left(\frac{\bar{E}^2 - L}{M}\right)^{(n-1)/n} \tag{9.35a}$$

$$S_{T_0} = \frac{\bar{E}}{2T_0}\left\{a - \frac{\eta}{\tau} - \frac{(\tau + \eta)}{\bar{E}^2}\left[\frac{L}{F(\tau)}\frac{\partial F}{\partial \tau} + \frac{(\bar{E}^2 - L)}{G(\tau)}\frac{\partial G}{\partial \tau}\right] - nb\frac{(\bar{E}^2 - L)}{\bar{E}^2}\right\}, \tag{9.35b}$$

where η is the recovery factor (see eqn (9.28)), and it has been assumed that the fluid properties vary with temperature according to

$$\frac{k}{k_r} = \left(\frac{T}{T_r}\right)^a, \quad \frac{\mu}{\mu_r} = \left(\frac{T}{T_r}\right)^b,$$

where T_r is some reference temperature, and the values of a and b are about 0.8, as discussed in Section 7.2.2.

For time-averaged measurements we get, from eqn (9.33b),

$$\overline{e'^2} = S_m^2\overline{(\rho u)'^2} + 2S_m S_{T_0}\overline{(\rho u)' T'_0} + S_{T_0}^2\overline{T'^2_0}. \tag{9.36}$$

In principle, it is possible to determine $\overline{(\rho u)'^2}$, $\overline{(\rho u)' T'_0}$, and $\overline{T'^2_0}$ by operating the hot-wire at three different overheat ratios. However, to obtain accurate results a large number of overheat ratios should be used in conjunction with a least-squares curve-fitting method (the fluctuation-diagram method first suggested by Kovasznay 1950a). A method for the rapid scanning of overheat ratios has been used by Walker *et al.* (1989), and a typical set of velocity measurements is shown in Fig. 9.13.

Kovasznay (1950a) has shown that the sensitivity ratio S_m/S_{T_0} depends

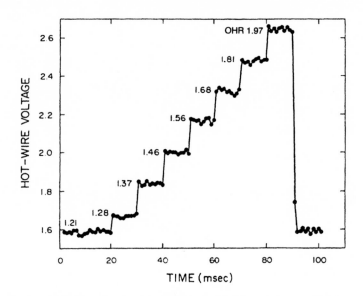

Fig. 9.13. The application of hot-wire signals for various stepped overheat ratios (OHR) (10 ms per overheat level) in a supersonic flow. (From Walker *et al.* 1989. Reproduced with permission of AIAA. © AIAA 1989.)

strongly on the overheat ratio, and the fluctuation-diagram method is usually carried out over the range $0.02 < \tau < 0.5$. At low overheat ratios, the hot-wire signal is primarily sensitive to T_0', and if the overheat ratio is sufficiently low the sensitivity to $(\rho u)'$ can be neglected. However, CT anemometers are difficult to use at small overheat ratios because serious nonlinear effects can occur (Smits *et al.* 1983) and a complex frequency-optimization procedure is required (Bonnet and Alziary de Roquefort 1980). At high overheat ratios the output is primarily sensitive to fluctuations in the mass flux, but it is difficult to increase τ to the point where the output is sensitive only to $(\rho u)'$. However, if the wire can be operated with a value of τ of about one, then Smits *et al.* (1983) have shown that the sensitivity of the wire to fluctuations in the stagnation temperature can be neglected.

As described by Spina *et al.* (1991) and by Donovan and Spina (1992), the fluctuating velocity component, u' can be estimated from the measured fluctuations in the mass flow rate, $(\rho u)'$, provided that the following approximations are valid:

If T_0' is small in comparison to T', the Strong Reynolds Analogy (SRA, Morkovin 1962) can be used to relate T' and u'; that is

$$\frac{T'}{\overline{T}} = - (\gamma - 1)M^2 \frac{u'}{\overline{U}}. \tag{9.37}$$

Furthermore, if the pressure fluctuations are small compared with the temperature fluctuations, then

$$\frac{p'}{\bar{p}} \ll \frac{T'}{\bar{T}},$$

which is typical of many flows without strong perturbations, and consequently

$$\frac{\rho'}{\bar{\rho}} = (\gamma - 1)M^2 \frac{u'}{\bar{U}}. \tag{9.38}$$

An expression relating u' and $(\rho u)'$ can be found as follows. A Reynolds decomposition is performed on $(\rho u)' = \rho U - \overline{\rho U}$, and ρ and U are expressed in terms of the mean and fluctuating components. A first-order series expansion is applied, and using the SRA results (eqns (9.37) and (9.38)) gives

$$\frac{u'}{\bar{U}} = \frac{(\rho u)'}{\overline{\rho U}} [1 + (\gamma - 1)M^2]^{-1}. \tag{9.39}$$

Further studies of the Reynolds-analogy concept in supersonic-boundary layer flows have been presented by Gaviglio (1987).

Smits (1990) has used two wires placed in close proximity to each other, with the first wire operated at a high overheat ratio, say 1.0, and the second wire operated at an overheat-ratio of about 0.4. Since the two wires will have different S_m/S_{T_0} ratios, $(\rho u)'$ and T_0' can be measured simultaneously. In this method, it is just as important to match the frequency responses of the two probes as it is to achieve the highest frequency response possible (Smits, personal communication). The two-wire technique has also been used in a comparative investigation by Walker *et al.* (1989) From consistency checks they concluded that the dual-wire probe produced results that were too high. They concluded that the most likely explanation was ill-conditioning in the flow variable separation process; that is although the flow seen by the two wires is very nearly the same, even small differences or small amounts of noise result in large errors in the computed results which increased both the indicated rms mass flux and the indicated rms total temperature.

SY and X-hot-wire probes Shear stress measurements with *CC*-anemometer equipment have been reported by Demetriades and Laderman (1973), Laderman (1976), and Laderman and Demetriades (1979). More recent investigations have been carried out with hot-wire probes operated in the *CT* mode. The response equations for inclined sensors can, for small fluctuations, be expressed in terms of the relationship between the fluctuating anemometer voltage, e', and fluctuations in the mass-flux, $(\rho u)'$, the flow angle, θ', and the total temperature, T_0'; that is

$$e' = \frac{\partial E}{\partial(\rho U)}(\rho u)' + \frac{\partial E}{\partial \theta}\theta' + \frac{\partial E}{\partial T_0}T_0' \qquad (9.40a)$$

or

$$e' = S_m(\rho u)' + S_{mv}\bar{\rho}v' + S_{T_0}T_0', \qquad (9.40b)$$

where S_m and S_{mv} are the longitudinal and transverse mass-flux sensitivities and S_{T_0} is the total-temperature sensitivity.

Extending the SN-probe work of Smits $et\ al.$ (1983) to SY and X-probes (Smits and Muck 1984; Donovan and Spina 1992) enables the response equation for an inclined sensor to be expressed as

$$E^2 = L + MG(\theta)(\rho\tilde{V})^n, \qquad (9.41a)$$

where $\tilde{V}$ and θ are the magnitude and flow angle of the (two-dimensional) flow vector (θ is defined as the angle between the instantaneous two-dimensional velocity vector and the mean-flow direction), and $G(\theta)$ represents the probe's angle response (with $G(0) = 1$). If the turbulence intensity is small, so that $\tilde{V} \simeq U$, then eqn (9.41) can be written as

$$E^2 = L + MG(\theta)(\rho U)^n. \qquad (9.41b)$$

Also, if the probe is aligned ($\bar{\theta} = 0°$) with the mean-flow direction, $\bar{U}$, then

$$\theta' \cong \frac{v'}{\bar{U}}, \qquad S_{mv} = \frac{1}{\bar{\rho}\bar{U}}\frac{\partial E}{\partial \theta}.$$

For a given overheat ratio, the values of L and M were determined from calibration data obtained at $\theta = 0°$, using a least-squares fit for $E^2 \sim (\overline{\rho U})^n$. The function $G(\theta)$ was obtained by a subsequent yaw calibration keeping $\overline{\rho U}$ constant. By applying a first-order perturbation analysis to eqn (9.41b), the functional form of S_m, S_{mv}, and S_{T_0} can be determined, as described by Smits $et\ al.$ (1983) by Smits and Muck (1984), and by Donovan and Spina (1992).

Measurement with an SY probe involves taking two successive measurements after a rotation of $180°$ around the probe stem. Provided that the probe is perfectly aligned, which may be difficult to achieve in practice (as observed by Morkovin and Phinney 1958; Smits and Muck 1984; Bonnet and Knani 1988), eqn (9.40b) can be applied to the two positions denoted 1 and 2. Denoting the sensitivities for position 1 as S_m, S_{mv}, and S_{T_0}, the value of the transverse mass-flux sensitivity will be $-S_{mv}$ for position 2. Consequently, for an aligned probe, we have (Smits and Muck 1984)

$$\overline{e_1'^2} - \overline{e_2'^2} = 4S_{mv}\bar{\rho}[S_m\overline{(\rho u)'v'} + S_{T_0}\overline{v'T_0}], \qquad (9.42a)$$

$$\overline{e_1'^2} + \overline{e_2'^2} = 2[S_m^2\overline{(\rho u)'^2} + S_{mv}^2\bar{\rho}^2\overline{v'^2} + S_{T_0}^2\overline{T_0'^2} + 2S_mS_{T_0}\overline{(\rho u)'T_0'}].$$
$$\qquad (9.42b)$$

In principle, the six unknowns in eqns (9.42a–b) can be determined by operating the wire at three different overheat ratios. Smits and Muck (1984) and Smits and Dussauge (1989) have discussed in detail the practical difficulties in implementing this method.

Measurements and signal-analysis procedures for X-probes have been reported by Fernando *et al.* (1987), by Bonnet and Knani (1988), by Donovan and Spina (1990), by Fernando and Smits (1990), by Spina *et al.* (1991), and by Donovan and Spina (1992).

If the two wires have identical sensitivities, then eqn (9.40b) applies to both wires except for the sign inversion in S_{mv} for wire 2. Because the two signals are sampled simultaneously it follows that

$$\overline{(e_1' - e_2')^2} = 4S_{mv}^2 \bar{\rho}^2 \overline{v'^2}, \tag{9.43a}$$

$$\overline{(e_1' + e_2')^2} = 4[S_m^2 \overline{(\rho u)'^2} + 2S_m S_{T_0} \overline{(\rho u)' T_0'} + S_{T_0}^2 \overline{T_0'^2}]. \tag{9.43b}$$

Bonnet and Knani (1988) have considered the response equations when the sensitivities of the two wires are different. Additional turbulent quantities can be obtained from response equations for $\overline{(e_1' - e_2')(e_1' + e_2')}$ (Bonnet and Knani 1988) or for $\overline{e_1' e_2'}$ (Demetriades and Laderman 1973).

To obtain direct measurements of the fluctuating mass-flow rates, $(\rho u)'$ and $\bar{\rho} v'$, Donovan and Spina (1992) operated the wires in their X-probe at such a high overheat ratio ($\tau \approx 1$) that the temperature sensitivity S_{T_0} in eqn (9.40b) could be ignored. The two wires in their X-probe were therefore only sensitive to ρU and $\bar{\rho} v'$. Four signal-analysis procedures were investigated, including the first-order response of eqn (9.40b) and a direct solution for ρU and θ based on eqn (9.41b) As expected they concluded from their measurements and analysis that the first-order response equation gave the least accurate results in high-turbulence regions near a wall.

9.3.2.1 Probe design

The probes used in supersonic flows have a somewhat different design from those used in subsonic flows. The prongs are tapered at the front of the probe body in streamlined wedge shape to minimize aerodynamic interference. Also, the prongs are usually kept short to reduce vibrations and deflection under load. Strain-gauging effects can also be a problem in supersonic flows, and it is recommended that the wire is mounted slack to avoid parasitic effects which considerably affect the high-frequency part of the signal. Checking that the parasitic strain-gauge effect can be neglected in a low-turbulence supersonic part of the flow is recommended (Bonnet, personnel communication). Typical SN, SY, and X-hot-wire probes are illustrated in Fig. 9.14. It is recommended that the active part of the wire is kept away from the prongs, which are surrounded by detached shock waves as shown in Fig. 9.11. For X-hot-wire probes, shocks from one wire

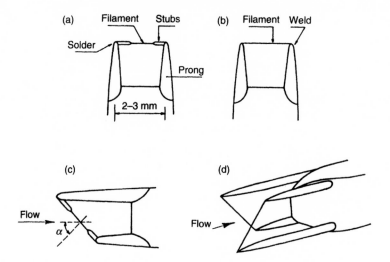

Fig. 9.14. Hot-wire probes used for supersonic flows: (a) an *SN* probe, 'etched' design; (b) an *SN* probe, 'welded' design; (c) an *SY* probe; (d) an *X*-probe. (From Smits and Dussauge 1989.)

and its support may interfere with the other wire, and the probes must be designed to avoid this interference at all Mach numbers (Kovasznay 1950a, Bradshaw 1971; Smits and Dussauge 1989).

Recent work suggests that commercially available probes can be used when the prongs are shortened and wedge-shaped fillets are added. The wire material is usually tungsten, gold-plated tungsten or platinum-plated tungsten, and the filament can be soldered or welded to the prongs. In the soldered design (Fig. 9.14(a)) the tungsten wire is first coated with copper to a diameter of 5–10 times the filament diameter. The coated wire is soldered to the prongs and the central portion of the coating is etched away using a dilute acid to expose the active portion of the wire. This procedure will reduce the aerodynamic interference caused by the relatively bulky prongs (Smits and Dussauge 1989).

10

VORTICITY MEASUREMENTS

This chapter describes the measurement of the components of the vorticity vector. The vorticity vector at a point is defined as the curl of the velocity vector. Using tensor notation this relationship can be expressed as

$$\omega_i = \varepsilon_{ijk} \frac{\partial U_k}{\partial x_j},$$
(10.1)

where ω_i ($i = 1$-3) are the components of the vorticity vector, U_j ($j = 1$-3) are the components of the velocity vector, and ε_{ijk} is the alternating tensor. As discussed by Wallace (1986), at each point in the flow field a spherical fluid particle can be imagined the motion of which can be decomposed into translation, expansion, and rotation. The vorticity can be interpreted as being twice the instantaneous solid-body rotation rate of particles along the principal axis in the fluid, where there is no shear deformation.

The importance of the vorticity for understanding turbulence follows from the dynamic equation for the transport of vorticity

$$\frac{\partial \omega_i}{\partial t} + U_j \frac{\partial \omega_i}{\partial x_j} = \omega_j \frac{\partial U_i}{\partial x_j} + \nu \frac{\partial^2 \omega_i}{\partial x_j \partial x_j}.$$
(10.2)

This equation shows that the rate of change of the vorticity of a particle is equal to the rate of deformation of the vortex lines of the particles plus the rate of viscous diffusion of its vorticity. Pressure does not enter directly into this equation unlike the transport equation for momentum; this leads to some simplification in interpretation and computation.

To explain the measurement of the vorticity vector in eqn (10.1), consider a cartesian (x, y, z)-coordinate system in which the components of the related velocity vector are denoted as (U, V, W). The corresponding vorticity components can be expressed as

streamwise component:
$$\omega_x = \frac{\partial W}{\partial y} - \frac{\partial V}{\partial z},$$
(10.3a)

The two cross-stream components are

normal component, and
$$\omega_y = \frac{\partial U}{\partial z} - \frac{\partial W}{\partial x},$$
(10.3b)

for the spanwise component:
$$\omega_z = \frac{\partial V}{\partial x} - \frac{\partial U}{\partial y}$$
(10.3c)

If the vorticity probe is aligned with the mean-flow direction so that $(U, V, W) = (\bar{U} + u, v, w)$ and the turbulence intensity is low, then eqns (10.3a–c) can be approximated by

$$\omega_x = \frac{\Delta w}{\Delta y} - \frac{\Delta v}{\Delta z}, \tag{10.4a}$$

$$\omega_y = \frac{\Delta U}{\Delta z} + \bar{U}^{-1} \frac{\Delta w}{\Delta t}, \tag{10.4b}$$

$$\omega_z = -\bar{U}^{-1} \frac{\Delta v}{\Delta t} - \frac{\Delta U}{\Delta y}, \tag{10.4c}$$

where the partial derivatives have been replaced by the corresponding finite-difference ratios, and Taylor's convective frozen-pattern hypothesis $(\partial/\partial x = -\bar{U}^{-1}(\partial/\partial t))$ has been used to convert the x-derivatives into time derivatives. The individual terms in eqns (10.4a–c) can, in principle, be evaluated by using probes containing single normal (SN) hot-wire sensors to measure the longitudinal velocity component, U; and X-arrays can determine (U, v) when placed in the (x, y)-plane or (U, w) when placed parallel to the (x, z)-plane. The streamwise vorticity in eqn (10.4a) requires the measurement of the transverse variation in the cross-flow velocity components v and w. This can be accomplished using two pairs of X-arrays with spatial wire separations in respectively, the y- and z-direction (the classical Kovasznay-type vorticity probe). To measure a cross-stream vorticity component ω_y or ω_z, given by eqn (10.4b) or (10.4c) the following wire arrangements can be used. For $\omega_z(\simeq -\bar{U}^{-1}\Delta v/\Delta t - \Delta U/\Delta y)$ the term $\Delta U/\Delta y$ can be evaluated from two parallel SN hot-wires separated a small distance in the y-direction. The term $\Delta v/\Delta t$, which is the temporal derivative of the cross-flow component, v, can be evaluated from an X-array placed parallel to the (x, y)-plane. From the format of eqns (10.4b) and (10.4c) it can be observed that ω_y can be obtained by rotating such a probe by 90 °. However, it should be noted that measurements of velocity gradients with parallel-wire probes are somewhat uncertain, because spatial-averaging errors are important when the wire separation is much larger than the Kolmogorov length scale and gradient evaluations become very sensitive to calibration uncertainties and noise when the separation distance approaches zero. The practical implementation of these vorticity-probe principles is described in the following subsections. Reviews of these techniques have also been given by Wallace (1986) and by Foss and Wallace (1989).

10.1 THE STREAMWISE VORTICITY COMPONENT

The hot-wire probe shown in Fig. 10.1 was developed by Kovasznay (1950*b*, 1954) to measure the longitudinal component of the vorticity $\omega_x(= \partial w/\partial y -$

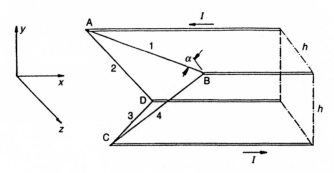

Fig. 10.1. A Kovasznay-type streamwise vorticity probe. (From Vukoslavčević and Wallace 1981.)

$\partial v/\partial z$). An ideal probe of this type contains four identical hot-wires, 1–4, which are all inclined at the same angle, α, relative to mean-flow direction. In the signal analysis developed by Kovasznay, the probe was considered as containing two X-arrays; wires 1 and 3, with a wire separation h, are parallel to the (x, y)-plane and wires 2 and 4, also with a wire separation h, are parallel to the (x, z)-plane. The four wires were operated in the constant current (CC) mode and they were connected to a single Wheatstone bridge, as shown in Fig. 10.2.

The following simplified assumptions were used in order to evaluate ω_x from the voltage difference E_{ω_x} in Fig. 10.2. (i) The signals from the four wires were evaluated using a simple X-array signal analysis for the two X-arrays; that is, for the wire pair 1,3 the effect of the w-component (in the z-direction) was ignored, and for the wire pair 2,4 the effect of the

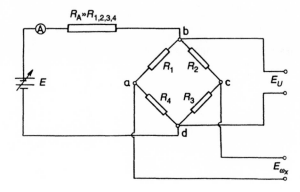

Fig. 10.2. A CC anemometer circuit for a Kovasznay-type streamwise vorticity probe. (From Kastrinakis *et al.* 1979.)

v-component (in the y-direction) was ignored; (ii) Denoting the velocity components at the centre of the wire configurations by (U_0, v_0, w_0), it was assumed that the longitudinal velocity component $U_0 = \bar{U} + u$ was constant within the probe volume, and the effect of spatial variations in v and w within the probe volume were only observed as a $\partial w/\partial y$ term for the wire configuration 1, 3 and as a $\partial v/\partial z$ term for the wire configuration 2, 4. In all other respects the flow field was assumed to be uniform within the probe volume. Provided that the wire separation h is small, so that a first-order series expansion can be applied, the effective velocity for the four identical wires can be expressed using a cosine-cooling law as

$$V_{e1} = (\bar{U} + u) \cos \alpha - w \left(y = \frac{h}{2} \right) \sin \alpha \simeq (\bar{U} + u) \cos \alpha$$

$$- \left(w_0 + \frac{h}{2} \frac{\partial w}{\partial y} \right) \sin \alpha, \tag{10.5a}$$

$$V_{e2} = (\bar{U} + u) \cos \alpha + v \left(z = -\frac{h}{2} \right) \sin \alpha \simeq (\bar{U} + u) \cos \alpha$$

$$+ \left(v_0 - \frac{h}{2} \frac{\partial v}{\partial z} \right) \sin \alpha, \tag{10.5b}$$

$$V_{e3} = (\bar{U} + u) \cos \alpha + w \left(y = -\frac{h}{2} \right) \sin \alpha \simeq (\bar{U} + u) \cos \alpha$$

$$+ \left(w_0 - \frac{h}{2} \frac{\partial w}{\partial y} \right) \sin \alpha, \tag{10.5c}$$

$$V_{e4} = (\bar{U} + u) \cos \alpha - v \left(z = \frac{h}{2} \right) \sin \alpha \simeq (\bar{U} + u) \cos \alpha$$

$$- \left(v_0 + \frac{h}{2} \frac{\partial v}{\partial z} \right) \sin \alpha, \tag{10.5d}$$

Taking the sum of the signals from the two X-arrays 1, 3 and 2, 4, then

$$V_{e1} + V_{e3} = 2(\bar{U} + u) \cos \alpha - h \frac{\partial w}{\partial y} \sin \alpha, \tag{10.6a}$$

$$V_{e2} + V_{e4} = 2(\bar{U} + u) \cos \alpha - h \frac{\partial v}{\partial z} \sin \alpha, \tag{10.6b}$$

and subtracting these two equations gives

$$(V_{e2} + V_{e4}) - (V_{e1} + V_{e3}) = h \sin \alpha \left(\frac{\partial w}{\partial y} - \frac{\partial v}{\partial z} \right) = h \sin \alpha \, \omega_x. \tag{10.7}$$

From the configuration of the Wheatstone bridge in Fig. 10.2 it follows for the voltages E_1, E_2, E_3, and E_4 across the four sensors that

$$(E_2 + E_4) - (E_1 + E_3) = 2E_{\omega_x}, \qquad (10.8)$$

where E_{ω_x} is the voltage difference between points a and c in Fig. 10.2. Equations. (10.7) and (10.8) are in a similar form, but due to the non-linear relationship between E and V_e the vorticity probe must be calibrated. A means of directly calibrating the probe by spinning it around its longitudinal axis in a rotation-free flow and transmitting the signal through a mercury-bath commutator was developed by Uberoi and Corrsin (1951). Such a calibration was performed by Kastrinakis *et al.* (1979) in a constant-velocity laminar flow. When the probe is rotated the fluid is in relative solid-body rotation with respect to it, so that the induced streamwise vorticity is simply twice the angular velocity with which the probe is rotated. Calibration curves for two flow velocities are shown in Fig. 10.3,

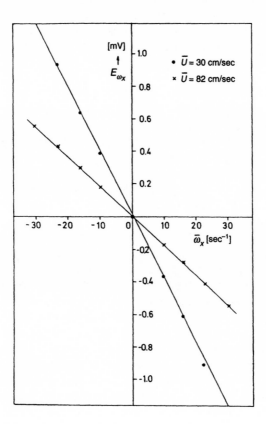

Fig. 10.3. Calibration curve of the Kovasznay-type vorticity probe with respect to ω_x. (E_{ω_x} is the voltage across the diagonal a–c in Fig. 10.2). (From Kastrinakis *et al.* 1979.)

demonstrating the necessity for a probe calibration when the flow velocity is changed.

Kistler (1952) analysed the operation of this probe and used it to measure one-dimensional-vorticity spectra and their decay in a turbulent grid flow. Corrsin and Kistler (1955) measured the rms streamwise vorticity distribution in a roughened boundary layer with a probe of this type. Also, in the first of many attempts to measure the vorticity which were undertaken at the Max-Plank-Institute für Strömmungsforchung in Göttingen, Kastrinakis (1976) measured the rms distribution of the streamwise vorticity across a turbulent channel flow using a Kovasznay-type probe with 2.8 mm long wires. Kastrinakis also used two Kovasznay-type probes to measure the two-point correlation coefficient for spanwise probe separation. Kastrinakis *et al.* (1977) used the same two probes to investigate wall-layer vorticical structures.

Kistler (1952) was the first to show that the simple analysis of eqn (10.7) for the evaluation of the streamwise vorticity component may be contaminated by parasitic velocity-component effects. For a probe with all its wires geometrically and electrically identical, Kistler found that the voltage, E_{ω_x}, across points a and c of the Wheatstone bridge in Fig. 10.2 can be expressed as

$$E_{\omega_x} = 2Ah(\sin \alpha)\omega_x + A(\cos \alpha)h^2 \left(\frac{\partial^2 U}{\partial y^2} - \frac{\partial^2 U}{\partial z^2} \right), \qquad (10.9)$$

where the coefficient A depends both on the constant operating current and on the streamwise velocity, which varies in turbulent flow. Wyngaard (1969) studied the spatial-resolution characteristics of this probe type and gaves a minimum value of the Kolmogorov length-scale to wire-length ratio of 0.3 as the criterion for adequate resolution. At this ratio, the probe measures approximately 86% of the true mean-square streamwise vorticity for an active wire-length to h ratio of unity.

From the discussion by Vukoslavčević and Wallace (1981) and the assumptions made for eqns (10.5a–d) it follows that it is implausible to assume that the Kovasznay-type probe is only sensitive to the velocity gradients of the streamwise vorticity and not to any of the other seven velocity gradients in the flow. The analysis presented by Kastrinakis *et al.* (1979) demonstrated that the value of ω_x, as measured by the Kovasznay-type probe, contains a considerable parasitic sensitivity to the streamwise and cross-stream velocity components. They found that contamination by the cross-stream velocity can be of the same order as that of the measured instantaneous vorticity. To account for the spatial variation in the flow field and the response of each wire to all three velocity components, Kastrinakis *et al.* (1979) suggested the construction of a probe with the same geometric arrangement as a Kovasznay-type probe, but with each wire operated independently. This, of course, has the disadvantage that it requires eight supporting prongs. Such a probe has been built and tested by Vukoslavčević and Wallace (1981), and the probe design is shown in Fig. 10.4. It has four wires forming two orthogonal X-arrays and each wire is heated by an independent CT anemometer.

The probe was calibrated and analysed assuming a power-law relation,

$$E^2 = A + BV_e^n,$$

between the voltage and the effective velocity for each wire, and the effective velocity was expressed in terms of Jørgensen's equation (eqn (3.2); that is,

$$V_e^2 = U_N^2 + k^2 U_T^2 + h^2 U_B^2,$$

where U_N, U_T, and U_B are, respectively, the normal, tangential, and binormal velocity components in the wire-fixed coordinate system. For simplicity, the yaw and pitch coefficients were assumed to be $k = 0$ and $h = 1$; that is, each wire was assumed to be cooled by the total normal velocity component. The analysis presented by Vukoslavčević and Wallace (1981) includes, to first order, the transverse variation in all three velocity components by a first-order Taylor series expansion. For this truncated velocity field representation they derived related exact expressions for the effective-velocity components normal to each wire. For a probe with an average wire angle of α and a characteristic probe dimension of h, the equations for the four cooling velocities, V_{ei} $(i = 1\text{-}4)$, are

$$V_{e1}^2 = \left[\left(U_0 + \frac{\partial U}{\partial y}\frac{h}{2}\right)\cos\alpha - \left(w_0 + \frac{\partial w}{\partial y}\frac{h}{2}\right)\sin\alpha\right]^2 + \left(v_0 + \frac{\partial v}{\partial y}\frac{h}{2}\right)^2,$$
$$(10.10a)$$

$$V_{e2}^2 = \left[\left(U_0 - \frac{\partial U}{\partial z}\frac{h}{2}\right)\cos\alpha + \left(v_0 - \frac{\partial v}{\partial z}\frac{h}{2}\right)\sin\alpha\right]^2 + \left(w_0 - \frac{\partial w}{\partial z}\frac{h}{2}\right)^2,$$
$$(10.10b)$$

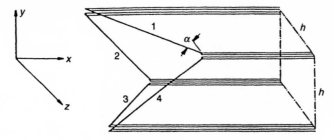

Fig. 10.4. A modified Kovasznay-type streamwise vorticity probe. Each wire is supported by a separate pair of prongs and operated in the CT mode. (From Vukoslavčević and Wallace 1981.)

$$V_{e3}^2 = \left[\left(U_0 - \frac{\partial U}{\partial y} \frac{h}{2} \right) \cos \alpha + \left(w_0 - \frac{\partial w}{\partial y} \frac{h}{2} \right) \sin \alpha \right]^2 + \left(v_0 - \frac{\partial v}{\partial y} \frac{h}{2} \right)^2,$$

(10.10c)

$$V_{e4}^2 = \left[\left(U_0 + \frac{\partial U}{\partial z} \frac{h}{2} \right) \cos \alpha - \left(v_0 + \frac{\partial v}{\partial z} \frac{h}{2} \right) \sin \alpha \right]^2 + \left(w_0 + \frac{\partial w}{\partial z} \frac{h}{2} \right)^2.$$

(10.10d)

Using the wire notation and the coordinate system shown in Fig. 10.4, eqns (10.10a–d) deviate slightly from those given by Vukoslavčević and Wallace (1981) and Wallace (1986). In the equations above, (U_0, v_0, w_0) represent the velocity components at the centre of the probe. Equations (10a–d) can be written in terms of $(V_{e1}^2 - v_0^2)^{1/2}$, etc., by assuming that the $(\partial v/\partial y)(h/2)$ and the $(\partial w/\partial z)(h/2)$ terms are small compared with, respectively, v_0 and w_0. Adding the rearranged eqn (10.10c) to (10.10a) and eqn (10.10d) to (10.10b) and subtracting the two sums, an approximate expression for the streamwise component of the vorticity can be obtained; that is,

$$\omega_x = \frac{1}{h \sin \alpha} \left[(V_{e4}^2 - w_0^2)^{1/2} + (V_{e2}^2 - w_0^2)^{1/2} - (V_{e3}^2 - v_0^2)^{1/2} \right.$$

$$\left. - (V_{e1}^2 - v_0^2)^{1/2} \right].$$

(10.11)

Equation (10.11) shows the explicit dependence of the streamwise vorticity on the cross-stream velocity components v_0 and w_0, and it is not of second order as previously thought.

Adding the rearranged eqns (10.10a–d) gives an approximate equation for U_0; that is,

$$U_0 = \frac{1}{4 \cos \alpha} \left[(V_{e1}^2 - v_0^2)^{1/2} + (V_{e2}^2 - w_0^2)^{1/2} + (V_{e3}^2 - v_0^2)^{1/2} \right.$$

$$\left. + (V_{e4}^2 - w_0^2)^{1/2} + h \sin \alpha \left(\frac{\partial w}{\partial y} + \frac{\partial v}{\partial z} \right) \right].$$

(10.12)

Expressions for the cross-stream velocity components, v_0 and w_0, can be obtained by subtracting the rearranged eqn (10.10d) from eqn (10.10b) and eqn (10.10c) from eqn (10.10a) yielding

$$v_0 = \frac{1}{2 \sin \alpha} \left[(V_{e2}^2 - w_0^2)^{1/2} - (V_{e4}^2 - w_0^2)^{1/2} + h \frac{\partial U}{\partial z} \cos \alpha \right], \quad (10.13a)$$

$$w_0 = \frac{1}{2 \sin \alpha} \left[(V_{e3}^2 - v_0^2)^{1/2} - (V_{e1}^2 - v_0^2)^{1/2} + h \frac{\partial U}{\partial y} \cos \alpha \right]. \quad (10.13b)$$

These expressions for v_0 and w_0 contain (in addition to the differences of the normal cooling-velocity components in the form in which they usually appear for X-arrays) the v_0- and w_0-components themselves as well as streamwise velocity gradients which must always be neglected because of the assumption that the velocity field, which is instantaneously sensed by the X-arrays, is uniform.

The probe was tested in a uniform, irrotational flow by pitching and yawing the probe through a range of angles and by varying the flow speed. A spurious vorticity, which has been explained by Kastrinakis *et al.* (1979), is detected if the v_0- and w_0-components of the equation are neglected. However, if the measured values of these components are included, the uncertainty in the measured vorticity in this gradient-free flow is reduced to random measurement errors alone. It is instructive to note that even this residual error can be about 50 s^{-1} from a measurement error of only $0.5 \, \text{mV}$ at the anemometer bridge output.

The instantaneous velocity gradients in turbulent shear flows can be large, and consequently the instantaneous vorticity components, for example, $\omega_x = \partial w/\partial y - \partial v/\partial z$, may also be large. However, in order to measure vorticity components with multi-array hot-wire probes, it is normally assumed that the cooling velocity along the length of each of the wire elements is uniform, and this may lead to significant signal-interpretation errors. Vukoslavčević and Wallace (1981) built a two-sensor gradient probe with parallel wires to measure the maximum instantaneous gradients of the fluctuating streamwise velocity components, $\partial u/\partial y$ and $\partial u/\partial z$. Near the wall, they found that the instantaneous error in the measured velocity components and, consequently, in the measured value of ω_x for the worst-case occurrences was larger than the evaluated maximum rms vorticity values obtained from eqn (10.11), which had been compensated for the parasitic effect of cross-stream velocity components.

Kastrinakis and Eckelmann (1983) used the modified Kovasznay-type probe to measure some of the basic statistics of the streamwise vorticity in a turbulent channel flow. The spacing between the sensors for this flow was about 11.5 viscous lengths or about 6.5 times their estimation of the Kolmogorov length scale in the buffer layer. They found that the v and w velocity components were systematically underestimated by about 20% when the probe was tested by pitching and yawing it in a uniform irrotational flow. They applied experimentally determined calibration factors to correct the data measured in turbulent flows. The same probe was used by Kastrinakis *et al.* (1983) to examine the percentage contribution to the mean-square streamwise vorticity component conditioned on the quadrant analysis of the uv-product time series. They also obtained the autocorrelation of ω_x as a function of the position across the channel, and from this they calculated the integral length scales for ω_x and ω_x^2.

10.2 CROSS-STREAM VORTICITY COMPONENTS

Foss (1976, 1978, 1981) and co-workers (Foss *et al.* 1986, 1987; Haw *et al.* 1989; Foss and Haw 1990*a*, 1993) have been developing arrays of hot-wire probes to measure the transverse vorticity component, $\omega_z = \partial V/\partial x - \partial U/\partial y$, in turbulent flows. As discussed in Section 10.1, the transverse vorticity is often evaluated from eqn (10.4c), that is, from $\omega_z \simeq -\bar{U}^{-1}(\Delta V/\Delta t) - (\Delta U/\Delta y)$ and this formulation enables ω_z to be determined from a probe containing an X-array parallel to the (x, y) plane and two *SN* sensors parallel to the z axis and separated by a small distance, h, in the y-direction. Figure 10.5 shows the initial probe developed and used by Foss and co-workers (Foss 1976, 1978, 1981; Foss *et al.* 1986, 1987). In the following discussion the two wires in the X-array will, using Foss's notation, be denoted by 1 and 2, and the two parallel *SN* hot-wires will be denoted by 3 and 4. It can be observed that in this early version of the probe the two *SN* wires 3 and 4 have a significant transverse spatial separation, Δz, from wires 1 and 2 in the X-array.

When the vorticity probe is aligned with the mean-flow direction and the turbulence intensity is low, then eqn (10.4c) can provide a good measure of ω_z. This signal-interpretation method was used by Foss (1976, 1978), and it was also applied by Antonia and Rajagopalan (1990) and by Rajagopalan and Antonia (1993) for the measurement of ω_z in the self-preserving turbulent wake of a cylinder.

Foss *et al.* (1987) have studied many of the operating principles of the vorticity probe shown in Fig. 10.5, and they found that the most important

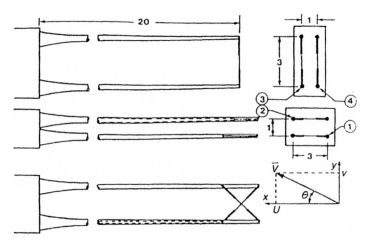

Fig. 10.5. The original four-sensor spanwise vorticity probe consisting of an X-array (sensors 1 and 2) and two *SN* hot-wires (sensors 3 and 4). The dimensions are in millimetres. (From Foss 1981.)

problem was that of the nonuniformity of the flow in the spanwise direction (the z-direction). They also discussed the low-pass filtering effect caused by the spatial separation between the sensors in the probe. Klewicki and Falco (1990) have examined the effect of the spacing, h, between the two parallel SN wires 3 and 4. This was done by using four sensors parallel to each other and the wall, and normal to the mean flow. The inner two wires were fixed, but the two other wires could be moved in the y-direction. Their results appear to be in good agreement with the predictions of Wyngaard (1969) that very little attenuation in the rms gradient is observed if the spacing over which the gradient is determined is no more than 3η, where η is the Kolmogorov length scale. They also observed a significant attenuation ($\sim 15\%$) when the spacing was increased to 6η. However the results presented were normalized by $(\Delta u/\Delta y)^2$ for the inner wires, and the normalization masks the noise problem, which becomes important when $\Delta y \to 0$ (Antonia, personal communications).

The spanwise vorticity probe (Fig. 10.5) has been used in a number of turbulent-flow studies. Foss et al. (1986) measured the velocity components U and V and the spanwise vorticity ω_z at the entraining boundary of a large plane shear layer. Falco (1983) constructed ensemble averages of the time series of the spanwise vorticity in a smoke-marked turbulent boundary layer. Klewicki and Falco (1990) have also used such a probe to measure the statistics of the spanwise vorticity component in a zero-pressure-gradient boundary layer over the range $1010 < \mathrm{Re}_\theta < 4850$ ($\mathrm{Re}_\theta = U_\infty \theta/\nu$, where U_∞ is the free-stream velocity and θ is the momentum thickness).

The inherent problems associated with the lateral displacement between the parallel SN wires and the X-array of the original probe stimulated the development of the compact probe geometry shown in Fig. 10.6. (In the work by Foss and co-workers, this probe is designated the 'Mitchell probe' in recognition of the contribution by Mitchell (1987) in threading and then attaching the final slant wire between the mounted parallel wires.) Results obtained with this probe have been presented by Foss and Haw (1990b) and by Haw et al. (1989). A compact probe of a similar type has also been developed and used by Antonia and Rajagopalan (1990) and by Rajagopalan and Antonia (1993).

In high turbulence intensity flows, and when the mean-flow direction deviates from the orientation of the probe stem, it may not be appropriate to use eqn (10.4c) for the evaluation ω_z. This is the problem addressed by Foss 1981, and in subsequent papers). In his method, the local transverse vorticity was determined by evaluating the microcirculation around a nominal $1\,\mathrm{mm}^2$ spatial domain, which then yields the average vorticity inside this domain. The basic principles of the velocity transformations in this technique can be explained by noting that the lateral vorticity, ω_z, can be expressed in either an (x, y) space-fixed coordinate system or an (s, n) streamwise coordinate system

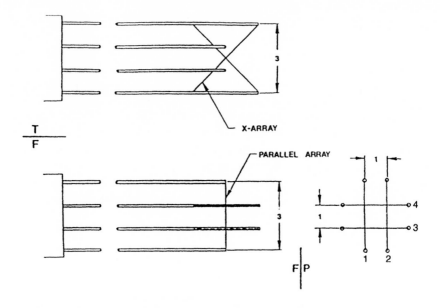

Fig. 10.6. A four-sensor (Mitchell) spanwise vorticity probe. The dimensions are in milli-metres. (From Haw *et al.* 1989. Reproduced with kind permission of Springer-Verlag. © Springer-Verlag.)

$$\omega_z = \frac{\partial V}{\partial x} - \frac{\partial U}{\partial y} = \frac{\partial U_n}{\partial s} - \frac{\partial U_s}{\partial n}, \qquad (10.14)$$

where (U_s, U_n) are the velocity components in the (s, n) coordinate system.

Consider first the relatively simple case of a low turbulence intensity flow with a two-dimensional mean velocity vector $(\bar{U}, \bar{V})$ at the centre of the vorticity probe in the space-fixed coordinate system (x, y). The magnitude of the mean velocity vector is denoted by

$$\bar{\bar{V}} = (\bar{U}^2 + \bar{V}^2)^{1/2} \qquad (10.15)$$

and the related (two-dimensional) mean flow angle, $\bar{\theta}$, specified by

$$\tan \bar{\theta} = \frac{\bar{V}}{\bar{U}}, \qquad (10.16)$$

determines the (s, n) coordinate system, as illustrated in Fig. 10.7. In the method proposed by Foss the X-array (wires 1 and 2) is used to evaluate the instantaneous (two-dimensional) flow angle, $\theta(t)$, and the corresponding instantaneous (two-dimensional) magnitude, $\tilde{V}(t)$, of the velocity vector is measured with the two *SN* wires 3 and 4. The theory assumes that $\theta(t)$ is constant throughout the probe volume and it ignores the response of the hot-wire sensors to the w velocity component. Introducing the angle

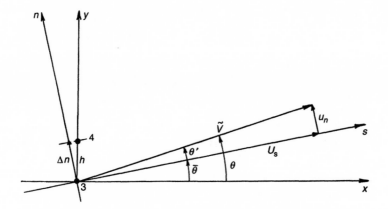

Fig. 10.7. The (s, n) coordinate system and the related notation.

$$\theta' = \theta(t) - \bar{\theta}, \tag{10.17}$$

it follows from Fig. 10.7 that

$$U_{n3}(t) = \tilde{V}_3(t) \sin \theta'(t), \tag{10.18a}$$

$$U_{n4}(t) = \tilde{V}_4(t) \sin \theta'(t), \tag{10.18b}$$

and

$$U_{s3}(t) = \tilde{V}_3(t) \cos \theta'(t), \tag{10.19a}$$

$$U_{s4}(t) = \tilde{V}_4(t) \cos \theta'(t), \tag{10.19b}$$

where $\tilde{V}_3(t)$ and $\tilde{V}_4(t)$ are the magnitudes of the instantaneous velocity vectors measured by the two (SN) wires 3 and 4. From eqns (10.18a–b) one may define

$$\frac{\Delta U_n}{\Delta t}(t_j) = \frac{\frac{1}{2}[\tilde{V}_3(t_{j+1}) + \tilde{V}_4(t_{j+1})] \sin \theta'(t_{j+1}) - \frac{1}{2}[\tilde{V}_3(t_j) + \tilde{V}_4(t_j)] \sin \theta'(t_j)}{t_{j+1} - t_j},$$

$$\tag{10.20}$$

where t_j and t_{j+1} are two consecutive times in a time-history record. Similarly, from eqns (10.19a–b)

$$\frac{\Delta U_s}{\Delta n}(t_j) = \frac{[\tilde{V}_4(t_j) - \tilde{V}_3(t_j)] \cos \theta'(t_j)}{h \cos \theta'}, \tag{10.21}$$

because the separation, Δn, between wires 3 and 4 in the (s, n)-coordinate system, (see Fig. 10.7) is

$$\Delta n = h \cos \theta'. \tag{10.22}$$

The (s, n)-format of eqn (10.14) can be expressed as

$$\omega_z \simeq -\bar{U}_s^{-1} \frac{\Delta U_n}{\Delta t} - \frac{\Delta U_s}{\Delta n} = -\bar{V}^{-1} \frac{\Delta U_n}{\Delta t} - \frac{\Delta U_s}{\Delta n}, \tag{10.23}$$

and by applying eqns (10.15), (10.20), and (10.21), ω_z can be evaluated in the (s, n) coordinate system.

When the turbulence intensity is high it may not be appropriate to use a fixed (s, n) coordinate system. To account for the large variations in the flow angle, θ, with time, Foss has introduced a variable (s, n) coordinate system, the orientation of which is only fixed for a few time steps. The above analysis can be modified to apply to this flow situation. Consider k data samples, covering a time period t_j to t_{j+k}. During this time period an average flow angle can be identified; that is,

$$\theta_k = \frac{1}{k+1} \sum_{i=0}^{k} \theta(t_{j+i}). \tag{10.24}$$

In the quasisteady (s, n) coordinate system defined by θ_k an average convection velocity, $U_{s,k}$, can be defined by

$$U_{s,k} = \frac{1}{k+1} \sum_{i=0}^{k} \tfrac{1}{2} [\tilde{V}_3(t_{j+i}) + \tilde{V}_4(t_{j+i})] \cos \theta'(t_{j+i}). \tag{10.25}$$

By applying a quasisteady Taylor hypothesis, $\omega_z(t_{j+i})$ can (during the time period t_j to t_{j+k}) be evaluated from

$$\omega_z(t_{j+i}) = -U_{s,k}^{-1} \frac{\Delta U_n}{\Delta t} - \frac{\Delta U_s}{\Delta n}, \tag{10.26}$$

with $\Delta U_n/\Delta t$ and $\Delta U_s/\Delta n$ being evaluated from equations similar to eqns (10.20) and (10.21). In the complete method developed by Foss (1981) and refined by Foss et al. (1986, 1987), by Haw et al. (1989), and by Foss and Haw (1990a, 1993), the above procedure is linked to the evaluation of the microcirculation around a small spatial domain, which yields the average vorticity within this domain.

Eckelmann et al. (1977) have reported the development and use of a five hot-film sensors probe to study the physical mechanism of turbulent shear flow. The probe, shown in Fig. 10.8, contains a V-array sensitive to (U, w), an X-array sensitive to (U, v) and a single SN sensor sensitive to U. The sensor arrangement enables the simultaneous measurement of (U, v, w) and the two cross-stream vorticity components. For ω_z, (see eqn (10.3c)), the single sensor (1) and the two sensors (2 and 3) of the V-array are used to evaluate $\partial U/\partial y$; and $\partial v/\partial x$ is obtained by applying Taylor's hypothesis to the time derivative of v measured with the X-array (sensors 4 and 5). In eqn (10.3b) for ω_y, the spanwise gradient, $\partial U/\partial z$, is evaluated from the streamwise velocities measured with the V- and X-arrays; $\partial w/\partial x$ is similarly

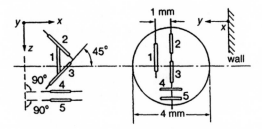

Fig. 10.8. A five-sensors hot-film probe used for the measurement of spanwise and normal vorticity components in a turbulent oil channel flow. (From Eckelmann *et al.* 1977.)

estimated from Taylor's hypothesis applied to the time derivative of w measured by sensors 2 and 3 of the V-array.

Measurements with the probe were carried out in an oil channel flow, which has a Kolmogorov length scale of about 1 mm in the wall region, giving good spatial resolution for velocity-component measurements (see, for example, Kastrinakis and Eckelmann 1983). According to Wyngaard (1969), the inequality

$$1 < \frac{h}{\eta} < {\sim}3, \tag{10.27}$$

where h is the spacing between the sensors, should be satisfied during gradient measurements. However, Eckelmann *et al.* (1977) reported considerable difficulties in obtaining accurate measurements of $\partial U / \partial y$ in the turbulent oil channel by taking differences in U over the 1 mm separation between sensor 1 and sensors 2,3, a distance which is approximately equal to the Kolmogorov length scale. The main problem was the evaluation of the gradient from a small difference of two large values of U obtained with different sensors. This difference was observed to be very sensitive to the calibration accuracy of sensors 1, 2, and 3. Noise also becomes a problem when $h \to 0$. Böttcher and Eckelmann (1985) have shown from calibration experiments in a known laminar velocity-gradient flow field that two closely spaced sensors will not correctly evaluate the velocity gradient; the most likely explanation is that this is due to sensor interference effects. They developed a correction scheme to compensate for this effect.

Antonia *et al.* (1988*a*) have measured, but not simultaneously, both the normal and the spanwise vorticity components, ω_y and ω_z in the self-preserving region of a turbulent wake 420 diameters downstream of a cylinder. The Reynolds number based on the free-stream velocity, U_∞, and the cylinder diameter, d, was 1170. The measurements were carried out using two X-hot-wire arrays separated in the appropriate cross-stream direction, and ω_y and ω_z were evaluated using eqns (10.4b–c); this assumes the validity of Taylor's hypothesis. It was found that the statistics of ω_y and

ω_z were essentially independent of which X-array was used to evaluate v or w. The spacing between the centre of the two X-arrays was about 1.6 mm, and the wires in each of the X-arrays were 1 mm apart. The Kolmogorov length scale was estimated to be about 0.45 mm at the centre line of the turbulent wake. Antonia *et al.* measured the first three statistical moments of the cross-stream vorticity fluctuations, and they estimated the statistics for the streamwise vorticity component, ω_x using measurements of the two corresponding velocity gradients. However, these gradients were not obtained simultaneously.

Kim and Fiedler (1989) have designed and used a six-sensor probe to measure ω_x and ω_z in a two-stream mixing layer. In the Technische Universität Berlin (TUB) probe, shown in Fig. 10.9, the separation of the wires in the X-array (I) placed in the (x, y)-plane is 2 mm. The vorticity probe in Fig. 10.8 has two SN sensors (3 and 4), whereas in the TUB probe these are replaced by two X-arrays (II and III) which are both parallel to the (x, z)-plane. The distance between the wires in arrays II and III is 1 mm, and the probe therefore provides a measure of the spatially averaged vorticity over a microcirculation domain of 2 mm $\times$ 2 mm (Foss 1981)).

The response equation for each wire was expressed, by a first-order Taylor-series expansion, in terms of the velocity (U_0, v_0, w_0) at the probe centre and the velocity gradients $\partial U/\partial y$, $\partial v/\partial z$, and $\partial w/\partial y$. In this analysis it was assumed that $\partial U/\partial z$ is small. Transverse-velocity corrections (that is, w for the (U, v) calculation, and v for the (U, w) calculation) are included in the TUB calculations. The spanwise vorticity, ω_z, was evaluated from eqn (10.4c) $(\omega_z \simeq -\bar{U}^{-1}\Delta v/\Delta t - \Delta U/\Delta y)$, with $\Delta v/\Delta t$ calculated from the v velocity component measured with X-array I and $\Delta U/\Delta y$ determined from X-arrays II and III. The streamwise vorticity ω_x $(= \partial w/\partial y - \partial v/\partial z)$ was also estimated with this probe. The value of $\partial w/\partial y$ was obtained from the two X-arrays II and III, and an estimate of $\partial v/\partial z$ was obtained from the two wires in X-array I using equations similar to eqns (10.29a–b) with $s = 0$.

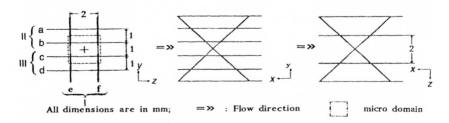

Fig. 10.9. A six-wire vorticity probe, containing three X-arrays, for the measurement of the spanwise and streamwise vorticity components. (From Kim and Fiedler 1989. Reproduced with kind permission of Springer-Verlag. © Springer-Verlag.)

10.2.1 Velocity-gradient measurements with closely spaced X-arrays

Consider two identical X-arrays parallel to the (x, y)-plane, as shown in Fig. 10.10, with a separation, s, between the centres of the two arrays and a distance b between the two wires in each array. Consequently the z-positions for the two hot-wires of array I are

$$\text{Wire 1:} \quad z = \frac{s}{2} + \frac{b}{2}, \qquad \text{Wire 2:} \quad z = \frac{s}{2} - \frac{b}{2},$$

and for the two hot-wires of array II are

$$\text{Wire 3:} \quad z = -\frac{s}{2} + \frac{b}{2}, \qquad \text{Wire 4:} \quad z = -\frac{s}{2} - \frac{b}{2}.$$

It will be assumed that: a cosine cooling law applies, the effect of the w component can be ignored for all four wires, and the probe geometry is compatible with a good spatial resolution (which requires the separation distances b and s to be comparable to the Kolmogorov length scale). Under these conditions, the effective velocities for the four wires can be expressed as

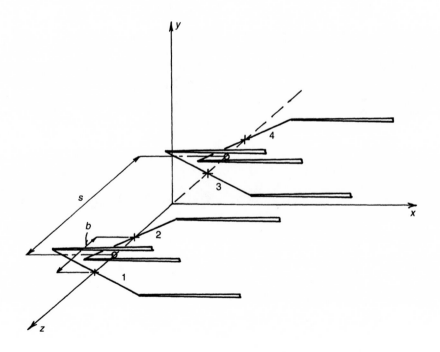

Fig. 10.10. The geometry and notation for two closely spaced X-arrays.

$$V_{e1} = U\left(z = \frac{s}{2} + \frac{b}{2}\right)\cos\alpha + v\left(z = \frac{s}{2} + \frac{b}{2}\right)\sin\alpha, \qquad (10.28a)$$

$$V_{e2} = U\left(z = \frac{s}{2} - \frac{b}{2}\right)\cos\alpha - v\left(z = \frac{s}{2} - \frac{b}{2}\right)\sin\alpha, \qquad (10.28b)$$

$$V_{e3} = U\left(z = -\frac{s}{2} + \frac{b}{2}\right)\cos\alpha + v\left(z = -\frac{s}{2} + \frac{b}{2}\right)\sin\alpha, \qquad (10.28c)$$

$$V_{e4} = U\left(z = -\frac{s}{2} - \frac{b}{2}\right)\cos\alpha - v\left(z = -\frac{s}{2} - \frac{b}{2}\right)\sin\alpha, \qquad (10.28d)$$

Denoting the velocity components at the centre of the coordinate system as by (U_0, v_0) and applying a first-order Taylor-series expansion gives

$$V_{e1} = \left[U_0 + \left(\frac{s}{2} + \frac{b}{2}\right)\frac{\partial U}{\partial z}\right]\cos\alpha + \left[v_0 + \left(\frac{s}{2} + \frac{b}{2}\right)\frac{\partial v}{\partial z}\right]\sin\alpha, \qquad (10.29a)$$

$$V_{e2} = \left[U_0 + \left(\frac{s}{2} - \frac{b}{2}\right)\frac{\partial U}{\partial z}\right]\cos\alpha - \left[v_0 + \left(\frac{s}{2} - \frac{b}{2}\right)\frac{\partial v}{\partial z}\right]\sin\alpha, \qquad (10.29b)$$

$$V_{e3} = \left[U_0 + \left(-\frac{s}{2} + \frac{b}{2}\right)\frac{\partial U}{\partial z}\right]\cos\alpha + \left[v_0 + \left(-\frac{s}{2} + \frac{b}{2}\right)\frac{\partial v}{\partial z}\right]\sin\alpha, \qquad (10.29c)$$

$$V_{e4} = \left[U_0 + \left(-\frac{s}{2} - \frac{b}{2}\right)\frac{\partial U}{\partial z}\right]\cos\alpha - \left[v_0 + \left(-\frac{s}{2} - \frac{b}{2}\right)\frac{\partial v}{\partial z}\right]\sin\alpha. \qquad (10.29d)$$

Equations (10.29a–d) represent four linear equations for the four unknowns: U_0, v_0, $\partial U/\partial z$, and $\partial v/\partial z$. Applying sum-and-difference methods to the signals from the two X-arrays gives for array I for the sum

$$V_{e1} + V_{e2} = \left(2U_0 + s\frac{\partial U}{\partial z}\right)\cos\alpha + b\frac{\partial v}{\partial z}\sin\alpha, \qquad (10.30a)$$

and for the difference

$$V_{e1} - V_{e2} = b\frac{\partial U}{\partial z}\cos\alpha + \left(2v_0 + s\frac{\partial v}{\partial z}\right)\sin\alpha, \qquad (10.30b)$$

and for array II for the sum

$$V_{e3} + V_{e4} = \left(2U_0 - s\frac{\partial U}{\partial z}\right)\cos\alpha + b\frac{\partial v}{\partial z}\sin\alpha, \qquad (10.31a)$$

and for the difference

$$V_{e3} - V_{e4} = b\frac{\partial U}{\partial z}\cos\alpha + \left(2v_0 - s\frac{\partial v}{\partial z}\right)\sin\alpha. \qquad (10.31b)$$

The velocity gradient $\partial U/\partial z$ can be evaluated by taking the difference bet-

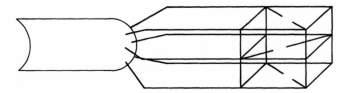

Fig. 10.11. The three wire $A_x I$-probe for derivative measurements. (From George and Hussein 1991. Reproduced with kind permission of Cambridge University Press from the *Journal of Fluid Mechanics*.)

ween eqns (10.30a) and (10.31a) and similarly $\partial v/\partial z$ can be obtained from eqns (10.30b) and (10.31b)

$$(V_{e1} + V_{e2}) - (V_{e3} + V_{e4}) = 2s \frac{\partial U}{\partial z} \cos \alpha, \qquad (10.32a)$$

$$(V_{e1} - V_{e2}) - (V_{e3} - V_{e4}) = 2s \frac{\partial v}{\partial z} \sin \alpha, \qquad (10.32b)$$

It should be observed that eqns (10.32a–b) only contain the spacing, s, between the two X-arrays and not the spacing, b, between the wires in the two arrays; this is the assumption that is normally made.

As described in George and Hussein (1991), a special $A_x I$-probe, Fig. 10.11, consisting of only three wires was developed for gradient measurements. It was originally hoped that the upper and lower pairs of wires could be treated as independent X-arrays, but this is an incorrect interpretation as pointed out by Hallbäck *et al.* (1991). The probe suggested by George and Hussain (1991) will give three simultaneous equations in the form of eqns (10.29a–c), which is insufficient information to evaluate simultaneously U_0, v_0, $\partial U/\partial z$, and $\partial v/\partial z$.

10.3 THE THREE COMPONENTS OF THE VORTICITY MEASURED SIMULTANEOUSLY

Wassmann and Wallace (1979) designed and constructed a nine-sensor hot-wire probe, which was capable of simultaneously estimating all the three vorticity components as well as the components of the velocity and strain-rate fields. Further developments and testing of this type of probe have been reported by Balint *et al.* (1987, 1991), Balint and Wallace (1989), and Vukoslavčević *et al.* (1989, 1991). A sketch of the latest version of the nine-sensor probe (Vukoslavčević *et al.* 1991), including the prong/sensor arrangement and the related dimensions, is shown in Fig. 10.12.

Probe specification The probe consists of three arrays of triple hot-wire

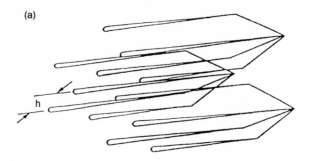

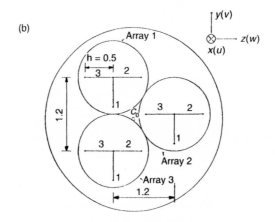

Fig. 10.12. A nine-sensor probe for simultaneous measurements of the three vorticity components. (a) a sketch of the prong and wire arrangement, (b) an end view giving the sensor arrangement and the dimensions (in mm). (From Vukoslavčević *et al.* 1991. Reproduced with kind permission of Cambridge University Press from the *Journal of Fluid Mechanics*.)

sensors mounted at 45° to the mean flow. The total dimensions of the probe sensing area are 1.7 mm vertically and 2.2 mm horizontally with an average distance between the array centres of about 1.2 mm and a distance h between the supporting prongs of any array of 0.5 mm. This is a considerably more compact arrangement than for the similar probe used by Vukoslavčević *et al.* (1989) in which the horizontal separation between the centres of arrays 2 and 3 was 1.3 mm and the vertical separation between the centres of arrays 1 and 3 was 1.5 mm. The diameter of the nine tungsten sensors is 2.5 μm and their length is about 0.7 mm, giving a length-to-diameter ratio of about 280. The prongs are nickel-plated tungsten wires with a diameter of about 255 μm, and they are tapered to about 75 μm at their tips.

Fabrication of the probe is difficult because of its small size, complex design, and fragility. The main problem is to fit twelve supporting prongs in the smallest feasible space while satisfying accuracy and resolution criteria. Vukoslavčević *et al.* (1991) have described the manufacturing process of this probe, which involves the use of small ceramic tubes to avoid electrical short circuits between the prongs or between the prongs and the probe body. The prong tips are manipulated under a microscope into their correct geometrical position, and are then epoxyed together at one end of the ceramic tubes. After nickel plating the prong tips, the 2.5 μm diameter unplated tungsten wire sensors are spot welded to the tips.

To obtain a compact probe design, each of the three arrays utilizes a common central prong, and two difficulties arose when this was first attempted as reported by Vukoslavčević *et al.* (1989). It was discovered that a resistance of the common ground prong greater than 0.3 Ω caused a feedback instability in the circuit, when they were operated simultaneously. Additionally, electrical crosstalk between the circuits resulted from this common resistance. The stability problem was overcome by nickel plating the tungsten wire prongs to reduce their resistance to about 0.1 Ω, which also substantially reduced the crosstalk. However, to achieve the required reduction in resistance, the nickel plating had to be relatively thick, which caused a substantial aerodynamical blockage for the latest, more compact, probe. To reduce the plating thickness, each common tungsten probe was first plated with copper. This significantly decreased the resistance to about 0.08 Ω with a much smaller increase in the prong diameter than for nickel plating. A simple electrical test demonstrated the absence of any electrical crosstalk when the resistance of the common prong was below 0.13 Ω.

Calibration tests showed no thermal contamination from array to array. However, there is thermal contamination within an individual three-sensor array, and to minimize this problem each hot-wire sensor was operated at a relatively low overheat ratio of 1.2. Due to the complex geometrical configuration of the nine-sensor probe it is difficult to separate thermal contamination and aerodynamical blockage effects, and these combined effects were accounted for by the evaluated calibration coefficients.

Spatial resolution The response equations developed for the nine sensors are based on a first-order Taylor-series expansion of the velocity at the probe centroid, and the accuracy with which second- and higher-order terms can be ignored increases with decreases in the array spacing. Also, in order to estimate accurately the cooling velocity for each sensor, its length should be comparable to the size of the smallest turbulent structure in the flow, that is, comparable to the Kolmogorov length scale, η.

In the study by Balint *et al.* (1991) measurements were carried out with the vorticity probe shown in Fig. 10.12 in a zero-pressure-gradient turbulent boundary layer at a momentum-thickness Reynolds number, $\text{Re}_\theta \, (= U_\infty \theta / v)$ of 2685. For the measurement position closest to the wall, $y^+ = 11.2$, the Kolmogorov length scale η was about 0.19 mm. Consequently, at this position the sensor lengths were 3.6 times the measured Kolmogorov length scale, and the distance over which velocity gradients were evaluated was 6.3η.

The incomplete spatial resolution of each sensor was evaluated by comparative spectral measurements with an *SN* hot-wire probe with a wire length of 0.3 mm, that is, of about 1.6η. The one-dimensional spectra, $E_u(f)$, measured by the two probes were found to be in close agreement, except for attenuation at the highest frequencies for the vorticity probe.

The probe must also be able to resolve adequately the velocity-gradient field in the flow. This resolution is primarily determined by the average spacing between sensor centres, which was equal to 6.3η at $y^+ = 11.2$. Wyngaard (1969) has theoretically investigated the resolution of a Kovasznay-type vorticity probe assuming isotropy and the form of the three-dimensional velocity spectrum given by Pao (1965). It was concluded that a reasonable resolution was obtained when $\eta/\ell \simeq 0.3$ and $h/\ell \simeq 0.7$, where ℓ is the active sensor length and h the spacing between the sensor elements. For these conditions, Wyngaard estimated that the Kovasznay-type probe measures about 86% of the true mean-square value of the vorticity. For the nine-sensor probe at $y^+ = 11.2$, $\eta/\ell = 0.27$ and the equivalent of h/ℓ for the nine-sensor probe is the ratio of the average spacing between sensor centres to the length of the sensors, which is about 1.7. Wyngaard (1969) has shown that additional attenuation of the vorticity spectrum occurs for $h/\ell > 0.7$, and the results from the nine-sensor probe will therefore contain some attenuation of the vorticity spectra at the highest frequencies.

Probe calibration Each sensor j of array i ($i = 1, 2, 3$; $j = 1, 2, 3$) was calibrated using a power law equation

$$E_{ij}^2 = A_{ij} + B_{ij}(V_{eij})^{n_{ij}}, \tag{10.33}$$

which relates the anemometer voltage E_{ij} for each sensor to the corresponding effective velocity, V_{eij}. For sensors in the (x, y)-plane of Fig. 10.11, the effective velocity can be expressed as

$$V_{eij}^2 = (U\cos\alpha_{eij} + V\sin\alpha_{eij})^2 + h_{ij}^2 W^2, \tag{10.34}$$

with similar expressions for the sensors in the (x, z)-plane. In eqn (10.34), the yaw dependence is expressed in terms of the effective angle, α_e, because this method does not require the measurement of the mean yaw angle, which is difficult to determine for a compact miniature probe.

During calibration, the probe was placed in the nominally irrotational free-stream core of the wind-tunnel and velocity, yaw, and pitch calibration procedures (as described in Chapters 4 and 5) were applied to each sensor.

The velocity calibration with the probe aligned with the mean-flow direction was used to determine A_{ij} and n_{ij} for each sensor by a least-squares curve-fitting procedure. By subsequently yawing and pitching the probe at constant velocities the corresponding values of B_{ij}, α_{eij}, and h_{ij} were calculated.

The calibration experiments demonstrated that the compactness of the probe resulted in a significant blockage effect, and it was also observed that the same response was not obtained for positive or negative angles in yaw or pitch. To account for these effects, the least-squares curve-fits of the calibration data were divided into three calibration ranges: high negative, pitch/yaw angles; near-zero pitch/yaw angles; and high positive, pitch/yaw angles. In each of these three ranges, separate effective yaw angles were determined, typically being $36.0°$, $39.9°$, and $46.1°$. The corresponding value for the pitch factor, h, was about 1.5, which is substantially higher than the value $h \simeq 1.2$ obtained for the slightly larger nine-sensor probe used by Vukoslavčević et al. (1989) and much larger than the typical value of about 1.05 for a plated SN hot-wire probe. The disadvantage of the multi-range calibration curve-fits is that the data-analysis program must choose between three sets of calibration constants each time it solves the response equation for the nine sensors.

Signal analysis The solution procedure is based on the effective velocity in eqn (10.34), with the velocity components at the centre of each wire in each array specified by applying a first-order Taylor-series expansion to the velocity components (U_0, V_0, W_0) at the centroid of the probe. Denoting the array number by the subscript $i (= 1, 2, 3)$, the effective velocity for the three sensors $j = 1, 2, 3$ in each array can be expressed as

$$V_{eij}^2 = \left[K_{ij1} \left(U_0 + C_{ij1} \frac{\partial U}{\partial z} + C_{ij2} \frac{\partial U}{\partial y} \right) + K_{ij2} \left(V_0 + C_{ij3} \frac{\partial v}{\partial z} + C_{ij4} \frac{\partial v}{\partial y} \right) \right]^2$$
$$+ K_{ij3} \left(W_0 + C_{ij5} \frac{\partial w}{\partial z} + C_{ij6} \frac{\partial w}{\partial y} \right)^2. \tag{10.35}$$

In eqn (10.35), K_{ij1} are the cosines of the effective angles α_{eij}, K_{ij2} are the sines of the effective angles, and K_{ij3} accounts for the aerodynamic blockage. The C_{ijl} ($l = 1, 2, 3, 4, 5, 6$) are constants which are positive or negative fractions of the projected prong spacing, h, for a given array geometry.

The vorticity probe provides nine simultaneous equations of the form given by eqns (10.35), which can be solved iteratively to provide simultaneous values for (U_0, V_0, W_0) and the six transverse gradients ($\partial U/\partial y$, $\partial U/\partial z$, $\partial v/\partial y$, $\partial v/\partial z$, $\partial w/\partial y$, and $\partial w/\partial z$). Of the three streamwise gradients, $\partial U/\partial x$ was evaluated from the continuity equation ($\partial U/\partial x + \partial v/\partial y + \partial w/\partial z = 0$), while the other two gradients, $\partial v/\partial x$ and $\partial w/\partial x$, were obtained using the Taylor hypothesis ($\partial/\partial x = -\bar{U}^{-1}(\partial/\partial t)$). The adequacy of this approximation was assessed by comparing the value of $\partial U/\partial x$ obtained from the continuity equation with the value of $-\bar{U}^{-1}(\partial U/\partial t)$.

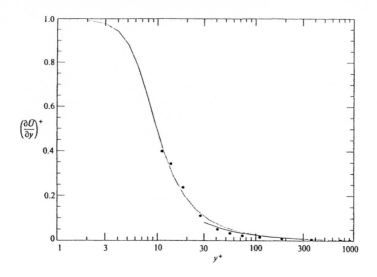

Fig. 10.13. A comparative evaluation of the mean velocity gradient $\partial \bar{U}/\partial y$ across the boundary layer: ($\cdots$) Spalding (1961), (——) Coles (1962), ($\bullet$) values directly measured by the nine-sensor probe. (From Vukoslavčević *et al.* 1991. Reproduced with kind permission of Cambridge University Press from the *Journal of Fluid Mechanics*.)

Eckelmann *et al.* (1977) and Böttcher and Eckelmann (1985) have shown that it is very difficult to obtain accurate instantaneous and mean-velocity gradients from the velocity differences measured with two or more sensors when they are measured over distances of only 1–2 Kolmogorov length scales. In a study by Vukoslavčević *et al.* (1991) the sensor separation at $y^+ = 11.2$ was 6.3η, and Fig. 10.13 contains a comparison between $\partial \bar{U}/\partial y$ measured with the nine-sensor probe and derivatives of fits of published mean-velocity measurements given by Spalding (1961) and the derivative of the logarithmic law for $y^+ > 30$ (shown as a solid line) with the constants given by Coles (1962). Fig. 10.13 gives a good indication of the ability of the probe to measure the instantaneous velocity gradients.

Velocity-gradient measurements Balint *et al.* (1991) have reported detailed measurements with the nine-sensor probe of statistical properties of both the velocity and vorticity fields in a zero-pressure-gradient turbulent boundary layer. The rms values of the individual velocity gradients, normalized with inner variables, are shown in Fig. 10.14 for the near-wall region. In and above the buffer layer, the rms values of the streamwise gradients measured by Balint *et al.* compare fairly well with those of Kim *et al.* (1987), as discussed by Piomelli *et al.* (1989). This fact supports the use of Taylor's hypothesis in evaluating the streamwise velocity gradients. The data in Fig. 10.14 also indicate the relatively smaller importance of these stream-

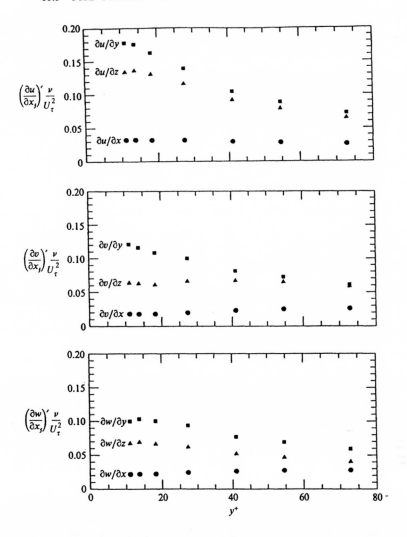

Fig. 10.14. Measured rms fluctuating velocity gradient components normalized with the inner scaling U_τ and ν. (From Balint *et al.* 1991. Reproduced with kind permission of Cambridge University Press from the *Journal of Fluid Mechanics*.)

wise gradients compared to the cross-stream gradients in the statistics of the ω_y and ω_z components.

A further vorticity probe development is the twelve-wire probe (Tsinober *et al.* 1992). The probe shown in Fig. 10.15 contains three arrays with four wires placed in a symmetrical orientation in each array, and it has the following improved features. Firstly, one of the possible error sources of

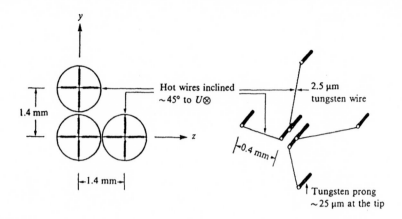

Fig. 10.15. A vorticity probe with twelve hot-wire sensors. (From Tsinober *et al.* 1992. Reproduced with kind permission of Cambridge University Press from the *Journal of Fluid Mechanics.*)

the nine-hot-wire probe (Fig. 10.12) is the presence of a common prong in each of the three wire arrays. The twelve hot-wire probe was developed without the use of common prongs. The problem was solved by producing a compound split prong consisting of several tungsten wires coated with Teflon and glued together. An electrochemical-erosion method was used to give a thickness of the tips of the prongs of less than $20 \,\mu m$. Secondly, problems were experienced in picking up sufficient velocity-component information from an array with three wires. This problem was resolved by adding a fourth wire making the array symmetrical, as shown in Fig. 10.15. Finally, a new calibration method was introduced by extending to three dimensions the least-squares polynomial approximation of Oster and Wygnanski (1982). For each combination of three wires in each array (that is, a total of $4 \times 3 = 12$) a calibration function was produced which gave the relationship between the three velocity components and the three voltages obtained from each wire in the array. These calibration functions for every velocity component in each array were constructed as three-dimensional polynomials of fourth order using Chebyshev orthorgonal polynomials.

A twelve hot-wire probe with a similar four-wire-array arrangement has also been developed by Marasli *et al.* (1993), who has developed an improved calibration procedure and has tested the probe in the wake of a circular cylinder.

11

CONDITIONAL SAMPLING TECHNIQUES

Distinct flow events are found in many types of turbulent flows, and they can often be identified by a *HWA* sensor placed in the flow field. In some fluid machinery, these events will occur periodically, and a phase-locked ensemble-averaging technique can be used to extract the periodic signal component from the fluctuating signal; this is discussed in Section 11.1. Coherent flow structures have also been observed in many turbulent flows, but their developments are often random in both space and time. To identify these large-scale structures, special signal-analysis procedures have been developed and applied to hot-wire probes; this is described in Section 11.2.

11.1 PHASE-LOCKED-AVERAGING TECHNIQUES

The phase-locked averaging (*PLA*) technique has been applied to a number of different flow situations, which all contain repetitive flow events; and the basic concepts of this technique are introduced with reference to rotating-wake phenomena.

11.1.1 Rotating-wake phenomena

The *PLA* method can be applied in principle to all types of rotating wakes (compressors, turbines, fans, pumps, windmills, propellors, and helicopter rotors). *HWA* measurements have been carried out behind single rotors and interstage between rotor and stator blade rows; and rotating traversing-gear mechanisms have been developed for measurements within the blade passage of a rotor (see, for example, Gorton and Lakshminarayana 1976. Lakshminarayana 1980, 1981). Instrumented rotor blades have also been developed. The fluctuating surface pressure distribution has been determined using miniature high-response pressure transducers (see, for example, Rothrock *et al.* 1982; Hardin *et al.* 1987; Hodson 1985; Sugeng and Fiedler 1986; and Ainsworth *et al.* 1991), and the mean and fluctuating surface shear stress has been measured by glue-on hot-film elements, as described in Section 8.8.2 (Hodson 1984; Sugeng and Fiedler 1986; Capece and Fleeter 1987; Hodson and Addision 1989; Addision and Hodson 1990*a,b*). Many turbomachinery investigations have also been carried out using conventional pressure probes and *LDA* techniques. Reviews of optical and nonoptical instrumentation for flow studies in turbomachinery have been given by Weyer (1980) and Lakshminarayana (1981), respectively.

The flow in rotating fluid machinery is inherently unsteady and it often displays semiregular flow phenomena associated with the blade-passing frequency of the rotor. Reported *PLA* studies (by *HWA*) aimed at determining the periodic flow component include those on: axial flow compressors (Evans 1975; Hirsch and Kool 1977; Schmidt and Okiishi 1977; Okiishi and Schmidt 1978; Lakshminarayana and Davino 1980; Hah and Lakshminarayana 1981; Butler and Wagner 1983; Shreeve and Neuhoff 1984; Pougare *et al.* 1985; Wisler *et al.* 1987; Capece and Fleeter 1987, 1989*a,b*; Schulz *et al.* 1990; Venkateswaran 1991; Li and Cumpsty 1991), axial flow fans (Whitfield *et al.* 1972; Raj and Lakshminarayana 1976; Larguier 1981), axial-flow inducers (Lakshminarayana and Poncet 1974; Lakshminarayana 1982*b*), axial-flow turbines (Joslyn *et al.* 1983; Hodson 1984, 1985; Binder *et al.* 1989 and Ainsworth *et al.* 1991), centrifugal fans (Raj and Swim 1981; Cau *et al.* 1987), propellors (Favier *et al.* 1977; Bennett 1978), and wind-turbines (Clausen and Wood 1988).

The *PLA* method has been used to identify many of the following flow phenomena which occur in turbomachinery: (i) the general flow field away from the hub and tip regions (this flow case will be used below to discuss the *PLA* method), (b) details of the flow in the tip-clearance region (Lakshminarayana *et al.* 1982; Hunter and Cumpsty 1982, Pandya and Lakshminarayana 1983; Lakshminarayana and Pandya 1984), (c) the use of compressor-casing treatment to improve the stall margins of wall stall (Greitzer *et al.* 1979; Smith and Cumpsty 1984; Johnson and Greitzer 1987), and (d) the inception of stall cells (Jackson 1987; McDougall *et al.* 1990; Inoue *et al.* 1991; Garnier *et al.* 1991), and the study of fully developed, rotating, stall cells (Day and Cumpsty 1978; Breugelmans *et al.* 1983; Das and Jiang 1984; Mathioudakis and Breugelmans 1985).

Consider a turbomachine consisting of a rotor and a downstream row of fixed stator blades. Assume initially that the rotating blade row can be approximated, at the radial position, R, of the hot-wire probe, by a linear row of equally spaced identical blades moving with a constant velocity $V_b = \omega R$ past the fixed stator blades. This flow situation is illustrated in Fig. 11.1. Three hot-wire probe positions are identified in this figure: A is downstream of the stator row, B is between the moving blade row and the stator row, and C is inside a blade passage and fixed to the moving blade row. Probe A is located at a fixed spatial position within a flow passage of fixed geometry, and the standard *HWA* signal-analysis procedure described in Chapters 3 to 6 can therefore be applied. A periodic component, caused by the rotor blade wakes, can often be observed in the signal from probe A, and the *PLA* technique can be used to extract this part of the signal. The signal analysis for the rotating probe C is also relatively simple (see Lakshminarayana 1981 for a review of rotating *HWA* probe methods in turbomachinery investigations). In a coordinate system moving

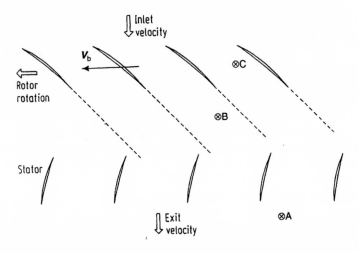

Fig. 11.1. The flow through an axial compressor represented by two linear rows of rotor and stator blades. The hot-wire-probe positions are denoted by A, B, and C.

with the velocity of the blade row, $V_b = \omega R$, the hot-wire probe will appear stationary, and probe C will therefore measure the velocity of the flow, V_r, relative to the moving blade row. The corresponding absolute flow velocity V, can be obtained by vector addition

$$V = V_r + V_b. \tag{11.1}$$

In this analysis, V_r may have three nonzero velocity components, while $V_b = (0, \omega R, 0)$. Equation (11.1) is also valid for the flow in front of and behind the rotor.

Consider in the same coordinate system, the flow-field downstream of the moving blade row. Because all the blades are assumed to be identical, and provided the flow is statistically stationary, the mean-velocity profile in the u-direction of the moving coordinate system will have a spatial periodicity $\Delta s = \Delta\theta R$, ($\Delta\theta = 2\pi/B$, where B is the number of blades). Superimposed on the mean-velocity profile will be random fluctuations. This rotating velocity field will be observed by the fixed probe at the point B as a time-varying velocity vector $U_i(t)$, ($i = 1\text{--}3$), containing a periodic component $\bar{U}_{i,E}(t)$ with a period $T = 2\pi/(\omega B)$ and a random component $u_{i,E}(t)$

$$U_i(t) = \bar{U}_{i,E}(t) + u_{i,E}(t), \tag{11.2}$$

and the following relationship is valid for the periodic component

$$\bar{U}_{i,E}(t) = \bar{U}_{i,E}(t + kT). \tag{11.3}$$

The axial flow velocity downstream of the rotor of an axial flow turbo-machine may look as shown in Fig. 11.2, with the velocity 'defect' corresponding to the wake regions. In this figure, the periodicity corresponding to each blade passage is $T = 2\pi/(\omega B)$, and the once-per-revolution periodicity $T^* = 2\pi/\omega$.

To avoid signal-analysis errors caused by variations between individual blade passages, the periodic component, $\bar{U}_{i,E}(t)$, is normally evaluated by using a once-per-revolution marker pulse, which ensures that the ensemble averaging is performed on data from the same blade passage. More generally, a (rotary) digital encoder can provide both the trigger event and the exact number of samples for either a once-per-revolution or once-per-blade signal analysis, (Lakshminarayana, personal communications). The digital data acquisition can either be carried out continuously or it can be initiated once-per-revolution (or once-per-blade). If the continuous mode is selected then the sampling rate, SR (samples s^{-1}), must be matched to the once-per-revolution periodicity of the rotor, T^* ($=2\pi/\omega$). Also, if M points are required for each blade passage then the following condition must be satisfied

$$SR = \frac{M}{T} = M\frac{\omega B}{2\pi}. \tag{11.4}$$

In the early investigations (for example Lakshminarayana and Poncet 1974), it was necessary to compensate for the available rather slow A/D sampling rate by first recording the hot-wire signals on a magnetic tape recorder, and then replaying the data at a lower speed. This procedure may still be necessary for high-speed flow investigations (Hah and Lakshminarayana, 1981),

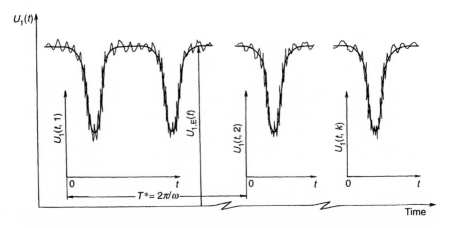

Fig. 11.2. The axial flow velocity downstream of the rotor of an axial-flow turbomachine. A selection of phase-locked sample records as used for ensemble averaging evaluations.

but this processing step can now be omitted in many turbomachinery investigations due to the development of modern A/D equipment.

The sampling-rate matching and the ensemble averaging procedure can be simplified by using 'burst'-mode data acquisition, which is initiated by a once-per-revolution marker pulse or a digital-encoder signal. This method will produce a number of digital sample records, $U_i(t, k)$, all phase-locked to the reference pulse, as illustrated in Fig. 11.2. The time of each sample record can be selected to correspond to one or several blade passages. Using a large number, K, of sample records (typically 100–300), the periodic component can be obtained by ensemble averaging

$$\bar{U}_{i,\mathrm{E}}(t) = \frac{1}{K} \sum_{k=1}^{K} U_i(t, k), \qquad (11.5)$$

because the contribution from the random component, $u_{i,\mathrm{E}}(t)$, goes to zero as $K \to \infty$. Using digital data acquisition with M points per blade passage, the temporal phase-locked mean-velocity variation,

$$\bar{U}_{i,\mathrm{E}}(t) = \bar{U}_{i,\mathrm{E}}(n\Delta t), \quad n = 1, M, \qquad (11.6)$$

can be transformed into the corresponding blade-to-blade spatial variation

$$\bar{U}_{i,\mathrm{E}}(\theta) = \bar{U}_{i,\mathrm{E}}(n\Delta\theta), \quad n = 1, M \qquad (11.7)$$

because

$$\Delta\theta = \omega\Delta t = \frac{2\pi}{BM}. \qquad (11.8)$$

Similarly, the phase-locked Reynolds normal and shear stresses can be obtained $(i, j = 1, 2, 3)$ from

$$\overline{u_i u_j}(\theta) = \frac{1}{K} \sum_{k=1}^{K} [U_i(\theta, k) - \bar{U}_{i,\mathrm{E}}(\theta)][U_j(\theta, k) - \bar{U}_{j,\mathrm{E}}(\theta)]. \qquad (11.9)$$

HWA techniques Several different hot-wire probes have been used for *PLA* measurements. Lakshminarayana and co-workers at the Pennsylvania State University have reported many studies of turbomachinery flow with triple ($3W$) hot-wire probes. The first published paper of flow inside a rotating blade passage was by Gorton and Lakshminarayana (1976), and Lakshminarayana and Poncet (1974) presented data obtained downstream of a rotor using a digital *PLA* technique. In this latter method, the voltage signals from the three sensors were acquired simultaneously and converted into the corresponding three simultaneous data records of the radial, tangential, and axial velocity components. (A general description of the voltage-to-velocity conversion for $3W$ probes is given in Section 6.3.) Using a once-per-revolution marker pulse, the *PLA* technique was used to evaluate the three phase-locked mean-velocity components and the related Reynolds stresses. In principle, this is the correct *HWA* procedure, but it

is computationally very laborious. In the study by Joslyn *et al.* (1983), this
method was compared with the simpler phase-locking of the linearized
anemometer signals prior to converting to velocity components. Their comparative results indicate that similar phase-locked mean-velocity profiles,
$\bar{U}_{i,\mathrm{E}}(t)$, can be obtained by both methods.

Cau *et al.* (1987) used an *X*-probe placed sequentially in two orthogonal
planes to evaluate the ensemble-averaged velocity vector. Evans (1975) and
Capece and Fleeter (1989*a,b*) placed an *X*-probe in the (θ, z)-plane downstream of the rotor. By assuming that the radial-velocity component was
small, and by applying a *PLA* technique, they evaluated the profiles of the
phase-locked circumferential and axial velocity components. A number of
investigators have also used a single yawed (*SY*) hot-wire probe with the
wire element positioned sequentially at three different spatial orientations
at each measuring point (Whitfield *et al.* 1972; Favier *et al.* 1977; Schmidt
and Okiishi 1977; Hirsch and Kool 1977; Okiishi and Schmidt 1978;
Lakshminarayana and Davino 1980; Larguier 1981; Raj and Swim 1981;
Wisler *et al.* 1987). Provided that the periodic velocity field is phase-locked
by a reference pulse, a separate *PLA* technique can be applied to the
time-history records from the three probe orientations. The phase-locked
mean-velocity components, $\bar{U}_{i,\mathrm{E}}(t)$, $(i = 1\text{--}3)$, were evaluated by performing 'simultaneous' data analysis on the three ensemble-averaged signals.
However, as the three time-history records were obtained sequentially, the
value of the evaluated velocity components was very sensitive to any degradation in amplitude or phase information by the *PLA* technique. Larguier
(1981) concluded that this method does not give results which are as
accurate as those obtained by the multi-sensor *PLA* technique. To improve
the accuracy of the evaluated phase-locked mean-velocity and Reynolds
stresses, Kuroumaru *et al.* (1982) used a twelve-position method in conjunction with a least-squares curve-fitting procedure. A two-dimensional adaptation of the *SY*-probe technique has been used inside a rotating blade
passage, where the radial velocity component was known to be small
(Hodson 1984, 1985). In this method the probe support was approximately
aligned with the mean-flow direction and the probe was rotated so that the
sensor element was placed sequentially in two symmetrical positions in the
(θ, z)-plane. The *SY*-probe technique can also be implemented by a
rotatable *SN*-probe with its sensor element inclined relative to the predominant two-dimensional flow vector (Evans 1975; Raj and Swim 1981). These
workers applied a *PLA* technique and used eqn (11.8) to transform the temporal information into measuring points relative to a blade passage. The
flow direction at each of these points was first determined by rotating the
probe and identifying the angular position with the minimum *PLA* signal,
which corresponds to the hot-wire being tangential to the phase-locked
mean-flow direction. The magnitude of the phase-locked two-dimensional
mean-velocity vector was obtained by repeating the measurements for each

of the blade passage points with the probe rotated by 90°, so that the wire was perpendicular to the phase-locked flow direction.

Axial compressors are subject to two distinct aerodynamic instabilities, rotating stall and surge, which restrict the operational range of a compressor. Rotating stall is characterized by one or more cells of reduced and often reversed flow, travelling about the circumference of the machine with an angular velocity which is much smaller than that of the rotor stages. Day and Cumpsty (1978) studied rotating stall cells with a rotating SN-probe method, which made use of the following features. Firstly, the front of the rotating stall cell was used to trigger a reference pulse. Secondly, there is a large variation in the flow direction between unstalled flow and stall cells, and this distinction can be used to determine the circumferential extent of the stall cell. Using a PLA technique in conjunction with eqn (11.8), they determined the circumferential variation in the flow angle and the related velocity magnitude from the maximum of the PLA signals, obtained by rotating the SN probe in steps of 10° through an angle of 180°. In general, it is not recommended that the flow direction is identified from the maximum signal of a rotating HWA probe, but it was an acceptable approach in this study because of the large variations in the flow angle. To resolve any ambiguity in the flow direction evaluated by the hot-wire probe, a total-pressure probe rotated through 360° was also used.

While stall and surge studies may provide insight into the compressor performance under such conditions, they are of little practical value because neither can be tolerated during compressor operation. The common practical procedure for avoiding these instabilities is to leave a safety margin, the so-called stall/surge margin, between the operational point of the compressor and the point at which full-scale instabilities occur. However, there can be considerable benefits in terms of reduced compressor weight and efficiency improvements if the stall/surge margin can be safely reduced. This has resulted in some recent studies of the inception of stall in axial compressors. As described by Garnier et al. (1991), the instability evolves as a small-scale amplitude wave in the axial velocity, which grows as it travels around the circumference of the compressor until, through a non-linear action, it causes a large-amplitude disturbance (stall or surge). HWA studies aimed at identifying the initiation and developments of these waves have been reported by Jackson (1987), McDougall et al. (1990), Inoue et al. (1991), and Garnier et al. (1991).

The flow inside and behind a rotor deviates from the simplified linear blade row approximation shown in Fig. 11.1. As discussed by Raj and Lakshminarayana (1976) 'the wake of a rotor blade, unlike a cascade or isolated airfoil wake, is three-dimensional in nature. The three-dimensionality is due to imbalance in pressure gradient and centrifugal forces inside the wake. While the pressure gradient is nearly the same across the wake, there is a variation in centrifugal forces inside the wake due to velocity defect and this results in radial flows.' Also, the results presented by, for example, Capece and Fleeter (1989*a*), for the instantaneous and ensemble-averaged rotor exit flow field (Fig. 11.3) demonstrate that the ensemble-averaging technique may discard information describing significant unsteady-flow phenomena. They observed that in the free-stream region, the instantaneous signals were analogous to one another and to the ensemble-averaged free-stream results, as expected. However, the instantaneous signals in the rotor-blade wake region were somewhat surprising. Some of the instantaneous signals were found to be analogous to one another and to the characteristics of the ensemble-averaged wake, but others differed significantly from the expected wake profiles. This phenomenon was interpreted as being due to the existence of a vortex-sheet structure in the rotor-blade wake, a process that is not phase-locked to the marker pulse. This unsteady-flow phenomenon is a source of total-pressure loss, flow unsteadiness, and acoustic excitation, and it may be as significant as the mean-velocity profile determined by the *PLA* method.

Looking at the detailed information contained in Fig. 11.3, it can be observed that, for example, the amplitude of the ensemble-averaged signal in the wake region is much smaller than the corresponding amplitude information in the individual wake events. It is therefore necessary to treat *PLA* results with caution. The *PLA* technique is clearly very useful in identifying

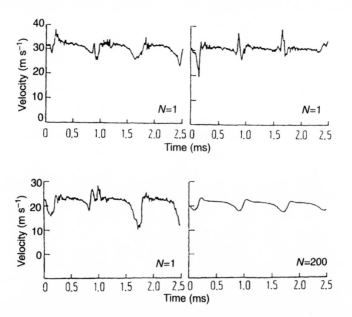

Fig. 11.3. Instantaneous and ensemble-averaged rotor-exit flow field. (From Capece and Fleeter 1989*a*.)

the existence and location of wake-flow events and it will give some indication of the blade-to-blade variation in the periodic component, $\bar{U}_{i,E}(t)$. However, unless the wake-flow events are strongly phase-locked, the *PLA* technique will smooth out significant flow events, and the magnitude of the evaluated phase-locked flow quantities may not be an accurate description of the wake-flow events.

11.1.2 Internal-combustion engines

The flow field within the cylinder of a reciprocating piston engine is the most important factor controlling the combustion process. As discussed by Heywood (1987) and by Arcoumanis and Whitelaw (1987), the flow field governs the flame propagation rate in homogeneous charge spark-ignition engines and it contributes to the control of the fuel–air mixing and burning rates in diesel engines. There are two main objectives of internal-combustion engine flow processes: (i) At maximum power the cylinder must be filled with as much fresh air as possible. The maximum engine power is limited by the air flow: only as much fuel can be burnt as there is oxygen available. (ii) A sufficiently fast combustion process must be achieved to release the fuel's chemical energy in a short crank-angle interval, typically about 25°, centred at the beginning of the power-producing stroke. The impact of the flow on the combustion process is different for spark-ignition and compression-ignition engines. In a conventional spark-ignition engine, the fuel and air are mixed together in the intake system, and the primary requirement during combustion is therefore a sufficiently turbulent flow field to ensure flame development and propagation at an appropriate rate. The highest velocities and turbulence are usually generated during the intake process and then they decay with time. Burn rates can be increased by means of turbulence enhancement just prior to combustion. One technique is to use bowl-in-piston engines as discussed by, for example, Fansler and French (1988) and by Fansler (1993). A second method, which has attracted much interest recently, is the creation of intake-generated 'tumble' (rotational air motion about an axis perpendicular to the cylinder axis) to convert mean-flow kinetic energy to turbulence kinetic energy via vortex breakdown and so augment mixing and combustion rates in spark-ignition engines. It has been found in several investigations that four-valve pent-roof configurations usually generate some degree of tumble (Khalighi 1990; Haworth *et al.* 1990; Le Coz *et al.* 1990; Arcoumanis *et al.* 1990, 1991*a*, 1993; Hu *et al.* 1992). A major driving force for these developments has been the move towards lean, or highly diluted, mixture combustion, which has the advantage of good fuel economy and reduced CO and NO_x emission. The associated slow laminar burning velocity of lean mixtures can be compensated for by an increased precombustion turbulence, but excessive turbulence levels may lead to ignition difficulties and quenching due to aerodynamic-flow straining (Bradley *et al.* 1988).

In diesel engines fuel is injected by the fuel injector into the engine cylinder towards the end of the compression stroke. The major problem is to achieve sufficiently rapid mixing between the injected fuel and the air in the cylinder to complete combustion in the selected crank-angle interval close to top centre. The fuel-injection system plays a critical role in the combustion efficiency, and studies have therefore been carried out both of the internal fluid flow in a (model) diesel-injector nozzle (Arcoumanis *et al.* 1992) and of the in-cylinder spray-tip penetration and droplet velocities (Arcoumanis *et al.* 1991*b*). The fuel–air mixing rate is the primary factor controlling the fuel-burning rate. The mixing rate, in turn, is controlled both by the fuel spray and by the turbulent flow field created by the intake process and piston motion.

Engine operating cycle and geometry In a reciprocating engine, a piston moves back and forth in a cylinder and transmits power through a connecting rod and crank mechanism to the drive shaft, as shown in Fig. 11.4. The majority of reciprocating engines operate on the *four-stroke cycle.*

An intake stroke, starting with the piston at the top dead centre (TDC) position, and ending with the piston at the bottom dead centre (BDC). During this stroke, a fresh air/fuel mixture is drawn into the cylinder via the

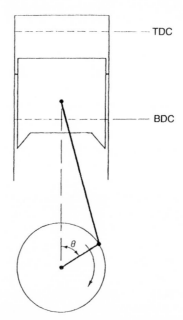

Fig. 11.4. Sketch of reciprocating internal-combustion engine showing Top Dead Centre (TDC) and Bottom Dead Centre (BDC) of the piston and the crank angle, θ.

inlet-valve system. To maximize the inducted mass, the inlet valve usually opens shortly before the stroke starts and closes after it ends.

A compression stroke, during which both the inlet and the exhaust valves are closed. The mixture inside the cylinder is compressed to a small fraction of the initial value, causing an increase in both the pressure and temperature. Near the end of the compression stroke, combustion is initiated, and the cylinder pressure rises more rapidly.

A power stroke, with both valves closed, which starts with the piston at TDC and ends at BDC. The high-temperature, high-pressure gases push the piston down and force the crank to rotate. For a well-designed engine, it is desirable to complete the combustion during the early part of the power stroke, but this may be difficult to achieve in lean or dilute spark-ignition engines because the combustion is quite slow. This explains the recent major interest in turbulence enhancement just prior to combustion. The exhaust valve opens as the piston approaches BDC, and the cylinder pressure drops to close to the exhaust pressure.

An exhaust stroke, during which the remaining burnt gas is being pushed out of the exhaust port by the cylinder moving from BDC to TDC. As the piston approaches TDC, the inlet valve opens and just after TDC the exhaust valve closes and the cycle starts again.

Flow through the inlet valve Reviews of the main effects of the intake system on the in-cylinder turbulent-flow parameters have been given by Heywood (1987) and Catania and Mittica (1987). As stated earlier, the maximum power of an internal-combustion engine at a given speed is proportional to the mass-flow rate of air. Thus, induction of the maximum possible air mass at full load during the intake stroke is a major objective. Also, to fully utilize the air in the combustion process it is necessary to set up the appropriate turbulent flow in the cylinder. The flow through the inlet valve enters the cylinder as high-shear intake jets which interact with the cylinder wall and the moving piston to form large-scale rotating-flow patterns. In general, swirl (defined as organized rotation of the flow around either the cylinder axis, axial swirl, or an axis parallel to the piston surface, 'tumble' or 'barrel' swirl) has a beneficial effect on the combustion process. The three-dimensional nature of the flow entering the cylinder during the intake stroke is therefore of basic importance for the turbulent combustion rate in reciprocating piston engines. The induction-system design determines the inlet-valve flow and thereby the development of the turbulent fluid motion within the cylinder, which in turn affects both the ignition process and the flame propagation rate. However, high turbulence levels, which are usually associated with fast burning and higher efficiency, can only be developed by high-pressure drops, and this means a reduced air-intake mass flow. A practical air-inlet system represents a compromise between these two conflicting requirements.

In engine-flow studies, it has until recently been assumed that the flow through the inlet valve is quasisteady. That is, the unsteady flow at any instant is approximately the same as the steady flow corresponding to the instantaneous valve lift and pressure drop. The validity of this approximation has been supported by unsteady and steady inlet-flow measurements by Bicen *et al.* (1984, 1985), by Arcoumanis *et al.* (1987*b*), and by Arcoumanis and Whitelaw (1991), and related *HWA* valve-flow investigations have been reported by Haghgooie *et al.* (1984), by Khalighi *et al.* (1986), and by Wagner and Kent (1988*b*). This approach provides boundary conditions for computational fluid dynamics (*CFD*) model work, which does not resolve the flow through the ports and valves. However, the assumption of quasisteady flow is no longer necessary due to the development of *CFD* codes for three-dimensional unsteady-flow calculations of the flow in the cylinder, the manifold, and the ports of production engines (see, for example, Haworth *et al.* 1990; Amsden *et al.* 1992). These developments require experimental inlet-manifold and valve flow results of adequate spectral and time resolution to be used as validation data (Arcoumanis *et al.* 1989).

In-cylinder flow-field studies An improved understanding of the fluid motion in internal-combustion engines is critical for further improvements in the fuel economy and emissions. However, it is very difficult to obtain accurate turbulence measurements in combusting flows, and it has therefore been common practice to correlate flow data acquired under motored conditions to combustion data acquired in the firing engine. This is often a valid approach, because similar turbulent-flow patterns before combustion under motored and firing conditions have been obtained by Ohigashi *et al.* (1971), by Rask (1979), and by Hall and Bracco (1987). It was also concluded from cycle-resolved analysis of in-cylinder velocity data under firing conditions that combustion did not greatly increase the turbulence intensity ahead of the flame (Witze and Mendes-Lopes 1986; Hall and Bracco 1987). However, the results of Witze *et al.* (1989) show that there are engine conditions where this approach must be treated with caution.

Many experimental investigations have been carried out on model or production internal-combustion engines using both *HWA* (Semenov 1958; Huebner and McDonald 1970; Ohigashi *et al.* 1971; Dent and Salama 1975; Lancaster 1976; Witze 1977, 1980; Catania 1980, 1982, 1985; Wakisaka *et al.* 1982; Tindal *et al.* 1982; Wakuri *et al.* 1983; Catania and Mittica 1985*a,b*, 1987, 1989, 1990; Khalighi *et al.* 1986; Dinsdale *et al.* 1988; Wagner and Kent 1988*b*. Subramaniyam *et al.* 1990; Catania *et al.* 1992) and *LDA* (Rask 1979, 1981, 1984; Arcoumanis *et al.* 1982, 1983, 1985, 1987*a,b*, 1991*a,b*; Liou and Santavicca 1982, 1985; Ball *et al.* 1983; Witze *et al.* 1984; Liou *et al.* 1984; Coghe *et al.* 1985; Bopp *et al.* 1986; Hall *et al.* 1986; Saxena and Rask 1987; Hall and Bracco 1987; Ikegami *et al* 1987; Fansler and French 1988; Glover *et al.* 1988; Namekawa *et al.* 1988; Fraser and Bracco

1988, 1989; Witze *et al.* 1990; Hu *et al.* 1992; Le Coz 1992; Fansler and French 1992; Fansler 1993).

Related cycle-resolved signal-analysis techniques apply to both the *HWA* and laser doppler anemometry (*LDA*) methods. Hot-wire anemometry has the advantage of providing a continuous signal, but there are many problems associated with the use of *HWA* in engines. The probe is intrusive and measurements are therefore usually restricted to the top part of the cylinder (above the TDC position of the piston in Fig. 11.4). Also, the probe cannot be used during combustion. To obtain accurate velocity measurements it is necessary to calibrate the probe and to compensate for the effect of the large variation in the temperature and the pressure during the engine cycle. This may be difficult to achieve in practice (Witze, personal communication). In a strongly three-dimensional flow field it is necessary to use a three-wire (3*W*) probe to obtain the three instantaneous velocity components. However most *HWA* engine studies have used a probe with a single sensor. Assuming that the cyclic variations are small, Catania (1982) inserted a probe radially and placed the wire at three spatial orientations at each measurement point in order to obtain the ensemble averaged mean-velocity components and the related Reynolds stresses. However, substantial cyclic variations occur in many engines, making the multi-position single-wire method difficult to use. Finally, a stationary hot-wire probe cannot detect flow reversal. For these reasons most of the recent in-cylinder-flow studies have been carried out with *LDA*.

The *LDA* technique, which requires expensive equipment and elaborate data analysis, can (due to its nonintrusive nature) provide flow information throughout the cylinder volume both under a motored condition and with combustion. An *LDA* system does not require calibration, and it can resolve the directional ambiguity inherent in *HWA*. For cycle-resolved analysis, *LDA* has the advantage of providing direct measurement of one or more velocity components. The signal analysis procedures (presented for a *HWA* system) are similar for *HWA* and *LDA*, except for the use of a small crank-angle window, $\Delta\theta$, for data processing with *LDA* due to the noncontinuous nature of the *LDA* signal. For completeness, it should also be mentioned that particle-image velocimetry (*PIV*), which can give a complete two-dimensional 'instantaneous'-vector presentation of the flow field, is now being applied to engine-flow studies (Reuss *et al.* 1989, 1990; Nino *et al.* 1993; Lee and Farrel 1993). As discussed by, for example Heywood (1987), Arcoumanis and Whitelaw (1987), and Catania and Mittica (1989), the in-cylinder turbulent-flow field is highly three-dimensional and nonstationary, due to the intake-system configuration and the time-dependent boundary conditions imposed by the piston motion and by valve transience. In order to specify cause-and-effect flow phenomena in engine performance, many investigations have attempted to identify which part of the velocity signal is due to turbulence and which part is due to the mean velocity. This task is difficult because changes in the mean flow occur on a time-scale of milliseconds; that is, on the same order of magnitude as typical time-scales of turbulence. To give an example, it was shown by Liou *et al.* (1984), with reference to a specific ported engine, that the question of whether the in-cylinder turbulence intensity without swirl is smaller or larger than it is with swirl could have different answers, depending on the procedure used to extract the turbulence intensity from the same velocity data, particularly for the no-swirl case.

The task of separating the flow motion into mean flow and turbulence is further complicated by the existence of cyclic variations in all engines. Three interrelated causes are considered to give combustion instability (Heywood 1989; Johansson 1993):

(1) the total amount of fuel, air and burned gas in the cylinder are not the same from cycle to cycle;

(2) the effect of an inhomogeneous air/fuel mixture, particularly close to the spark plug;

(3) the flow field in the engine varies from cycle to cycle.

The relative importance of these three causes is not well-understood, but it is generally accepted that mixture motion is a principal cause of cyclic variation in the combustion process. Reynolds (1980) suggested that the cyclic-gas-motion variations occur because the location and size of the recirculating regions formed during the intake process are very sensitive to minor variations in the inducted flow field. For example, swirl inlet ports often produce swirl patterns with off-axis centres of rotation and even with multiple-vortex patterns, and the swirl-centre motion need not be phase-locked to the engine's crank angle (Fansler, personal communications). For single-point velocity analysis, it is a major problem that the frequencies of the larger-scale turbulent velocity fluctuations may overlap the frequencies associated with the unsteady mean motion and with any instability-related phenomena (for example flapping jets, precessing swirl) that may occur. There is currently no consensus as to whether cyclic variations are a feature of the mean-flow motion or whether they correspond to some large-scale coherent structure of the turbulence, and as a consequence there is no generally accepted definition of mean-flow field and turbulence for internal combustion engines.

The in-cylinder flow field has been analysed using either phase-locked ensemble-averaging techniques or cycle-resolved reduction procedures.

The phase-locked-averaging (PLA) method The most common method for evaluating the periodic nature of the engine cycle has been the phase-locked-averaging (*PLA*) technique. For engine configurations where the cycle-to-cycle variation in the mean flow is relatively small, the decomposition into mean and fluctuating components can be achieved by phase-locking the velocity signal to a specific crank-angle position and ensemble averaging the corresponding instantaneous velocity signal. Depending on the hot-wire probe type and wire orientation, one or more of the components of the velocity vector can be evaluated. Applying the *PLA* method described in Section 11.1.1, the instantaneous velocity components (specified by the crank angle, θ, position) $U_i(\theta, k)$, ($i = 1\text{-}3$) in the kth-cycle can be separated into two components,

$$U_i(\theta, k) = \bar{U}_{i,\mathrm{E}}(\theta) + u_{i,\mathrm{E}}(\theta, k), \qquad (11.10)$$

and the (phase-locked) ensemble-averaged mean velocity components, $\bar{U}_{i,\mathrm{E}}(\theta)$, are obtained from eqn 11.5 as

$$\bar{U}_{i,\mathrm{E}}(\theta) = \frac{1}{K} \sum_{k=1}^{K} U_i(\theta, k),$$

where K is the number of cycles considered. The corresponding rms velocity fluctuations are defined as

$$u'_{i,\mathrm{E}}(\theta) = \left[\frac{1}{K} \sum_{k=1}^{K} u^2_{i,\mathrm{E}}(\theta, k) \right]^{1/2} = \left\{ \frac{1}{K} \sum_{k=1}^{K} [U_i(\theta, k) - \bar{U}_{i,\mathrm{E}}(\theta)]^2 \right\}^{1/2}.$$

$$(11.11)$$

Cycle-resolved reduction procedures It has been observed in many internal combustion engine flow studies that significant cycle-to-cycle variations

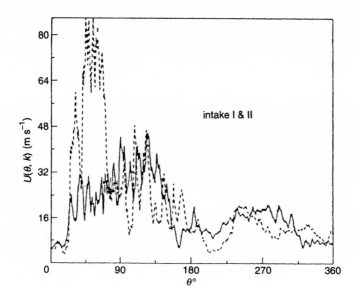

Fig. 11.5. Cyclic variations in the in-cylinder flow field of the internal-combustion engine shown in Fig. 11.6. (From Catania and Mittica 1989.)

occur in the mean flow. The existence of cyclic variations in an in-cylinder flow field is illustrated in Fig. 11.5 by individual cycle-velocity patterns measured with a hot-wire probe positioned in the top part of the internal diesel combustion engine shown in Fig. 11.6. The engine had a 75.5 mm bore, a 83.5 mm stroke, and a compression ratio of 16. The induction system was made up of two swirl tangential ducts, and valves of the same size and type. The (assumed) predominant swirl-velocity component was measured with the hot-wire probe, shown in Fig. 11.6, placed with its sensor parallel to the cylinder axis ($\alpha = 0°$), at the position indicated by the dot,

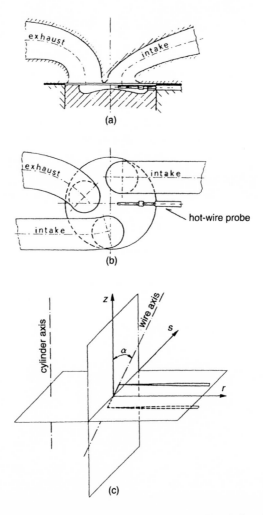

Fig. 11.6. Schematic of: test engine and probe set up (a and b); (c) the reference frame and wire orientation for multi-position measurements. (From Catania and Mittica 1987.)

3.5 mm underneath the flat cylinder and 17.5 mm from the cylinder axis. The hot-wire probe was placed upstream of intake valve I and downstream of intake valve II, so that the main swirling flow was sensed on induction. The data sampled from the probe was acquired and analysed using an advanced *HWA* technique by Catania (1982) and validated in nonstationary as well as stationary conditions, at different gas pressures and temperatures (Catania 1985; Catania and Mittica 1987).

The combustion process in a reciprocating engine occurs within each individual cycle and it is therefore influenced by the mean- and turbulent-flow pattern in each cycle. It has therefore been proposed that the analysis of engine turbulence should be studied on a single-cycle basis, and several reduction techniques of in-cylinder velocity data have been developed to characterize the mean velocity and turbulence in each cycle, and to permit statistical analysis of the turbulent flow parameters determined from cycle-resolved results. The various methods used for the evaluation of the individual cycle mean velocities (as described below) are essentially forms of local time averaging (with an averaging interval T) or equivalently, of low-pass filtering (with a cut-off filter of approximately $1/T$), despite the differences in their implementation. It is therefore not surprising that they yield similar results when applied with similar values of T (Fansler, personal communication). Using the same set of velocity data, Catania and Mittica (1989) have studied the effectiveness of cycle-resolved data reduction techniques. For comparative purposes, the cycle-resolved methods were identified as: Nonstationary Time Averaging (NTA, Lancaster 1976), Cubic Spline Fitting (CSF, Rask 1981), Inverse Fast Fourier Transform (IFT, Liou and Santavicca 1985), Time-Averaging Filtering (TAF, Catania and Mittica 1985a,b) and Linear Trent Removal (LTR, Daneshyar and Fuller 1986). Further cycle-resolved studies include: direct bandpass filtering (Yianneskis et al. 1989; Lorenz and Prescher 1990; Murakami et al. 1990), the IFT method (zur Loye et al. 1989; Le Coz 1992) and the NTA procedure (Witze et al. 1990). Enotiadis et al. (1990) carried out a comparative cycle-resolved study using high-pass filtering and cubic spline fitting. Hilton et al. (1991) developed an autocorrelation based analysis of the in-cylinder flow field. A phase-stability criterion was used to distinguish between turbulence and cyclic variations, which were assumed to be statistically independent.

In the cycle-resolved method the instantaneous velocity during each cycle (at a selected measuring point) is separated into a mean nonstationary component and a 'turbulent' component. In each method, it is necessary to select a low-pass filter cut-off (or equivalently a local averaging time, T). The selection of the cut-off frequency is clearly important. One method used is to low-pass filter the ensemble-mean-velocity trace, $\bar{U}_{i,E}(\theta)$, and to raise the cut-off frequency until no statistically significant structure in $\bar{U}_{i,E}(\theta)$ is attenuated or smoothed out. This leads to a value of T which is comparable with the integral time-scale of the 'turbulence'. This criterion is conservative in as much as it ensures that the unsteadiness of the mean motion will not contribute to the 'turbulent' part of the signals. However, for the turbulence evaluation, the method will exclude contributions which have characteristic frequencies below the cut-off frequency (Fansler and French 1988; Fansler (1993).

In the cycle-resolved methods, for the kth engine cycle

$$U_i(\theta, k) = \bar{U}_i(\theta, k) + u_i(\theta, k), \tag{11.12}$$

where $\bar{U}_i(\theta, k)$ are the mean-velocity components and $u_i(\theta, k)$ are the velocity fluctuations about the mean. Each of the above methods uses a different technique to evaluate $\bar{U}_i(\theta, k)$; as summarized by Catania and Mittica (1989).

The in-cycle mean-velocity components $\bar{U}_i(\theta, k)$ can be expressed as

$$\bar{U}_i(\theta, k) = \bar{U}_{i,E}(\theta) + u_{i,LF}(\theta, k), \tag{11.13}$$

where $\bar{U}_{i,E}(\theta)$ is the (phase-locked) ensemble-averaged mean-velocity components (see eqn 11.5) and $u_{i,LF}(\theta, k)$ is the low-frequency cyclic fluctuating components of the mean. From the definitions of the terms in eqn (11.13) it follows that ensemble averaging of the cycle-resolved data will give the same mean-velocity results as the *PLA* method; that is,

$$\bar{U}_{i,E}(\theta) = \frac{1}{K} \sum_{k=1}^{K} \bar{U}_i(\theta, k). \tag{11.14}$$

The corresponding low-frequency rms values $u'_{i,LF}(\theta)$ of the components of the in-cycle mean velocity about the ensemble mean velocity can be evaluated, by ensemble averaging, as

$$u'_{i,LF}(\theta) = \left[\frac{1}{K} \sum_{k=1}^{K} u^2_{i,LF}(\theta, k)\right]^{1/2} = \left\{\frac{1}{K} \sum_{k=1}^{K} [\bar{U}_i(\theta, k) - \bar{U}_{i,E}(\theta)]^2\right\}^{1/2}. \tag{11.15}$$

The high-frequency cycle-resolved, ensemble-averaged, turbulence-intensity components, $u'_{i,HF}(\theta)$, during the time period $t \pm T/2$ used to evaluate $\bar{U}_i(\theta, k)$ are defined as

$$u'_{i,HF}(\theta) = \left[\frac{1}{K} \sum_{k=1}^{K} u_i^2(\theta, k)\right]^{1/2} = \left\{\frac{1}{K} \sum_{k=1}^{K} [U_i(\theta, k) - \bar{U}_i(\theta, k)]^2\right\}^{1/2}. \tag{11.16}$$

From these definitions it follows to first order that

$$u'^2_{i,E}(\theta) \cong u'^2_{i,LF}(\theta) + u'^2_{i,HF}(\theta). \tag{11.17}$$

The *PLA* method will therefore overestimate the turbulence intensity when there are significant low-frequency fluctuations in the mean flow.

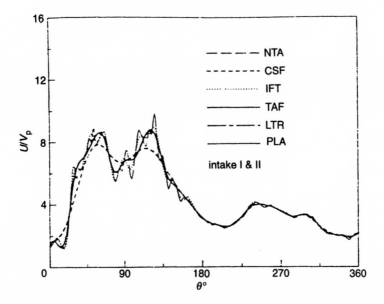

Fig. 11.7. Normalized ensemble-averaged mean-velocity results obtained with cycle-resolved and the *PLA* methods. (From Catania and Mittica 1989.)

Using the same hot-wire data (from the engine shown in Fig. 11.6) Catania and Mittica (1989) have carried out a comparison of ensemble-averaged velocity distributions obtained with both cycle-resolved reduction procedures and the *PLA* technique. In this comparison, the hot-wire cooling velocity, which corresponds primarily to the predominant swirling velocity, will be denoted by U. The mean-velocity data, normalized by the mean piston velocity, V_p, are presented in Fig. 11.7 and show that good agreement is obtained with all methods except for U_{CSF}, which deviates from the other distributions during the intake stroke. The corresponding rms value of the low-frequency cyclic fluctuation of the mean flow, u'_{LF}, is presented in Fig. 11.8, which demonstrates that large cycle-to-cycle variations occur in the mean motion during the early part of the intake stroke. When large values of u'_{LF} occur, eqn (11.17) indicates that the turbulence intensity, u'_E, evaluated by the *PLA* method will be significantly larger than the cycle-resolved turbulence intensity u'_{HF}. This conclusion is confirmed by the turbulence-intensity results shown in Fig. 11.9.

A main advantage of cycle-resolved analysis is that it can identify two interacting processes: convection by the mean coherent large-scale velocity field and mixing by the chaotic small-scale turbulence. This separation provides a method for identifying both causes for cyclic variations in the

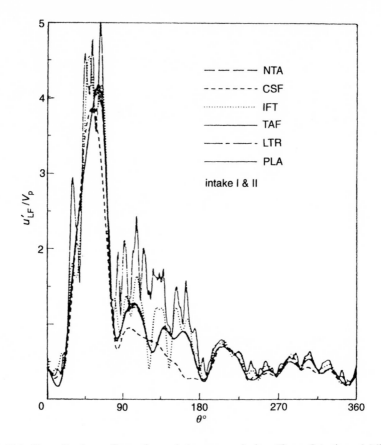

Fig. 11.8. Normalized rms fluctuations of the mean velocity. (From Catania and Mittica 1989.)

cylinder air motion and the corresponding effects on the combustion process. For example, Le Coz (1992) has shown that cyclic variations of large-scale fluid motion in the electrode gab contribute strongly to the cyclic variations of the initial flame. Cycle-resolved analysis has also been used by Fansler (1993) to obtain details of turbulence production near the top dead centre (TDC) for different geometries of bowl-in-piston engines.

Length-scale measurements To characterize the turbulence of the flow and to formulate/validate models of turbulence (see Section 1.6) it is necessary to determine both the intensity of the turbulent fluctuations and the eddy-size range, which vary from the large energy-containing eddies to the smallest dissipating eddies (see Section 2.4.2).

In internal combustion engines, large energy-containing eddies are initially formed by the inlet jet flow. Their sizes are roughly equal to the thickness

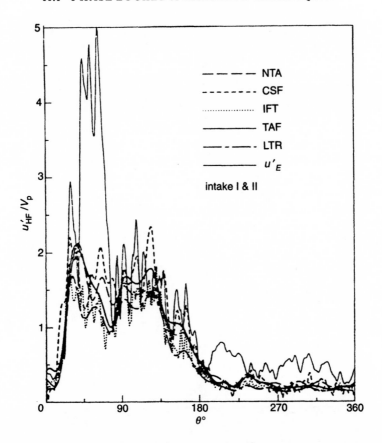

Fig. 11.9. Normalized turbulence-intensity distributions. (From Catania and Mittica 1989.)

of the local jet, and they are typically about 0.5–3.0 mm in size depending on the valve/engine configuration (Fraser and Bracco 1988; Reuss *et al.* 1989). Furthermore, other flow processes such as vortex breakdown (for example, a tumbling vortex crushed by the piston as it approaches TDC) can lead to generation of new turbulent eddy structures unrelated to the inlet-shear-layer thickness.

Velocity measurements made at two spatially separated points can be used to identify the correlation distance of the turbulent flow. The positions of the two points will be denoted by A, (x, y, z), and B, $(x + r_1, y + r_2, z + r_3)$, and the velocity fluctuation in the r_1-direction will be denoted by u. A measure of the size of the energy-containing eddies is the (longitudinal) integral length-scale, L, defined by eqn (2.66)

$$L = \int_0^\infty f(r_1) \, dr_1,$$

where $f(r_1)$ is the longitudinal-correlation coefficient defined by eqn (2.65a) for a statistically stationary turbulent flow. The corresponding transverse integral scale, L_g, is defined by eqn (2.67)

$$L_g = \int_0^\infty g(r_2)\, dr_2,$$

where $g(r_2)$ is the lateral correlation coefficient. Typical variations for $f(r_1)$ and $g(r_2)$ are shown in Fig. 2.26; For homogeneous isotropic turbulence the relationship between the two integral length-scales is $L = 2L_g$. However, the difference between L and L_g can be much larger in turbulent shear flows, as was discussed in Section 2.4.2.

Because the flow in an internal-combustion engine is nonstationary, it is necessary to evaluate the longitudinal and lateral correlation coefficients, $f(r_1)$ and $g(r_2)$, by ensemble averaging

$$f(r_1) = \frac{1}{K-1} \sum_{k=1}^{K} \frac{u(x,k)u(x+r_1,k)}{u'(x)u'(x+r_1)}, \tag{11.18}$$

$$g(r_2) = \frac{1}{K-1} \sum_{k=1}^{K} \frac{u(y,k)u(y+r_2,k)}{u'(y)u'(y+r_2)}. \tag{11.19}$$

The corresponding integral length-scales can be evaluated using eqns (2.66) and (2.67).

It is a very complex task to carry out spatial-correlation measurements inside a motored internal-combustion engine, and only a few investigations have been reported: Wakuri *et al.* (1983) used *SN* hot-wire probes with variable separation; Collings *et al.* (1987) and Dinsdale *et al.* (1988) obtained measurements with a flying hot-wire probe; Ikegami *et al.* (1987) applied the laser homodyne technique; Glover *et al.* (1988) applied a scanning *LDA*; Fraser and Bracco (1988, 1989) used a two-point, single-probe volume *LDA* technique; and Reuss *et al.* (1989) deduced the length scale from particle-image velocity measurements. A two-point *LDA* technique has been applied by Belmabrouk *et al.* (1991) to a steady-flow rig which simulates the induction stroke of an engine.

As explained in Section 2.4.2, the integral length scale may also be estimated from the auto-correlation coefficient function $\rho_u(\tau)$, obtained from a single point measurement. For a statistically steady turbulent flow, $\rho_u(\tau)$ is defined by eqn (2.69), but for an internal-combustion engine an ensemble-averaging definition is required; namely,

$$\rho_u(\theta,\tau) = \frac{1}{K-1} \sum_{k=1}^{K} \frac{u(\theta,k)u(\theta+\tau,k)}{u'(\theta)u'(\theta+\tau)}. \tag{11.20}$$

An integral time-scale, T_1, corresponding to eqn (2.70), can be defined by

$$T_1 = \int_0^\infty \rho_u(\theta, \tau)\mathrm{d}\tau,$$

and by assuming the validity of Taylor's hypothesis for the convection of a 'frozen' turbulent structure past the measurement point with the mean velocity, the integral length scale, L, can be estimated from eqn (2.71) as

$$L = \bar{U}T_1.$$

This method has been used in several investigations including those by Witze (1977), Wakisaka *et al.* (1982), Liou and Santavicca (1985), Fansler and French (1988), Subramaniyam *et al.* (1990), Catania and Mittica (1990), and Catania *et al.* (1992). However, estimating the integral length scale from a time-scale via Taylor's hypothesis is, of course, fraught with danger for such a highly nonstationary situation as in-cylinder flow. Such results should be treated as an order-of-magnitude estimates.

Cyclic variations in the mean flow will, in general, have larger correlation lengths than the large turbulent eddies, and some investigators have therefore sought to distinguish between the corresponding length scales. As an example, Fraser and Bracco (1988) demonstrated that the (low-frequency) integral length scale determined by the *PLA* method could be more than twice the value of the turbulence integral length scale obtained by a cycle-resolved procedure.

11.2   THE IDENTIFICATION OF TURBULENT FLOW PHENOMENA

Our physical picture of turbulent shear flows has changed significantly over the last fifty to sixty years. During the period 1920 to 1950, turbulence was in general viewed as a random process, which could only be described statistically. The flow field was assumed to contain a well-defined and repeatable mean on which random-velocity fluctuations were superimposed. The effect of turbulence on the mean motion was usually expressed in terms of Reynolds stresses, as explained in Chapter 1. The turbulence was characterized by a wide range of randomly interacting scales ranging from large energy-containing eddies to small dissipating eddies. The concept of an eddy was useful in formulating turbulence models, but it was essentially an abstraction, and there was no real attempt at identifying any significant real turbulent motion.

Most of the subsequent significant changes to our understanding of turbulent shear flows have come from a dual approach of high-quality flow-visualization experiments and related conditional-sampling techniques applied, in particular, to hot-wire probe signals. Flow-visualization experiments in many different types of shear flows, usually at moderate Reynolds numbers, have demonstrated features which are inconsistent with the notion of turbulence as a random process. Subsequently, conditional sampling

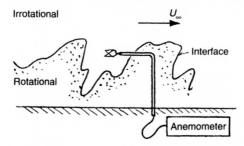

Fig. 11.10. A hot-wire probe placed in a turbulent boundary layer. Note the relatively sharp interface between the rotational turbulent flow and the irrotational free-stream flow.

techniques have been developed and applied to hot-wire probe signals to provide quantitative data for the turbulent-flow processes identified.

The first significant new feature added to the physical picture of turbulence was the observation by Corrsin (1943) and Townsend (1947) that there exists a sharp interface between turbulent and non-turbulent flow in the outer part of turbulent shear flows. A hot-wire probe placed in such a flow, as illustrated in Fig. 11.10, will identify the intermittent nature of turbulence, as confirmed by Corrsin and Kistler (1955). The developing picture of turbulence, attributed primarily to Townsend (1956), was of a turbulent fluid of nearly uniform intensity moved about by the slow convective motion of a system of large eddies with sizes comparable to the width of the shear layer and much larger than the eddies containing most of the turbulent energy. These large eddies were primarily responsible for the entrainment of irrotational potential flow from outside the turbulent interface.

Townsend produced a picture of the large-eddy motion based on the conventional correlation tensor; but, in general, the picture of turbulence was still that of a random process. The conditionally sampled hot-wire-probe measurements made by Kovasznay *et al.* (1970) first demonstrated that there was a relationship between the interface and coherent large-scale motions in the outer part of a turbulent-boundary-layer flow.

11.2.1 Turbulent boundary layers

The study of coherent flow structures will first be considered for a turbulent boundary layer on a flat plate, due to its importance for both fundamental turbulence research and engineering applications. Over the last three decades, an enormous research effort has been expended on this flow case, but there is still considerable uncertainty as to the detailed features of this type of flow, as summarized by Robinson (1991). In this section, a physical process will be outlined which can link the dynamics of the large-scale flow structures in the outer region to the 'burst' phenomena in the near-wall region; this is based primarily on an assessment of previously published results. The following notation will be used. The coordinates (x, y, z) are, respectively, the longitudinal, transverse, and spanwise directions, and the corresponding fluctuating velocity components will be denoted by (u, v, w). Mainly u and v velocity data have been selected to show the feasibility of the proposed physical model. The discussion of the boundary layer flow has been separated into two parts: (a) the outer flow region and (b) the near-wall region linked by a dynamic-flow mechanism.

11.2.1.1 Outer flow region

The first type of conditional sampling, carried out in the outer part of turbulent shear flows, was based on the observation that these flows have an intermittent nature with turbulent regions interspersed with non-turbulent regions. A relatively sharp interface separates the turbulent flow (which is highly rotational) from the non-turbulent flow (which can, for practical

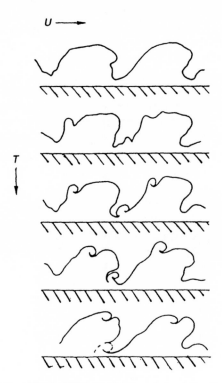

Fig. 11.11. The formation of 'typical (Falco) eddies' on the back interface of turbulent bulges. (From Falco 1977.)

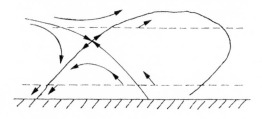

Fig. 11.12. The flow field related to the large-scale flow structure in a coordinate system moving with the saddle point. (From Falco 1977.)

purposes, be considered to be irrotational). Figure 11.10 indicates that the turbulent region forms large bulges which are also referred to as large-scale turbulent eddies. The part of the bulge (facing downstream), which first reaches the probe is normally referred to as the 'front', and the upstream part of the interface is referred to as the 'back'. Flow-visualization results at moderate Reynolds numbers by Falco (1977) have also identified the development of 'typical eddies' (which will be referred to as Falco eddies) on the back interface, as shown in Fig. 11.11. The corresponding motion of the entrainment flow is illustrated in Fig. 11.12. In a coordinate system moving with the velocity of the saddle point, it is observed that the flow below the saddle point is directed towards the wall and this causes the motion of the Falco-eddies towards the wall as indicated by Fig. 11.11. It is proposed, as will be discussed in the near-wall section, that these Falco-eddies may be the link mechanism for the observed 'burst' events in the near-wall regions. This is a conceptually attractive and simple physical model. (However, it should be noted that the Falco eddies are highly dependent on the Reynolds numbers, and the proposed link mechanism must therefore be treated with some caution (Bradshaw, Tiederman, personal communication).) The conceptual link mechanism must be modified for the cases of fully developed turbulent pipe and channel flows, which do not contain any irrotational entrainment fluid. In these cases it is still possible to identify large-scale coherent motions in the outer flow and interfaces between regions of high and moderate turbulence intensities. It is possible that similar Falco-type eddies may form on these interfaces and be convected towards the wall.

The intermittency function, I Since turbulent shear flows contain two distinct states, it is useful to introduce an intermittency function, I, which can be used to discriminate between the turbulent and non-turbulent regions. The intermittency function $I(r, t)$ at a point, r, denoted below simply by $I(t)$, is defined by

$$I(t) = \begin{cases} 1 \text{ for turbulent flow} \\ 0 \text{ for non-turbulent flow.} \end{cases} \tag{11.21}$$

The time-averaged value of $I(t)$ is the intermittency factor, $\gamma \ (= \bar{I})$, which represents the fraction of time for which the flow at the selected position in space, r, is turbulent.

Expressed in terms of a digital time-history record with N data points, the mathematical definition of γ is

$$\gamma = \lim_{N \to \infty} \sum_{i=1}^{N} \frac{I(t_i)}{N}. \tag{11.22}$$

The intermittency function, I, which identifies the turbulent and non-turbulent regions has to be generated by a signal (or combination of signals) from, for example, a hot-wire probe. The method must be able to: (i) identify the flow as turbulent when the front of a turbulent region passes the probe, (ii) give a turbulent flow region identification without drop out until the back of the turbulent region reaches the probe, and (iii) classify the flow as non-turbulent until the next turbulent region reaches the probe. The discrimination between turbulent and non-turbulent fluid can be made on the presence or absence of vorticity, and a vorticity probe would therefore be the natural choice except for the requirement of a very complex probe capable of spatial differentiation (see Chapter 10). For this reason, simpler, mainly *ad hoc*, discrimination methods based on a hot-wire probe containing a single normal wire, two single normal wires, or an X-array have been used. As discussed by Antonia (1981), the signal algorithms have included linear operations (for example, differentiation, smoothing), nonlinear operations (for example, rectification) as well as combinations of

TABLE 11.1 *Turbulence-detector functions.*

Townsend (1949)	$|u|$, $|\partial u/\partial t|$
Corrsin and Kistler (1955)	$|u|$, $|\partial u/\partial t|$
Heskestad (1965)	u^2
Gartshore (1966)	$|\partial u/\partial t|$
Fiedler and Head (1966)	$|\partial u/\partial t|$
Kaplan and Laufer (1968)	$(\partial u/\partial t - \langle \partial u/\partial t \rangle)^2$
Wygnanski and Fiedler (1970)	$(\partial u/\partial t)^2 + (\partial^2 u/\partial t^2)^2$
Kovasznay *et al.* (1970)	$|\partial^2 u/\partial y \partial t|$
Antonia and Bradshaw (1971)	$(\partial u/\partial t)^2$
Antonia (1972)	$(\partial uv/\partial t)^2$
Thomas (1973)	$|\partial u/\partial t|$ filtered
Bradshaw and Murlis (1974)	$|\partial uv/\partial t|$ or $|\partial^2 uv/\partial t^2|$

From Hedley and Keffer (1974). Reprinted with the kind permission of Cambridge University Press from the *Journal of Fluid Mechanics*.

velocity components. A list of turbulence-detector functions used is given in Table 11.1.

The principle of a typical detector system is shown in Fig. 11.13. The signal from the hot wire (1) may be passed through a low-pass filter (2) to remove low-frequency fluctuations in the non-turbulent regions. Feeding this signal through a rectifier and noise chipper (3) creates a signal with very different amplitude characteristics in the turbulent and non-turbulent regions. However, it should be noted that the turbulent signal occasionally approaches zero, which would cause a 'drop-out' problem if a simple threshold level was applied to this signal. To overcome this problem, the signal may be fed through a smoothing filter (4) before a threshold

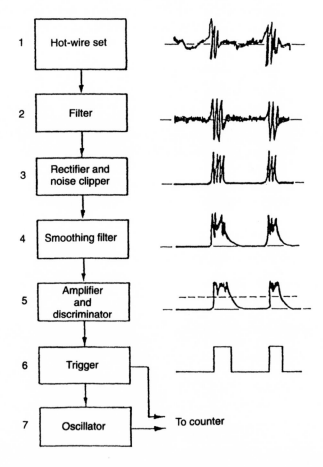

Fig. 11.13. A block diagram of an intermittency measuring system. (From Corrsin and Kistler 1955.)

discriminator (5) is applied to create the intermittency function (6). In practice, most detector systems based on velocity components contain two variable parameters: a threshold level, C, and a hold time, τ_H which represent the length of the time interval over which the threshold criterion is applied; as a consequence, the intermitting function will not change state for short events with durations of less than τ_H. The hold time should not be much larger than the characteristic time-scale associated with the fine-scale structure of the turbulence within the turbulent region. The optimization of C and τ_H has been considered in several studies (for example, Bradshaw and Murlis 1974; Kibens *et al.* 1974; Antonia and Atkinson 1974).

Major difficulties in the determination of the intermittency function, I, with detector systems based on velocity fluctuations have been reported. These problems are due to:

1. All the detector systems assume very different amplitude characteristics in the turbulent and non-turbulent regions. However there is no physical reason why the many *ad hoc* detector functions (listed in Table 11.1) should all indicate a correct change of state when an interface moves past a detector probe.

2. The difficulties in identifying a highly re-entrant interface. The smoke photographs of Fiedler and Head (1966), Falco (1977, 1980), and Head and Bandyopadhyay (1981) have revealed a highly contorted interface, rather than the simplistic configuration indicated in Fig. 11.10. (This is partly because the Schmidt number of smoke and dye is very high. The vorticity interface is much less re-entrant because of viscous diffusion (Bradshaw, personal communication).)

3. Mathematical and operational difficulties with the detection system. Within the turbulence region, the discriminator signal should remain positive, but it can occasionally have zero values. The discriminator system may respond to high wave number intermittency as well as to interfacial intermittency.

To minimize the problems related to 3, Bradshaw and Murlis (1974) and Murlis *et al.* (1982) employed the following two definitions for the velocity-field intermittency

1. *Retail intermittency*, derived from the unsmoothed criterion function without the application of any 'hold time' other than the digital-sampling time interval (which is small).

2. *Wholesale intermittency*, in which short turbulent regions far from any long turbulent regions are ignored altogether as being due to noise, eccentricities of the algorithm, or unimportant glancing encounters with turbulent regions; while short turbulent regions adjacent to a long turbulent region are regarded as merging with it, and short irrotational regions within

a turbulent zone are ignored, so that the scheme is 'turbulence-biased'. The principle of this intermittency-detection scheme, based on the shear stress $uv(t)$ signal, is shown in Fig. 11.14.

It has been observed in several investigations that the identification of turbulent and non-turbulent zones by temperature as a passive scalar can be operationally superior to the use of detection systems based on a velocity or a vorticity component. This method has been utilised in several different turbulent flow studies: in boundary layers (Chen and Blackwelder 1978; Subramanian *et al.* 1982), in the wake of a heated circular cylinder (LaRue

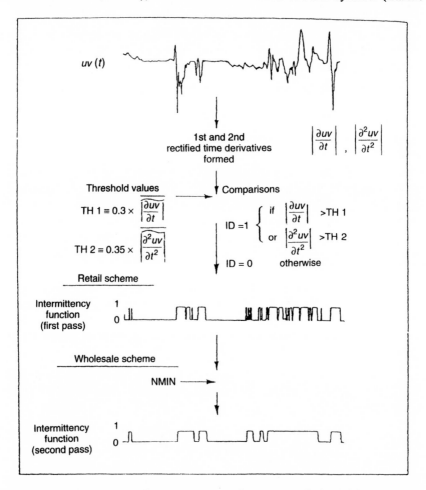

Fig. 11.14. The velocity-intermittency scheme. The first pass ('retail') intermittency function, ID, is set equal to 0 or 1. For the second pass ('wholesale'), intervals less than NMIN times sampling interval are set equal to 1 if they are irrotational and to 0 if they are turbulent. (From Murlis *et al.* 1982. Reproduced with kind permission of Cambridge University Press from the *Journal of Fluid Mechanics*.)

1974; LaRue and Libby 1974; Fabris 1979), in a circular heated jet (Antonia *et al.* 1975; Chevray and Tutu 1978; Komori and Ueda 1985), in a plane heated jet (Davies *et al.* 1975; Moum *et al.* 1979). Weir and Bradshaw (1977) and Weir *et al.* (1981) studied the merging of two plane shear layers, one of which was marked by slight heating. In these flow studies, the fluctuating fluid temperature was measured with a sensor operated as a resistance-wire (see Section 7.4.3 for the operational principles of resistance-wires). Antonia *et al.* (1975) were able to dispense with the hold time, τ_H, and the value of the intermittency factor, γ, showed only a slight dependence on the threshold level, C, except when γ approached unity. LaRue (1974), however, retained a non-zero hold time, but also found that the statistics of I were nearly independent of the values of the detection parameters. The constant-threshold method used in most studies will fail if the cold-zone temperature varies with time, as was the case in the investigation by Weir *et al.* (1981). A detection method was therefore developed which was based on inspection of the magnitude of $d\theta/dt$ rather than of the temperature fluctuation, θ, itself. Finally, hot-wire measurements have also been carried out in smoke-filled boundary layers, with the intermittency function being determined by flow visualization (Fiedler and Head 1966; Falco 1977, 1980; Head and Bandyopadhyay 1981).

Turbulent and non-turbulent zone averages   Having introduced the intermittency function, $I(t)$, in eqn 11.21, and the intermittency factor, γ, in eqn 11.22, it is now possible to define turbulent and non-turbulent zone averages. Consider a flow quantity, $Q(t)$, at a point r, for which a digital time-history record with N data points is available. The overall time mean value is

$$\bar{Q} = \sum_{i=1}^{N} \frac{Q(t_i)}{N} \tag{11.23}$$

provided N is sufficiently large, as discussed in Chapter 12. Following Kovasznay *et al.* (1970), the corresponding conditional averages in the turbulent and non-turbulent zones can be expressed, for the turbulent fluid as

$$\bar{Q}_t = \sum_{i=1}^{N} \frac{I(t_i)Q(t_i)}{N\gamma} = \frac{\sum_{i=1}^{N} I(t_i)Q(t_i)}{\sum_{i=1}^{N} I(t_i)} \tag{11.24}$$

and as

$$\bar{Q}_n = \sum_{i=1}^{N} \frac{[1 - I(t_i)]Q(t_i)}{N[1 - \gamma]} = \frac{\sum_{i=1}^{N} [1 - I(t_i)]Q(t_i)}{\sum_{i=1}^{N} [1 - I(t_i)]} \tag{11.25}$$

for the non-turbulent fluid.

It follows from eqns (11.23) to (11.25) that the conventional and conditional zone averages are related by

$$\bar{Q} = \gamma \bar{Q}_t + (1 - \gamma)\bar{Q}_n. \tag{11.26}$$

On the basis of the definitions above, conventional and zonal fluctuations can be introduced as:

$$q(t) = Q(t) - \bar{Q},$$
$$q_t(t) = Q(t) - \bar{Q}_t;$$
$$q_n(t) = Q(t) - \bar{Q}_n.$$

For large values of N, the corresponding time-mean-square values are

$$\overline{q^2} = \sum_{i=1}^{N} \frac{q^2(t_i)}{N}, \tag{11.27}$$

$$\overline{q_t^2} = \sum_{i=1}^{N} \frac{I(t_i)\, q_t^2(t_i)}{N\gamma}, \tag{11.28}$$

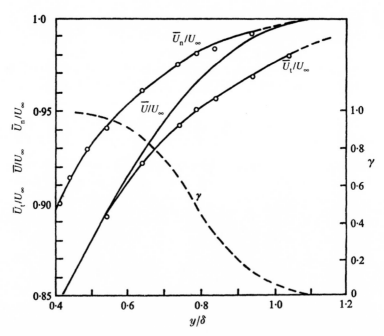

Fig. 11.15. Zone averages of the streamwise mean-velocity component. The intermittency factor, γ, is given for reference. (From Kovasznay *et al.* 1970. Reproduced with kind permission of Cambridge University Press from the *Journal of Fluid Mechanics.*)

$$\overline{q_n^2} = \sum_{i=1}^{N} \frac{[1 - I(t_i)]\, q_n^2(t_i)}{N[1 - \gamma]}. \tag{11.29}$$

The relationship between the three mean-square values defined by eqns (11.27) to (11.29) is given (Antonia (1972); Hedley and Keffer 1974) by

$$\overline{q^2} = \gamma \overline{q_t^2} + (1 - \gamma)\overline{q_n^2} + \gamma(1 - \gamma)(\bar{Q}_t - \bar{Q}_n)^2. \tag{11.30}$$

Figures 11.15 and 11.16 show zonal-averaged results obtained in the outer part of a turbulent boundary-layer flow. Figure 11.15, which contains results for the mean longitudinal velocity component, demonstrates a clear difference between the average mean velocity in the two regions, with the non-turbulent flow moving faster on average than the turbulent flow. The corresponding zonal averages for the Reynolds shear stress, $\overline{uv}$, in Fig. 11.16, demonstrate very different values in the two zones. For the shear stress, $\overline{uv}$, it can be shown (Chevray and Tutu (1978)) that

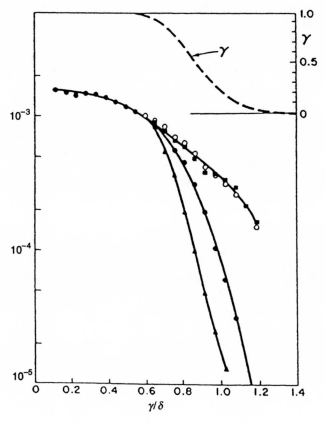

Fig. 11.16. The zone average of the Reynolds stress: ($\bullet$) $\overline{uv}/U_\infty^2$, ($\blacksquare$) $\overline{u_t v_t}/U_\infty^2$, ($\blacktriangle$) $\overline{u_n v_n}/U_\infty^2$, and ($\bigcirc$) $\overline{uv}/\gamma U_\infty^2$. (From Blackwelder and Kaplan 1972.)

$$\overline{uv} = \gamma\,\overline{u_t v_t} + (1-\gamma)\overline{u_n v_n} + \gamma(\gamma-1)(\bar{U}_t \bar{V}_t + \bar{U}_n \bar{V}_n - \bar{U}_t \bar{V}_n - \bar{U}_n \bar{V}_t). \tag{11.31}$$

If the shear stress $\overline{uv}$ is carried completely by the turbulent region and as the last term on the right-hand side, which contains mean transverse velocities, is usually small then

$$\overline{uv} = \gamma\,\overline{u_t v_t}. \tag{11.32}$$

The quantity $\overline{uv}/(\gamma U_\infty^2)$ is also shown in Fig. 11.16, and it compares quite well with $\overline{u_t v_t}/U_\infty^2$, which supports the hypothesis that the Reynolds stress, $\overline{uv}$, is carried almost entirely by the turbulent region.

Conditional point averages Conditional point averages have been used extensively in the study of coherent flow phenomena within turbulent shear layers. In general, these techniques require simultaneous measurements of an eduction signal, $Q_a(t)$, at a spatial position r_a and a detector signal $Q_b(t)$ at another spatial position, r_b. The signals Q_a and Q_b may represent the same physical quantity, for example, the longitudinal velocity component, or any combination of the fluctuating components of the velocity, the shear stress, the vorticity, the temperature, the wall shear stress, the pressure, etc. For self-eduction from one probe only, $r_a = r_b$ and $Q_a = Q_b$.

The basic assumption in all eduction techniques is that it is possible to link a coherent flow phenomenon to a specific signal feature in the detector signal, $Q_b(t)$. These events may occur randomly in time, and it is therefore necessary to introduce a second type of detector function, which can identify the temporal positions t_i (in the detector signal) when the flow phenomenon occur. A discussion of the detector function used for this purpose is presented in the near-wall region Section 11.2.1.2. Having identified the t_i values, the corresponding conditional (point) average of the eduction signal $Q_a(t)$, is defined as

$$\langle Q_a(\tau)\rangle_b = \frac{1}{N}\sum_{i=1}^{N} Q_a(t_i + \tau), \tag{11.33}$$

where N is the number of identified events in the detector signal. The subscript b denotes that the identified events are detected with a probe placed at r_b. (Note that in many applications the indices a and b are omitted.) The individual realizations of the eduction sensor signal, $Q_a(t_i + \tau)$ are assumed to contain two components $A(\tau)$ and $B(t_i + \tau)$, where $A(\tau)$ represents a part of the signal which is uniquely related to the specified event in the detector signal and $B(t_i + \tau)$ corresponds to random local flow events. For N separate detection events we have

$$Q_a(t_1 + \tau) = A(\tau) + B(t_1 + \tau),$$
$$Q_a(t_2 + \tau) = A(\tau) + B(t_2 + \tau),$$
$$\vdots$$

$$Q_a(t_N + \tau) = A(\tau) + B(t_N + \tau).$$

Because the local event $B(t_i + \tau)$ is assumed to be a random function, it follows that for large values of N

$$\frac{1}{N} \sum_{i=1}^{N} B(t_i + \tau) \to 0 \quad \text{for } N \to \infty,$$

and the conditionally averaged or educed signal therefore becomes

$$\langle Q_a(\tau) \rangle_b \cong A(\tau) \tag{11.34}$$

In practice 'phase jitter' may severely affect the averaging process, causing a smearing effect and a reduction in the amplitude of the educed signal. Variation in the arrival time at position r_a of events detected at r_b is a common problem. Related correction procedures have been used in several investigations leading to significant enhancement of both the amplitude and general signal features of the educed signals (Blackwelder and Kaplan 1976; Zilbermann et al. 1977; Johansson et al. 1987a, b).

For measurements in the outer part of a turbulent boundary layer, the simplest point averages are those determined by the instants that the fronts or the backs of turbulent bulges pass over the detector probe. Figure 11.17 show the average convection velocities of these fronts and backs, and the fronts are observed to move faster than the backs indicating significant deformation processes within the large-scale eddies. Also, as summarized by Cantwell (1981), it has been found that the upstream-facing (back) portion of the turbulent bulges is the most active, and this activity is associated with a saddle-point, in a convected frame of reference. This observation is of course consistent with the previously described developments of Falco eddies (Fig. 11.11). A wide variety of flows, not just turbulent boundary layers, seem to exhibit this property.

It has been concluded in many papers that coherent flow structures are easier to identify from their spatial coherence than from the temporal variation in a signal from a single sensor. Many turbulent shear flow investigations have therefore been carried out with a rake of sensors (hot-wire, hot-film, or resistance wires) usually placed along a line in space. The rake may typically span (most of) the boundary layer in order to identify related coherent motions at various vertical positions, or the rake can be placed parallel to the wall to measure the spanwise extent of the coherent flow structures. Other common conditional sampling approaches have used the Reynolds shear stress, $\overline{uv}$, as measured by an X-probe. The corresponding simultaneous $u(t)$, $v(t)$, and $uv(t)$ signals may be used in any combination for detection and eduction. Alternatively a second sensor may provide the detector signal.

Coherent flow structures in the outer part of the flow have been studied with both the rake and X-probe methods. The spatial extent of coherent

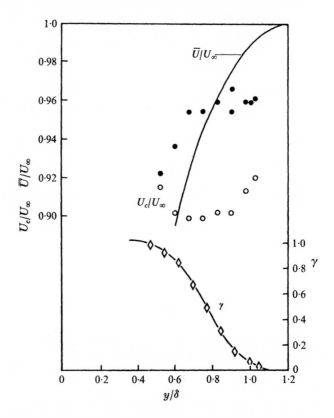

Fig. 11.17. Local convection velocities of the interface: (●) fronts, and (○) backs. The conventional mean-velocity profile and the intermittency factor are given for reference. (From Kovasznay *et al.* 1970. Reproduced with kind permission of Cambridge University Press from the *Journal of Fluid Mechanics*.)

flow events has been determined using temperature as a passive contaminant in conjunction with a rake of resistance-wires spanning the height of the boundary layer (Chen and Blackwelder 1978; Subramanian *et al.* 1982). Figure 11.18 shows a set of simultaneous temperature signals from a ten-wire rake. The arrows in the figure identify a 'cold' temperature front, specified by a rapid decrease in temperature, which corresponds to the sensors moving from a 'warm' to a 'cold' part of the fluid. The temperature front was detected by the variable-interval time-averaged (VITA) method, details of which is given in Section 11.2.1.2 (for the near-wall region). Broadly speaking, the VITA method identifies events which have a steep gradient over a time period which is comparable to the integral time-scale of the VITA method, and it will detect cold fronts when $d\theta/dt < 0$ and warm fronts when $d\theta/dt > 0$. In the study by Subramanian *et al.* (1982) about 80

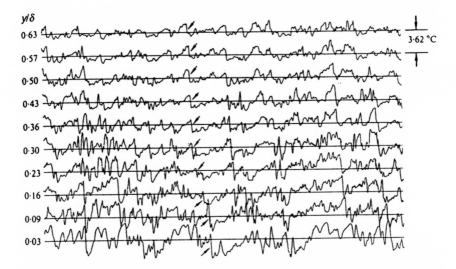

Fig. 11.18. Simultaneous temperature signals from a ten-wire rake in the turbulent regions. The horizontal time span is $18.7 U_\infty \Delta t / \delta$. One particular 'cold' temperature front is denoted by the arrows. (From Chen and Blackwelder 1978. Reproduced with kind permission of Cambridge University Press from the *Journal of Fluid Mechanics*.)

per cent of all detected fronts were cold fronts and about 20 per cent were warm fronts. As indicated by Fig. 11.18, the cold fronts can often be identified across the whole boundary layer, while the warm fronts are usually identifiable over a limited number of sensors. Temperature front results have also been linked by conditional sampling to the velocity field measured by an X-probe. (Subramanian *et al.* 1982). Corresponding temperature and velocity results obtained at a vertical position, $y/\delta = 0.32$, are shown in Fig. 11.19.

The results presented in Figs 11.18 and 11.19 are consistent with the following events at the back and front of the turbulent bulges (large-scale eddies) in the outer flow region, taking into account that the turbulent regions are warm due to wall heating:

1. The cold-front events corresponds to events at the back of a bulge, as shown in Fig. 11.20(a). Using Taylor's hypothesis to change $\langle \theta(t) \rangle$ into $\langle \theta(x) \rangle$, it follows that the sensor moves from warm turbulent fluid to cold entraining (irrotational) fluid. This is consistent with the value of $\langle uv \rangle$ being highest in the turbulent region. The longitudinal velocity component $\langle u \rangle$, is lower in the turbulent region than in the entrainment region, which is consistent with the zonal averages in Fig. 11.15. The vertical motion of the flow is outwards in the turbulent region and inwards in the entrainment flow close to the back of the bulge. Figure 11.18 shows that coherent

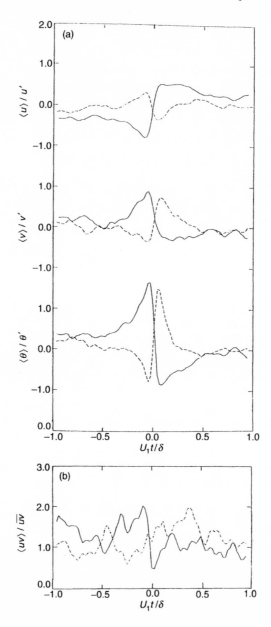

Fig. 11.19. Conditional averages obtained in a turbulent boundary layer, at $y/\delta = 0.32$ by the VITA method with the detection criterion based on θ and different signs of $\dot{\theta}$: ——, $\dot{\theta} < 0$ ($N = 100$); (---), $\dot{\theta} > 0$ ($N = 97$). (a) Individual signals, and (b) $\langle uv \rangle$. (From Subramaniam *et al.* 1982. Reproduced with kind permission of Cambridge University Press from the *Journal of Fluid Mechanics*.)

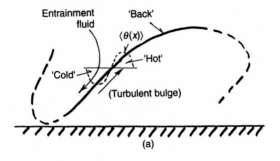

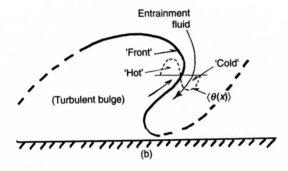

Fig. 11.20. An interpretation of flow events at the back and front of turbulent bulges using conditional-sampled data from Figs 11.18 and 11.19. Note that $\langle \theta(t) \rangle$ has been converted into $\langle u(\theta) \rangle$ using the Taylor's hypothesis. (a) Back, and (b) Front.

regions of the entrainment flow often extend across most of the boundary layer. A significant shift in the temporal location of the temperature front (the back of the bulge) is observed at $y/\delta \simeq 0.2$, which is consistent with the possible presence of a Falco eddy at this position of the back interface.

2. The warm fronts can similarly be shown to correspond to events at the front of a bulge (Figure 11.20(b)). It follows from the $\langle \theta(x) \rangle$ signal that for a warm front, the sensor moves from a cold (irrotational) region into a warm (turbulent) region. This is consistent with the value of $\langle uv \rangle$ being highest in the turbulent region. The value of $\langle u \rangle$ is higher in the cold (irrotational) flow region than in the warm (turbulent) region, and the $\langle v \rangle$ result demonstrates that the irrotational flow is directed inwards and the warm (turbulent) flow outwards. Both methods therefore, respectively, detect entrainment of 'cold' irrotational fluid behind the back (cold front) and ahead (warm front) of selected turbulent bulges. Of these events, the flow near the back is by far the most coherent. These observations are basically consistent with the definitions and results obtained with the inter-

mittency technique, the main difference is that the intermittency method identifies all backs and fronts, while the VITA method selects particularly well defined events.

11.2.1.2   *The near-wall region*

Considerable effort has been spent on the identification of flow phenomena in the near-wall region and on the structural relationship between the inner and outer flow. As summarized by Cantwell (1981) the results obtained by Klebanoff (1954) show that a sharp peak in the rate of production of turbulent energy (production $= \overline{uv}\partial\bar{U}/\partial y$) occurs at the outer edge of the viscous sublayer. Measurements in pipe flow by Laufer (1954) show a similar effect. Integration over the thickness of the boundary layer leads to the result that the first 5 per cent of the boundary layer contributes to over half of the total production of turbulent energy. This important result was the primary motivation for the early work by Kline and co-workers at Stanford University, and it has remained a primary motivation for much of the subsequent work on boundary layer structure.

Beginning in the late 1950s, a series of experiments was carried out at Stanford University using flow visualization to study the turbulent boundary layer (Rundstadler *et al.* 1963; Schraub and Kline 1965; Kline *et al.* 1967; Kim *et al.* 1971; Offen and Kline 1974). The flow in a low-Reynolds-number turbulent boundary layer was visualized by a hydrogen-bubble wire placed parallel to, and at various distances above, the wall. In the region $0 \leqslant y^+ \leqslant 10$ these workers observed a semiregular distributed spanwise structure composed of alternating arrays of high- and low-speed regions described as 'streaks'. They noticed that occasionally one of the streaks of low streamwise momentum would lift away from the wall and interact with the outer flow. This process was named 'bursting' because the interaction was quite sudden and it occurred randomly in space and time. A considerable amount of turbulence production occurred during this sequence. The bursting was described as being composed of three stages: the lifting of a low-speed streak, followed by an oscillatory growth which was terminated after a few cycles by the onset of a more chaotic motion called breakup. Flow-visualization studies have also been reported by Corino and Brodkey (1969) in the wall region of a fully developed turbulent pipe flow and in the outer region of a turbulent boundary layer on a flat plate by Nychas *et al.* (1973).

The proposed link mechanism between the flow in the outer and near-wall regions will be discussed using (some of) the sequence of events presented by Corino and Brodkey (1969) as being representative of the 'burst' phenomena; these events appear to be similar for a turbulent boundary layer and for a fully developed turbulent pipe or channel flow. In Fig. 11.21(a) a fast moving large-scale disturbance (*LSD*) is coming from

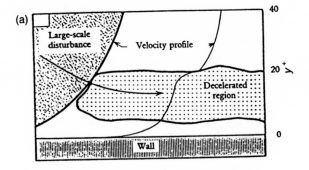

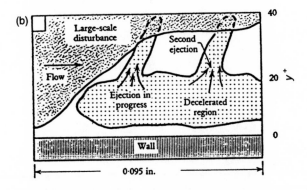

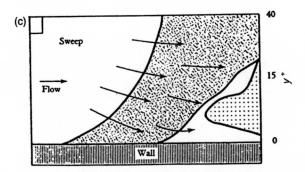

Fig. 11.21. The sequence of events in a burst: (a) a fast moving large-scale disturbance (*LSD*) interacts with fluid in a decelerated region near the wall, (b) the *LSD* is moving over the decelerated region causing one òr more ejections. (c) the burst event is terminated by a 'cleansing sweep' by the *LSD*. (Based on Corino and Brodkey 1969. Reproduced with kind permission of Cambridge University Press from the *Journal of Fluid Mechanics*.)

the upstream direction and it interacts with the fluid in the decelerated region near the wall. As discussed in Section 11.2.1.1, it is proposed that this large-scale coherent fluid is a Falco eddy, because it has been shown that Falco eddies move in towards the wall on the back of the large-scale eddies. The fast moving LSD flows over the decelerated region and causes one or more ejections from this region, as shown in Fig. 11.2l(b). The complete sequence of ejection events is normally referred to as a 'burst'. The burst process is terminated by the LSD sweeping over the remainder of the decelerated fluid and it sometimes removes all of the retarded flow, a process often referred to as a cleansing sweep. (Fig. 11.21c). The ejections into the LSD will reduce its momentum, but because an LSD is large compared to the dimensions of the burst events it will continue to move downstream and cause further bursts until it loses its excess momentum. The proposed flow model can, in principle, satisfy all length and time-scale criteria, and it presents a mechanism for linking the large-scale eddies in the outer flow to the burst phenomena in the near-wall region.

One aspect of the near-wall flow which has received considerable attention is the scaling law for the average time period, T_B, between burst events. This information can provide valuable information about the dynamic relationship between the inner region of intense turbulence production and the large-scale eddies in the outer region. Prior to about 1981, the general consensus was that T_B was more likely to scale on the outer time-scale $T_{out} = \delta/U_\infty$ where U_∞ is the free-stream velocity and δ is the boundary layer thickness (Rao *et al.* 1971; Laufer and Badri Narayanan 1971; Willmarth 1975; Cantwell 1981). Following a spatial-resolution study by Blackwelder and Haritonidis (1983) it has been suggested in several investigations (for example Chambers *et al.* 1983; Willmarth and Sharma 1984; Luchik and Tiederman 1987) that inner scaling is more appropriate, with an inner time-scale of $T_{in} = \nu/U_\tau^2$ (ν is the kinematic viscosity and U_τ is the friction velocity). However other investigators (Alfredsson and Johansson 1984a; and Shah and Antonia 1989) have proposed the use of a mixed scaling law involving both T_{out} and T_{in}. One major problem in identifying scaling laws by conditional-sampling techniques is that, in general, they detect ejections and not complete burst events, which can contain multiple ejections. This problem has been studied by Bogard and Tiederman (1986) and by Luchik and Tiederman (1987), who used a probability-distribution method to identify ejections belonging to the same burst. This method has also been used by Shah and Antonia (1989). The proposed flow model suggests that the trigger mechanism for the bursts is the appearance of Falco eddies in the near-wall region. For the length, l_f, and velocity, U_f scales of a typical Falco eddy

$$\nu/U_\tau < l_f < \delta \qquad U_\tau < U_f < U_\infty,$$

which suggests that neither inner nor outer scaling are appropriate, and that the best scaling law is likely to involve a mixture of inner and outer parameters.

Conditional sampling techniques A large number of mainly *ad-hoc* methods have been used to study coherent flow events in the near-wall region. Reviews and comparative studies of conditional sampling techniques have been given by Antonia (1981), Subramanian *et al.* (1982), Alfredsson and Johansson (1984*a*, *b*), Bogard and Tiederman (1986), Luchik and Tiederman (1987), and Morrison *et al.* (1989, 1992). The principles of the two most common methods, the *uv*-quadrant and the VITA (variable-interval time-averaging) techniques are outlined below, and typical results are also presented.

The uv-quadrant method The quadrant-analysis method was used by Willmarth and Lu (1972), Wallace *et al.* (1972), Lu and Willmarth (1973), Brodkey *et al.* (1974) and Sabot and Comte-Bellot (1976) to sort out contributions to the Reynolds shear stress, $\overline{uv}$, (measured with an X-probe) from the four quadrants of the (u,v)-plane, as identified in Table 11.2. Figure 11.22 shows the corresponding typical results for a fully developed turbulent oil channel flow. The results demonstrate that ejections ($u < 0$, $v > 0$) which have been identified as being part of the burst event, and sweeps ($u > 0$, $v < 0$) make similar large contributions to the Reynolds shear stress, $\overline{uv}$. This method has been modified by Lu and Willmarth (1973) to identify 'violent' events. In the quadrant-hole technique the (u,v) plane is divided into five regions as shown in Fig. 11.23. The cross-hatched region is called the 'hole' and it is bounded by curves for which $|uv| =$ constant. The four quadrants, excluding the hole, are the other four regions. The contributions to $\overline{uv}$ from the five regions can be specified by introducing a parameter H and setting $|uv| = Hu'v'$ where u' and v' are the local root-mean square values of the u- and v-signals. The parameter H is called the hole size or the threshold value. The quadrant-hole method identifies large-amplitude contributions to $\overline{uv}$ from each of the four quadrants. Corresponding results from a fully developed turbulent water-channel flow

TABLE 11.2 *The uv-quadrant notation*

Sign of u	Sign of v	Sign of uv	Type of motion
−	+	−	Ejection
+	−	−	Sweep
−	−	+	Interaction (wallward)
+	+	+	Interaction (outward)

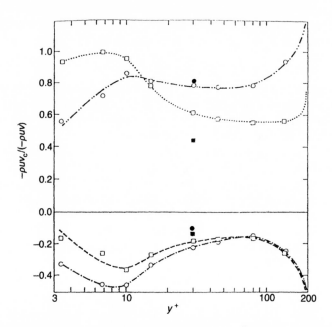

Fig. 11.22. The classified Reynolds stresses (see Table 11.2) normalized with the local average Reynolds stress: (●, ■) the results of Willmarth and Lu (1972). (· · ·), sweep, (——·——) ejection, (——·——), outward interaction, and (– – –) wallward interaction. (From Brodkey *et al.* 1974. Reproduced with kind permission of Cambridge University Press from the *Journal of Fluid Mechanics*.)

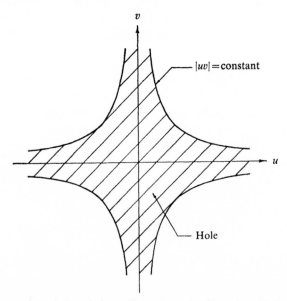

Fig. 11.23. The quadrant-hole technique. (From Lu and Willmarth 1973. Reproduced with kind permission of Cambridge University Press from the *Journal of Fluid Mechanics*.)

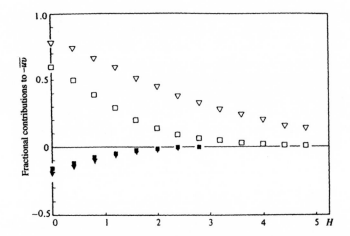

Fig. 11.24. The contributions to $\overline{uv}$ from the different quadrants (q1 to q4) at $y^+ = 50$, as function of threshold level: ($\blacksquare$) q1, ($\triangledown$) q2, ($\blacktriangledown$) q3, ($\square$) q4). (From Alfredsson and Johansson 1984*b*. Reproduced with kind permission of Cambridge University Press from the *Journal of Fluid Mechanics*.)

are shown in Fig. 11.24. When $H \geqslant 2$, the quadrant 1 and 3 contributions are so small that the corresponding conditional $uv(t)$ signal only consists of ejection and sweep events. For $H \geqslant 4$ only ejection events are detected.

The main disadvantage of the basic (uv)-quadrant technique, as observed by Blackwelder and Kaplan (1976), is the elimination of phase information, which prevents the identification of coherent flow structures. However, the quadrant-hole technique can be extended to the study of either ejection (quadrant 2) or sweep (quadrant 4) events. In this method, all ejection (or sweep) events with amplitudes greater than the threshold parameter, H, are identified. For self-eduction from an X-probe, the reference times, t_i, for the ejection events are usually selected as those corresponding to the maximum values of each peak in the truncated $uv(t)$ signal. Knowing the value of t_i for N identified events, eqn (11.33) can be used to evaluate the corresponding conditional averages: $\langle u(\tau) \rangle$, $\langle v(\tau) \rangle$, and $\langle uv(\tau) \rangle$. Alternatively, the reference times, t_i can be identified by a separate detector probe. This principle was applied by Willmarth and Lu (1972) and by Lu and Willmarth (1973) by placing as a detector probe an *SN* hot-wire probe at the edge of the viscous sublayer. In the study by Nakagawa and Nezu (1981), the detector and eduction probes were respectively a *V*- and an X-probe, and the detection method was based on applying the quadrant-hole technique to the $uv(t)$ signal from the detector probe.

The VITA method The most widely used detector method is the variable-interval time-averaging (VITA) technique. Blackwelder and Kaplan (1976) were the first to apply the VITA technique, in conjunction with a rake of hot-wire probes, to detect bursts in the wall region ($y^+ < 100$). This method is similar to the use of a detection function with a hold time to identify the turbulent/non-turbulent interface in the outer part of a shear layer, as described in Section 11.2.1.1.

The method will be explained with reference to the fluctuating component of a signal $q_b(t)$ from a detector probe placed at a spatial position r_b. Mathematically, the variable-interval time-average is defined as

$$\hat{q}_b(t, T) = \frac{1}{T} \int_{t - \frac{1}{2}T}^{t + \frac{1}{2}T} q_b(s)\,ds, \tag{11.35}$$

where T is the averaging time. Essentially, this is a moving average, which gives the local mean value of the quantity q_b. The short-time variance (or VITA variance) of the fluctuating detector signal, $q_b(t)$, is defined as

$$\mathrm{var}(q_b) = \frac{1}{T} \int_{t - \frac{1}{2}T}^{t + \frac{1}{2}T} q_b^2(s)\,ds - \left(\frac{1}{T} \int_{t - \frac{1}{2}T}^{t + \frac{1}{2}T} q_b(s)\,ds \right)^2.$$

In the detection criterion used by Blackwelder and Kaplan (1976), an event was deemed to occur when the short-time variance exceeded $k(q_b')^2$, where k is the chosen threshold and q_b' is the root-mean-square value of $q_b(t)$. The corresponding detection function, $D(t)$, was defined as

$$D(t) = \begin{cases} 1 & \text{if } \mathrm{var}(q_b) > k(q_b')^2 \\ 0 & \text{otherwise.} \end{cases} \tag{11.37}$$

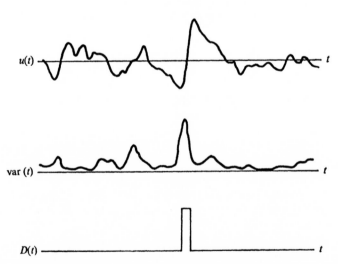

Fig. 11.25. A schematic diagram of the VITA detection process. (From Blackwelder and Kaplan 1976. Reproduced with kind permission of Cambridge University Press from the *Journal of Fluid Mechanics*.)

A schematic diagram of their detection process applied to the longitudinal velocity component, $u(t)$, at $y^+ = 15$ is shown in Fig. 11.25. The time positions, t_i, of the detected events is usually taken as corresponding to the time positions of the maximum values of the short-time variance, but other definitions such as the time of first detection have been used (see, for example, Subramanian $et\,al.$ 1982; and Chambers $et\,al.$ 1983).

The application of the VITA technique to flow-structure detection requires simultaneous measurements of a detector signal and one or more eduction signals. The probes used for both purposes usually have SN or X-wire configurations. To determine the spatial extent of the coherent flow structures, the probes may be fixed on a rake arrangement or on separate traversing mechanisms. The conditionally averaged or educed signal $\langle Q_a(\tau) \rangle_b$ corresponding to the signal $Q_a(t)$ from an eduction wire placed at a spatial position r_a, is obtained from eqn 11.33, as explained in Section 11.2.1.1.

The VITA technique has two adjustable parameters, the integrating time, T, and the threshold, k; and the amplitude and the shape of the educed signal and the frequency of occurrence of the detected events depends on the values selected for T and k. The effect of the value of the integration time, T, can be estimated by assuming that the detected event has a sinusoidal shape with an angular frequency, ω. Johansson and Alfredsson (1982) have studied the effect of the value of T on the short-time variance of a sinusoidal signal. Their results, shown in Fig. 11.26, demonstrate a type

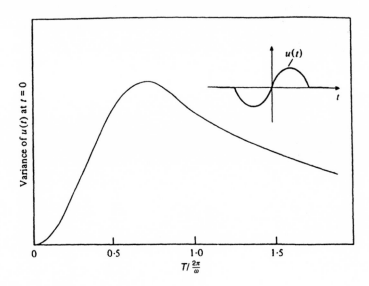

Fig. 11.26. The short-time variance (at $t = 0$) of $u(t) = \sin \omega t$ $(-\pi/\omega < t < \pi/\omega)$, 0 otherwise. (From Johansson and Alfredsson 1982. Reproduced with kind permission of Cambridge University Press from the *Journal of Fluid Mechanics*.)

of bandpass filter character, with the maximum of the short-time variance located at an integration time of 72% of the period. It is observed that the short-time variance tends to zero when $T \to 0$, and the VITA method therefore does not respond to spikes of a duration which is much shorter than the integration time, T.

The VITA method was originally applied to the longitudinal velocity component, $u(t)$, and it was assumed that the burst events were related to strong accelerations, as indicated by Fig. 11.25. However, the basic VITA method does not distinguish between signals with strong positive or negative slopes, and it has been shown (for example, Johansson and Alfredsson 1982; Luchik and Tiederman 1987) that the two types of events are both significant and that they represent different flow aspects. It is therefore necessary to add a gradient criteria

$$D(t) = \begin{cases} 1 & \text{if } \text{var}(q_b) > k(q_b')^2, \, dq_b/dt > 0 \text{ or } dq_b/dt < 0 \\ 0 & \text{otherwise.} \end{cases} \qquad (11.38)$$

in order to identify these specific flow events. An example of the number of detected events with positive or negative slopes is shown in Fig. 11.27

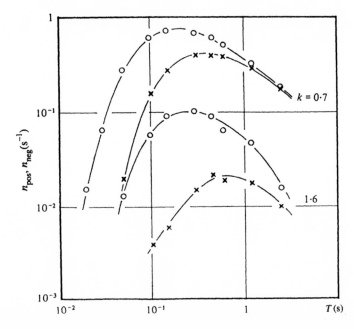

Fig. 11.27. The number of events with (○) a positive slope and with (x) a negative slope detected per unit time as function of the integration time; Re = 13 800, y^+ = 12.9. (From Johansson and Alfredsson 1982. Reproduced with kind permission of Cambridge University Press from the *Journal of Fluid Mechanics*.)

as function of the integration time, T, and of the threshold level, k, demonstrating the importance of both events.

The VITA technique has been applied to the following fluctuating quantities: the velocity components, u (Blackwelder and Kaplan 1976; Blackwelder and Eckelmann 1979; Johansson and Alfrèdsson 1982; Blackwelder and Haritonidis 1983; Willmarth and Sharma 1984; Alfredsson and Johansson 1984a,b; Bogard and Tiederman 1986; Luchik and Tiederman 1987; and Shah and Antonia 1989); v (Alfredsson and Johansson 1984b), uv (Alfredsson and Johansson 1984b; Morrison et al. 1989, 1992); temperature, θ (Chen and Blackwelder 1978; Subramanian et al. 1982; Morrison et al. 1989); wall shear stress, τ_w (Chambers et al. 1983); and pressure, p (Johansson et al. 1987b).

Much of the near-wall turbulence research has been aimed at identifying flow processes which cause the production of turbulent energy. For this work, the $uv(t)$ signal is the obvious choice for conditional-sampling techniques, but X-probes are difficult to use in the near-wall region. Most detector probes, placed at the position of maximum turbulence production ($y^+ \approx 15$), have therefore been SN probes, which provide a $u(t)$ signal for analysis. The early VITA studies assumed, based on separate flow-visualization studies, that burst phenomena were related to strong accelerations in the $u(t)$ signal. However, it was the simultaneous flow-visualization and conditional-sampling studies by Bogard and Tiederman (1986) and by Luchik and Tiederman (1987), and more recently by Tubergen and Tiederman (1993) that related the VITA and slope method to specific parts of the burst event. (The main technique used by Tiederman and co-workers

TABLE 11.3   *Ejection categories*

1 The ejection is in the middle stage of development, it is clearly distinguishable, and it is still strongly lifting (moving away from the near wall) as it passes through the probe.
2 The development is in the later stages. It is clearly distinguishable and still lifting (but not strongly) at the probe.
3 The development is in the early stage. The ejections originate very close to the probe but they are not clearly distinguishable because of the short distance they can be viewed before they reach the probe.
4 The ejection development has finished upstream of the probe, with no apparent lifting when the probe is reached.
5 The head of the ejection passes over the top of the probe with only the tail contacting the probe. The lifting or nonlifting of this tail is ambiguous.
6 The head of the ejection passes over the top of the probe, and the tail of the ejection clearly does not lift as it contacts the probe.

From Bogard and Tiederman (1986). Reprinted with the kind permission of Cambridge University Press from the *Journal of Fluid Mechanics*.

was actually a $-u$ level and a slope technique, but comparative studies have shown that for practical purposes both the VITA and the $-u$ level techniques with slope criteria, will identify similar events). They also concluded that the detection wire observed ejections in different stages of their developments and they classified the ejection events which intersected the wire into the six groups shown in Table 11.3. Their results demonstrated that the VITA or $-u$ level techniques detect the leading edge of an ejection when $\partial u/\partial t < 0$ (at the midpoint of the detected event) and detect the trailing edge of an ejection when $\partial u/\partial t > 0$. A burst may have several ejections, and results for the conditionally averaged results at $y^+ = 30$ of $\langle u \rangle$, $\langle v \rangle$, and $\langle uv \rangle$ for the leading edge of the first ejection and for the trailing edge of the last ejection in a burst are shown in Fig. 11.28. Luchik and Tiederman (1987) concluded that the magnitude of the positive $\langle u \rangle$ and negative $\langle v \rangle$ levels following the burst are larger than those leading the burst. The $\langle uv \rangle$ signal following the burst is consistently negative and it is in fact a significant quadrant 4 uv-sweep ($u > 0$, $v < 0$), whereas the sweep leading the burst gives a smaller contribution to $\overline{uv}$. It is important to notice that, if the conditional sampling corresponding to $\partial u/\partial t > 0$ or $\partial u/\partial t < 0$ had been carried out over all ejections, the events at the beginning and end of the burst would not have been apparent.

Conditionally sampled VITA results for an X-probe placed at $y^+ = 50$ are shown in Fig. 11.29. Using the event identification developed by Tiederman and co-workers based on simultaneous flow visualization and conditional sampling, it can be concluded that the results in Fig. 11.29(c) correspond to the leading edge of an ejection, while Fig. 11.29(b) corresponds to the trailing edge of an ejection. It can also be observed that the VITA method without any slope criteria, Fig. 11.29(a), does not provide any useful information. It is observed that the VITA technique identifies a regular large negative peak for $\langle uv \rangle$ for both events, with the $\langle uv \rangle$ peak occurring after the leading edge in Fig. 11.29(c), and the $\langle uv \rangle$ peak occurring before the trailing edge in Fig. 11.29(b); that is both detection methods identify turbulence production within an ejection. However, the leading- and trailing-edge events identified by the VITA technique occupy only a small fraction of the ejection process. Consequently, a basic VITA technique based on the $u(t)$ signal cannot identify the major contributions to the turbulence production. To study this flow aspect, Morrison *et al.* (1989, 1992) used the $uv(t)$ signal and introduced a VITA + LEVEL method, with a 'level' criterion added to avoid drop-out problems during the period of an identified ejection process.

Wall-sensor signals Investigations have been carried out to establish whether the coherent structures in the near-wall region give rise to identifiable signals from either a microphone (pressure) or a shear-stress sensor mounted flush with the wall. Johansson *et al.* (1987*b*) demonstrated, using the VITA technique, that there is a strong relationship between large

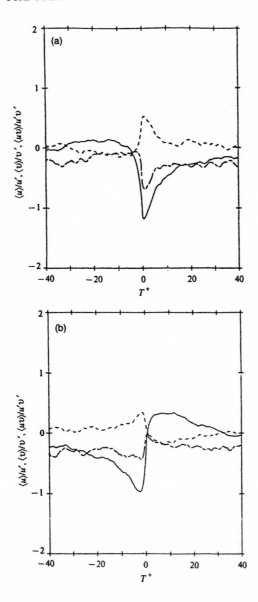

Fig. 11.28. Conditionally averaged velocity signals in a fully developed channel flow for $Re_h = 17\,800$: (a) centred on the leading edge of the first ejection in a burst (b) centred on the trailing edge of the last ejection in a burst. (——) $\langle u \rangle$, (– – –) $\langle v \rangle$, (— · —) $\langle uv \rangle$. (From Luchik and Tiederman 1987. Reproduced with kind permission of Cambridge University Press from the *Journal of Fluid Mechanics*.)

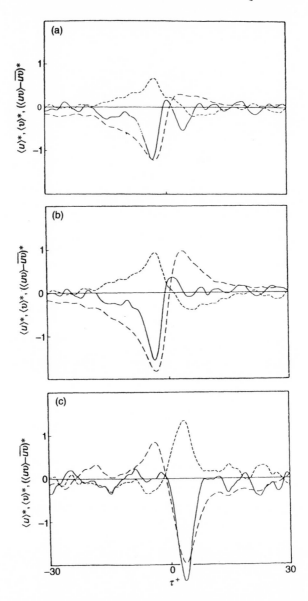

Fig. 11.29. Conditional averages at $y^+ = 50$ of (— —) u, (---) v, and (—) uv for events detected in the u-signal the VITA method: (a) All events (b) events with a positive slope, and (c) events with a negative slope. (From Alfredsson and Johansson 1984b. Reproduced with kind permission of Cambridge University Press from the *Journal of Fluid Mechanics*.)

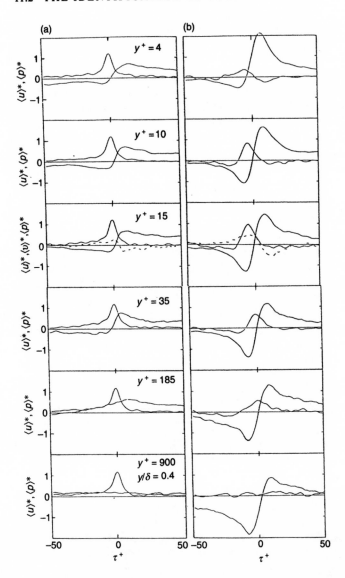

Fig. 11.30. Conditionally averages of the wall pressure, p_w, and u at various y^+ positions, and (- - -) v at $y^+ = 15$, for (a) detection of positive pressure peaks ($k = 2.5$), and (b) VITA detection of accelerating events ($k = 1.0$, $T^+ = 24$). $\langle u \rangle^* = \langle u \rangle / u_{rms}$, and $\langle p \rangle^* = \langle p \rangle / k p_{rms}$ in (a) and $\langle p \rangle / p_{rms}$ in (b). Note that τ^+ denotes the time in wall units relative to the detection time. $x = 4$ m. (From Johansson *et al.* 1987b. Reproduced with kind permission of Cambridge University Press from the *Journal of Fluid Mechanics*.)

positive pressure peaks and strong positive gradients in the u-velocity signals, particularly for flow phenomena occurring in the buffer region $10 \leqslant y^+ \leqslant 35$ (Fig. 11.30). In interpreting the results in this figure, it is necessary to take into account the spatial separation (2 mm) between the microphone and the hot-wire probe, and the pressure peak should therefore be shifted about $5t^*$, resulting in the identified p- and u-events occurring at the same time. Using the interpretation of the u-velocity signal of Luchik and Tiederman (1987), it follows that the results shown in Fig. 11.30 are predominantly related to the termination of a burst, and the subsequent initiation of a cleansing sweep. This change in flow phenomena has, in many investigations, been interpreted as a shear-layer effect.

The practicality of using the wall shear stress, τ_w, as a detector signal originated with Eckelmann (1974), who showed that τ_w correlated with velocity fluctuations for a hot-wire probe placed near the wall ($y^+ \leqslant 25$). The shear stress signal has, in particular, been used to identify sweep events ($u > 0$, $v < 0$), but this definition is too general for event detection. Several investigations have concentrated on sweeps with fronts of strongly accelerated fluids. These fronts are associated with a large, positive, temporal velocity gradient, $\partial U/\partial t$, and with a large, positive, spatial velocity gradient, $\partial U/\partial y$. From the discussion of the wall pressure effects above, it follows that these events will be cleansing sweeps at the end of bursts.

The wall shear stress can be expressed as

$$\tau_w = \mu \left. \frac{\partial U}{\partial y} \right|_w \tag{11.39}$$

where $(\partial U/\partial y)|_w$ is the gradient at the wall of the $U(y)$-velocity profile. Since $\partial U/\partial y$ is positive during a sweep event, it follows from eqn (11.39) that $\partial \tau_w/\partial t$ is also positive, and this latter slope criterion was used by Chambers et al. (1983) for their VITA detection of sweep events. Black-welder and Eckelmann (1979) applied the VITA technique to an SN hot-wire probe placed at $y^+ = 15$ to study the relationships between events in the u-velocity and wall-shear-stress signals.

Randolph et al. (1987, 1989) have investigated the detection of cleansing sweeps by simultaneous measurements of the temporal variation in $\partial U/\partial t$ and $\partial U/\partial y$. As described above, a cleansing sweep should correspond to simultaneous large positive values of both these gradients. Measurements were carried out in the Göttingen oil channel which has been described by Eckelmann (1974), and the gradient measurements were performed with a TSI-1244 20w two-sensor gradient probe. The spacing, h, between the two sensors was 3 mm, corresponding to $\Delta h^+ = hU_\tau/\nu = 5$. The gradient probe was introduced into the flow in such a way that the two sensors were parallel to the wall and perpendicular to the mean flow direction. The two sensors measured the instantaneous velocities $U_1(y - h/2)$ and $U_2(y + h/2)$ and the spatial velocity gradient was evaluated as

$$\frac{\partial U}{\partial y} \simeq \frac{U_2 - U_1}{h}. \tag{11.40}$$

Randolph *et al.* (1987, 1989) have discussed calibration and interpretation difficulties related to the use of eqn (11.40). One operational difficulty with this method is that the separation distance $h^+ = 5$ will produce significant differences in the time histories of $\partial U_1/\partial t$ and $\partial U_2/\partial t$ in the near-wall region. Nevertheless, the results in Fig. 11.31 (obtained at $y^+ = 15$ and 27.8) show a strong correlation between the two temporal gradients and

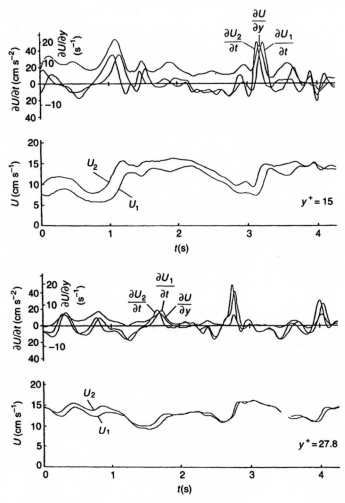

Fig. 11.31. The time histories of velocities and the space and time derivatives at wall distances of $y^+ = 15$ and $y^+ = 27.8$. (From Randolph *et al.* 1987. Reproduced with kind permission of Springer-Verlag. © Springer-Verlag.)

$\partial U/\partial y$. A correlation method based on the simultaneous occurrence of large values for these gradients was used as a criterion for recognizing the sweep events.

11.2.2   Free shear layers

The concept of turbulent shear flow has also been greatly influenced by the flow visualization of a two-stream mixing layer by Brown and Roshko (1974), which demonstrated that regular spanwise vortex structures persisted down the length of their test facility. These motions originated in the transitional part of the layer, and they appeared to be a permanent feature of the flow even at high Reynolds numbers. A detailed study of the vortex pairing process, by which the turbulent mixing layer grows was carried out by Winant and Browand (1974). These coherent vortex features were shown by Dimotakis and Brown (1976) to exist for values of $Re_x (= xU_1/\nu)$ of at least 3×10^6. The persistence of these regular vortex structures, however, remains a subject of some controversy (Chandrsuda *et al.* 1978; Bell and Mehta 1992). Reviews of coherent structures in the free shear layers have been given by Hussain (1983, 1986), Wygnanski and Petersen (1987), and Liu (1989).

Perturbed free shear layers A primary objective of low-level forcing has been the overcoming of the problem of 'jitter' in conditional sampling and the introduction of a phase reference for the coherent structures. This approach is supported by flow-visualization experiments which have demonstrated the existence of phase-locked regular large-scale coherent structures under specific exitation conditions. These observations have led to the concept of 'preferred modes'; that is undisturbed shear layers contain a latent order, which can be enhanced and determined by a 'preferred mode' forcing. Only later (as summarized by, for example, Ho and Huerre 1984) was it realized that the evolution of mixing layers is highly susceptible to even a very low-amplitude disturbance. High-quality data obtainable from forced free shear layers can, nevertheless, greatly enhance the basic understanding of shear-layer dynamics.

Investigations of many types of perturbed turbulent flows have been reported, in particular for free round jets (Crow and Champagne 1971; Chan 1974; Bechert and Pfizenmaier 1975; Moore 1977; Hussain and Zaman 1980; 1981; Zaman and Hussain 1980; Bradley and Ng 1989) and plane turbulent mixing layers (Oster and Wygnanski 1982; Ho and Huang 1982; Fiedler and Mensing 1985; Disimile 1986; Weisbrot and Wygnanski 1988; Wygnanski and Weisbrot 1988; Latigo 1989).

For a regular phase-locked large-scale flow structure, it is in principle possible to determine the spatial and temporal development by simultaneously recording the forcing signal and the signal from one or more hot-wire probe(s) placed sequentially at measuring points throughout the flow

region of interest. In the study by Wygnanski and Weisbrot (1988), measurements were taken with a rake of seven X-wire probes which were
traversed across the flow. The velocities measured at each point were
sampled simultaneously with the forcing signal, which were used to obtain
a phase reference for averaging the data. The phase-locked data were used
to compute the coherent velocity fluctuations, as well as the coherent spanwise component of vorticity, without the need of Taylor's hypothesis. The
high spatial and temporal resolutions of the data were used to compare calculated particle paths and streaklines with corresponding flow-visualization
pictures. The results of such a comparison are shown in Fig. 11.32, demonstrating a very good agreement between the two methods.

Natural free shear layers The large-scale flow structures in natural, turbulent, shear flows occur somewhat irregularly, and significant variations
have been observed in their temporal and spatial developments. To identify
these large-scale structures, conditional signal-analysis techniques have
been used to link the structures to the occurrence of a specified type of event
in a detector signal. One common method, which has been applied both
to a circular jet (Lau and Fisher 1975; Bruun 1977, 1979b; Yule 1978) and
to a two-dimensional free shear layer (Browand and Weidman 1976; and

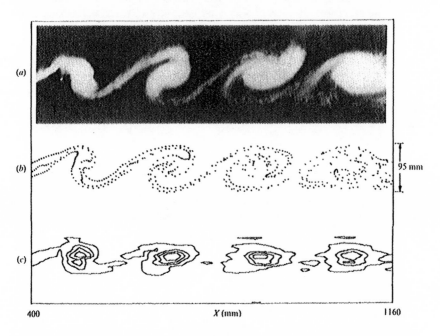

Fig. 11.32. Phase-locked structures in a perturbed plane mixing layer: (a) smoke illuminated
stroboscopically (exposure time = 0.8 s), (b) calculated streaklines, and (c) calculated iso-
dynes. (From Weisbrot and Wygnanski 1988. Reproduced with kind permission of Cambridge
University Press from the *Journal of Fluid Mechanics*.)

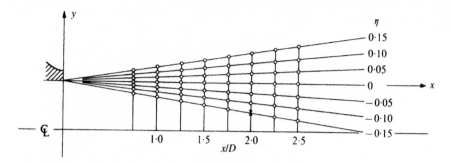

Fig. 11.33. The air-jet co-ordinate system and hot-wire positions: (x) the fixed (detector) wire, ($\bigcirc$) the eduction wire. Re $= 10^4$, $\eta = y/x$.

Browand and Trout 1980), relies on a large-scale flow structure in the shear layer creating a specific signal pattern ('footprint') in the irrotational flow surrounding the shear layer. The principles of this method will be demonstrated using the approach and results of Bruun (1977, 1979b) for a circular jet. In these studies, an SN detector probe placed in the potential flow ($x/D = 2.0$, $\eta = -0.15$) provided the detector signal and a second SN probe (the eduction wire) was moved sequentially to the specified positions in the flow field, as shown in Fig. 11.33 for a circular jet.

The eduction technique requires the selection of a detector criterion which is directly related to a specific type of large-scale flow event. In

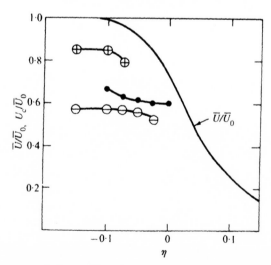

Fig. 11.34. The mean velocity profile and convection velocity of: ($\bullet$) cross-correlation and positive ($\oplus$) and negative ($\ominus$) educed signals at $x/D = 2.0$. Re $= 10^4$.

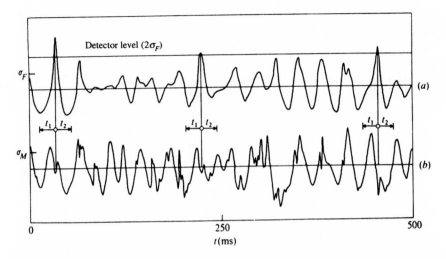

Fig. 11.35. The eduction principle, Re $= 10^4$: (a) a fixed (detector) wire, at $x/D = 2.0$, $\eta = -0.15$, and (b) a moving (eduction) wire at, $x/D = 2.0$, $\eta = -0.025$.

investigations by Bruun (1977, 1979b) the criterion was either large positive or negative peaks in the fluctuating u-velocity signal, because initial convection-velocity results for a jet Reynolds number Re$_D = 10^4$ (Fig. 11.34) had demonstrated that positive and negative peaks in the u-velocity signal did correspond to fast and slow moving flow events, respectively. An example of the identification of detector events for positive peaks is shown in Fig. 11.35; which also contains the related signal from an eduction wire. The corresponding educed signal, evaluated using eqn 11.33, is shown in

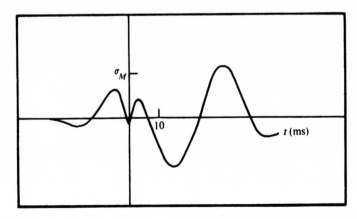

Fig. 11.36. The educed signal, for a moving wire at $x/D = 2.0$, $\eta = -0.025$.

Fig. 11.36; it contains several distinct positive and negative peaks. Corresponding peaks can be identified in the three separate time-history events (t_1, t_2) in Fig. 11.35. However, considerable variation can be seen to exist in both the amplitude and in the relative position of the individual peaks, and as a result the magnitude of the regular peaks in the educed signal will usually be significantly smaller than the peaks in the separate events. *It is therefore not possible to relate directly the magnitude of the educed signal to an 'average' large-scale flow event.* Constructions of vector or contour plots of coherent structures based on the amplitude information in educed signals are therefore likely to be highly inaccurate.

The evaluation of large-scale flow structure In the study of a circular jet at a moderate Reynolds number of 10^4 (Bruun 1977), it was observed from smoke flow-visualization experiments that the shear layer contained regular coherent flow structures. To describe the corresponding large-scale features identified by the eduction process, the development of the regular

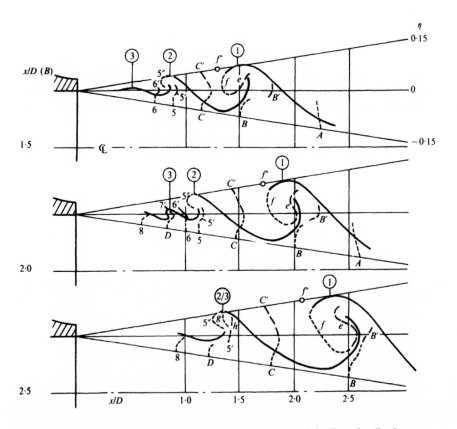

Fig. 11.37. A comparison of the educed large-scale structure and the flow-visualization structure (——).

peaks in the educed signals was studied along constant-η lines. The spatial and temporal developments of these peaks are compared with flow-visualization results in Fig. 11.37. Having demonstrated the ability of the method above to identify large-scale flow structures in a transitional, moderate-Reynolds-number jet, the technique was also applied to a higher-Reynolds-number jet ($\mathrm{Re}_D = 2 \times 10^5$), where it has not been possible to demonstrate the existence of regular coherent structures with flow visualization. Also, initial measurements had showed that the jet flow is highly Reynolds-number dependent, as demonstrated in Fig. 11.38 for the variation with the Reynolds number of the Strouhal number and the turbulence intensity in the potential core.

The conditionally sampled results for the higher Reynolds number did not provide evidence for an axisymmetric coherent flow structure spanning the width of the shear layer. With the detector wire placed in the potential core, the educed signals provided some evidence of regular events on the inside of the mixing region. Although some small amplitude peaks were identified in the outer part of the mixing region their magnitudes were so small that it was not justifiable to relate them to any significant coherent flow structure.

Additional conditionally sampled measurements were carried out using either an SN detector probe placed at $\eta = 0$ (the middle of mixing layer) or an X probe placed at various radial positions at $x/D = 2$. These measurements identified two types of flow events, referred to as 'bursts' ($u > 0$, $v < 0$) and 'sweeps' ($u < 0$, $v > 0$) which contributed significantly to the measured Reynolds shear stress, $\overline{uv}$. Eduction measurements showed

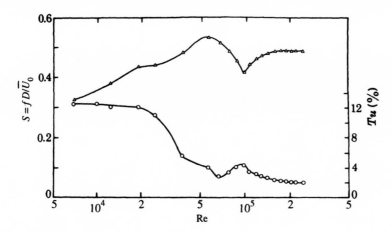

Fig. 11.38. The Reynolds-number dependence of the Strouhal number and the turbulence intensity in the potential core, for a hot-wire probe at $x/D = 2$, $\eta = -0.15$. $\triangle$, Strouhal number; $\bigcirc$, turbulence intensity.

that these events take place as circumferential narrow tongues of ejected and entrained fluid. The two types of events occur at all circumferential positions in the jet, but the eduction technique did not establish any underlying axisymmetric large-scale flow structures. It has been observed in many investigations that a low (but significant) correlation can be detected over large axial distances. In the author's opinion, this observation cannot be interpreted as the existence of a semiregular axisymmetric coherent flow structure, its identity being masked by local fluctuations of high magnitude. The original semiregular flow structure formed near the jet exit is continuously transformed by the sweep and burst events. The transformation of the flow structure is rapid, and it soon loses any resemblance of axisymmetry. However, the flow will retain smeared-out regions of low and high vorticity, which are often referred to as large eddies (see, for example, Yule 1978). The measured low correlation is a manifestation of the rather slow decay of these vorticity regions.

12

TIME-SERIES ANALYSIS

The output from a hot-wire anemometer (HWA) sensor will in general be a time-varying signal representing typically, a velocity component. Other measurable quantities include the vorticity, the temperature, the concentration in gas mixtures, etc. In a turbulent flow, the anemometer signal will be of a random nature, and a statistical description of the content of the signal is therefore necessary. In the early statistical turbulent-flow studies, much emphasis was placed on spectral analysis and statistics based on statistically independent samples. However, the development of computational fluid dynamics (CFD) has caused a change of emphasis in much of the current experimental HWA work. As described in Chapter 1, the validation of CFD codes requires accurate measurements of the mean-velocity components and Reynolds stresses, and the measurement of these quantities and the related error estimates will therefore be described in some detail. Most contemporary experimental work is carried out using digital computers, and the corresponding evaluation procedures and problems are also highlighted in this chapter, which reflects the accumulated expertise in this field. Noticeable contributions include those by Bendat and Piersol (1966, 1971, 1986), Jenkins and Watts (1968), Lumley (1970), Tennekes and Lumley (1972), Van Atta (1974), Rabiner and Gold (1975), George *et al.* (1978), Williams (1981), Castro (1989), and Tropea (1991). The author has particulary found presentations by Bendat and Piersol (1971, 1986) both clear and useful. Consequently, for many of the statistical quantities described in this chapter, the notation and derivations of Bendat and Piersol will be used. Tennekes and Lumley (1972) and George *et al.* (1978) have also demonstrated that the integral time-scale T_I of the fluctuating part of the signal is a key factor in evaluating both the optimum sampling rate and error estimates for the measured statistical quantities. This chapter contains an integration of these two approaches to random data analysis.

12.1   THE CLASSIFICATION OF PHYSICAL DATA

Any observed data representing a physical phenomenon can broadly speaking be classified as being either deterministic or random. Deterministic data are those data that can be described by an explicit mathematical relationship. Data representing a random physical phenomenon cannot be described by an explicit mathematical relationship because there is no way of predicting an exact value at a future instant in time. Random data may be

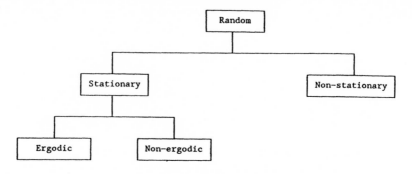

Fig. 12.1. The classification of random data.

classified as shown in Fig. 12.1, and the following notation will be used: when a random process is observed over a finite time interval it shall be referred to as a single *time-history record* or a *sample record*. The collection of all possible time-history records which the random phenomenon might have produced is called a *random process* or a *stochastic process*.

12.1.1 Stationary random processes

Using the notation above, the collection of time-history records, also called the ensemble, (as for example, shown in Fig. 12.2) will form a random process. The mean value of the random process at a time t_1 can be evaluated by taking the instantaneous value of each time-history record at t_1, summing the values, and dividing by the number of sample records. In a similar manner, an autocorrelation function representing the correlation between the values of the random process at two different times can be calculated by taking the ensemble average of the product of the instantaneous values at two times, t_1 and $t_1 + \tau$. For a random process, the mean value, $\mu_x(t_1)$, and the autocorrelation function, $R_x(t_1, t_1 + \tau)$, are given by

$$\mu_x(t_1) = \lim_{N \to \infty} \frac{1}{N} \sum_{k=1}^{N} x_k(t_1), \tag{12.1a}$$

$$R_x(t_1, t_1 + \tau) = \lim_{N \to \infty} \frac{1}{N} \sum_{k=1}^{N} x_k(t_1) x_k(t_1 + \tau), \tag{12.1b}$$

where the summations assumes that each record is equally likely.

For the general case, where $\mu_x(t_1)$ and $R_x(t_1, t_1, + \tau)$ defined by eqns (12.1a–b) vary as the time t_1 varies, the random process is said to be *non-stationary*. For the special case where $\mu_x(t_1)$ and $R_x(t_1, t_1 + \tau)$ do not vary as the time t_1 varies (that is, $\mu_x(t_1) = \mu_x$ and $R_x(t_1, t_1 + \tau) = R_x(\tau)$), the random process is said to be *weakly stationary*. When all possible moments

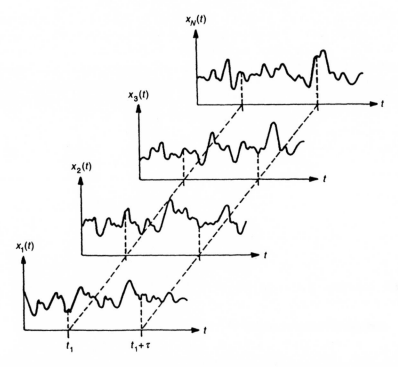

Fig. 12.2. An ensemble of time-history records defining a random process. (From Bendat and Piersol 1980. Reproduced with kind permission of John Wiley & Sons Inc. © John Wiley & Sons Inc.)

and joint moments are time invariant, the random process is said to be *strongly stationary* or stationary in the strictest sense.

12.1.2 Ergodic random processes

In the previous section it was shown how the properties of a random process can be evaluated using *ensemble averages* at specific instances in time. However, in most cases, it is also possible to describe the properties of a stationary random process by calculating *time averages* over a specific sample record in the ensemble. This can be demonstrated by considering the kth time-history record of the random process illustrated in Fig. 12.2. The corresponding mean value, $\mu_x(k)$, and autocorrelation function, $R_x(\tau, k)$, are given by

$$\mu_x(k) = \lim_{T \to \infty} \frac{1}{T} \int_0^T x_k(t)\, \mathrm{d}t, \tag{12.2a}$$

$$R_x(\tau, k) = \lim_{T \to \infty} \frac{1}{T} \int_0^T x_k(t)\, x_k(t + \tau)\, \mathrm{d}t. \tag{12.2b}$$

If the random process is stationary and $\mu_x(k)$ and $R_x(\tau, k)$ (defined in eqns (12.2a–b)) do not differ when evaluated over different time-history records, the random process is said to be *ergodic*. For ergodic random processes the time-averaged mean value and autocorrelation function (as well as all other time-averaged properties) are equal to the corresponding ensemble-averaged values. That is, $\mu_x(k) = \mu_x$ and $R_x(\tau, k) = R_x(\tau)$. In the rest of the chapter it will be assumed that the random process under consideration is ergodic.

This chapter considers basic statistical properties used to describe either a single time-history record, $X(t)$, or two simultaneous time-history records, $X(t)$ and $Y(t)$, representing different physical processes. For evaluation purposes, the instantaneous values $X(t)$ and $Y(t)$ can be expressed in terms of time-mean values and fluctuating components; that is,

$$X(t) = \bar{X} + x(t), \qquad (12.3a)$$
$$Y(t) = \bar{Y} + y(t). \qquad (12.3b)$$

Overbars are used to denote a time-averaged value.

The statistical quantities have been grouped according to the type of information provided:

1. *Amplitude-domain statistics* The quantities which can be described by amplitude-domain statistics provide information related to the amplitude distribution of the signal, but they provide no time-history information. For all these quantities, an optimum sampling criterion can be defined, based on the integral time-scale of the time history record. Typical statistical properties include: (a) for a *single time-history record $X(t)$*: (the mean, $\bar{X}$, mean-square value, $\overline{x^2}$, higher-order moments, $\overline{x^m}$, and the probability density function, $p(x)$), and (b) for *two simultaneous time-history records $X(t)$ and $Y(t)$*: (the mean values $\bar{X}$ and $\bar{Y}$, and for the fluctuating components: $\overline{xy}$ (corresponding to Reynolds stresses if x and y represent two fluctuating velocity components at a point), spatial correlations (when the signals $x(t)$ and $y(t)$ are obtained from two spatially separated sensors), higher-order moments $\overline{x^m y^p}$ and the joint-probability density function, $p(x, y)$).

2. *Time-domain statistics* The autocorrelation function, $R_x(\tau)$, for a single time-history record $x(t)$ and the cross-correlation function, $R_{xy}(\tau)$, for two time-history records $x(t)$ and $y(t)$. The autocorrelation function can provide information on the time interval over which the signals are correlated and on the rate of decay of this correlation. The cross-correlation function can, for spatially separated sensors, measure the average convection velocity of disturbances in the flow.

3. *Frequency-domain statistics* An alternative, but complementary way of describing the time-domain nature of the signal is by way of its energy spectrum. Physically, this is simply a measure of how much energy is con-

tained within the signal in each frequency band. Due to the many features which are involved in spectral measurements, this chapter will, for clarity, only describe the autospectral density function from a single time-history record.

12.2 UNCERTAINTY SPECIFICATION FOR MEASURED STATISTICAL QUANTITIES

The probability density function (pdf) of the signal from many physical processes approximates a Gaussian distribution, and the uncertainty specification given in this chapter is based on the assumption that this approximation is valid. Even when the pdf of the signal deviates significantly from a Gaussian distribution, the analysis presented can often be used as an order-of-magnitude estimate of the uncertainty in the statistical quantities evaluated.

Consider a number of statistically independent samples, $x(k)$, from a random (ergodic) process. The corresponding probability density function $p(x(k))$ will be centred around the true ensemble average, μ_x, with a standard deviation σ_x (for some statistical quantities the centre of the measured probability density function will be offset with a bias b from the true ensemble average; the fixed error, b, is additional to the random error represented by the corresponding standard deviation, σ). Introducing the standardized variable

$$z = \frac{x - \mu_x}{\sigma_x},\qquad(12.4)$$

the standardized Gaussian probability density function, $p(z)$, can be expressed as

$$p(z) = \frac{1}{\sqrt{2\pi}}\, e^{-z^2/2}\qquad(12.5)$$

The corresponding probability, $P(z)$, will be

$$P(z) = \int_{-\infty}^{z} p(\xi)\, d\xi.\qquad(12.6)$$

To illustrate the uncertainty specification, consider $x(k)$ to be an estimate for the ensemble average, μ_x. The uncertainty specification for $x(k)$ being an estimate for μ_x can be expressed as

$$-z_{\alpha/2} < \frac{x(k) - \mu_x}{\sigma_x} < z_{\alpha/2},\qquad(12.7)$$

where $z_{\alpha/2}$ is the value of z for which $P(z_{\alpha/2}) = 1 - \alpha/2$, that is $x(k)$ will, statistically, fall within the interval

TABLE 12.1 *A Gaussian probability density distribution*

$z_{\alpha/2}$	$(1 - \alpha)\%$
1.65	90
1.96	95
2.33	98
2.57	99

$$\mu_x - z_{\alpha/2}\,\sigma_x < x(k) < \mu_x + z_{\alpha/2}\sigma_x \qquad (12.8)$$

with a probability of $(1 - \alpha)\%$.

Table 12.1 contains the relationships between typical values of $z_{\alpha/2}$ and $1 - \alpha$. For an ergodic (stationary) random process the uncertainty specification above can also be applied to the analysis of a time-history record, $x(t)$. In particular, it should be noted that the random variable, x, used in eqn (12.8) can, in principle, represent both the sample record, $x(t)$, and any related statistical quantity, provided that the appropriate standard deviation, σ, is substituted for σ_x. Using carets, $\hat{}$, to describe an *estimated* (*measured*) *value* and overbars to denote time-mean values, this point can be illustrated by the following uncertainty specifications:

1. The measured mean value, $\hat{\bar{X}}$ (true value, $\bar{X}$), will statistically fall within the interval

$$\bar{X} - z_{\alpha/2}\,\sigma[\hat{\bar{X}}] < \hat{\bar{X}} < \bar{X} + z_{\alpha/2}\sigma[\hat{\bar{X}}] \qquad (12.9a)$$

with a probability of $(1 - \alpha)$ %
The variance of $\hat{\bar{X}}$ is

$$\text{var}\,[\hat{\bar{X}}] = \sigma^2[\hat{\bar{X}}], \qquad (12.9b)$$

where $\sigma[\hat{\bar{X}}]$ is the standard deviation of the measured mean value, $\hat{\bar{X}}$.

2. The measured mean-square value of the fluctuating part of the signal, $\hat{\overline{x^2}}$ (true value $\overline{x^2}$), will fall within the interval

$$\overline{x^2} - z_{\alpha/2}\,\sigma[\hat{\overline{x^2}}] < \hat{\overline{x^2}} < \overline{x^2} + z_{\alpha/2}\sigma[\hat{\overline{x^2}}] \qquad (12.10a)$$

with a probability of $(1 - \alpha)\%$
The variance of $\hat{\overline{x^2}}$ is

$$\text{var}\,[\hat{\overline{x^2}}] = \sigma^2[\hat{\overline{x^2}}], \qquad (12.10b)$$

where $\sigma[\hat{\overline{x^2}}]$ is the standard deviation of the measured mean square value of the fluctuating signal, $\hat{\overline{x^2}}$.

These examples illustrate that the uncertainty of a measured (unbiased) statistical quantity can be specified once the variance of that quantity has been determined. For statistically independent samples, it is shown in Section

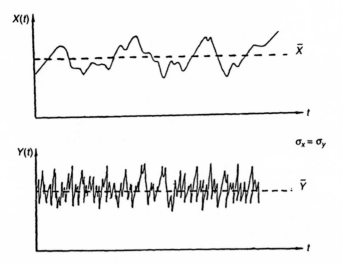

Fig. 12.3. Two sample records $X(t)$ and $Y(t)$ with the same standard deviation, $\sigma_x = \sigma_y$, but with different frequency contents.

12.3 that $\sigma[\hat{\bar{X}}] = \sigma_x/\sqrt{N}$ and $\sigma[\hat{\bar{x^2}}] = \sigma_x^2/\sqrt{N}$ where N is the number of independent samples and σ_x is the standard deviation of the random process. It is consequently possible to evaluate the uncertainty corresponding to a given number of independent samples, N, or to evaluate the number of independent samples required for a given uncertainty specification.

For the basic statistical quantities considered in this chapter, the following aspects will be discussed: (i) their definition based on an infinitely long sample record, (ii) the formulation of digital evaluation procedures for finite time history records, (iii) error estimation and (iv) specification of optimum-sampling-rate criteria. The importance of the sampling criteria can be illustrated with reference to the two sample records, $X(t)$ and $Y(t)$, shown in Fig. 12.3, which both have the same standard deviation, $\sigma_x = \sigma_y$. The two signals clearly have different frequency contents, and intuitively one would expect that different sampling criteria should be applied to the two time-history records to achieve the same accuracy for similar statistical quantities evaluated from them. The criterion for optimum sampling is explained in Section 12.3.1.1.

12.3 AMPLITUDE-DOMAIN ANALYSIS

12.3.1 Single time-history record statistics

12.3.1.1 Mean value

Consider a time-history record $X(t)$ which consists of a mean, $\bar{X}$, and a fluctuating component $x(t)$, related by

$$X(t) = \bar{X} + x(t).$$

For an infinitely long time-history record, the mean value, $\bar{X}$, is simply the average of all values, which can be expressed as

$$\bar{X} = \lim_{T \to \infty} \frac{1}{T} \int_0^T X(t) \, dt. \tag{12.11}$$

Figure 12.4 represents a finite time-history record of such a signal. When using *digital* data analysis, the continuous signal $X(t)$, $0 \leqslant t \leqslant T$ is replaced by a corresponding digital sample record $X(n)$, $n = 1, 2, \ldots, N$, which is indicated by the dots in Fig. 12.4. The initial point, t_0, is arbitrary and does not enter into the following equations, unless the sequence is assumed to start at $n = 0$ and to end at $N - 1$. The total sample time is T and N is the corresponding number of samples. N and T are related by the sampling rate, SR (samples s^{-1}), by eqn (3.6); that is,

$$N = SR \times T$$

and the time interval between samples, Δt, is given by

$$\Delta t = 1/SR = T/N \tag{12.12}$$

In digital form, the estimate (or measured value), for the mean value of a finite sample record, $X(n)$, $n = 1, 2, \ldots, N$ can be expressed as

$$\hat{\bar{X}} = \frac{1}{N} \sum_{n=1}^{N} X(n), \tag{12.13}$$

and $\hat{\bar{X}}$ will be an unbiased estimate of the true mean value, $\bar{X}$.

Uncertainty estimate for $\hat{\bar{X}}$ The uncertainty in $\hat{\bar{X}}$ can be estimated by evaluating the variance of $\hat{\bar{X}}$. The following time-series evaluation of var $[\hat{\bar{X}}]$ is based on the derivations given by Bendat and Piersol (1971, 1986). The evaluation of the error estimation for the statistical quantities discussed in subsequent sections is also primarily based on these references. The

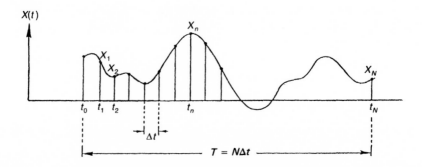

Fig. 12.4. A digital representation of a sample record.

accuracy of a parameter estimate based on measured sample values can be described by its mean square error. Since $\hat{X}$ is an unbiased estimate, the mean square error will be equal to the variance of $\hat{X}$, which is defined as

$$\text{var} [\hat{X}] = \sigma^2[\hat{X}] = E[(\hat{X} - \bar{X})^2]. \tag{12.14}$$

Derivations for a continuous sample record give

$$\text{var} [\hat{X}] = E\left[\left(\frac{1}{T}\int_0^T X(t)\,dt - \bar{X}\right)^2\right],$$

$$= \frac{1}{T^2} E\left[\left(\int_0^T (X(t) - \bar{X})\,dt\right)^2\right].$$

Introducing the fluctuating component, $x(t) = X(t) - \bar{X}$,

$$\text{var} [\hat{X}] = \frac{1}{T^2} E\left[\left(\int_0^T x(t)\,dt\right)^2\right],$$

$$= \frac{1}{T^2} \int_0^T \int_0^T E[x(\xi)\,x(\eta)]\,d\eta\,d\xi.$$

The autocorrelation function, $R_x(\tau)$ (see Section 12.4) for the fluctuating component $x(t)$ of a stationary random process can be expressed as

$$R_x(\tau) = E[x(t)\,x(t + \tau)] \tag{12.15}$$

Introducing the autocorrelation coefficient function, $\rho_x(\tau) = R_x(\tau)/\sigma_x^2$, the variance can be expressed as

$$\text{var} [\hat{X}] = \frac{\sigma_x^2}{T^2} \int_0^T \int_0^T \rho_x(\eta - \xi)\,d\eta\,d\xi = \frac{\sigma_x^2}{T^2} \int_0^T \int_{-\xi}^{T-\xi} \rho_x(\tau)\,d\tau\,d\xi,$$

$$= \frac{\sigma_x^2}{T} \int_{-T}^T \left(1 - \frac{|\tau|}{T}\right)\rho_x(\tau)\,d\tau, \tag{12.16}$$

where the last expression was obtained by reversing the orders of integration between τ and ξ and carrying out the ξ integration. When $\tau \gg 0$, $\rho_x(\tau) \to 0$, and because $\rho_x(\tau) = \rho_x(-\tau)$

$$\text{var} [\hat{X}] = \sigma^2[\hat{X}] \approx \frac{2\sigma_x^2}{T} \int_0^T \rho_x(\tau)\,d\tau. \tag{12.17}$$

Finally, introducing the integral time-scale defined by

$$T_I = \int_0^T \rho_x(\tau)\,d\tau,$$

(see eqn (12.71)), the expression for the variance becomes (Tennekes and Lumley 1972)

$$\text{var} [\hat{X}] = \sigma^2[\hat{X}] \approx \frac{2T_I \sigma_x^2}{T}. \tag{12.18}$$

In this expression, σ_x^2 is the variance of the sample record, $X(t)$, as described in Section 12.3.2. The uncertainty in the measured mean value, $\hat{\bar{X}}$, can, using inequality (12.9a), be specified as

$$\bar{X} - z_{\alpha/2}\sigma[\hat{\bar{X}}] < \hat{\bar{X}} < \bar{X} + z_{\alpha/2}\sigma[\hat{\bar{X}}],$$

with a probability of $(1 - \alpha)$ %. As described by George et al. (1978), a related analysis can also be carried out using ensemble averaging. To evaluate digitally the ensemble mean, the individual realizations of the random process are added and this sum is divided by the number of realizations, N, that is

$$X_N = \frac{1}{N} \sum_{n=1}^{N} X_n. \tag{12.19}$$

Denoting the true mean value by $\bar{X}$, the variance of X_N can be expressed as

$$\text{var}\,[X_N] = E[\,(X_N - \bar{X})^2\,]. \tag{12.20}$$

If the samples are evenly distributed and *statistically independent* then it can be shown that

$$\text{var}\,[X_N] = \frac{1}{N}\,\text{var}\,[X], \tag{12.21}$$

where var [X] is the mean-square value about the mean and it is given by

$$\text{var}\,[X] = \sigma_x^2,$$

and consequently

$$\text{var}\,[X_N] = \frac{\sigma_x^2}{N}, \tag{12.22}$$

for N *independent* realizations.

For an ergodic process, a statistical quantity can be evaluated by either ensemble or time-mean averaging. Therefore, var $[X_N]$ = var $[\hat{\bar{X}}]$, and by equating eqns (12.18) and (12.22)

$$\text{var}\,[\hat{\bar{X}}] = \sigma^2[\hat{\bar{X}}] = \frac{\sigma_x^2}{N} \approx \frac{2T_I\sigma_x^2}{T}, \tag{12.23}$$

from which it follows that

$$N \simeq \frac{T}{2T_I}. \tag{12.24}$$

The N independent realizations in eqn (12.22) can therefore also be treated for signal-analysis purposes as N consecutive *independent* samples in a time-history record. Consequently, to first order, samples separated by two

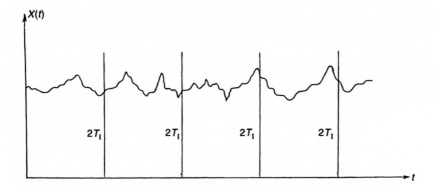

Fig. 12.5. Segments of a stationary random process, $X(t)$, illustrating pieces of record which contribute to the mean as if they were statistically independent. (From George *et al.* 1978.)

integral time-scales, $2T_I$, therefore contribute to the evaluated quantities as statistically independent samples. This principle is illustrated in Fig. 12.5. For digital measurements, the sample interval

$$\Delta t = 2T_I, \tag{12.25}$$

(or the sampling rate $SR = 1/(2T_I)$) represents the *optimum sampling criterion for amplitude-domain analysis.* For a given number of samples, N, sampling much more slowly will extend the measuring time without any significant increase in accuracy, and a faster sampling rate will increase the uncertainty because the samples will no longer be statistically independent. Finally, introducing the normalized rms error

$$\varepsilon[\hat{\bar{X}}] = \frac{\sigma[\hat{\bar{X}}]}{\bar{X}} = \frac{1}{\sqrt{N}}\frac{\sigma_x}{\bar{X}} \approx \left(\frac{2T_I}{T}\right)^{1/2}\frac{\sigma_x}{\bar{X}}, \tag{12.26}$$

the uncertainty in the measured mean value $\hat{\bar{X}}$, can be expressed in terms of $\varepsilon[\hat{\bar{X}}]$ as

$$1 - z_{\alpha/2}\varepsilon[\hat{\bar{X}}] < \frac{\hat{\bar{X}}}{\bar{X}} < 1 + z_{\alpha/2}\varepsilon[\hat{\bar{X}}], \tag{12.27}$$

with a probability of $(1 - \alpha)$ %.

To illustrate the principle of uncertainty estimation for practical flow situations, three fluid-flow examples will be considered. In each case the time-history record, $X(t)$, will be assumed to represent the longitudinal velocity component, $U(t)$.

Example A
Consider a typical turbulent boundary layer flow on a flat plate in a wind-tunnel, as shown in Fig. 12.6(a). The free-stream velocity, $U_\infty = 10$ m and

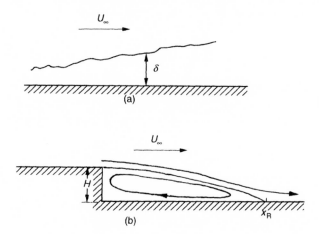

Fig. 12.6. (a) A turbulent boundary layer on a flat plate, where U_∞ is the free-stream velocity and δ is the boundary layer thickness. (b) The flow over a backward-facing step. U_∞ is the free-stream velocity, H is the step height, and x_R is the reattachment length.

the boundary layer thickness, δ, at the measurement position is 5 cm. Measurements are carried out in the outer part of the boundary layer (where $\bar{U} \simeq U_\infty$) and the local turbulence intensity $\sigma_u/\bar{U} = 10\%$. Determine the optimum sampling rate, the number of (independent) samples, and the total sampling time in order to obtain an accuracy for $\hat{\bar{U}}$ of $\pm 1\%$ with a 98% confidence level (probability).

Solution From Table 12.1 it follows that a 98% confidence level corresponds to $z_{\alpha/2} = 2.33$. As the required accuracy is given as a percentage eqn (12.27) can be used to obtain

$$\frac{\hat{\bar{U}}}{\bar{U}} = 1 \pm 2.33 \frac{\sigma[\hat{\bar{U}}]}{\bar{U}} = 1 \pm 0.01,$$

$$N = \left[\frac{T_u}{a} \left(\frac{z_u}{2} \right) \right]^2$$

from which it follows, using eqn (12.26), that

$$\frac{\sigma[\hat{\bar{U}}]}{\bar{U}} = \frac{1}{\sqrt{N}} \frac{\sigma_u}{\bar{U}} = \frac{0.01}{2.33} = 0.0043.$$

Since the turbulence intensity is 10%, the number of independent samples required is

$$N = \frac{1}{0.0043^2} \left(\frac{\sigma_u}{\bar{U}} \right)^2 = \frac{0.01}{0.0043^2} = 540 \text{ samples}.$$

To evaluate the corresponding total sampling time, T, the integral time-scale, T_I, must be measured or estimated. In this example, the integral time-scale is estimated from the time-scale of the large-scale turbulent eddies as $T_I \simeq \delta/U_\infty = 5 \times 10^{-2}/10 = 5$ ms. The corresponding optimum time between samples is, using eqn (12.25),

$$\Delta t = 2T_I = 10 \text{ ms}$$

and the total sampling time

$$T = N\Delta t = 540 \times 10 \text{ ms} = 5.4 \text{ s}.$$

Example B
Air with a free-stream velocity $U_\infty = 10$ ms^{-1} flows over a backward-facing step (Fig. 12.6(b)). Measurements are to be carried out in the shear layer surrounding the separated flow region at positions where $\bar{U} \simeq U_\infty$, and where the turbulence intensity, $\sigma_u/\bar{U}$, may reach 20%. From autocorrelation measurements the integral time-scale, T_I, has been evaluated as being about 50 ms. Determine the optimum sampling rate, the number of (independent) samples and the total sampling time to achieve an accuracy for $\bar{U}$ of $\pm 1\%$ with a 99% confidence level.

Solution From Table 12.1 it follows that a 99% confidence level corresponds to $z_{\alpha/2} = 2.57$. As the required accuracy is given as a percentage it follows, combining eqns (12.26) and (12.27) that

$$\frac{\sigma[\hat{\bar{U}}]}{\bar{U}} = \frac{1}{\sqrt{N}} \frac{\sigma_u}{\bar{U}} = \frac{0.01}{2.57} = 0.0039.$$

Since the (maximum) turbulence intensity is 20% the number of independent samples, N, will be

$$N = \frac{0.2^2}{0.0039^2} = 2600 \text{ samples}.$$

In this example, the integral time-scale, T_I, is assumed to have been measured, and the optimum time between samples is

$$\Delta t = 2T_I = 0.1 \text{ s}.$$

The total sampling time is evaluated using

$$T = N\Delta t = 2600 \times 0.1 = 260 \text{ s}.$$

The existence of low-frequency phenomena in the turbulent flow, as is reflected in the value of the integral time-scale, is observed to result in a much longer sampling time. This principle is also illustrated in the next example.

Example C
Water flows over a backward-facing step. The details of the flow are as follows (Tropea 1994). The free-stream velocity is $U_\infty = 2$ ms^{-1}, and at the measurement point $\bar{U} \simeq U_\infty$ and $\sigma_u^2 = 0.2$ m^{2}s^{-2}. The step height, H,

(Fig. 12.6(b)) is 5 cm and the reattachment length $x_R = 8H$. It is estimated that the integral time-scale T_I is dominated by low-frequency phenomena which scale on x_R.

Evaluate the measurement conditions which will give an accuracy in $\hat{\bar{U}}$ of $\pm 0.01\ ms^{-1}$ with a 95% confidence level.

Solution The integral time-scale may, in this case, be estimated as

$$T_I = \frac{x_R}{U_\infty} = \frac{8 \times 0.05}{2} = 0.2\ s.$$

From Table 12.1, a confidence level of 95% corresponds to $z_{\alpha/2} = 1.96$. Since the accuracy is specified in absolute terms, eqn (12.9a) will be used, giving

$$\hat{\bar{U}} = \bar{U} \pm 1.96\sigma[\hat{\bar{U}}] = \bar{U} \pm 0.01,$$

and therefore

$$\sigma[\hat{\bar{U}}] = \frac{0.01}{1.96} = 0.005.$$

Then applying eqn (12.23)

$$\mathrm{var}\,[\hat{\bar{U}}] = \sigma^2[\hat{\bar{U}}] = \frac{\sigma_u^2}{N} = 0.005^2 = 25 \times 10^{-6}.$$

The corresponding total number of (independent) samples is

$$N = \frac{0.2}{25 \times 10^{-6}} = 8000\ \text{samples},$$

and the total sampling time

$$T = N\Delta t = N2T_I = 8000 \times 0.4 = 3200\ s \simeq 53\ min.$$

In practice, it may be necessary to select the sampling rate $SR\ (=1/\Delta t)$ as the nearest even integer value, in this case 4 Hz. The corresponding sampling time would be about 33 min.

12.3.1.2 *The mean-square value and higher-order moments*

The mean-square value ψ_x^2 For completeness, the mean-square value will also be described, although this quantity is not used very frequently in turbulent flow studies. The mean-square value, ψ_x^2, is simply the average of the squared values of the time-history record $X(t)\ (=\bar{X} + x(t))$. In equation form

$$\psi_X^2 = \lim_{T \to \infty} \frac{1}{T}\int_0^T X^2(t)\ dt, \tag{12.28}$$

and for a digital time-history record the estimated (or measured) value $\hat{\psi}_X^2$ is

$$\hat{\psi}_X^2 = \frac{1}{N} \sum_{n=1}^{N} X^2(n). \tag{12.29}$$

The variance, σ_x^2 The variance of a time-history record, $X(t)$ usually denoted σ_x^2, is the mean-square value about the mean. This can be expressed as

$$\sigma_x^2 = \lim_{T \to \infty} \frac{1}{T} \int_0^T (X(t) - \bar{X})^2 \, \mathrm{d}t = \lim_{T \to \infty} \frac{1}{T} \int_0^T x^2(t) \, \mathrm{d}t. \tag{12.30}$$

The positive root σ_x is denoted the *standard deviation*. From eqns (12.28) and (12.30) it follows that

$$\sigma_x^2 = \psi_X^2 - \bar{X}^2. \tag{12.31}$$

For a digital time-history record, an unbiased estimate for σ_x^2 can be obtained using

$$\hat{\bar{x^2}} = \frac{1}{N-1} \sum_{n=1}^{N} (X(n) - \hat{\bar{X}})^2. \tag{12.32}$$

However, application of eqn (12.32) is not computer efficient because the measured mean value, $\hat{\bar{X}}$, has to be subtracted from each value of $X(n)$ before squaring the signal. A simpler computational procedure can be obtained by expanding the right-hand side of eqn (12.32), giving

$$\hat{\bar{x^2}} = \frac{1}{N-1} \left[\sum_{n=1}^{N} X^2(n) - 2\hat{\bar{X}} \sum_{n=1}^{N} X(n) + N\hat{\bar{X}}^2 \right].$$

Since $\Sigma_{n=1}^{N} X(n) = N\hat{\bar{X}}$ the above equation can be reduced to

$$\hat{\bar{x^2}} = \frac{1}{N-1} \left[\sum_{n=1}^{N} X^2(n) - N\hat{\bar{X}}^2 \right], \tag{12.33}$$

in which the main computational step is a running average of $X^2(n)$.

Uncertainty estimate for $\hat{\bar{x^2}}$ The uncertainty principle described for $\hat{\bar{X}}$ in Section 12.3.1 can also be applied to $\hat{\bar{x^2}}$. Since the true value $\bar{x^2} = \sigma_x^2$, the uncertainty in $\hat{\bar{x^2}}$ can, using eqn (12.10a), be expressed as

$$\sigma_x^2 - z_{\alpha/2}\sigma[\hat{\bar{x^2}}] < \hat{\bar{x^2}} < \sigma_x^2 + z_{\alpha/2}\sigma[\hat{\bar{x^2}}], \tag{12.34}$$

with a probability of $(1 - \alpha)\%$, provided the probability distribution for $\hat{\bar{x^2}}$ is Gaussian.

The variance of $\hat{\bar{x^2}}$ can be expressed as

$$\mathrm{var}[\hat{\bar{x^2}}] = \sigma^2[\hat{\bar{x^2}}] = E[(\hat{\bar{x^2}} - \sigma_x^2)^2], \tag{12.35}$$

where σ_x^2 is the true value. Following a set of derivations similar to those for the mean value $\hat{\bar{X}}$ it can be shown that

$$\sigma^2[\hat{\overline{x^2}}] \approx \frac{2}{T} \int_{-\infty}^{\infty} R_x^2(\tau)\, d\tau.$$

Introducing the autocorrelation coefficient function $\rho_x(\tau) = R_x(\tau)/\sigma_x$,

$$\sigma^2[\hat{\overline{x^2}}] \approx \frac{4\sigma_x^4}{T} \int_0^T \rho_x^2(\tau)\, d\tau. \tag{12.36}$$

In eqn (12.17) for $\sigma^2[\hat{\bar{X}}]$ we introduced an integral time-scale $T_I = \int_0^T \rho_x(\tau)\, d\tau$. For $\rho_x(\tau)$ we have $\rho_x(0) = 1$ and $\rho_x(\tau) \to 0$ for $\tau \gg 0$. If it is assumed that $\rho_x(\tau) = \exp(-\tau/T_I)$ then it can be shown that $\int_0^T \rho_x^2(\tau)\, d\tau$ is equal to $0.5 T_I$, and eqn (12.36) becomes

$$\sigma^2[\hat{\overline{x^2}}] \approx \frac{2T_I \sigma_x^4}{T}, \tag{12.37a}$$

and introducing eqn (12.24)

$$\sigma^2[\hat{\overline{x^2}}] \approx \frac{\sigma_x^4}{N}, \tag{12.37b}$$

corresponding to N independent samples.

For the related normalized rms error, $\varepsilon[\hat{\overline{x^2}}]$

$$\varepsilon[\hat{\overline{x^2}}] = \frac{\sigma[\hat{\overline{x^2}}]}{\sigma_x^2} \approx \left(\frac{2T_I}{T}\right)^{1/2} \simeq \frac{1}{\sqrt{N}}. \tag{12.38a}$$

As discussed by, for example, Bendat and Piersol (1986) the corresponding error for $(\overline{x^2})^{1/2}$ can be evaluated from

$$\varepsilon[\hat{\overline{x^2}}] \simeq 2\varepsilon[(\hat{\overline{x^2}})^{1/2}]. \tag{12.38b}$$

It is illustrative to compare $\varepsilon[\hat{\overline{x^2}}]$ (in eqn (12.38a)) and $\varepsilon[\hat{\bar{X}}]$ (in eqn (12.26)). Noting that the integral time-scale, T_I, is the same in the two equations for the error estimates, and denoting the total sampling time for $\hat{\bar{X}}$ as $T_{\bar{X}}$ and for $\overline{x^2}$ as $T_{\overline{x^2}}$, then

$$\frac{\varepsilon[\hat{\overline{x^2}}]}{\varepsilon[\hat{\bar{X}}]} \approx \left[\frac{T_{\bar{X}}}{T_{\overline{x^2}}}\right]^{1/2} \left[\frac{\sigma_x}{\bar{X}}\right]^{-1}. \tag{12.39}$$

Consider Example A in Section 12.3.1.1 in which $\sigma_u/\bar{U} = 0.1$. If the same sampling criteria is used for $\hat{\bar{X}}$ and $\hat{\overline{x^2}}$, that is, $T_{\bar{X}} = T_{\overline{x^2}}$, then $\varepsilon[\hat{\overline{x^2}}]$ will be ten times larger than $\varepsilon[\hat{\bar{X}}]$. The uncertainty in the turbulent quantities is therefore always much larger than the uncertainty in the mean value, when the same sampling criteria are applied to both statistical quantities.

Higher-order moments The mth-moment, $\overline{x^m}$, of the fluctuating component $x(t)$ of the sample record is defined as

$$\overline{x^m} = \lim_{T \to \infty} \frac{1}{T} \int_0^T x^m(t)\, dt, \tag{12.40}$$

where m has a positive integer value. The measured third moment can be expressed in digital form as

$$\widehat{\overline{x^3}} = \frac{1}{N} \sum_{n=1}^{N} (X(n) - \hat{\bar{X}})^3, \tag{12.41}$$

and by expansion the following more computer efficient form can be obtained

$$\widehat{\overline{x^3}} = \frac{1}{N} \left(\sum_{n=1}^{N} X^3(n) - 3\hat{\bar{X}} \sum_{n=1}^{N} X^2(n) + 2N\hat{\bar{X}}^3 \right). \tag{12.42}$$

The third moment, which is normally presented in nondimensional form as the skewness,

$$S = \overline{x^3} / \sigma_x^3, \tag{12.43}$$

is an indicator of a lack of statistical symmetry in the signal. An example of a signal with positive skewness is shown in Fig. 12.7. The fourth moment, nondimensionalized by σ_x^4 is called the kurtosis, or the flatness factor; it is represented by the symbol K, where

$$K = \overline{x^4} / \sigma_x^4. \tag{12.44}$$

Two functions, one with a relatively small and the other with a relatively

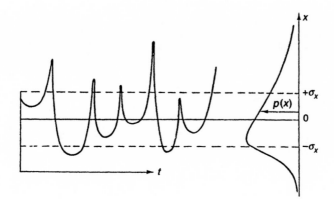

Fig. 12.7. A time-history record with positive skewness. (From Tennekes and Lumley 1972.)

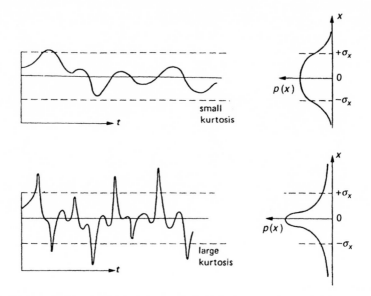

Fig. 12.8. Time-history records with small and large kurtosis. (From Tennekes and Lumley 1972.)

large kurtosis, are shown in Fig. 12.8. The value of the kurtosis is large if the values of $p(x)$ in the tails of the probability density function (defined in the following section) are relatively large.

Uncertainty estimate for $\widehat{x^m}$ Castro (1989) has discussed the standard deviations of higher-order moments, and has given the following standard deviations of $\sigma_x^3(6/N)^{1/2}$, $\sigma_x^4(96/N)^{1/2}$, $\sigma_x^5(720/N)^{1/2}$, and $\sigma_x^6(10\,170/N)^{1/2}$ for estimates of the third, fourth, fifth and sixth moment for a signal with a normal probability-density-function distribution. Consequently, even when N is large the standard deviations become very large for the higher-order moments.

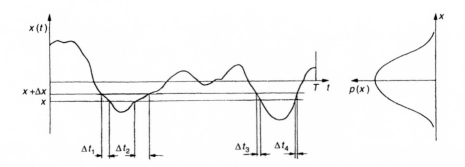

Fig. 12.9. The probability density function.

12.3.1.3 The probability density function

Consider a random stationary signal, $x(t)$, which is assumed to have a zero mean. At any particular instant there is a specific probability that the signal amplitude lies between x and $x + \Delta x$. With reference to Fig. 12.9, if $T_x = \Sigma \Delta t_i$ is the total amount of time that $x(t)$ falls inside the range $(x, x + \Delta x)$ during the total signal time T, then this probability is just T_x/T. This ratio will approach an exact probability description as T approaches infinity; that is

$$\text{Prob}\,[x < x(t) \leqslant x + \Delta x] = \lim_{T \to \infty} \frac{T_x}{T}. \tag{12.45}$$

A *probability density function*, $p(x)$, can be defined by

$$p(x) = \lim_{\Delta x \to 0} \frac{\text{Prob}\,[x < x(t) \leqslant x + \Delta x]}{\Delta x} = \lim_{\Delta x \to 0} \lim_{T \to \infty} \left[\frac{1}{\Delta x} \frac{T_x}{T} \right]. \tag{12.46}$$

The probability density function, $p(x)$, is always a real-valued, nonnegative function.

The probability that the instantaneous value, $x(t)$, is less than or equal to some value x is defined by $P(x)$, which is equal to the integral of the probability density function from $-\infty$ to x. This function $P(x)$ is known as the *probability distribution function*, and it should not be confused with the probability density function, $p(x)$. Specifically,

$$P(x) = \text{Prob}\,[x(t) \leqslant x] = \int_{-\infty}^{x} p(\xi)\, d\xi. \tag{12.47}$$

For a signal $x(t)$ with a zero mean we have using $p(x)$

$$0 = \int_{-\infty}^{\infty} x p(x)\, dx, \tag{12.48}$$

and the variance, σ_x^2 can be evaluated from

$$\sigma_x^2 = \overline{x^2} = \int_{-\infty}^{\infty} x^2 p(x)\, dx. \tag{12.49}$$

Similarly the higher-order moments can be expressed as

$$\overline{x^m} = \int_{-\infty}^{\infty} x^m p(x)\, dx. \tag{12.50}$$

For digital data analysis of a finite record, an estimate (measured value) of the probability density function can be obtained from

$$\hat{p}(x) = \frac{N_x}{NW}, \tag{12.51}$$

where N is the total number of samples and N_x is the number of data points with values of $x(t)$ between $x - W/2$ and $x + W/2$. It should be noted that the estimate $\hat{p}(x)$ is not unique since it clearly depends on the width, W, of the amplitude intervals.

Uncertainty estimate for $p(x)$ The estimate $\hat{p}(x)$ is in general a biased estimate of $p(x)$, and both the bias and variance of $\hat{p}(x)$ must be considered in evaluating the mean square error

$$E[(\hat{p}(x) - p(x))^2] = \text{var}[\hat{p}(x)] + b^2[\hat{p}(x)]. \tag{12.52}$$

It can be shown that

$$\text{var}[\hat{p}(x)] \approx \frac{p(x)}{NW}, \tag{12.53}$$

where N corresponds to statistically independent samples (spaced at least two integral time-scales apart), and the bias

$$b[\hat{p}(x)] \approx \frac{W^2}{24} p''(x),$$

where $p''(x)$ is the second derivative of $p(x)$ with respect to x. The total mean square error of the probability density estimate $\hat{p}(x)$ can therefore be expressed as

$$E[(\hat{p}(x) - p(x))^2] \approx \frac{p(x)}{NW} + \left[\frac{W^2 p''(x)}{24}\right]^2. \tag{12.54}$$

and the normalized mean square error is

$$\varepsilon^2[\hat{p}(x)] \approx \frac{1}{NWp(x)} + \frac{W^4}{576}\left[\frac{p''(x)}{p(x)}\right]^2. \tag{12.55}$$

Equation (12.55) demonstrates that there are conflicting requirements on the window width, W, in probability-density measurements. A large value of W is desirable to reduce the random error, but a small value is needed to reduce the bias error. In practice, values of $W \leqslant 0.2\sigma_x$ will usually limit the normalized bias error to less than 1% for approximately Gaussian random data (Bendat and Piersol 1986).

It is illustrative to compare $\varepsilon[\hat{p}(x)]$ (eqn (12.55)) and $\varepsilon[\hat{\bar{X}}]$ (eqn (12.26)). In this comparison, N in eqn (12.55) is set equal to $T/(2T_1)$ and it should be noted that the integral time-scale is the same in the equations for both error estimates. The total sampling time for $\hat{\bar{X}}$ is denoted as $T_{\bar{X}}$ and for $\hat{p}(x)$ it is denoted as $T_{p(x)}$. If the width, W, is selected as corresponding to 32 intervals covering the amplitude range $\pm 3\sigma_x$, that is, $W = \sigma_x/6$, and the bias error is ignored, then

$$\frac{\varepsilon[\hat{p}(x)]}{\varepsilon[\hat{\bar{X}}]} \approx \left[\frac{T_{\bar{X}}}{T_{p(x)}}\right]^{1/2}\left[\frac{\sigma_x}{\bar{X}}\right]^{-1}\frac{2.5}{(\sigma_x p(x))^{1/2}}. \tag{12.56}$$

Consider Example A in Section 12.3.1.1 in which $\sigma_u/\bar{U} = 0.1$ and $\sigma_u = 1$ m s^{-1}. If the probability density function, $p(x)$, is a normalized Gaussian function then $p(0) = 1/(2\pi)^{1/2} \simeq 0.4$. If the same sampling criteria are selected for $\hat{\bar{X}}$ and for $\hat{p}(x)$ then $T_{\hat{\bar{X}}} = T_{p(x)}$ and eqn (12.56) predicts that $\varepsilon[\hat{p}(0)]$ is about 40 times larger than $\varepsilon[\hat{\bar{X}}]$. Comparing $\varepsilon[\hat{p}(x)]$ in eqn (12.55) and $\varepsilon[\widehat{x^2}]$ in eqn (12.38a), then for similar sampling criteria, and with $W = \sigma_x/6$

$$\frac{\varepsilon[\hat{p}(x)]}{\varepsilon[\widehat{x^2}]} \approx \frac{2.5}{(\sigma_x p(x))^{1/2}}. \qquad (12.57)$$

Consequently, for the conditions in example A, $\varepsilon[\hat{p}(0)]$ is about four times larger than $\varepsilon[\widehat{x^2}]$.

12.3.2 The joint statistics of two sample records

This section deals with the joint amplitude statistics of two simultaneous sample records $x(t)$ and $y(t)$, which are both assumed to have zero means.

12.3.2.1 The mean value, the spatial correlation, and higher moments

The mean value $\overline{xy}$ The mean value of the product $x(t)\,y(t)$ is defined as

$$\overline{xy} = \lim_{T \to \infty} \frac{1}{T}\int_0^T x(t)\,y(t)\,\mathrm{d}t. \qquad (12.58)$$

The corresponding equation for the estimate $\widehat{\overline{xy}}$ obtained by digital sampling is

$$\widehat{\overline{xy}} = \frac{1}{N}\sum_{n=1}^{N} x(n)\,y(n). \qquad (12.59)$$

The quantity $\overline{xy}$ is important in turbulent-flow studies because it typically represents a shear stress, $\overline{uv}$, or a velocity temperature correlation, $\overline{u\theta}$.

Uncertainty estimate for $\widehat{\overline{xy}}$ The mean value $\overline{xy}$ is a special case ($\tau = 0$) of the cross-correlation function $R_{xy}(\tau)$ discussed in Section 12.4.2. Using eqn (12.82), a conservative estimate for the variance var$[\widehat{\overline{xy}}] = \sigma^2[\widehat{\overline{xy}}]$ will be

$$\mathrm{var}[\widehat{\overline{xy}}] = \mathrm{var}[\hat{R}_{xy}(0)] \approx \frac{T_1\sigma_x^2\sigma_y^2}{T}[1 + \rho_{xy}^2(0)]. \qquad (12.60)$$

Provided that $\sigma_x \simeq \sigma_y$ then the error estimate for $\widehat{\overline{xy}}$ is similar to that for $\widehat{x^2}$ (see eqn (12.37a)).

Spatial correlation When the two fluctuating signals $x(t)$ and $y(t)$ are obtained from spatially separated sensors, the term $\overline{xy}$ represents a spatial correlation, and information can be obtained about typical length scales in turbulent flows. As discussed in Section 2.4.2, a spatial correlation coefficient function may be introduced

$$\rho_{ij}(r) = \frac{\overline{u_i(0)\, u_j(r)}}{\sigma_{u_i}\sigma_{u_j}}, \qquad (12.61)$$

where u_i and u_j are fluctuating velocity components and the variable r is the spatial separation between the two sensors. $\rho_{ij}(r)$ can be determined by measuring $\overline{u_i(0)\, u_j(r)}$ for various spatial separations. The use of this function for the evaluation of spatial integral and microscales is described in Section 2.4.2.

The *higher-order moments* $\overline{x^m y^p}$, where m and p are positive integers, can be defined as

$$\overline{x^m y^p} = \lim_{T \to \infty} \frac{1}{T} \int_0^T x^m(t)\, y^p(t)\, \mathrm{d}t. \qquad (12.62)$$

and the corresponding digital estimate is given by

$$\widehat{\overline{x^m y^p}} = \frac{1}{N} \sum_{n=1}^{N} x^m(n)\, y^p(n). \qquad (12.63)$$

12.3.2.2 Joint-probability density function

The joint-probability density distribution $p(x, y)$ can be defined with reference to Fig. 12.10, which contains a pair of simultaneous time-history

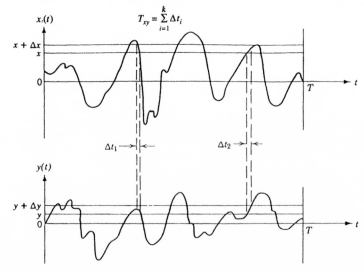

Fig. 12.10. Joint-probability measurement. (From Bendat and Piersol 1966. Reproduced with kind permission of John Wiley & Sons Inc. © John Wiley & Sons Inc.)

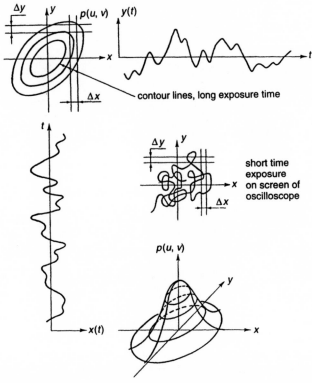

Fig. 12.11. A visual display of a joint-probability density function. (From Tennekes and Lumley 1972.)

records, $x(t)$ and $y(t)$, which both have zero means. Denoting as $T_{x,y}$ the amount of time that $x(t)$ and $y(t)$ simultaneously fall inside the ranges $(x, x + \Delta x)$ and $(y, y + \Delta y)$ respectively, during a total sample time, T, then

$$p(x,y) = \lim_{\substack{\Delta x \to 0 \\ \Delta y \to 0}} \lim_{T \to \infty} \left[\frac{T_{x,y}}{\Delta x \Delta y T} \right]. \qquad (12.64)$$

As described by Tennekes and Lumley (1972), a simple way of visualizing the joint-probability density distribution is to display one variable, $x(t)$, on the x-axis of an oscilloscope, while the other variable, $y(t)$, is displayed on the y-axis, as shown in Fig. 12.11. The joint-probability density function $p(x, y)$ is proportional to the fraction of the time that the moving spot in Fig. 12.11 spends in a small window between x and $x + \Delta x$, y and $y + \Delta y$.

For digital data analysis an estimate, $\hat{p}(x, y)$ can be obtained from

$$\hat{p}(x, y) = \frac{N_{x,y}}{NW_x W_y},$$

(12.65)

where W_x and W_y are narrow intervals centred around x and y, respectively, and $N_{x,y}$ is the number of pairs of data that simultaneously fall within these intervals.

Uncertainty estimate for $\hat{p}(x, y)$ Assuming that the windows W_x and W_y are sufficiently small so that the bias error is negligible, then the mean square error associated with the estimate $\hat{p}(x, y)$ will be given by the variance, var$[\hat{p}(x, y)]$. A first-order estimate of this variance is given by

$$\text{var}\,[\hat{p}(x,y)] \approx \frac{A\,p(x,y)}{NW_x W_y},$$

(12.66)

where A is an unknown constant (to be determined from experimental data) and N corresponds to statistically independent samples.

12.4 TIME-DOMAIN ANALYSIS

12.4.1 The autocorrelation function

The autocorrelation function describes the general dependence of the data at one time on the values at another time. Consider a time-history record, $x(t)$, with a zero mean, as illustrated in Fig. 12.12. For an infinitely long time record, the autocorrelation between the values of $x(t)$ at times t and $t + \tau$ can be expressed as

$$R_x(\tau) = \lim_{T \to \infty} \frac{1}{T} \int_0^T x(t)\,x(t + \tau)\,dt.$$

(12.67)

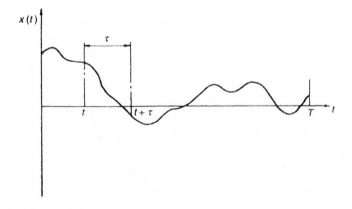

Fig. 12.12. Autocorrelation measurements.

The quantity $R_x(\tau)$ is always a real-valued even function so that

$$R_x(-\tau) = R_x(\tau). \tag{12.68}$$

The maximum value of $R_x(\tau)$ occurs at $\tau = 0$ with

$$\sigma_x^2 = R_x(0). \tag{12.69}$$

The autocorrelation coefficient function, $\rho_x(\tau)$, is defined as

$$\rho_x(\tau) = \frac{R_x(\tau)}{\sigma_x^2}, \tag{12.70}$$

and a typical autocorrelation coefficient function, $\rho_x(\tau)$, is shown in Fig. 12.13. The corresponding integral time-scale,

$$T_I = \int_0^\infty \rho_x(\tau)\, d\tau, \tag{12.71}$$

is a measure of the time separation over which the signals $x(t)$ and $x(t + \tau)$ are correlated. As discussed in Section 12.3.1.1, the two signals can be considered to be statistically uncorrelated when $\tau \geqslant 2T_I$. For digital data analysis, $\tau = 2T_I$ is the optimum time interval between samples for the amplitude-domain statistics described in Section 12.3. However, if this sampling rate is selected for autocorrelation analysis then the correlated part of $R_x(\tau)$ (or $\rho_x(\tau)$) would be missed. Assume that p equally spaced points are required to specify the autocorrelation function in the time-lag interval $0 \leqslant \tau \leqslant 2T_I$. In this case, the interval between samples will be $\Delta t = 2T_I/p$, the sampling rate $SR = 1/\Delta t = p/(2T_I)$, and for a time-history record with a total number of samples, N, only $N_p = N/p$ can be considered to be statistically independent.

A time-history record, $x(t)$, will, when sampled digitally at equally spaced intervals Δt, produce a digital time record $x(n\Delta t)$, $n = 0, 1, \ldots,$

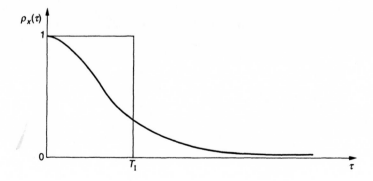

Fig. 12.13. The autocorrelation coefficient function, $\rho_x(\tau)$, showing the integral time-scale, T_I.

$N - 1$. An unbiased estimate of the autocorrelation function at the time delay $\tau = r\Delta t$ can be obtained (see, for example, Castro 1989) from

$$\hat{R}_x(r\Delta t) = \frac{1}{N-r} \sum_{n=0}^{N-r} x(n\Delta t) x(n\Delta t + r\Delta t), \quad r = 0, 1, 2, \ldots, m, \quad (12.72)$$

where r is called the lag number and m is the maximum lag number. To obtain a similar uncertainty in the estimate $\hat{R}_x(r\Delta t)$ for all values of r, it is usual to select $m \ll N$.

Uncertainty estimate for $\hat{R}_x(\tau)$ It can be shown for large values of the sampling time, T, that the variance of the autocorrelation function can be expressed as

$$\text{var}[\hat{R}_x(\tau)] \approx \frac{1}{T} \int_{-\infty}^{\infty} [R_x^2(\xi) + R_x(\xi + \tau) R_x(\xi - \tau)] \, d\xi, \quad (12.73)$$

and introducing the autocorrelation coefficient function $\rho_x(\tau) = R_x(\tau)/\sigma_x^2$ then

$$\text{var}[\hat{R}_x(\tau)] \approx \frac{\sigma_x^4}{T} \int_{-\infty}^{\infty} [\rho_x^2(\xi) + \rho_x(\xi + \tau) \rho_x(\xi - \tau)] \, d\xi, \quad (12.74)$$

At $\tau = 0$, eqn (12.74) becomes

$$\text{var}[\hat{R}_x(0)] \approx \frac{4\sigma_x^4}{T} \int_0^{\infty} \rho_x^2(\xi) \, d\xi \approx \frac{2T_1\sigma_x^4}{T}. \quad (12.75)$$

Since $R_x(0) = \sigma_x^2$, eqn (12.75) is of course identical to the variance of $\overline{x^2}$, eqn (12.37). For large τ where $\rho_x(\tau)$ approaches zero, it follows that

$$\rho_x^2(\xi) \gg \rho_x(\xi + \tau) \rho_x(\xi - \tau)$$

and

$$\text{var}[\hat{R}_x(\tau)] \approx \frac{2\sigma_x^4}{T} \int_0^{\infty} \rho_x^2(\xi) \, d\xi \approx \frac{T_1\sigma_x^4}{T}. \quad (12.76)$$

A general conservative estimate for the variance of $\hat{R}_x(\tau)$ can be given by

$$\text{var}[\hat{R}_x(\tau)] \approx \frac{T_1\sigma_x^4}{T} [1 + \rho_x^2(\tau)]. \quad (12.77)$$

It is instructive to compare the sampling criteria and standard deviations for $\overline{x^2}$ and $\hat{R}_x(\tau)$. From eqn (12.37a)

$$\sigma[\overline{x^2}] \approx \left(\frac{2T_1\sigma_x^4}{T} \right)^{1/2}$$

and optimum measurements are based on N independent samples taken two integral time-scales $(2T_1)$ apart; that is $N \approx T/(2T_1)$ (see eqn (12.24)). From eqn (12.77) we obtain for $\sigma[\hat{R}_x(\tau)]$

$$\sigma[\hat{R}_x(\tau)] \approx \left\{\frac{T_1\sigma_x^4}{T}\left[1 + \rho_x^2(\tau)\right]\right\}^{1/2}.$$

If we evaluate both quantities from a time-history record length, T, then for $\tau = 0$, at which $\rho_x(0) = 1$,

$$\frac{\sigma[\hat{R}_x(0)]}{\sigma[\hat{x^2}]} \approx 1.$$

This is, of course, consistent with $R_x(0) = \overline{x^2} = \sigma_x^2$. However, to specify p points of the autocorrelation $R_x(\tau) = R_x(p\Delta\tau)$, in the interval $0 \leqslant \tau \leqslant 2T_1$, a sampling rate $SR = p/(2T_1)$ is required. Therefore, although the required total sampling time is the same for the $\overline{x^2}$- and $R_x(\tau)$-evaluations, the $R_x(\tau)$ measurements must be sampled p-times faster and the total number of samples required will therefore also be p times larger.

12.4.2 The cross-correlation function

The cross-correlation function describes the general dependence of the values from one set of data on the other. Consider the pair of time-history records $x(t)$ and $y(t)$ shown in Fig. 12.14. Both are assumed to have zero means. For infinitely long time records, the cross-correlation between $x(t)$ and $y(t + \tau)$ can be expressed as

$$R_{xy}(\tau) = \lim_{T \to \infty} \frac{1}{T}\int_0^T x(t)\,y(t + \tau)\,dt. \qquad (12.78)$$

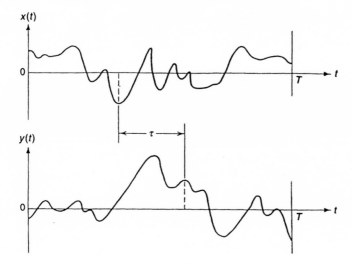

Fig. 12.14. Cross-correlation measurements. (From Bendat and Piersol 1966. Reproduced with kind permission of John Wiley & Sons Inc. © John Wiley & Sons Inc.)

For a digital representation of finite time records, an unbiased estimate of the cross-correlation function at the time delay $\tau = r\Delta t$ can be obtained from

$$\hat{R}_{xy}(r\Delta t) = \frac{1}{N - r} \sum_{n=0}^{N-r} x(n\Delta t)\, y(n\Delta t + r\Delta t), \quad r = 0, 1, 2, \ldots, m,$$

(12.79)

Uncertainty estimate for $\hat{R}_{xy}(\tau)$ For large values of the sampling time, T, it can be shown that an approximate expression for the variance of the cross-correlation function is

$$\mathrm{var}[\hat{R}_{xy}(\tau)] \approx \frac{1}{T} \int_{-\infty}^{\infty} [R_x(\xi)\, R_y(\xi) + R_{xy}(\xi + \tau)\, R_{yx}(\xi - \tau)]\, \mathrm{d}\xi,$$

(12.80)

which, introducing the auto- and cross-correlation coefficient functions, can be expressed as

$$\mathrm{var}[\hat{R}_{xy}(\tau)] \approx \frac{\sigma_x^2 \sigma_y^2}{T} \int_{-\infty}^{\infty} [\rho_x(\xi)\, \rho_y(\xi) + \rho_{xy}(\xi + \tau)\, \rho_{yx}(\xi - \tau)]\, \mathrm{d}\xi.$$

(12.81)

Provided that $\rho_x(\tau) \approx \rho_y(\tau)$, a conservative estimate for the variance of $\hat{R}_{xy}(\tau)$ can be given by

$$\mathrm{var}[\hat{R}_{xy}(\tau)] \approx \frac{T_1 \sigma_x^2 \sigma_y^2}{T} [1 + \rho_{xy}^2(\tau)],$$

(12.82)

where $\rho_{xy}(\tau) = R_{xy}(\tau)/\sigma_x \sigma_y$ is the cross-correlation coefficient function.

The cross-correlation function does not necessarily have a maximum at $\tau = 0$, as was the case for the autocorrelation function. If $x(t)$ and $y(t)$ represent signals from two sensors separated a distance Δx in the mean-flow direction then the time delay, τ_m, corresponding to the maximum value of $R_{xy}(\tau)$ (or $\rho_{xy}(\tau)$) can be used to evaluate the average convection velocity $U_c = \Delta x/\tau_m$ of disturbances in the flow. A typical variation in $\rho_{xy}(\tau, \Delta x)$ for spatially separated sensors is shown in Fig. 12.15. It should be observed that the peak value of ρ_{xy} decreases with increasing values of Δx, reflecting the deformation of the turbulent structures in the flow. For each cross-correlation, $\rho_{xy}(\tau, \Delta x)$, the time-delay range must be selected so that

$$-n \times 2T_1 < \tau - \tau_m < n \times 2T_1,$$

where n is typically between 1 and 2 depending on how much of the correlation tail needs to be determined. If p points are required to cover the time interval $2T_1$ then the sampling rate must be $SR = p/(2T_1)$.

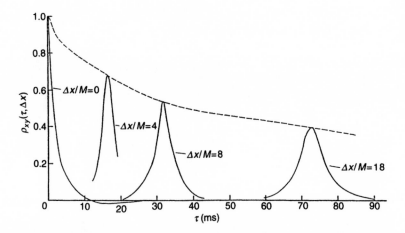

Fig. 12.15. Measured space–time cross-correlations in grid-generated turbulence where Δx is the spatial separation and M is the mesh size. (From Comte-Bellot and Corrsin 1971. Reproduced with kind permission of Cambridge University Press from the *Journal of Fluid Mechanics*.)

12.5 SPECTRAL DENSITY FUNCTIONS

Spectral density functions can be defined in three different equivalent ways:

(1) via a narrow-band filtering method;
(2) via a Fourier transform of the correlation function;
(3) via a finite Fourier transform of the original time-history record.

For simplicity of presentation, the next section will only discuss the definition and measurements of the autospectral density function from a single time-history record, $x(t)$. The methodology presented can be extended to the cross-spectral density function between two simultaneous time-history records $x(t)$ and $y(t)$, as described by, for example, Bendat and Piersol (1986).

12.5.1 Spectra by the narrow-band filtering method

Physically, the autospectral density function, $G_x(f)$, is a measure of how much energy is contained within the signal in each frequency band. For a bandpass filter with sharp cut-off characteristics covering the frequency range $(f, f + \Delta f)$, an estimate of the mean square value of $x(t)$ within the bandwidth Δf centred at f can be obtained from

$$\hat{\sigma}_x^2(f, \Delta f) = \frac{1}{T}\int_0^T x^2(t, f, \Delta f)\, \mathrm{d}t. \tag{12.83}$$

Equation (12.83) is a consistent and unbiased estimate of the true mean square value, and as T tends to infinity

$$\sigma_x^2(f, \Delta f) = \lim_{T \to \infty} \frac{1}{T} \int_0^T x^2(t, f, \Delta f) \, dt. \tag{12.84}$$

Mathematically, the autospectral density funtion, $G_x(f)$, can be defined by

$$G_x(f) = \lim_{\Delta f \to 0} \frac{\sigma_x^2(f, \Delta f)}{\Delta f} = \lim_{\substack{T \to \infty \\ \Delta f \to 0}} \frac{1}{(\Delta f)T} \int_0^T x^2(t, f, \Delta f) \, dt, \tag{12.85}$$

and an estimated (measured) value, $\hat{G}_x(f)$, for a finite bandwidth, Δf, and a finite record length, T, can be obtained from

$$\hat{G}_x(f) = \frac{1}{(\Delta f)T} \int_0^T x^2(t, f, \Delta f) \, dt = \frac{\hat{\sigma}_x^2(f, \Delta f)}{\Delta f}. \tag{12.86}$$

A spectrum analyser is a physical realization of the narrow-band-filter concept, which requires the following analysis steps (see Fig. 12.16):

(1) frequency filtering of the signal, $x(t)$, by a narrow-band filter with a central frequency f and a bandwidth Δf. The output signal from the filter unit will be $x(t, f, \Delta f)$;

(2) squaring of the instantaneous value of the filtered signal to obtain $x^2(t, f, \Delta f)$;

(3) averaging of the squared instantaneous value over the record length, T, to obtain an estimate $\hat{\sigma}_x^2(f, \Delta f)$ of the mean-square value of the filtered signal within the bandwidth Δf;

(4) division by the filter bandwidth, Δf, to obtain the estimate $\hat{G}_x(f)$;

(5) repeating steps 1–4 for selected values of the central frequency, f, throughout the frequency range of interest.

To cover the complete frequency range, a tuneable narrow-band filter can be used. Alternatively, the spectrum analyser may contain a bank of narrow-

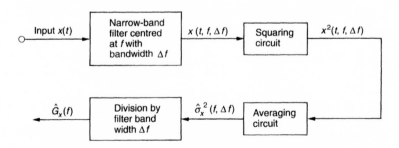

Fig. 12.16. The analysis steps in a power spectral-density analyser.

band filters, which together cover the frequency range of interest. The bandwidth, Δf, of the filters is usually selected so that it has either a constant value or a constant percentage of the central frequency, f.

Uncertainty estimate for $\hat{G}_x(f)$ The estimate $\hat{G}_x(f)$ defined by eqn (12.86) is a biased estimate, and the mean square error of the estimate $\hat{G}_x(f)$ must therefore be evaluated from

$$E[(\hat{G}_x(f) - G_x(f))^2] = \text{var}[\hat{G}_x(f)] + b^2[\hat{G}_x(f)]. \quad (12.87)$$

The bias can, provided that $\Delta f^2 G_x''(f) < G_x(f)$, be evaluated from

$$b[\hat{G}_x(f)] \approx \frac{\Delta f^2}{24} G_x''(f), \quad (12.88)$$

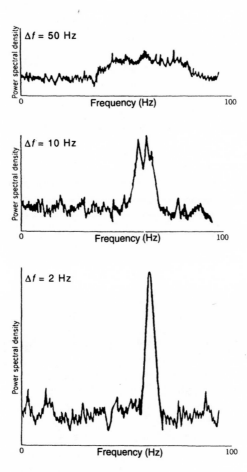

Fig. 12.17. The effect of the filter bandwidth on peak detection in power spectral-density measurements. (From Bendat and Piersol 1966. Reproduced with kind permission of John Wiley & Sons Inc. © John Wiley & Sons Inc.)

where $G_x''(f)$ is the second derivative of $G_x(f)$ with respect to f, and the variance of $\hat{G}_x(f)$ can be expressed as

$$\text{var}[\hat{G}_x(f)] \approx \frac{G_x^2(f)}{\Delta f T}. \qquad (12.89)$$

The normalized mean square error, $\varepsilon^2[\hat{G}_x(f)]$, can therefore be expressed as

$$\varepsilon^2[\hat{G}_x(f)] = \frac{E[(\hat{G}_x(f) - G_x(f))^2]}{G_x^2(f)} \approx \frac{1}{\Delta f T} + \frac{\Delta f^4}{576}\left(\frac{G_x''(f)}{G_x(f)}\right)^2. \qquad (12.90)$$

It follows from eqn (12.90) that there are conflicting requirements for the resolution bandwidth, Δf. A large value of Δf is required to reduce the random error, while a small value of Δf is needed to suppress the bias part of the error. The suppression of the bias error can be very critical since spectra often display sharp peaks with large second derivatives. Such peaks can be completely missed if too large a value is selected for Δf, as shown in Fig. 12.17.

12.5.2 Spectra via correlation functions

The autospectral density function can also be defined in terms of a single Fourier transform of a previously calculated correlation function $R_x(\tau)$. This approach gives a two-sided spectral density function usually denoted by $S_x(f)$, which is defined for f over $(-\infty, \infty)$. The two functions $S_x(f)$ and $R_x(\tau)$ form an exact Fourier transform pair, given by

$$S_x(f) = \int_{-\infty}^{\infty} R_x(\tau)\,e^{-i2\pi f \tau}d\tau \qquad (12.91)$$

and

$$R_x(\tau) = \int_{-\infty}^{\infty} S_x(f)\,e^{i2\pi f \tau}df. \qquad (12.92)$$

Equations (12.91) and (12.92) and the corresponding equations for the cross-correlations and cross-spectral density functions are often referred to as the *Wiener–Khintchine relations* in honour of the two mathematicians, N. Wiener of the United States and A. I. Khintchine of the former USSR, who independently proved the Fourier transform relationship between correlation functions and spectral density functions in the early 1930s.

The two-sided autospectral density function $S_x(f)$, where f varies over $(-\infty, \infty)$, is related to the one-sided autospectral density function $G_x(f)$, $0 \leqslant f < \infty$, as measured by the narrow-band filtering method (Section 12.5.1), by

$$G_x(f) = \begin{cases} 2S_x(f), & 0 \leqslant f < \infty \\ 0 & \text{otherwise.} \end{cases} \qquad (12.93)$$

Since the autocorrelation $R_x(\tau)$ is an even function, the relationships between $G_x(f)$ and $R_x(\tau)$ can also be expressed as

$$G_x(f) = 4 \int_0^\infty R_x(\tau) \cos(2\pi f \tau) \mathrm{d}\tau, \qquad (12.94)$$

$$R_x(\tau) = \int_0^\infty G_x(f) \cos(2\pi f \tau) \mathrm{d}f. \qquad (12.95)$$

This method of evaluating $G_x(f)$ is very time-consuming in practice since it requires the evaluation of $R_x(\tau)$ first. It is usually only applied to short time series and it will therefore not be considered further.

12.5.3 Spectra via finite Fourier transforms

Bendat and Piersol (1986) have given an excellent description of the related definitions and evaluations of both autospectral and cross-spectral density functions, and this reference is the main source for the following description.

The autospectral density function can also be defined and evaluated from a finite Fourier transform of the original data record. Consider a stationary random process, with the time-history record $x_k(t)$ representing one possible realization of this process. For a finite time interval $0 \leqslant t \leqslant T$ define

$$S_x(f, T, k) = \frac{1}{T} X_k^*(f, T) X_k(f, T), \qquad (12.96a)$$

where

$$X_k(f, T) = \int_0^T x_k(t) e^{-i2\pi f t} \mathrm{d}t. \qquad (12.96b)$$

The quantity $X_k(f, T)$ represents a finite Fourier transform of $x_k(t)$, and $X_k^*(f, T)$ is the complex conjugate of $X_k(f, T)$. A finite time interval, T, has been chosen, since the corresponding Fourier transform will exist for a stationary record, whereas the Fourier transform does not exist if T is infinite.

It is a common mistake to define the two-sided autospectral density function from a single realization, $x_k(t)$, by letting the time record, T, tend to infinity. This procedure defines

$$S_x(f, k) = \lim_{T \to \infty} S_x(f, T, k), \qquad (12.97)$$

which is an unsatisfactory definition of the autospectral density function, since it will be shown that the relative error in the spectral estimate does not improve as T tends to infinity. The correct way of defining $S_x(f)$ is by the expression

$$S_x(f) = \lim_{T \to \infty} E[S_x(f, T, k)], \qquad (12.98)$$

where $E[S_x(f, T, k)]$ is the expected value obtained by taking the ensemble average. It can be shown that in the limit when T tends to infinity

$$\lim_{T \to \infty} E[S_x(f, T, k)] = \int_{-\infty}^{\infty} R_x(\tau) e^{-i2\pi f \tau} d\tau. \qquad (12.99)$$

Replacing $S_x(f)$ by the corresponding one-sided autospectral density function, $G_x(f)$ gives

$$G_x(f) = 2 \lim_{T \to \infty} \frac{1}{T} E[|X_k(f, T)|^2], \qquad (12.100)$$

where $|X_k(f, T)|^2$ is the sum of the square of the amplitudes of the real and imaginary parts of the Fourier integral.

Spectral estimates from finite time records An estimate for $G_x(f)$, based on a single time-history record of a finite time, T, is given by

$$\tilde{G}_x(f) = \frac{2}{T} |X(f, T)|^2. \qquad (12.101)$$

However, the direct application of eqn (12.101) has two distinct disadvantages: spectral leakage and an unacceptably high random error.

Spectral leakage (windowing) Analysis of a time-history record of a finite time, T, will result in spectral 'leakage', and windowing techniques can be applied to minimize this phenomenon. A time-history record, $x_k(t)$, specified for a finite time, T, can be viewed mathematically as an unlimited time-history record, $x(t)$, viewed through a window, $w(t)$, where

$$w(t) = \begin{cases} 1 & 0 \leqslant t \leqslant T \\ 0 & \text{otherwise.} \end{cases} \qquad (12.102)$$

Consequently, for the complete time interval $(-\infty, \infty)$

$$x(t)\,w(t) = \begin{cases} x_k(t) & 0 \leqslant t \leqslant T \\ 0 & \text{otherwise.} \end{cases} \qquad (12.103)$$

The finite Fourier transform of the function $x(t)\,w(t)$ is given by

$$X(f, T) = \int_{-\infty}^{\infty} x(t)\,w(t) e^{-i2\pi f t} dt. \qquad (12.104)$$

The convolution theorem states that the Fourier transform of a product of two functions is equivalent to the transform of one of the functions convolved with the transform of the other; that is

$$X(f, T) = \int_{-\infty}^{\infty} X(\xi) W(f - \xi) \, d\xi. \qquad (12.105)$$

For the rectangular function, $w(t)$ defined by eqn (12.102), the Fourier transform, $W(f)$, is given by

$$W(f) = T \frac{\sin \pi f T}{\pi f T}. \qquad (12.106)$$

A plot of $W(f)$ is shown in Fig. 12.18. The large side lobes of $W(f)$ allow leakage of power at frequencies which are well separated from the main lobe of the spectral window, and this may introduce a significant distortion of the estimated spectrum, particularly when the data is narrow band in nature. If it is considered necessary to suppress the leakage problem then it is common practice to introduce a data window that tapers the time-history data to eliminate the discontinuities at the beginning and the end of the record being analysed. Many of the early windows (Bartlett, Tukey and Parzen, see for example, Jenkins and Watts 1968) were developed for spectral estimates via correlation functions. However, spectral evaluations are now normally performed using the Fast Fourier Transform (FFT) technique. Harris (1978) has presented an in-depth study of modern windows and the special conditions which apply to their use in conjunction with

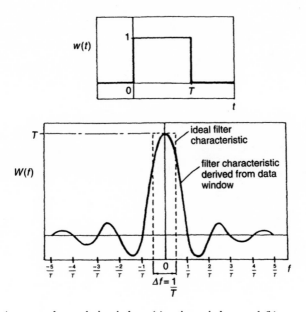

Fig. 12.18. A rectangular analysis window: (a) a time window, and (b) a spectral window.

FFT routines. One of the earliest (and it is still commonly employed) is the cosine squared or Hanning window, shown in Fig. 12.19, and defined as

$$w_h(t) = \begin{cases} \dfrac{1}{2}\left[1 - \cos\left(\dfrac{2\pi t}{T}\right)\right] = 1 - \cos^2\left(\dfrac{\pi t}{T}\right) & 0 \leqslant t \leqslant T \\ 0 & \text{otherwise.} \end{cases} \qquad (12.107a)$$

The Fourier Transform of eqn (12.107a) is

$$W_h(f) = \tfrac{1}{2}W(f) - \tfrac{1}{4}W(f - f_1) - \tfrac{1}{4}W(f + f_1), \qquad (12.107b)$$

where $f_1 = 1/T$ and $W(f)$ is defined in eqn (12.106).

Comparing Figs 12.18 and 12.19 it can be observed that the Hanning window has smaller side-lobes and a larger bandwidth for the main lobe than the rectangular window. In general, the use of any tapering operation to suppress side-lobe leakage will increase the bandwidth of the main lobe of the spectral window in an autospectral density analysis. However, this can be countered by: (i) increasing the block duration, T, for each *FFT* computation to bring the main lobe bandwidth back to what it would have been without tapering, and then (ii) overlapping the blocks so that the total duration of the data analysed remains the same (Piersol, personal communication). As described by Welch (1967), the result of this overlapping procedure will provide spectral estimates for tapered data with the same frequency resolution and only a slightly greater random error than an analysis without tapering.

Finally, if a tapered window, $w(t)$, is used in eqn (12.104) to evaluate a

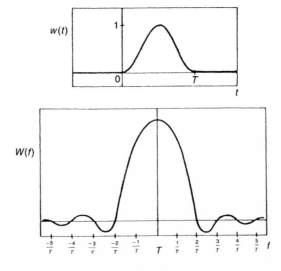

Fig. 12.19. A Hanning analysis window: (a) a time window, and (b) a spectral window.

spectral estimate then a loss factor must be introduced. As discussed by Bendat and Piersol (1986), when a Hanning window is used then a scale factor of $(8/3)^{1/2}$ should be introduced to obtain the correct magnitude of the spectral density estimate. Harris (1978) and Nuttall (1981) have discussed the effects of other tapering data windows.

Uncertainty estimate for $\tilde{G}_x(f)$ For a single sample record it can be shown (see, for example Jenkins and Watts 1968; George *et al.* 1978) that the normalized standard deviation, which defines the random portion of the estimation error, is

$$\varepsilon_r[\tilde{G}_x(f)] = \frac{\sigma[\tilde{G}_x(f)]}{G_x(f)} = 1; \qquad (12.108)$$

that is, the standard deviation is as big as the spectral signal itself.

In practice, the random error of the autospectrum estimate can be reduced by computing an ensemble of estimates from n_d different subrecords each of length, T, as shown in Fig. 12.20. By ensemble averaging the results a 'smooth' estimate can be obtained from

$$\hat{G}_x(f) = \frac{2}{n_d T} \sum_{j=1}^{n_d} |X_j(f, T)|^2. \qquad (12.109)$$

Provided the individual spectral estimates are statistically independent, then it follows by applying the principle of eqn (12.21) to $\hat{G}_x(f)$ that

$$\text{var}[\hat{G}_x(f)] = \frac{\text{var}[\tilde{G}_x(f)]}{n_d} = \frac{G_x^2(f)}{n_d}. \qquad (12.110)$$

Consequently, using the uncertainty specification in Section 12.2, the uncertainty in $\hat{G}_x(f)$ can he expressed as

$$G_x(f)(1 - z_{\alpha/2}/\sqrt{n_d}) < \hat{G}_x(f) < G_x(f)(1 + z_{\alpha/2}/\sqrt{n_d}), \qquad (12.111)$$

with a probability of $(1 - \alpha)\%$. From eqn (12.110) the normalized random rms error for $\hat{G}_x(f)$ is

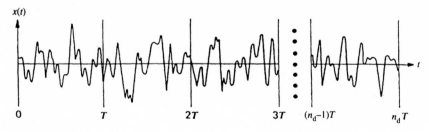

Fig. 12.20. The subdivision of data into n_d records of individual length, T. (From Bendat and Piersol 1986. Reproduced with kind permission of John Wiley & Sons Inc. © John Wiley & Sons Inc.)

$$\varepsilon_r[\hat{G}_x(f)] = \frac{1}{\sqrt{n_d}} \, . \tag{12.112}$$

12.5.4 Digital spectral analysis

Spectral evaluations via correlations or finite Fourier transforms can both be implemented digitally. However, the first method, which requires the initial evaluation of the autocorrelation function is computationally very laborious except for short time-history records, and the digital implementation of this method will therefore not be discussed.

When spectral density functions are evaluated by digital techniques, the discretization of the original continuous time-history record will result in an additional problem, which is usually referred to as aliasing.

Aliasing Consider, as shown in Fig. 12.4, the discretization of a continuous signal $x(t)$ using a sampling rate $SR = 1/\Delta t$, so that the individual samples are a time Δt apart. It follows, from the general sampling theorem, that at least two samples per cycle are required to define a frequency component in the original data. Consequently, the highest frequency that can be defined uniquely by a sampling rate of $SR = 1/\Delta t$ is $1/(2\Delta t)$ Hz. Any energy in the original signal above $1/(2\Delta t)$ will be 'folded back' and interpreted by the digital sampling as lower frequency components in the energy spectrum. This principle is illustrated in Fig. 12.21. The band limiting frequency,

$$f_c = \frac{1}{2\Delta t}, \tag{12.113}$$

is called the Nyquist frequency, or folding frequency. As discussed by, for example, Castro (1989), all data at frequencies of $f \pm nf_c$ have the same cosine function as data at a frequency of f, and they are therefore indis-

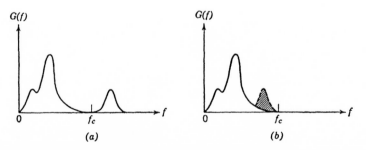

Fig. 12.21. An aliased autospectrum due to folding: (a) the true spectrum. (b) the aliased spectrum. (From Bendat and Piersol 1966. Reproduced with kind permission of John Wiley & Sons Inc. © John Wiley & Sons Inc.)

tinguishable from f. Consequently, measurements of the signal's energy content at frequencies f which are less than f_c will be contaminated by the energy at frequencies $2nf_c \pm f$.

The usual methods of surmounting the aliasing problem are either to low-pass filter the input signal at f_c, to remove the energy at higher frequencies, or to choose a sufficiently high sampling rate to ensure that energy levels in the signal above the aliasing frequency are negligible. Castro (1989) concluded that unless the original signal was fed through an anti-aliasing filter prior to digitization, distorted spectral data can be expected at frequencies higher than about one sixth of the digital sampling frequency, with the degree of distortion depending on the spectral shape beyond the aliasing frequency.

Discrete Fourier Transform (DFT) The discrete version of the finite Fourier transform described in Section 12.5.3 is called the Discrete Fourier Transform (DFT). Consider a digital time-record (Fig. 12.20) subdivided into n_d segments of equal time, T. Each record segment, $x_j(t)$, is represented by N data points, and the values of the total record (time $T_r = N \times T$) can be specified as $x_j(n\Delta t)$, $(n = 0, 1, \ldots, N - 1; j = 1, 2, \ldots, n_d)$; $\Delta t = T/N$ is the time interval between individual samples.

The finite Fourier transform (see eqn (12.109)) will produce values at discrete frequencies

$$k\Delta f = \frac{k}{T} = \frac{k}{N\Delta t}, \tag{12.114}$$

with the lowest frequency being $f_{\min} = \Delta f = 1/T$. This is also the frequency resolution of the analysis.

The Fourier component for each segment can, using the digital equivalent of eqn (12.96), be expressed as

$$X_j(k\Delta f) = \Delta t \sum_{n=0}^{N-1} x_j(n\Delta t)e^{\frac{-i\,2\pi kn}{N}}, \quad k = 0, 1, \ldots, N - 1. \tag{12.115}$$

The corresponding estimate of the one-sided autospectral density function, $\hat{S}_x(k\Delta f)$, based on averaging over the n_d segments, can be evaluated as

$$\hat{S}_x(k\Delta f) = \frac{1}{n_d N\Delta t} \sum_{j=1}^{n_d} |X_j(k\Delta f)|^2, \quad k = 0, 1, \ldots, N - 1. \tag{12.116}$$

As described by Bendat and Piersol (1986) the values of k above the Nyquist frequency, $N/2$, can be interpreted as negative frequency values, and the estimate of the one-sided spectrum, $\hat{G}_x(k\Delta f)$, can therefore be expressed as

$$\hat{G}_x(k\Delta f) = \frac{2}{n_d N\Delta t} \sum_{j=1}^{n_d} |X_j(k\Delta f)|^2, \quad k = 1, \ldots, \frac{N}{2} - 1. \tag{12.117}$$

Note, that for $k = 0$ and $N/2$ the multiplier in eqn (12.117) is 1 since the value of $S_x(f)$ is not doubled at zero frequency or at the Nyquist frequency, when the negative frequencies are folded over to obtain $G_x(f)$ (Piersol, personal communication; see also Castro 1989, p. 102).

As discussed above, the lowest frequency of the *DFT* analysis is $f_{min} = 1/T$. However, the spectral value at zero frequency, $G_x(0)$, has an important physical meaning. From the correlation definition of $G_x(f)$ in eqn (12.94) it follows that

$$G_x(0) = 4\int_0^\infty R_x(\tau)\,d\tau, \tag{12.118}$$

and introducing eqns (12.70) and (12.71) we have

$$G_x(0) = 4\sigma_x^2 T_I, \tag{12.119}$$

where σ_x is the standard deviation and T_I is the integral time-scale of the time-history record. It is therefore common in most *DFT* and *FFT* routines to include evaluations for $f = 0$. In this analysis, it is important to remove the *DC* component of the time-history record for each segment, since it will appear as an addition to the spectral estimate, $G_x(0)$.

The computational disadvantage of the *DFT* method is that, for each subinterval, it requires N^2 multiply–add operations, which is very time consuming, except for low values of N.

Cooley and Tukey (1965) have developed an algorithm for the fast computation of the *DFT* which has become known as the Fast Fourier Transform (*FFT*). In this algorithm, there are about $4N\log_2(2N)$ complex multiplications (Castro 1989), which is smaller than N^2 when N is large.

There are a number of forms of the *FFT* algorithm, but most require that the input data consists of a number of points which are powers of 2. The *FFT* algorithm is now readily available in commercial software packages. Most *DFT* and *FFT* packages perform the numerical spectral operations on N numbers and the user has to introduce the appropriate value of Δt in eqns (12.115) to (12.117) to obtain the one-sided spectrum, $\hat{G}_x(k\Delta f)$ (Castro 1989). For details of the *FFT* procedure and its practical implementation, the reader is referred to Bergland (1969), Bracewell (1978), Brigham (1974), Cooley and Tukey (1965), Jenkins and Watts (1968), Newland (1984), Bendat and Piersol (1986), and Roberts and Mullis (1987).

Sampling criteria for $\hat{G}_x(f)$ The evaluation of the spectral estimate, $\hat{G}_x(f)$, requires the selection of an appropriate number of subrecords, n_d, the sampling interval, Δt, and the number of samples, N, in each subrecord. The sampling time for each subrecord is $T = \Delta t N$.

1. The time interval, Δt, between samples and the corresponding sampling rate, $SR = 1/\Delta t$, is determined by the maximum frequency of interest. The Nyquist frequency, $f_c = 1/(2\Delta t)$, is the maximum frequency that

can be interpreted uniquely. A common method of dealing with the aliasing problem is to feed the signal through a low-pass filter which is set to f_c prior to digitization.

2. The lowest resolvable frequency (and also the bandwidth of the spectral analysis) is $\Delta f = 1/T = 1/(N\Delta t)$. Having specified Δt by the maximum frequency of interest, the frequency resolution will depend on the number of samples, N, selected in each subrecord. If there is likely to be one or more narrow peaks in the spectra, then $N\Delta t$ must be sufficiently large to ensure that $\Delta f = 1/(N\Delta t)$ is small enough to resolve the peak satisfactorily. To achieve this, it may be necessary to set $\Delta f < 0.1f_p$, where f_p is the frequency of the lowest narrow band peak. When a tapered data window is used, the increase in the effective bandwidth should be considered, and a scaling factor is required in the spectral evaluation to compensate for the loss caused by the tapering.

3. The uncertainty in the spectral estimate, $\hat{G}_x(f)$, as specified by the normalized random rms error, $\varepsilon_r[\hat{G}_x(f)]$, in eqn (12.112), varies inversely with the square root of the number of subrecords, n_d. Therefore to keep ε_r below 10% requires 100 blocks (subrecords) of data.

REFERENCES

Abdel-Rahman, A., Tropea C., Slawson, P., and Strong, A. (1987). On temperature compensation in hot-wire anemometry. *J. Phys. E.: Sci. Instr.*, **20**, 315-319.

Abdel-Rahman, A. A., Hitchman, G. J., Slawson, P. R., and Strong, A. B. (1989). An X-array hot-wire technique for heated turbulent flows of low velocity. *J. Phys. E.: Sci. Instr.*, **22**, 638-644.

Abel, R. and Resch, F. J. (1978). A method for the analysis of hot-film anemometer signals in two-phase flows. *Int. J. Multiphase Flow*, **4**, 523-533.

Acrivlellis, M. (1978). An improved method for determining the flow field of multidimensional flows of any turbulence intensity. *DISA Info.*, No. 23, 11-16.

Acrivlellis, M. (1980). Measurements by means of triple-sensor probes. *J. Phys. E.: Sci. Instr.*, **13**, 986-992.

Acrivlellis, M. (1982). Hot-wire signal analysis. *J. Phys. E.: Sci. Instr.*, **15**, 289-297.

Acrivlellis, M. (1989). Determination of the magnitudes and signs of flow parameters by hot-wire anemometry, Part 2, Measurements using a triple hot-wire probe. *Rev. Sci. Instr.*, **60**, 1281-1285.

Acrivlellis, M. and Felsch, K. O. (1979). A new method of determining the flow field with low and high turbulence intensity. *Proc. Symp. Turbulent Boundary Layers*, Niagara Falls, NY, pp. 169-177.

Adams, E. W. and Johnston, J. P. (1988). Effects of the separating shear layer on the reattachment flow structure, Part One, Pressure and turbulence quantities. *Exp. in Fluids*, **6**, 400-408.

Adams, E. W., Johnston, J. P., and Eaton, J. K. (1984). Experiments on the structure of turbulent reattaching flow. Report MD-43, Department of Mechanical Engineering, Stanford University, Stanford, CA.

Addison, J. S. and Hodson, H. P. (1990a). Unsteady transition in an axial-flow turbine, Part 1, Measurements on the turbine rotor. *ASME, J. Turbomachinery*, **112**, 206-214.

Addison, J. S. and Hodson, H. P. (1990b). Unsteady transition in an axial-flow turbine, Part 2, Cascade measurements and modelling. *ASME, J. Turbomachinery*, **112**, 215-221.

Adler, D. (1972). A hot wire technique for continuous measurement in unsteady concentration fields of binary gaseous mixtures. *J. Phys. E.: Sci. Instr.*, **5**, 163-169.

Adler, D. and Zvirin, Y. (1975). The time response of a hot wire concentration transducer. *J. Phys. E.: Sci. Instr.*, **8**, 185-188.

Adrian, R. J., Johnson, R. E., Jones, B. G., Merati, P., and Tung, A. T. C. (1984). Aerodynamic disturbances of hot-wire probes and directional sensitivity. *J. Phys. E.: Sci. Instr.*, **17**, 62-71.

Ahmed, S. A. and So, R. M. C. (1986). Concentration distributions in a model combustor. *Exp. in Fluids*, **4**, 107-113.

Ainsworth, R. W., Dietz, A. J., and Nunn, T. A. (1991). The use of semi-conductor sensors for blade surface pressure measurement in a model turbine stage. *ASME, J. Eng. for Gas Turbines and Power*, **113**, 261-268.

Ajagu, C. O., Libby, P. A., and LaRue, J. C. (1982). Modfied gauge for time-resolved skin-friction measurements. *Rev. Sci. Instr.*, **53**, 1920-1926.

Alfredsson, P. H. and Johansson, A. V. (1984a). Time scales for turbulent channel flow. *Phys. Fluids*, **27**, 1974–1981.

Alfredsson, P. H. and Johansson, A. V. (1984b). On the detection of turbulence-generating events. *J. Fluid Mech.*, **139**, 325–345.

Alfredsson, P. H., Johansson, A. V., Haritonidis, J. H., and Eckelmann, H. (1988). The fluctuating wall-shear stress and the velocity field in the viscous sublayer. *Phys. Fluids*, **31**, 1026–1033.

Ali, S. F. (1975). Hot-wire anemometry in moderately heated flow. *Rev. Sci. Instr.*, **46**,185–191.

Al-Kayiem, H. H. (1989). Separated flow on a high lift wing. Ph.D. thesis, Department of Mechanical and Manufacturing Engineering, University of Bradford.

Al-Kayiem, H. H. and Bruun, H. H. (1991). Evaluation of a flying X hot-wire probe system. *Meas. Sci. Technol.*, **2**, 374–380.

Almquist, P. and Legath, E. (1965). The hot-wire anemometer at low air velocities. *DISA Info.*, No. 2, 3–4.

Amsden, A. A., O'Rourke, P. J., Butler, T. D., Meintjes, K., and Fansler, T. D. (1992). Comparisons of computed and measured three-dimensional velocity fields in a motored two-stroke engine. SAE Paper No. 920418.

Andreas, E. L. (1979a). The calibration of cylindrical hot-film velocity sensors. *ASME, J. Appl. Mech.*, **46**, 15–20.

Andreas, E. L. (1979b). Analysis of crossed hot-film velocity data. *DISA Info.*, No. 24, 15–23.

Andreopoulos, J. (1981). Comparison test of the response to pitch angles of some digital hot wire anemometry techniques. *Rev. Sci. Instr.*, **52**, 1376–1381.

Andreopoulos, J. (1983a). Improvements of the performance of triple hot wire probes. *Rev. Sci. Instr.*, **54**, 733–740.

Andreopoulos, J. (1983b). Statistical errors associated with probe geometry and turbulence intensity in triple hot-wire anemometry. *J. Phys. E.: Sci. Instr.*, **16**, 1264–1271.

Andreopoulos, J., Durst, F., Zaric, Z., and Jovanovic, J. (1984). Influence of Reynolds number on characteristics of turbulent wall boundary layers. *Exp. in Fluids*, **2**, 7–16.

Andrews, G. E., Bradley, D., and Hundy, G. F. (1972). Hot wire anemometer calibration for measurements of small gas velocities. *Int. J. Heat Mass Transf.*, **15**, 1765–1786.

Anhalt, J. (1973). Device for in-water calibration of hot-wire and hot-film probes. *DISA Info.*, No. 15, 25–26.

Antonia, R. A. (1972). Conditionally sampled measurements near the outer edge of a turbulent boundary layer. *J. Fluid Mech.*, **56**, 1–18.

Antonia, R. A. (1981). Conditional sampling in turbulence measurement. *Ann. Rev. Fluid Mech.*, **13**, 131–156.

Antonia, R. A. and Atkinson, J. D. (1974). Use of a pseudo-turbulent signal to calibrate an intermittency measuring circuit. *J. Fluid Mech.*, **64**, 679–699.

Antonia, R. A. and Bradshaw, P. (1971). Conditional sampling of turbulent shear flows. Imperial College Aero. Report No 71-04, London.

Antonia, R. A. and Browne, L. W. B. (1983). The destruction of temperature fluctuations in a turbulent plane jet. *J. Fluid Mech.*, **134**, 67–83.

Antonia R. A. and Mi, J. (1993a). Corrections for velocity and temperature derivatives in turbulent flows. *Exp. in Fluids*, **14**, 203–208.

Antonia, R. A. and Mi, J. (1993a). Corrections for velocity and temperature derivatives in turbulent flows. *Exp. in Fluids*, **14**, 203–208.

Antonia, R. A. and Mi, J. (1993b). Temperature dissipation in a turbulent round jet. *J. Fluid Mech.*, **250**, 531–551.

Antonia, R. A. and Rajagopalan, S. (1990). Performance of lateral vorticity probe in a turbulent wake. *Exp. in Fluids*, **9**, 118–120.

Antonia, R. A., Prabhu, A. and Stephenson, S. E. (1975). Conditionally sampled measurements in a heated turbulent jet. *J. Fluid Mech.*, **72**, 455–480.

Antonia, R. A., Danh, H. Q., and Prabhu, A. (1977). Response of a turbulent boundary layer to a step change in surface heat flux. *J. Fluid Mech.*, **80**, 153–177.

Antonia, R. A., Chambers, A. J., and Hussain, A. K. M. F. (1980). Errors in simultaneous measurements of temperature and velocity in the outer part of a heated jet. *Phys. Fluids*, **23**, 871–874.

Antonia, R. A., Browne, L. W. B., and Chambers, A. J. (1981a). Determination of time constants of cold wires. *Rev. Sci. Instr.*, **52**, 1382–1385.

Antonia, R. A., Chambers, A. J., Sokolov, M., and Van Atta, C. W. (1981b). Simultaneous temperature and velocity measurements in the plane of symmetry of a transitional turbulent spot. *J. Fluid Mech.*, **108**, 317–343.

Antonia, R. A., Browne, L. W. B., and Shah, D. A. (1988a). Characteristics of vorticity fluctuations in a turbulent wake. *J. Fluid Mech.*, **189**, 349–365.

Antonia, R. A., Krishnamoorthy, L. V., and Fulachier, L. (1988b). Correlation between the longitudinal velocity fluctuation and temperature fluctuation in the near-wall region of a turbulent boundary layer. *Int. J. Heat Mass Transf.*, **31**, 723–730.

Antonia, R. A., Zhu, Y., and Kim, J. (1993). On the measurement of lateral velocity derivatives in turbulent flows. *Exp. in Fluids*, **15**, 65–69.

Arcoumanis, C. and Whitelaw, J. H. (1987). Fluid mechanics of internal combustion engines – a review. *Proc. I. Mech. Engrs.*, **201**, 57–74.

Arcoumanis, C. and Whitelaw, J. H. (1991). In-cylinder flow measurements in motored internal combustion engines. *I. Mech. E. Conf.* paper C433/060.

Arcoumanis, C., Bicen, A. F., and Whitelaw, J. H. (1982). Measurements in a motored four-stroke reciprocating model engine. *ASME, J. Fluids Eng.*, **104**, 235–241.

Arcoumanis, C., Bicen, A. F., and Whitelaw, J. H. (1983). Squish and swirl-squish interaction in motored model engines. *ASME, J. Fluids Eng.*, **105**, 105–112.

Arcoumanis, C., Bicen, A. F., Vlachos, N. S., and Whitelaw, J. H. (1985). Effects of flow and geometry boundary conditions on fluid motion in a motored IC model engine. *Proc. Inst. Mech. Eng.*, **196**, 1–10.

Arcoumanis, C., Whitelaw, J. H., and Flamang, P. (1989). Flow in the inlet manifold of a production diesel engine. *Proc. Inst. Mech. Eng.*, **203**, 39–49.

Arcoumanis, C., Hadjiapostolou, A., and Whitelaw, J. H. (1987a). Swirl center precession in engine flows. SAE Paper No. 870370.

Arcoumanis, C., Vafidis, C., and Whitelaw, J. H. (1987b). Valve and in-cylinder flow generated by a helical port in a production diesel engine. *ASME, J. Fluids Eng.*, **109**, 368–375.

Arcoumanis, C., Hu, Z., Vafidis C., and Whitelaw, J. H. (1990). Tumbling motion: A mechanism for turbulence enhancement in spark-ignition engines. SAE Paper 900060.

Arcoumanis, C., Enotiadis, A. C., and Whitelaw, J. H. (1991a). Frequency analysis of tumble and swirl in motored engines. *Proc. Inst. Mech. Eng.*, **205D**, 177–184.

Arcoumanis, C., Hadjipostolou, A., and Whitelaw, J. H. (1991b). Flow and combustion in a hydra direct-injection diesel engine. SAE Paper No. 910177.

Arcoumanis, C., Nouri, J. M., and Andrews, R. J. (1992). Measurement of the internal fluid flow in a diesel engine injector using refractive index matching. Proc. I. Mech. Eng., Seminar on 'Diesel fuel injection systems' pp. 27–33.

Arcoumanis, C., Hu, Z., and Whitelaw, J. H. (1993). Steady flow characterization of tumble-generating four-valve cylinder heads. Proc. Inst. Mech. Eng., 207, 203–210.

Arya, S. P. S. and Plate, E. J. (1969). Hot-wire measurements in non-isothermal flow. Inst. and Cont. Syst., 42, 87–90.

Arzoumanian, E. and Debieve, J. F. (1989). Un processus programme en anémometrie a fil chaud en écoulement supersonique. Report Institut de Mecanique Statistique de la Turbulence, Marseille, France.

Aydin, M. and Leutheusser, H. J. (1980). Very low velocity calibration and application of hot-wire probes. DISA Info., No. 25, 17–18.

Azad, R. S. (1983). Corrections to measurements by hot wire anemometer in proximity of a wall. Report MET-7, Department of Mechanical Engineering, University of Manitoba.

Azad, R. S. and Burhanuddin, S. (1983). Measurements of some features of turbulence in wall-proximity. Exp. in Fluids, 1, 149–160.

Azad, R. S. and Kassab, S. Z. (1989). A new method of obtaining dissipation. Exp. in Fluids, 7, 81–87.

Badran, O. O. (1993). A flying hot-wire study of separated flows. Ph.D. Thesis, Department of Mechanical and Manufacturing Engineering, University of Bradford.

Baille, A. (1973). Hot wire cooling relationship for low flow velocities (In French). Bull. Dir. Etud. Rech., 3, Series A.

Baldwin, L. V., Sandborn, V. A., and Laurence, J. C. (1960). Heat transfer from transverse and yawed cylinders in continuum, slip and free molecule air flows. ASME, J. Heat Transf., 82, 77–86.

Balint, J. L., Wallace, J. M. (1989). The statistical properties of the vorticity field of a two-stream turbulent mixing layer. Advance in Turbulence 2, (ed. H. H. Fernholz and H. E. Fiedler), pp. 74–78. Springer, Berlin.

Balint, J. L., Vukoslavčević, P., and Wallace, J. M. (1987). A study of vortical structure of the turbulent boundary layer. In Advances in Turbulence, (ed. G. Comte-Bellot and J. Mathiew), pp. 456–464. Springer, Berlin.

Balint, J. L., Wallace, J. M., and Vukoslavčević, P. (1991). The velocity and vorticity vector fields of a turbulent boundary layer, Part 2, Statistical properties. J. Fluid Mech., 228, 53–86.

Ball, W. F., Pettifer, H. F., and Waterhouse, C. N. F. (1983). Laser doppler velocimeter measurements of turbulence in a direct-injection diesel combustion chamber. Inst. Mech. Eng. Paper No. C52/83.

Bank, N. and Gauvin, W. H. (1977). Inclined hot-wire response equations for a flow field having a dominant tangential velocity component. Can. J. Chem. Eng., 55, 516–520.

Barre, S., Dupont, P., and Dussauge, J. P. (1992). Hot-wire measurements in turbulent transonic flows. Eur. J. Mech., B/Fluids, 11, 439–454.

Barrett, R. V. (1987). Measurements in 3-dimensional boundary layers and narrow wakes using a single sensor hot wire probe. ICIASF '87 record. Proc. Int.

Congr. Instrumentation in Aerospace Simulation Facilities, Williamsburg, USA, pp. 356-368.

Barrett, R. V. (1989). A single-sensor hot-wire anemometer for 3-dimensional boundary layers and narrow wakes. *DANTEC Info.*, No. 8, 6-10.

Bartenwerfer, M. (1979). Remarks on hot-wire anemometry using 'squared signals'. *DISA Info.*, No. 24, 4, 40.

Batt, R. G. and Kubota, T. (1968). Experimental investigation of laminar near wakes behind 20° wedges at $M_\infty = 6$, *AIAA J.*, **6**, 2077-2083.

Bauer, A. B. (1965). Direct measurement of velocity by hot-wire anemometry. *AIAA J.*, **3**, 1189-1191.

Bearman, P. W. (1971). Corrections for the effect of ambient temperature drift on hot-wire measurements in incompressible flow. *DISA Info*, No. 11, 25-30.

Bechert, D. and Pfizenmaier, E. (1975). On wavelike pertubations in the free jet travelling faster than the mean flow in the jet. *J. Fluid Mech.*, **72**, 341-352.

Becker, H. A. and Brown A. P. G. (1974). Response of Pitot probes in turbulent streams. *J. Fluid Mech.*, **62**, 85-114.

Beguier, C., Rey, C., Dumas, R., and Astier, M. (1973). Une novelle sonde anémométrique. *C. R. Acad. Sci. Paris*, **A277**, 475-478.

Beguier, C., Dekeyser, I., and Launder, B. E. (1978a). Ratio of scalar and velocity dissipation time scales in shear flow turbulence. *Phys. Fluids*, **21**, 307-310.

Beguier, C., Fulachier, L., and Keffer, J. F. (1978b). The turbulent mixing layer with an asymmetrical distribution of temperature. *J. Fluid Mech.*, **89** 561-587.

Behrens, W. (1971). Total temperature thermocouple probe based on recovery temperature of circular cylinder. *Int. J. Heat Mass Transf.*, **14**, 1621-1630.

Behrens, W. and Ko, D. R. S. (1971). Experimental stability studies in wakes of two-dimensional slender bodies at hypersonic speeds. *AIAA J.*, **9**, 851-857.

Bell, J. H. and Mehta, R. D. (1992). Measurements of the streamwise vortical structures in a plane mixing layer. *J. Fluid Mech.*, **239**, 213-248.

Bellhouse, B. J. and Rasmussen, C. G. (1968). Low-frequency characteristics of hot-film anemometers. *DISA Info.*, No. 6, 3-10.

Bellhouse, B. J. and Schultz, D. L. (1966). Determination of mean and dynamic skin friction, separation and transition in low-speed flow with a thin-film heated element. *J. Fluid. Mech.*, **24**, 379-400.

Bellhouse, B. J. and Schultz, D. L. (1967). The determination of fluctuating velocity in air with heated thin film gauges. *J. Fluid Mech.*, **29**, 289-295.

Bellhouse, B. J. and Schultz, D. L. (1968). The measurement of fluctuating skin friction in air with heated thin-film gauges. *J. Fluid Mech.*, **32**, 675-680.

Belmabrouk, H., Lance, M., Grosjean, N., and Michard, M. (1991). Turbulence length scale measurements by two-point laser doppler anemometry in a steady flow. SAE Paper No. 910474.

Bendat, J. S. and Piersol, A. G. (1966). *Measurement and Analysis of Random Data*. John Wiley & Sons, New York.

Bendat, J. S. and Piersol, A. G. (1971). *Random Data: Analysis and Measurement Procedures*. John Wiley & Sons, New York.

Bendat, J. S. and Piersol, A. G. (1980). *Engineering Appplications of Correlation and Spectral Analysis*. John Wiley & Sons, New York.

Bendat, J. S. and Piersol, A. G. (1986). *Random Data: Analysis and Measurement Procedures*, (2nd ed). John Wiley & Sons, New York.

Bennett, J. C. (1978). High response measurements of prop-fan flow fields. *Proc. Dynamic Flow Symposium*, Johns Hopkins University, Baltimore, Maryland.

Berger, E. (1964). Bestimmung der hydrodynamischen Grössen einer Kármánschen Wirbelstrasse aus Hitzdrahtmessungen bei kleinen Reynoldschen Zahlen. *ZFW* **12**, 41–59.

Berger, E., Freymuth, P., and Froebel, E. (1963). Theorie und konstruktion von konstant-temperatur-hitzdrahtanemometern. *Konstruktion*, **15**, 495–497.

Bergland, G. D. (1969). A guided tour of the fast Fourier transform. *IEEE Spectrum*, **6**, 41–52.

Bergström, H. and Högström, U. (1987). Calibration of a three-axial fibre-film system for meteorological turbulence measurements. *DANTEC Info.*, No. 5, 16–20.

Bertrand, J. and Couderc, J. P. (1978). Hot-film probe calibration in liquids. *DISA Info.*, No. 23, 28–32.

Bestion, D., Gaviglio, J., and Bonnet, J. P. (1983). Comparison between constant-current and costant-temperature hot-wire anemometers in high-speed flows. *Rev. Sci. Instr.*, **54**, 1513–1524.

Betchov, R. (1948*a*). L'influence de la conduction thermique sur les anemometres a fils chauds. *Proc. Kon. Ned. Akad. Wet.*, **51**, 721–730.

Betchov, R. (1948*b*). L'inertie thermique des anemometres a fil chaud et le calcul approche de leurs caracteristiques. *Proc. Kon. Ned. Akad. Wet.*, **51**, 224–233.

Beuther, P. D., Shabbir, A., and George, W. K. (1987). X-wire response in turbulent flows of high intensity turbulence and low mean velocities. *ASME, Symp. on Thermal Anemometry, ASME Fluid Eng. Div. Spring Meeting, FED* Vol. 53, Cincinnati, 39–42.

Bhatia, J. C., Durst, F., and Jovanovic, J. (1982). Corrections of hot-wire anemometer measurements near walls. *J. Fluid Mech.*, **122**, 411–431.

Bicen, A. F., Vafidis, C., and Whitelaw, J. H. (1984). Steady and unsteady airflow through an intake valve of a reciprocating engine. In *Flow in Internal Combustion Engines - II, FED* Vol. 20, 47–55.

Bicen, A. F., Vafidis, C., and Whitelaw, J. H. (1985). Steady and unsteady airflow through the intake valve of a reciprocating engine. *J. Fluids Eng.*, **107**, 413–420.

Binder, A., Schroeder, Th., and Hourmouziadis, J. (1989). Turbulence measurements in a multistage low-pressure turbine. *ASME, J. Turbomachinery*, **111**, 153–161.

Birch, A. D., Brown, D. R., Dodson, M. G., and Swaffield, F. (1986). Aspects of design and calibration of hot-film aspirating probes used for the measurement of gas concentration. *J. Phys. E.: Sci. Instr.*, **19**, 59–63.

Bissonnette, L. R. and Mellor, G. L. (1974). Experiments on the behaviour of an axisymmetric turbulent boundary layer with a sudden circumferential strain. *J. Fluid Mech.*, **63**, 369–413.

Blackshear, P. L. and Fingerson, L. M. (1962). Rapid-response heat flux probe for high temperature gases. *J. Amer. Rocket Soc.*, **32**, 1709–1715.

Blackwelder, R. F. (1981). Hot-wire and hot-film anemometers. In *Methods of Experimental Physics: Fluid Dynamics*, (ed. R. J. Emrich), Vol. 18, Part A, pp. 259–314. Academic Press.

Blackwelder, R. F. and Eckelmann, H. (1979). Streamwise vortices associated with the bursting phenomenon. *J. Fluid Mech.*, **94**, 577–594.

Blackwelder, R. F. and Haritonidis, J. H. (1983). Scaling of the bursting frequency in turbulent boundary layers. *J. Fluid Mech.*, **132**, 87–103.

Blackwelder, R. F. and Kaplan, R. E. (1976). On the wall structure of the turbulent boundary layer. *J. Fluid Mech.*, **76**, 89–112.

Blackwelder, R. F. and Kovasznay, L. S. G. (1972). Time scales and correlations in a turbulent boundary layer. *Phys. Fluids*, **15**, 1545–1554.

Blair, M. F. and Bennett, J. C. (1987). Hot-wire measurements of velocity and temperature fluctuations in a heated turbulent boundary layer. *J. Phys. E.: Sci. Instr.*, **20**, 209–216.

Blinco, P. H. and Sandborn, V. A. (1973). Use of the split-film sensor to measure turbulence in water near a wall. *Proc. Symp. on Turbulence in Liquids*, University of Missouri, Rolla, pp. 403–413.

Blinco, P. H. and Simons, D. B. (1974). Characteristics of turbulent boundary shear stress. *J. Eng. Mech. Div.*, *ASCE*, **100**, No. EM2, 203–220.

Boerner, Th. and Leutheusser, H. J. (1984). Calibration of split-fibre probe for use in bubbly two-phase flow. *DISA Info.*, No. 29, 10–13.

Bogard, D. G. and Tiederman, W. G. (1986). Burst detection with single-point velocity measurements. *J. Fluid Mech.*, **162**, 389–413.

Boman, U. R. (1992). Hot-wire calibration over a large temperature range. *Exp. in Fluids*, **12**, 427–428.

Bond, A. D. and Porter, A. M. (1967). Self-aligning hot-wire probe. *J. Roy. Aero. Soc.*, **71**, 657–658.

Bonis, M. and van Thinh, N. (1973). A heat transfer law for a conical hot-film probe in water. *DISA Info.*, *No.* 14, 11–14.

Bonnet, J. P. and Alziary de Roquefort, T. (1980). Determination and optimization of frequency response of constant temperature hot-wire anemometers in supersonic flows. *Rev. Sci. Instr.*, **51**, 234–239.

Bonnet, J. P. and Knani, M. A. (1988). Calibration and use of inclined hot wires in a supersonic turbulent wake. *Exp. in Fluids*, **6**, 179–199.

Bonnet, J. P., Debieve, J. F., and Dussauge, J. P. (1986). Methodes de l'anemometrie par fil chaud pour l'etude des ecoulements turbulents a vitesse supersonique. Colloque DRET/ONERA, Ecoulements Turbulents Compressibles, Poitiers, pp. 195–214.

Bopp, S., Vafidis, C., and Whitelaw, J. H. (1986). The effect of engine speed on the TDC flowfield in a motored reciprocating engine. SAE Paper No. 860023.

Borgos, J. A. (1980). A review of electrical testing of hot-wire and hot-film anemometers. *TSI Quart.*, **VI**, No. 3, 3–9.

Börner, T. and Leutheusser, H. J. (1984). Calibration of split-fibre probe for use in bubbly two-phase flow. *DISA Info.*, No. 29, 10–13.

Börner, T., Franz, K., and Buchholz, R. (1980). Eichprobleme bei der anwendung der heissfilm-anemometertechnik zur bestimmung von turbulenzgrössen in zweiphasenströmungen. *Chem. Ing. Tech.*, **52**, 764–765.

Börner, T., Martin, W. W., and Leutheusser, H. J. (1982). Comparative measurements in bubbly two-phase flow using laser-doppler and hot-film velocimetry. *Proc. Annual Meeting A.I.Ch.E.*, LA, California.

Böttcher, J. and Eckelmann, H. (1985). Measurement of the velocity gradient with hot-film probes. *Exp. in Fluids*, **3**, 87–91.

Bourke, P. J. and Pulling, D. J. (1970). A turbulent heat flux meter and some measurements of turbulence in air flow through a heated pipe. *Int. J. Heat Mass Transf.*, **13**, 1331–1338.

Bowers, C. G., Willits, D. H., and Bowen, H. D. (1988). Comparison of temperature correction methods for hot wire anemometers. *Trans. ASAE*, **31**, 1552–1555.

Bracewell, R. N. (1978). *The Fourier Transform and its Applications* (2nd ed). McGraw-Hill, New York.

Bradbury, L. J. S. (1969). A pulsed-wire technique for measurements in highly-turbulent flow. NPL Aero Report 1284.

Bradbury, L. J. S. (1976). Measurements with a pulsed-wire and a hot-wire anemometer in the highly turbulent wake of a normal flat plate. *J. Fluid Mech.*, **77**, 473–497.

Bradbury, L. J. S. (1978). Examples of the use of the pulsed wire anemometer in higly turbulent flow. *Proc. Dynamic Flow Conf.*, Marseille, pp. 489–509.

Bradbury, L. J. S. and Castro, I. P. (1971). A pulsed-wire technique for measurements in highly turbulent flows. *J. Fluid Mech.*, **49**, 657–691.

Bradbury, L. J. S. and Castro, I. P. (1972). Some comments on heat-transfer laws for fine wires. *J. Fluid Mech.*, **51**, 487–495.

Bradley, T. A. and Ng, T. T. (1989). Phase-locking in a jet forced with two frequencies. *Exp. in Fluids*, **7**, 38–48.

Bradley, D., Hynes, J., Lawes, M., and Sheppard, C. G. W. (1988). Limitations to turbulence-enhanced burning rates in lean burn engines. Paper C46/88. *Proc. I. Mech. E. Int. Conf. on Combusion in Engines—Technology and Applications*, London, pp. 17–24.

Bradshaw, P. (1971). *An introduction to turbulence and its measurement.* Pergamon Press, Oxford.

Bradshaw, P. (1972). The understanding and prediction of turbulent flows. *Aero. J.*, **76**, 403–418.

Bradshaw, P. and Gregory, N. (1961). The determination of local turbulent skin friction from observations in the viscous sublayer. ARC R&M No. 3202.

Bradshaw, P. and Murlis J. (1974). On the measurement of intermittency in turbulent flow. Imperial College Aero Report, 74–04. Imperial College, London.

Bradshaw, P. and Terrell, M. G. (1969). A response of a turbulent boundary layer on an 'infinite' swept wing to the sudden removal of pressure gradient. NPL Aero. Report 1305. A. R. C. 31 514.

Bradshaw, P., Cebeci, T., and Whitelaw, J. H. (1981). *Engineering calculation methods for turbulent flow.* Academic Press, London.

Bragg, G. M. and Tevaarwerk, J. (1974). The effect of a liquid droplet on a hot wire anemometer probe. In *Flows, Its Measurement and Control in Science and Industry*, Vol. 1, ISA, pp. 599–603. Pittsburgh.

Bremhorst, K. (1978). Response to stream temperature perturbations at higher frequencies of finite aspect ratio hot-wire anemometers. *J. Phys. E.: Sci. Instr.*, **11**, 812–814.

Bremhorst, K. (1981). Effect of mounting systems on heat transfer from inclined cylinders in cross-flow. *Int. J. Heat Mass Transf.*, **24**, 243–250.

Bremhorst K. (1985). Effect of fluid temperature on hot-wire anemometers and an improved method of temperature compensation and linearisation without use of small signal sensitivities. *J. Phys. E.: Sci. Instr.*, **18**, 44–49.

Bremhorst, K. (1988). Flow interference effect with parallel array hot-wire anemometer probes. *J. Phys. E.: Sci. Instr.*, **21**, 722–724.

Bremhorst, K. and Gilmore, D. B. (1976a). Comparison of dynamic and static hot wire anemometer calibrations for velocity perturbation measurements. *J. Phys. E.: Sci. Instr.*, **9**, 1097–1100.

Bremhorst, K. and Gilmore, D. B. (1976b). Response of hot wire anemometer probes to a stream of air bubbles in a water flow. *J. Phys. E.: Sci. Instr.*, **9**, 347–352.

Bremhorst, K. and Gilmore, D. B. (1978). Influence of end conduction on the sensitivity to stream temperature fluctuations of a hot-wire anemometer. *Int. J. Heat Mass Transf.*, **21**, 145–154.

Bremhorst, K. and Graham, L. J. W. (1990). A fully compensated hot/cold wire anemometer system for unsteady flow velocity and temperature measurements. *Meas. Sci. Technol.*, **1**, 425–430.

Bremhorst, K. and Krebs, L. (1976). Reconsideration of constant current hot wire anemometers for the measurement of fluid temperature fluctuations. *J. Phys. E.: Sci. Instr.*, **9**, 804–806.

Bremhorst, K. and Listijono, J. (1987). Static pressure effects on calibration of velocity transducers at the nozzle exits. *Exp. in Fluids*, **5**, 344–348.

Bremhorst, K., Krebs, L., and Gilmore, D. B. (1977). The frequency response of hot-wire anemometer sensors to heating current fluctuations. *Int. J. Heat Mass Transf.*, **20**, 315–22.

Breugelmans, F. A. E., Mathiodakis, K., and Casalini, F. (1983). Flow in rotating stall cells of a low speed axial flow compressor. *6th Int. Symp. on Air-Breathing Engines*, Paris, pp. 632–642.

Brigham, E. O. (1974). *The Fast Fourier Transform*. Prenctice-Hall, Englewood Cliffs, New Jersey.

Brodkey, R. S., Wallace, J. M., and Eckelmann, H. (1974). Some properties of truncated turbulence signals in bounded shear flows. *J. Fluid Mech.*, **63**, 209–224.

Browand, F. K. and Troutt, T. R. (1980). A note on spanwise structure in the two-dimensional mixing layer. *J. Fluid Mech.*, **97**, 771–781.

Browand, F. K. and Weidman, P. D. (1976). Large scales in the developing mixing layer. *J. Fluid Mech.*, **76**, 127–144.

Brown, G. L. (1967). Theory and application of heated films for skin friction measurement. *Proc. Heat Transfer and Fluid Mechanics Institute*, Stanford University, Stanford, CA, pp. 361–381.

Brown, G. L. and Rebollo, M. R. (1972). A small, fast-response probe to measure composition of a binary gas mixture. *AIAA J.*, **10**, 649–652.

Brown, G. L. and Roshko, A. (1974). On density effects and large structure in turbulent mixing layers. *J. Fluid Mech.*, **64**, 775–816.

Browne, L. W. B. and Antonia, R. A. (1987). The effect of wire length on temperature statistics in a turbulent wake. *Exp. in Fluids*, **5**, 426–428.

Browne, L. W. B., Antonia, R. A., and Shah, D. A. (1988). Selection of wires and wire spacing for X-wires. *Exp. in Fluids*, **6**, 286–288.

Browne, L. W. B., Antonia, R. A., and Chua, L. P. (1989a). Calibration of X-probes for turbulent flow measurements. *Exp. in Fluids*, **7**, 201–208.

Browne, L. W. B., Antonia, R. A., and Chua, L. P. (1989b). Velocity vector cone angle in turbulent flows. *Exp. in Fluids*, **8**, 13–16.

Browne, L. W. B., Zhu, Y., and Antonia, R. A. (1991). Dissipation estimates in turbulent flows using the zero-wire-length technique. *Exp. in Fluids*, **11**, 197–199.

Bruun, H. H. (1971). Interpretation of a hot wire signal using a universal calibration law. *J. Phys. E.: Sci. Instr.*, **4**, 225–231.

Bruun, H. H. (1972). Hot-wire data corrections in low and high turbulence intensity flows. *J. Phys. E.: Sci. Instr.*, **5**, 812–818.

Bruun, H. H. (1975a). On the temperature dependence of constant temperature hot-wire probes with small wire aspect ratio. *J. Phys. E.: Sci. Instr.*, **8**, 942–951.

Bruun, H. H. (1975b). Interpretation of X-hot-wire signals. *DISA Info.*, No. 18, 5–10.

Bruun, H. H. (1976a). A note on static and dynamic calibration of constant-temperature hot-wire probes. *J. Fluid Mech.*, **76**, 145–155.

Bruun, H. H. (1976b). A digital comparison of linear and nonlinear hot wire data evaluation. *J. Phys. E.: Sci. Instr.*, **9**, 53–57.

Bruun, H. H. (1977). A time-domain analysis of the large-scale flow structure in a circular jet. Part 1. Moderate Reynolds number. *J. Fluid Mech.*, **83**, 641–671.

Bruun, H. H. (1979a). Review article: Interpretation of hot-wire probe signals in subsonic airflows. *J. Phys. E.: Sci. Instr.*, **12**, 1116–1128.

Bruun, H. H. (1979b). A time-domain evaluation of the large-scale flow structure in a turbulent jet. *Proc. Roy. Soc. Lond.*, **A367**, 193–218.

Bruun, H. H. (1987). Use of computers in HWA. Advanced HWA Course. University of Manchester Institute of Science and Technology.

Bruun, H. H. and Davies, P. O. A. L. (1972). Measurements of turbulent quantities by single hot-wires and X hot-wires using digital evaluation techniques. *Proc. Conf. on Fluid Dynamic Measurements in the Industrial and Medical Environment*, Leicester University, 163–166.

Bruun, H. H. and Farrar, B. (1988). Hot-film probe studies of kerosene/water and gas/liquid flows. *Proc, 1st World Conf. on Experimental Heat Transfer, Fluid Mechanics and Thermodynamics*, Dubrovnik, Yugoslavia, pp. 371–379.

Bruun, H. H. and Tropea, C. (1980). Calibration of normal, inclined and X-array hot-wire probes. Techical Report SFB80/M/170 Sonderforschungs-bereich 80, Universität Karlsruhe.

Bruun, H. H. and Tropea, C. (1985). The calibration of inclined hot-wire probes. *J. Phys. E.: Sci. Instr.*, **18**, 405–413.

Bruun, H. H., Khan, M. A., Al-Kayiem, H. H., and Fardad, A. A. (1988). Velocity calibration relationships for hot-wire anemometry. *J. Phys. E.: Sci. Instr.*, **21**, 225–232.

Bruun, H. H., Farrar, B., and Watson, I. (1989). A swinging arm calibration method for low velocity hot-wire probe calibration. *Exp. in Fluids*, **7**, 400–404.

Bruun, H. H., Al-Kayiem, H. H., and Badran, O. O. (1990a). Flying hot-wire anemometry with X-wire probes. *ASME, Proc. on the Heuristics of Thermal Anemometry, FED*, Vol. 97, University of Toronto, pp. 81–86.

Bruun, H. H., Nabhani, N., Al-Kayiem, H. H., Fardad, A. A., Khan, M.A., and Hogarth, E. (1990b). Calibration and analysis of X hot-wire probe signals. *Meas. Sci. Technol.*, **1**, 782–785.

Bruun, H. H., Nabhani, N., Fardad, A. A., and Al-Kayiem, H. H. (1990c). Velocity component measurements by X hot-wire anemometry. *Meas. Sci. Technol.*, **1**, 1314–1321.

Bruun, H. H., Nabhani, N., and Fardad, A. A. (1991). A comparative study of signal analysis procedures for X hot-wire probes and multi-position single yawed probes. *Proc. Second World Conf. on Experimental Heat Transfer, Fluid Mechanics and Thermodynamics*, Dubrovnik, Yugoslavia, pp. 443–451.

Bruun, H. H., Fitouri, A., and Khan, M. K. (1993). The use of a multiposition single yawed hot-wire probe for measurements in swirling flow. *ASME, Third Int. Symp. on Thermal Anemometry, FED.*, Vol. 167, Washington D. C., pp. 57–65.

Bryer, D. W. and Pankhurst, R. C. (1971). *Pressure probe methods for determining wind speed and flow direction.* HMSO, London.

Buddhavarapu, J. and Meinen, D. (1988). A new data analysis scheme and a three-sensor probe for measurements in a 3-D flow. *TSI, Flow lines*, Spring, 3–5, 16–19.

Burchill, W. E. and Jones, B. G. (1971). Interpretation of hot-film anemometer response in a non-isothermal field. *Proc. Symp. on Turbulence in Liquids,* University of Missouri, Rolla, 26–34.

Buresti, G. and Di Cocco, N. R. (1987). Hot-wire measurement procedures and their appraisal through a simulation technique. *J. Phys. E.: Sci. Instr.,* **20,** 87–99.

Buresti, G. and Talamelli, A. (1992). On the error sensitivity of calibration procedures for normal hot-wire probes. *Meas. Sci. Technol,* **3,** 17–26.

Butler, T. L. and Wagner, J. W. (1982). An improved method for calibration and use of a three sensor hot-wire probe in turbomachinery flows. *Proc. AIAA 20th Aerospace Sciences Meeting,* Orlando, Florida, USA, Paper AIAA-82-0195.

Butler, T. L. and Wagner, J. W. (1983). Application of a three-sensor hot-wire probe for incompressible flow. *AIAA J.,* **21,** 726–732.

Cantwell, B. J. (1976). A flying hot wire study of the turbulent near wake of a circular cylinder at a Reynolds number of 140000. Ph.D. Thesis, California Institute of Technology.

Cantwell, B. J. (1981). Organized motion in turbulent flow. *Ann. Rev. Fluid Mech.,* **13,** 457–515.

Cantwell, B. J. and Coles, D. E. (1983). An experimental study of entrainment and transport in the turbulent near wake of a circular cylinder. *J. Fluid Mech.,* **136,** 321–374.

Capece, V. R. and Fleeter, S. (1987). Unsteady a aerodynamic interactions in a multistage compressor. *ASME, J. Turbomachinery,* **109,** 420–428.

Capece, V. R. and Fleeter, S. (1989a). Measurement and analysis of unsteady flow structures in rotor blade wakes. *Exp. in Fluids,* **7,** 61–67.

Capece, V. R. and Fleeter, S. (1989b). Experimental investigation of multistage interaction gust aerodynamics. *ASME, J. Turbomachinery,* **111,** 409–417.

Cardell, G. (1993). A note on the temperature-dependent hot-wire calibration method of Cimbala and Park. *Exp. in Fluids,* **14,** 283–285.

Carr, L. W. and McCroskey, W. J. (1979). A directionally sensitive hot-wire probe for detection of flow reversal in highly unsteady flows. *ICIASF '79 Record, Proc. Int. Congr. Instrumentation in Aerospace Simulation Facilities,* pp. 154–162.

Carraway, D. L. (1991). The use of silicon microsensors in smart skins for aerodynamic research. *ICIASF '91, Record, Proc. Int. Congr. Instrumentation in Aerospace Simulation Facilities,* pp. 413–422.

Cartellier, A. (1990). Optical probes for local, void fraction measurements: Characterization of performance. *Rev. Sci. Instr.,* **61,** 874–886.

Cartellier, A. (1992). Simultaneous void fraction measurement, bubble velocity and size estimate using a single optical probe in a gas-liquid two-phase flow. *Rev. Sci. Instr.,* **63,** 5442–5453.

Cartellier, A. and Achard, J. L. (1991). Local phase detection probes in fluid/fluid two-phase flow. Review Article. *Rev. Sci. Instr.,* **62,** 279–303.

Castro, I. P. (1985). Time-domain measurements in separated flows. *J. Fluid. Mech.,* **150,** 183–201.

Castro, I. P. (1986). The measurement of Reynold's stresses. In *Encyclopedia of Fluid Mechanics,* Chapter 37. Gulf Publishing Company, Texas, USA.

Castro, I. P. (1989). *An Introduction to the Digital Analysis of Stationary Signals.* Adam Hilger, Bristol.

Castro, I. P. (1991). Pulsed wire anemometry. *Proc. 2nd World Conf. on Experimental Heat Transfer, Fluid Mechanics and Thermodynamics,* Dubrovnik, Yugoslavia, pp. 202–211.

Castro, I. P. and Cheun, B. S. (1982). The measurement of Reynolds stresses with a pulsed-wire anemometer. *J. Fluid Mech.*, **118**, 41–58.

Castro, I. P. and Dianat, M. (1983). Surface flow patterns on rectangular bodies in thick boundary layers. *J. Wind Eng. and Industr. Aerodyn.*, **11**, 107–119.

Castro, I. P. and Dianat, M. (1987). A pulsed hot wire probe for near wall measurements. *6th Symp. on Turbulent Shear Flows*, Toulouse, Paper 6–3.

Castro, I. P. and Dianat, M. (1990). Pulsed wire velocity anemometry near walls. *Exp. in Fluids*, **8**, 343–352.

Castro, I. P. and Haque, A. (1987). The structure of a turbulent shear layer bounding a separation region. *J. Fluid Mech.*, **179**, 439–468.

Castro, I. P. and Haque, A. (1988). The structure of a shear layer bounding a separation region. Part 2: Effects of free-stream turbulence. *J. Fluid Mech.*, **192**, 577–595.

Castro, I. P., Dianat, M., and Bradbury, L. J. S. (1987). The pulsed wire skin-friction technique. in *Turbulent Shear Flow 5*, (ed. F. Durst *et al.*) pp. 278–290. Springer, Berlin.

Castro, I. P., Dianat, M., and Haque, A. (1989). Shear layers bounding separated regions. In *Turbulent Shear Flows 6*, (ed. J. C. André *et al.*), pp. 299–312. Springer, Berlin.

Catania, A. E. (1980). Air flow investigation in the open combustion chamber of a high-speed, four-stroke diesel engine. ASME, Paper No. 80-FE-5.

Catania, A. E. (1982). 3-D swirling flows in an open-chamber automotive diesel engine with different induction systems. In *Flows in Internal Combustion Engines*, (ed. T. Uzkan) pp. 53–66. ASME, New York.

Catania, A. E. (1985). Induction system effects on the fluid-dynamics of a D. I. automotive diesel engine. ASME, Paper No. 85-DGP-11.

Catania, A. E. and Mittica, A. (1985*a*). A contribution to the definition and measurement of turbulence in a reciprocating I. C. engine. ASME, Paper No. 85-DGP-12.

Catania, A. E. and Mittica, A. (1985*b*). Cycle-by-cycle, correlation and spectral analysis of I. C. engine turbulence. In *Flows in Internal Combustion Engines—III, FED*, (ed. T. Uzkan, W. G. Tiederman, and J. M. Novak), Vol. 28, ASME, New York.

Catania, A. E. and Mittica, A. (1987). Induction system effects on small-scale turbulence in a high-speed diesel engine. *ASME, J. Eng. for Gas Turbines and Power*, **109**, 491–502.

Catania, A. E. and Mittica, A. (1989). Extraction techniques and analysis of turbulence quantities from in-cylinder velocity data. *ASME, J. Eng. for Gas Turbines and Power*, **111**, 466–478.

Catania, A. E. and Mittica, A. (1990). Autocorrelation and autospectra estimation of reciprocating engine turbulence. *ASME, J. Eng. for Gas Turbines and Power*, **112**, 357–368.

Catania, A. E., Dongiovanni, C., and Mittica, A. (1992). Time-frequency spectral structure of turbulence in an automotive engine. SAE Paper No 920153.

Cau, G., Mandas, N., Manfrida, G., and Nurzia, F. (1987). Measurements of primary and secondary flows in an industrial forward-curved centrifugal fan. *ASME, J. Fluids Eng.*, **109**, 353–358.

Chambers, F. W., Murphy, H. D., and McEligot, D. M. (1983). Laterally converging flow, Part 2, Temporal wall shear stress. *J. Fluid Mech.*, **127**, 403–428.

Champagne, F. H. (1978). The temperature sensitivity of hot wires. *Proc. Dynamic Flow Conf.*, Marseille, pp. 101-113.

Champagne, F. H., Sleicher, C. A., and Wehrmann, O. H. (1967). Turbulence measurements with inclined hot-wires. Part 1, Heat transfer experiments with inclined hot-wire. *J. Fluid Mech.*, **28**, 153-175.

Chan, Y. Y. (1974). Spatial waves in turbulent jets, Part II., *Phys. Fluids*, **17**, 1667-1670.

Chandrsuda, C., Mehta, R. D., Weir, A. D., and Bradshaw, P. (1978). Effect of free-stream turbulence on large structure in turbulent mixing layers. *J. Fluid Mech.*, **85**, 693-704.

Chang, P. H., Adrian, R. J., and Jones, B. G. (1983). Comparison between triple-wire and X-wire measurement techniques in high intensity shear flow. *Proc. 8th Bien. Symp. on Turbulence*, University of Missouri, Rolla, pp. 206-220.

Cheesewright, R. (1972). The application of digital techniques to hot-wire anemometry in highly turbulent flows. *Proc. Conf. on Fluid Dynanmic Measurements in the Industrial and Medical Environment*, Leicester University, pp. 145-151.

Chen, C. H. P. and Blackwelder, R. F. (1978). Large-scale motion in a turbulent boundary layer: a study using temperature contamination. *J. Fluid Mech.*, **89**, 1-31.

Chevray, R. and Tutu, N. K. (1972). Simultaneous measurements of temperature and velocity in heated flows. *Rev. Sci. Instr.*, **43**, 1417-1421.

Chevray, R. and Tutu, N. K. (1978). Intermittency and preferential transport of heat in a round jet. *J. Fluid Mech.*, **88**, 133-160.

Chew, Y. T. and Ha, S. M. (1988). The directional sensitivities of crossed and triple hot-wire probes. *J. Phys. E.: Sci. Instr.*, **21**, 613-620.

Chew, Y. T. and Ha, S. M. (1990). A critical evaluation of the explicit data analysis algorithm for a crossed wire anemometer in highly turbulent isotropic flow. *Meas. Sci. Technol.*, **1**, 775-781.

Chew, Y. T. and Simpson, R. L. (1988). An explicit non-real time data reduction method of triple sensors hot-wire anemometer in three-dimensional flow. *ASME, J. Fluids Eng.*, **110**, 110-119.

Cho, S. H. and Becker, H. A. (1985). Response of static pressure probes in turbulent streams. *Exp. in Fluids*, **3**, 93-102.

Christiansen, T. and Bradshaw, P. (1981). Effect of turbulence on pressure probes. *J. Phys. E.: Sci. Instr.*, **14**, 992-997.

Christiansen, W. H. (1961). Development and calibration of a cold wire probe for use in shock tubes. California Institute of Technology, Pasadena, CA. GALCIT Memo No. 62.

Christman, P. J. and Podzimek, J. (1981). Hot-wire anemometer behaviour in low velocity-air flow. *J. Phys. E.: Sci. Instr.*, **14**, 46-51.

Chue, S. H. (1975). Pressure probes for fluid measurement. In *Progress in Aerospace Sciences*, Vol. 16 (ed. D. Küchemann), pp. 147-223.

Cimbala, J. M. and Park, W. J. (1990). A direct hot-wire calibration technique to account for ambient temperature drift in incompressible flow. *Exp. in Fluids*, **8**, 299-300.

Clausen, P. D. and Wood, D. H. (1988). An experimental investigation of blade element theory for wind turbines, Part 2, Phase-locked averaged results. *J. Wind Eng. and Industr. Aerodyn.*, **31**, 305-322.

Clausen, P. D. and Wood, D. H. (1989). The correction of X-probe results for transverse contamination. *ASME, J. Fluids Eng.*, **111**, 226-229.

Clauser, F. H. (1954). Turbulent boundary layers in adverse pressure gradients. *J. Aero. Sci.*, **21**, 91–108.

Coghe, A., Gamma, F., Mauri, M., Brunello, G., Calderini, F., and Ferri Degli Antoni, L. (1985). In-cylinder air motion measurements by laser velocimetry under steady-state flow conditions. SAE Paper No. 850123.

Coles, D. E. (1956). The law of the wake in the turbulent boundary layer. *J. Fluid Mech.*, **1**, 191–226.

Coles, D. E. (1962). The turbulent boundary layer in compressible fluid. Rand Corporation Report R-403R-PR.

Coles, D. E. (1968). The young person's guide to the data. In *Proc. Computation of Turbulent Boundary Layers*. AFOSR-IFB Stanford Conference, Vol. 2, (ed. D. E. Coles and E. A. Hirst), pp. 1–45.

Coles, D. and Wadcock, A. J. (1979). Flying-hot-wire study of flow past a NACA 4412 airfoil at maximum lift. *J. AIAA*, **17**, 321–329.

Colin, P. and Olivari, D. (1971). Three applications of hot wire techniques for fluid dynamic measurements. *ICIASF '71 Record. Proc. Int. Congr. on Instrumentation in Aerospace Simulation* Facilities, Rhode-st-Genese, pp. 173–184.

Collings, N., Roughton, A. W., and Ma, T. (1987). Turbulence length scale measurements in a motored internal combustion engine. SAE Paper No. 871692.

Collis, D. C. and Williams, M. J. (1959). Two-dimensional convection from heated wires at low Reynolds numbers. *J. Fluid Mech.*, **6**, 357–384.

Comte-Bellot, G. (1976). Hot-wire anemometry. *Ann. Rev. Fluid Mech.*, **8**, 209–231.

Comte-Bellot, G. (1977). The physical background for hot-film anemometry. *Proc. Symp. on Turbulence in Liquids*, University of Missouri, Rolla, pp. 1–13.

Comte-Bellot, G. and Corrsin, S. (1971). Simple Eulerian time correlation of full- and narrow-band velocity signals in grid-generated, 'isotropic' turbulence. *J. Fluid Mech.*, **48**, 273–337.

Comte-Bellot, G. and Schon, J. P. (1969). Harmoniques crees par excitation parametrique dans les anemometres a fil chaud a intensite constante. *Int. J. Heat Mass Transf.*, **12**, 1661–1677.

Comte-Bellot, G., Strohl, A., and Alcaraz, E. (1971). On aerodynamic disturbances caused by single hot-wire probes. *ASME, J. Appl. Mech.*, **38**, 767–774.

Cook, N. J. and Redfearn, D. (1976). Calibration and use of a hot-wire probe for highly turbulent and reversing flows. *J. Ind. Aerodyn.*, **1**, 221–231.

Cooley, J. W. and Tukey, J. W. (1965). An algorithm for the machine calculation of complex Fourier series. *Mathematics of Computation*, **19**, 297–301.

Corino, E. R. and Brodkey, R. S. (1969). A visual investigation of the wall region in turbulent flow. *J. Fluid Mech.*, **37**, 1–30.

Corrsin, S. (1943). Investigation of flow in an axially symmetric heated jet of air. NACA Report No. W-94.

Corrsin, S. (1947). Extended applications of the hot-wire anemometer. *Rev. Sci. Instr.*, **18**, 469–471.

Corrsin, S. (1949). Extended applications of the hot-wire anemometer. NACA Technical Note 1864.

Corrsin, S. (1963). Turbulence: experimental methods. In *Handbuch der Physik*, Vol. 8.2, pp. 523–590. Springer, Berlin.

Corrsin, S. and Kistler, A. L. (1955). Free stream boundaries of turbulent flows. NACA Report No. 1244.

Corrsin, S. and Kovasznay, L. G. (1949). On the hot-wire length correction. *Phys. Rev.*, **75**, 1954.

Corsiglia, V. R., Schwind, R. G. and Chigier, N. A. (1973). Rapid scanning, three-dimensional hot-wire anemometer surveys of wing-tip vortices. *J. Aircraft*, **10**, 752–757.

Cowell, T. A. and Heikal, M. R. (1988). The calibration of constant temperature hot-wire anemometer probes at low velocities. *Proc. Second UK National Conf. on Heat Transfer*, University of Strathclyde, pp. 1607–1622.

Croom, C. C., Manuel, G. S., and Stack, J. P. (1987). In-flight detection of Tollmien-Schlichting instabilities in laminar flow. SAE Paper 871016.

Crouch, J. D. and Saric, W. S. (1986). Oscillating hot-wire measurements above an FX63-137 airfoil. *AIAA 24th Aerospace Sciences Meeting*, January 6–9, Nevada. AIAA-86-0012.

Crow, S. C. and Champagne, F. H. (1971). Orderly structure in jet turbulence. *J. Fluid Mech.*, **48**, 547–591.

Cutler, A. D. and Bradshaw, P. (1991). A crossed hot-wire technique for complex turbulent flows. *Exp. in Fluids*, **12**, 17–22.

Dahm, M. and Rasmussen, C. G. (1969). Effect of wire mounting system on hot-wire probe characteristics. *DISA Info.*, No. 7, 19–24.

Daneshyar, H. and Fuller, D. E. (1986). Definition and measurement of turbulence parameters in reciprocating I. C. engines. SAE Paper No. 861529.

Das, D. K. and Jiang, H. K. (1984). An experimental study of rotating stall in a multistage axial-flow compressor. *ASME, J. Eng. for Gas Turbines and Power*, **106**, 542–551.

Davidov, B. I. (1961). On the statistical dynamics of an incompressible turbulent flow. *Dokl. Akad. Nauk., SSSR*, **136**, 47–50.

Davies, A. E., Keffer, J. F., and Baines, W. D. (1975). Spread of a heated plane turbulent jet. *Phys. Fluids*, **18**, 770–775.

Davies, P. O. A. L. and Bruun, H. H. (1968). The performance of a yawed hot wire. *Proc. Symp. Instrumentation and Data Processing for Industrial Aerodynamics*, National Physical Laboratory, 10.1–10.15.

Davies, P. O. A. L. and Fisher, M. J. (1964). Heat transfer from electrically heated cylinders. *Proc. Roy. Soc.*, **A280**, 486–527.

Davies, T. W. and Patrick, M. A. (1972). A simplified method of improving the accuracy of hot-wire anemometry. *Proc. Conf. on Fluid Dynamic Measurements in the Industrial and Medical Environment*, Leicester University, pp. 152–155.

Davis, M. R. (1970). The dynamic response of constant resistance anemometers. *J. Phys. E.: Sci. Instr.*, **3**, 15–20.

Day, I. J. and Cumpsty, N. A. (1978). The measurement and interpretation of flow within rotating stall cells in axial compressors. *J. Mech. Eng. Sci.*, **20**, 101–114.

De Grande, G. and Kool, P. (1981). An improved experimental method to determine the complete Reynolds stress tensor with a single rotating slanting hot wire. *J. Phys. E.: Sci. Instr.*, **14**, 196–201.

De Haan, R. E. (1971). Dynamic theory of a short hot-wire normal to an incompressible air-flow, constant resistance operation. *Appl. Sci. Res.*, **24**, 231–260.

Dekeyser, I. and Launder, B. E. (1983). A comparison of triple-moment temperature-velocity correlations in the asymmetric heated jet with alternative closure models. *Proc. 4th Symp. on Turbulent Shear Flow*, University of Karlsruhe, 14.1–14.8.

Delhaye, J. M. (1969). Hot-film anemometry in two-phase flow. *Proc. 11th Nat.*

ASME/AIChE Heat Transfer Conf. on Two-Phase Flow Instrumentation, Minneapolis, Minnesota, pp. 58-69.

Demetriades, A. and Laderman, A. J. (1973). Reynolds stress measurements in a hypersonic boundary layer. *AIAA J.*, **11**, 1594-1596.

Demin, V. S. and Zheltukhin, N. A. (1973). Interpretation of hot-wire anemometer readings in a flow with velocity, pressure and temperature fluctuations. *Fluid Mech.—Sov. Res.*, **2**, 64-75.

Dengel, P. and Fernholz, H. H. (1990). An experimental investigation of an incompressible turbulent boundary layer in the vicinity of separation. *J. Fluid Mech.*, **212**, 615-636.

Dengel, P., Fernholz, H. H., and Vagt, J. D. (1981). Turbulence and mean flow measurements in an incompressible axisymmetric boundary layer with incipient separation. *Proc. 3rd Int. Symp. on Turbulent Shear Flow*, University of California, Davis, CA.

Dengel, P., Fernholz, H. H., and Vagt, J. D. (1982). Turbulent and mean flow measurements in an incompressible axisymmetric boundary layer with incipient separation. In *Turbulent Shear Flows* 3, (ed. L. J. S. Bradbury *et al.*), pp. 225-236. Springer, Berlin.

Dengel, P., Fernholz, H. H. and Hess, M. (1987). Skin-friction measurements in two- and three-dimensional highly turbulent flows with separation. In *Advances in Turbulence*, (ed. G. Comte-Bellot and J. Mathieu), pp. 470-479. Springer, Berlin.

Dent, J. C. and Salama, N. S. (1975). The measurement of turbulence characteristics in an internal combustion engine cylinder. SAE Paper No. 750886.

Derksen, R. W. and Azad, R. S. (1983). An examination of hot-wire length corrections. *Phys. Fluids.*, **26**, 1751-1754.

Devenport, W. J. and Sutton P. (1991). Near-wall behaviour of separated and reattaching flows. *AIAA J.*, **29**, 25-31.

Devenport, W. J. and Sutton, E. P. (1992). The effects of a centrebody on an axisymmetric flow through a sudden expansion. *30th Aerospace Sciences Meeting and Exhibition*, Reno, NV, AIAA paper 92-0431.

Devenport, W. J., Evans, G. P., and Sutton, E. P. (1990). A traversing pulsed-wire probe for velocity measurements near a wall. *Exp. in Fluids*, **8**, 336-342.

Dewey, C. F. (1961). Hot wire measurements in low Reynolds number hypersonic flows. *J. Amer. Rocket Soc.*, **28**, 1709-1718.

Dewey, C. F, (1965*a*). Near wake of a blunt body at hypersonic speeds. *AIAA J.*, **3**, 1001-1010.

Dewey, C. F. (1965*b*). A correlation of convective heat transfer and recovery temperature data for cylinders in compressible flow. *Int. J. Heat and Mass Transf.*, **8**, 245-252.

Dianat, M. and Castro, I. P. (1989). Measurements in separating boundary layers. *AIAA J.*, **27**, 719-724.

Dianat, M. and Castro, I. P. (1991). Turbulence in a separated boundary layer. *J. Fluid Mech.*, **226**, 91-123.

Dimotakis, P. E. and Brown, G. L. (1976). The mixing layer at high Reynolds number: large-structure dynamics and entrainment. *J. Fluid Mech.*, **78**, 535-560.

Dinsdale, S., Roughton, A., and Collings, N. (1988). Length scale and turbulence intensity measurements in a motored internal combustion engine. SAE Paper No. 880380.

DISA (1965). Measurements of flow velocity in liquids using a DISA constant temperature anemometer. *DISA Info.*, No. 1, 1-5.

DISA (1971). Improvement in frequency response. *DISA Info.*, No. 11, 42.

DISA (1977). Instruction manual. DISA 55M1O System.

DISA (1982). Triple-split fibre probe. DISA Publication No. 2203E.

Disimile, P. J. (1986). Phase averaged transverse vorticity measurements in an excited, two-dimensional mixing layer. *AIAA J.*, **24**, 1621–1627.

Dixon, S. L. (1978). Measurement of flow direction in a shear flow. *J. Phys. E.: Sci. Inst.*, **11**, 31–34.

Döbbeling, K., Lenze, B., and Leuckel, W. (1990a). Basic considerations concerning the construction and usage of multiple hot-wire probes for highly turbulent three-dimensional flows. *Meas. Sci. Technol.*, **1**, 924–933.

Döbbeling, K., Lenze, B., and Leuckel, W. (1990b). Computer-aided calibration and measurements with a quadruple hotwire probe. *Exp. in Fluids*, **8**, 257–262.

Döbbeling, K., Lenze, B., and Leuckel, W. (1992). Four-sensor hot-wire probe measurements of the isothermal flow in a model combustion chamber at different levels of swirl. *Exp. Thermal and Fluid Sci.*, **5**, 381–389.

Donovan, J. F. and Spina, E. F. (1990). An improved analysis of crossed-wire signals obtained in supersonic flow. *ASME, Proc. Symp. on Heuristics of Thermal Anemometry*, FED, Vol. 97, Toronto University, pp. 41–51.

Donovan, J. F. and Spina, E. F. (1992). An improved analysis method for cross-wire signals obtained in supersonic flow. *Exp. in Fluids*, **12**, 359–368.

Doughman, E. L. (1972). Development of a hot-wire anemometer for hypersonic turbulent flows. *Rev. Sci. Instr.*, **43**, 1200–1202.

Downing, P. M. (1972). Reverse flow sensing hot wire anemometer. *J. Phys. E.: Sci. Instr.*, **5**, 849–851.

Dring, R. P. (1982). Sizing criteria for laser anemometry particles. *ASME, J. Fluids Eng.*, **104**, 15–17.

Dring, R. P. and Gebhart, B. (1969). Hot-wire anemometer calibration for measurements at very low velocity. *ASME, J. Heat Transf.*, **91**, 241–244.

Drubka, R. E., Tan-atichat, J., and Nagib, H. M. (1977). Analysis of temperature compensating circuits for hot-wires and hot-films. *DISA Info.*, No. 22, 5–14.

Dryden, H. L. and Kuethe, A. M. (1929). The measurement of fluctuations of air speed by the hot-wire anemometer. NACA Technical Report 320.

Dryden, H. L., Schubauer, G. B., Mock, W. C., and Skramstad, H. K. (1937). Measurements of intensity and scale of windtunnel turbulence and their relation to the critical Reynolds number of spheres. NACA Technical Report 581.

D'Souza, G. J., Montealegre, A., and Weinstein, H. (1968). Measurement of turbulent correlations in a coaxial flow of dissimilar fluids. NASA CR-970.

Duncan, A. and Hartrmann, U. (1985). Some experiments on the dynamic behaviour of a split film probe. Herman Föttinger Institut für Thermo- und Fluiddynamik, T. U. Berlin, Institutbericht Nr 01/85.

Dupont, P. and Debieve, J. F. (1990). A hot-wire method for measuring turbulence in transonic or supersonic heated flows. *12th Symp. on Turbulence*, University of Missouri-Rolla.

Dupont, P. and Debieve, J. F. (1992). A hot wire method for measuring turbulence in transonic or supersonic heated flows. *Exp. in Fluids*, **13**, 84–90.

Durão, D. F. G., Laker, J., and Whitelaw, J. H. (1980). Bias effects in laser doppler anemometry. *J. Phys. E.: Sci. Instr.*, **13**, 442–445.

Durst, F. (1977). Hot-wire and laser-doppler techniques in turbulence research. Report SFB80/EM/119. Sonderforchungsbereich 80. Karlsruhe University.

Durst, F., Jovanovic, J., and Kanevce, Lj. (1987). Probability density distribution in turbulent wall boundary-layer flows. *Turbulent Shear Flows* 5, (ed. F. Durst *et al.*), pp. 197–220. Springer, Berlin.

Dvorak, K. and Syred, N. (1972). The statistical analysis of hot-wire anemometer signals in complex flow fields. *Proc. Conf. on Fluid Dynamic Measurements in the Industrial and Medical Environment*, Leicester University, pp. 136–144.

Eaton, J. K., Jeans, A. H., Ashjaee, J., and Johnston, J.P. (1979). A wall-flow-direction probe for use in separating and reattaching flows. *ASME, J. Fluids. Eng.*, **101**, 364–366.

Eaton, J. K. and Johnston, J. P. (1980). Turbulent flow reattachment: an experimental study of the flow and structure behind a backward-facing step. Report MD-39, Department of Mechanical Engineering, Stanford University.

Eaton, J. K. and Johnston, J. P. (1983). Low frequency unsteadiness of a reattaching turbulent shear layer. In *Turbulent Shear Flow* 3, (ed. L. J. S. Bradbury *et al.*), pp. 162–170. Springer, Berlin.

Eaton, J. K., Westphal, R. V., and Johnston, J. P. (1981). Two new instruments for flow direction and skin-friction measurements in separated flows. *ISA Trans.*, **21**, 69–78.

Eckelmann, H. (1972). Hot-wire and hot-film measurements in oil. *DISA Info.*, No. 13, 16–21.

Eckelmann, H. (1974). The structure of the viscous sublayer and the adjacent wall region in a turbulent channel flow. *J. Fluid Mech.*, **65**, 439–459.

Eckelmann, H. and Reichardt, H. (1971). An experimental investigation in a turbulent channel flow with a thick viscous sublayer. *Proc. Symp. on Turbulence in Liquid*, University of Missouri, Rolla, pp. 144–148.

Eckelmann, H., Nychas, S. G., Brodkey, R. S., and Wallace, J. M. (1977). Vorticity and turbulence production in pattern recognized turbulent flow structures. *Phys. Fluids*, **20**, S225–S231.

Elsenaar, A. and Boelsma, S. H. (1974). Measurements of the Reynolds stress tensor in a three-dimensional turbulent boundary layer under infinite swept wing conditions. NLR Technical Report No. 74095U.

Elsner, J. and Gundlach, W. R. (1973). Some remarks on the thermal equilibrium equation of hot-wire probes. *DISA Info.*, No. 14, 21–24.

Elsner, J. W. (1972). An analysis of hot-wire sensitivity in non-isothermal flow. *Proc. Dynamic Flow Conf.*, Marseille, pp. 156–159.

Elsner, J. W., Domagala, P., and Elsner, W. (1993). Effect of finite spatial resolution of hot-wire anemometry on measurements of turbulence energy dissipation. *Meas. Sci. Technol.*, **4**, 517–523.

Enotiadis, A. C., Vafidis, C., and Whitelaw, J. H. (1990). Interpretation of cyclic flow variations in motored internal combustion engines. *Exp. in Fluids*, **10**, 77–86.

Erm, L. P., Smits, A. J., and Joubert, P. N. (1987). Low Reynolds number turbulent boundary layers on a smooth flat surface in a zero pressure gradient. In *Turbulent Shear Flow* 5, (ed. F. Durst *et al.*), pp. 186–196. Springer, Berlin.

Evans, G. P. (1973). Separation bubble at a pipe entrance. Ph.D. Thesis, University of Cambridge.

Evans, R. L. (1975). Turbulence and unsteadiness measurements downstream of a moving blade row. *ASME, J. Eng. for Power*, **97**, 131–139.

Ezekwe, C. I., Pierce, F. J., and McAllister, J. E. (1978). Measured Reynolds stress tensors in a three-dimensional turbulent boundary layer. *AIAA J.*, **16**, 645–646.

Ezraty, R. (1970). Sur la mesure des caractéristiques turbulentes dans les écoulements d'eau. Thèse de Docteur Ingénieur, Marseille.

Ezraty, R. and Coantic, M. (1970). Sur le mesures des tenstions de frottement turbulent dans un écoulement d'eau. Note, Compte-rendus de l' Académie des Sciences de Paris, Séance de 16 Février.

Fabris, G. (1978). Probe and method for simultaneous measurements of 'true' instantaneous temperature and three velocity components in turbulent flow. *Rev. Sci. Instr.*, **49**, 654–664.

Fabris, G. (1979). Conditional sampling study of the turbulent wake of a cylinder, Part I. *J. Fluid Mech.*, **94**, 673–709.

Fabris, G. (1983*a*). Third-order conditional transport correlations in the two-dimensional turbulent wake. *Phys. Fluids*, **26**, 422–427.

Fabris, G. (1983*b*). Higher-order statistics of turbulent fluctuations in the plane wake. *Phys. Fluids*, **26**, 1437–1445.

Fage, A. and Falkner, V. M. (1931). On the relation between heat transfer and surface friction for laminar flow. ARC R&M No. 1408.

Falco R. E. (1977). Coherent motions in the outer region of turbulent boundary layers. *Phys. Fluids*, **20**, S124-S132.

Falco, R. E. (1980). Combined simultaneous flow visualization/hot-wire anemometry for the study of turbulent flows. *ASME, J. Fluids Eng.*, **102**, 174–182.

Falco, R. E. (1983). New results, a review and synthesis of the mechanism of turbulence production in boundary layers and its modification. AIAA Paper No. 83-0377.

Fand, R. M. and Keswani, K. K. (1972). A continuous correlation equation for heat transfer from cylinders to air in cross-flow for Reynolds numbers from 10^{-2} to 2×10^5. *Int. J. Heat Mass Transf.*, **15**, 559–562.

Fansler, T. D. (1993). Turbulence production and relaxation in bowl-in-piston engines. SAE, Paper No. 930479.

Fansler, T. D. and French, D. T. (1988). Cycle-resolved laser-velocimetry measurements in a reentrant-bowl-in-piston engine. SAE Paper No. 880377.

Fansler, T. D. and French, D. T. (1992). The scavenging flow field in a crankcase-compression two-stroke engine – A three dimensional laser-velocimetry survey. SAE, Paper No. 920417.

Fardad, A. A. (1989). Measurement of three-dimensional flow quantities inside a curved duct. Ph.D. Thesis, Department of Mechanical and Manufacturing Engineering, University of Bradford.

Fardad, A. A. and Bruun, H. H. (1991). A traversing mechanism for hot-wire probe measurements inside a curved duct. *DANTEC Info.*, No. 10, 9–11.

Farrar, B. (1988). Hot-film anemometry in dispersed oil-water flows. Ph.D. Thesis, Department of Mechanical and Manufacturing Engineering, University of Bradford.

Farrar, B. and Bruun, H. H. (1989). Interaction effects between a cylindrical hot-film anemometer probe and bubbles in air/water and oil/water flows. *J. Phys. E.: Sci. Instr.*, **22**, 114–123.

Fattahi, B. and Hsu, C. T. (1977). Feasibility of the use of three-dimensional anemometers in a simulated tornado flow measurement. *J. Phys. E.: Sci. Instr.*, **10**, 73–79.

Favier, D., Rebont, J., and Maresca, C. (1977). An experimental study of three-dimensional characteristics of propeller wakes under stalling conditions. *ASME, J. Fluids Eng.*, **99**, 745-752.

Favre, A. J., Gaviglio, J. J., and Dumas, R. (1957). Space-time double correlations and spectra in a turbulent boundary layer. *J. Fluid Mech.*, 2, 313-342.

Fedyaev, A. A., Mironov, A. K., and Sergievskiy, E. D. (1990). An experimental investigation of binary turbulent boundary layer structures. *DANTEC Info.*, No. 9, 13-15.

Fernando, E. M. and Smits, A. J. (1990). A supersonic turbulent boundary layer in an adverse pressure gradient. *J. Fluid Mech.*, 211, 285-307.

Fernando, E. M., Donovan, J. F., and Smits, A. J. (1987). The calibration and operation of a constant-temperature crossed-wire probe in supersonic flow. *Proc. Symp. on Thermal Anemometry, ASME Fluids Eng. Div. Spring Meeting*, FED, Vol. 53, Cincinatti, Ohio.

Fernholz, H. H. and Finley, P. J. (1980). A critical commentary on mean flow data for two-dimensional compressible turbulent boundary layers. AGARDograph No. 253.

Feyzi, F., Kornberger, M., Rachor, N., and Ilk, B. (1989). Development of two multi-sensor hot-film measuring techniques for free-flight experiments. *ICIASF, '89 Record, Proc. Int. Congr. on Instrumentation in Aerospace Simulation Facilities*, pp. 443-449.

Fiedler, H. (1978). On data acquisition in heated turbulent flows. *Proc. Dynamic Flow Conf.*, Marseille, pp. 81-100.

Fiedler, H. (1982). A note on the resistance characteristics of cold-wire probes. *DISA Info.*, No. 27, 38-39.

Fiedler, H. and Head, M. R. (1966). Intermittency measurements in the turbulent boundary layer. *J. Fluid Mech.*, 25, 719-735.

Fiedler, H. E. and Mensing, P. (1985). The plane turbulent shear layer with periodic excitation. *J. Fluid Mech.*, 150, 281-309.

Fingerson, L. M. (1968). Practical extensions of anemometer techniques. *Proc. Symp. on Advances in Hot-wire Anemometry*, University of Maryland, pp. 258-275.

Fingerson, L. M. and Freymuth, P. (1983). Thermal anemometers. In *Fluid Mechanics Measurements*, (ed. R. J. Goldstein), pp. 99-154. Hemisphere, Washington.

Foss, J. F. (1976). Accuracy and uncertainty of transverse vorticity measurements. *Bull. Am. Phys. Soc.*, 21, 1237.

Foss, J. F. (1978). Transverse vorticity measurements. *Proc. Dynamic Flow Conf.*, Marseille, pp. 983-1001.

Foss, J. F. (1981). Advanced techniques for transverse vorticity measurements. *Proc. Seventh Biennial Symp. on Turbulence*, (ed. Zakin and G. K. Patterson), University of Missouri, Rolla, pp. 208-218.

Foss, J. F. and Haw, R. C. (1990a). Transverse vorticity measurements using a compact array of four sensors. *ASME, Proc. Symp. on the Heuristics of Thermal Anemometry*, FED, Vol. 97, Toronto University, pp. 71-76.

Foss, J. F. and Haw, R. C. (1990b). Vorticity and velocity measurements in a 2:1 mixing layer. *ASME, Forum on Turbulent Flow, FED*, Vol. 94, University of Toronto, pp. 115-120.

Foss, J. F. and Haw, R. C. (1993). Transverse vorticity and velocity measurements from a compact four-wire probe, Part 1, Probe design and algorithms. (Submitted for publication.)

Foss, J. F. and Wallace, J. M. (1989). The measurement of vorticity in transitional and fully developed turbulent flows. In *Advances in Fluid Mechanics Measurements*, (ed. M. Gad-el-Hak), Lecture Notes in Engineering, Vol. 45, pp. 263-321. Springer, Berlin.

Foss, J. F., Klewicki, C. L., and Disimile, P. J. (1986). Transverse vorticity measurements using an array of four hot-wire probes. NASA CR 178098.

Foss, J. F., Ali, S. K., and Haw, R. C. (1987). A critical analysis of transverse vorticity measurements in a large plane shear layer. In *Advances in Turbulence*, (ed. G. Comte-Bellot and J. Mathieu), pp. 446–455. Springer, Berlin.

Francis, M. S., Kennedy, D. A. and Butler, G. A. (1978). Technique for the measurement of spatial vorticity distributions. *Rev. Sci. Instr.*, **49**, 617–623.

Franke, P. G. and Preuss, K. (1970). A suitable method for calibration of hot-film probes. *DISA Info.*, No. 10, 28–29.

Fraser, R. A. and Bracco, F. V. (1988). Cycle-resolved LDV integral length scale measurements in an I. C. engine. SAE Paper No. 880381.

Fraser, R. A. and Bracco, F. V. (1989). Cycle-resolved LDV integral length scale measurements investigating clearance height scaling, isotropy and homogeneity in an I.C. engine. SAE Paper No. 890615.

Frenkiel, F. N. (1949). The influence of the length of a hot-wire on the measurement of turbulence. *Phys. Rev.*, **75**, 1263–1264.

Frenkiel, F. N. (1954). Effects of wire length in turbulence investigations with a hot-wire anemometer. *Aero. Q.*, **5**, 1–24.

Freymuth, P. (1967). Feedback control theory for constant-temperature hot-wire anemometers. *Rev. Sci. Instr.*, **38**, 677–681.

Freymuth, P. (1968). Noise in hot-wire anemometers. *Rev. Sci. Instr.*, **39**, 550–557.

Freymuth, P. (1969). Non-linear control theory for constant-temperature hot-wire anemometers. *Rev. Sci. Instr.*, **40**, 258–262.

Freymuth, P. (1970). Hot-wire anemometer thermal calibration errors. *Instr. and Contr. Systems*, **43**, 82–83.

Freymuth, P. (1972). Improved linearization for hot-wire anemometers. *J. Phys. E.: Sci. Instr.*, **5**, 533–534.

Freymuth, P. (1977a). Frequency response and electronic testing for constant-temperature hot-wire anemometers. *J. Phys. E.: Sci. Instr.*, **10**, 705–710.

Freymuth, P. (1977b). Further investigation of the nonlinear theory for constant-temperature hot-wire anemometers. *J. Phys. E.: Sci. Instr.*, **10**, 710–713.

Freymuth, P. (1978a). A bibliography of thermal anemometry. *TSI Quart.*, **IV**, No. 4, 3–26.

Freymuth, P. (1978b). A comparative study of the signal-to-noise ratio for hot-film and hot-wire anemometers. *J. Phys. E.: Sci. Instr.*, **11**, 915–918.

Freymuth, P. (1978c). Extension of the nonlinear theory to constant-temperature hot-film anemometers. *TSI Quart.*, **IV**, No. 3, 3–6.

Freymuth, P. (1979). Engineering estimate of heat conduction loss in constant temperature thermal sensors. *TSI Quart.*, **V**, No. 3, 3–9.

Freymuth, P. (1981). Calculation of square wave test for frequency optimised hot-film anemometers. *J. Phys. E.: Sci. Instr.*, **14**, 238–240.

Freymuth, P. (1982). Off-optimum operation of hot-wire anemometers and analysis of an anemometer with shaped bridge impedance. *Proc. Sensors & Systems '82 Pasadena Conference*, Vol. 3, Various sensor applications, Pasadena, CA, USA., pp. 38–51.

Freymuth, P. (1983a). History of thermal anemometry. In *Handbook of Fluids in Motion*, (ed. N.P. Cheremisinoff and R. Gupta), pp. 79–91. Ann Arbor Science Publishers, Ann Arbor.

Freymuth, P. (1983b). Addendum to '*A Bibliography of Thermal Anemometer*', published by TSI Inc.

Freymuth, P. (1992). *Bibliography of Thermal Anemometry*, published by TSI Inc.

Freymuth, P. and Fingerson, L. M. (1977). Electronic testing of frequency response for thermal anemometers. *TSI Quart.*, **III**, No. 4, 5–12.

Friehe, C. A. and Schwarz, W. H. (1968). Deviations from the cosine law for yawed cylindrical anemometer sensors. *ASME, J. Appl. Mech.*, **90**, 655–662.

Frota, M. N. and Moffat, R. J. (1982). Triple hot-wire technique for measurements of turbulence in heated flows. *Proc. 7th Int. Heat Transfer. Conf.*, München, **4**, 491–496.

Frota, M. N. and Moffat, R. J. (1983). Effect of combined roll and pitch angels on triple hot-wire measurements of mean and turbulence structure. *DISA Info.*, No. 28, 15–23.

Frota, M. N., Moffat, R. J., and Honami, S. (1983). Flow disturbance induced by the DISA triaxial hot-wire probe 55P91. *Disa Info.*, No. 28, 24–26.

Fujita, H. and Kovasznay, L. S. G. (1968). Measurement of Reynolds stress by a single rotated hot wire anemometer. *Rev. Sci, Instr.*, **39**, 1351–1355.

Fulachier, L. (1978). Hot-wire measurements in low speed heated flow. *Proc. Dynamic Flow Conf.*, Marseille, pp. 465–487.

Fulachier, L. and Dumas, R. (1976). Spectral analogy between temperature and velocity fluctuations in a turbulent boundary layer. *J. Fluid Mech.*, **77**, 257–277.

Fulachier, L., Arzoumanian, E., and Dumas, R. (1982). Effect on a developed turbulent boundary layer of a sudden local wall motion. *Proc. IUTAM Symp. on Three-dimensional Turbulent Boundary Layers*, pp. 188–198. Springer, Berlin.

Furth, W. F. (1956). Hot-wire response to a parabolic velocity. M.S. thesis, The Johns Hopkins University.

Garnier, V. H., Epstein, A. H., and Greitzer, E. M. (1991). Rotating waves as a stall inception indication in axial compressors. *ASME, J. Turbomachinery*, **113**, 290–302.

Gartshore, I. S. (1966). An experimental examination of the large-eddy equilibrium hypothesis. *J. Fluid Mech.*, **24**, 89–98.

Gasser, D., Thomann, H., and Dengel, P. (1993). Comparison of four methods to measure wall shear stress in a turbulent boundary layer with separation. *Exp. in Fluids*, **15**, 27–32.

Gaster, M. and Bradbury, L. J. S. (1976). The measurement of spectra of highly turbulent flows by a randomly triggered pulsed-wire anemometer. *J. Fluid Mech.*, **77**, 499–509.

Gaulier, C. (1977). Measurement of air velocity by means of a triple hot-wire probe. *DISA Info.*, No. 21, 16–20.

Gaviglio, J. (1987). Reynolds analogies and experimental study of heat transfer in the supersonic boundary layer. *Int. J. Heat Mass Transf.*, **30**, 911–926.

George, W. K. (1988). Governing equations, experiments and the experimentalist. *Proc. 1st World Conf. on Experimental Heat Transfer, Fluid Mechanics and Thermodynamics*, Dubrovnik, Yugoslavia, pp. 230–240.

George, W. K. (1990). Governing equations, experiments and the experimentalist. *Exp. Thermal and Fluid Sci.*, **3**, 557–566.

George, W. K. and Hussein, H. J. (1991). Locally axisymmetric turbulence. *J. Fluid Mech.*, **233**, 1–23.

George, W. K., Beuther, P. D. and Lumley, J. L. (1978). Processing of random signals. *Proc. Dynamic Flow Conf.*, Marseille, pp. 757–799.

George, W. K., Beuther, P. D., and Ahmad, M. (1981). Polynomial calibration and quasi-linearization of hot-wires. Turbulence Research Laboratory Report, Sunny Buffalo, USA.

George, W. K., Beuther, P. D., and Shabbir, A. (1989*a*). Polynomial calibrations for hot wires in thermally varying flows. *Exp. Thermal and Fluid Sci.*, **2**, 230–235.

George, W. K., Hussein, H. J., and Woodward, S. H. (1989*b*). An evaluation of the effect of a fluctuating convection velocity on the validity of Taylor's hypothesis. *Proc. 10th Australasian Fluid Mechanics Conf.*, Melbourne.

Geremia, J. O. (1972). Experiments on the calibration of flush mounted sensors. *DISA Info.*, No. 13, 5–10.

Gerich, R. (1975). A study of nonstationary relative velocities at the outlet of a Francis turbine impeller, using a rotating hot-film probe. *DISA Info.*, No. 19, 19–25.

Gessner, F. B. and Arterberry, S. H. (1982). A method of flow field determination for moderately skewed, three-dimensional flows. *Proc. IUTAM Symp. on Three-dimensional Turbulent Boundary Layer*, Berlin, pp. 79–93.

Gessner, F. B. and Moller, G. L. (1971). Response behaviour of hot wires in shear flow. *J. Fluid Mech.*, **47**, 449–468.

Gherson, P. and Lykoudis, P. S. (1984). Local measurements in two-phase liquid-metal magneto-fluid-mechanic flow. *J. Fluid Mech.*, **147**, 81–104.

Gibson, M. M. and Verriopoulos, C. A. (1984). Turbulent boundary layer on a mildly, curved convex surface. Part 2: Temperature field measurements. *Exp. in Fluids*, **2**, 73–80.

Gieseke, T. J. and Guezennec, Y. G. (1993). An experimental approach to the calibration and use of triple hot-wire probes. *Exp. in Fluids*, **14**, 305–315.

Gilmore, D. C. (1967). The probe interference effect of hot-wire anemometers. Mechanical Engineering Research Laboratory, McGill University, Technical Note 67-3.

Ginder, R. B. and Bradbury, L. J. S. (1973). Preliminary investigation of a pulsed-gauge technique for skin friction measurements in highly turbulent flows. ARC Report 34448.

Giovanangeli, J. P. (1980). A non-dimensional heat transfer law for a slanted hot-film in water flow. *DISA Info.*, No. 25, 6–9.

Glover, A. R., Hundleby, G. E. and Hadded, O. (1988). An investigation into turbulence in engines using scanning LDA. SAE Paper No. 880379.

Goldschmidt, V. W. (1965). Measurement of aerosol concentrations with a hot wire anemometer. *J. Colloid Sci.*, **20**, 617–634.

Goldschmidt, V. W. and Eskinazi, S. (1966). Two-phase turbulent flow in a plane jet. *ASME, J. Appl. Mech.*, **33**, 735–747.

Goldschmidt, V. W. and Householder, M. K. (1969). The hot wire anemometer as an aerosol droplet size sampler. *Atmos. Envir.*, **3**, 643–651.

Gorton, C. A. and Lakshminarayana, B. (1976). A method of measuring the three-dimensional mean flow and turbulence quantities inside a rotating turbomachinery passage. *ASME, J. Eng. for Power*, **98**, 137–146.

Gourdon, C., Costes, J., and Domenech, S. (1981). Triple hot-film probe calibration in water. *DISA Info.*, No. 26, 19–22.

Graham, L. J. W. and Bremhorst, K. (1990). A linear compensation technique for inclined hot-wire anemometers subjected to fluid temperature changes. *Meas. Sci. Technol.*, **1**, 1322–1325.

Graham, L. J. W. and Bremhorst, K. (1991). Instantaneous time-constant adjustment of cold-wires acting as resistance thermometers when using multi-wire anemometer probes. *Meas. Sci. Technol.*, **2**, 238–241.

Grant, H. P. and Kronauer, R. E. (1962). Fundamentals of hot-wire anemometry. *ASME Symp. on Measurements in Unsteady Flow*, pp. 44–53.

Greitzer, E. M., Nikkanen, J. P., Haddad, D. E., Mazzawy, R. S., and Joslyn, H. D. (1979). A fundamental criterion for the application of rotor casing treatment. *ASME, J. Fluids Eng.*, 101, 237–243.

Guitton, D. E. (1969). A transient technique for measuring skin-friction using a flush-mounted heated film and its application to a wall jet in still air. McGill University Report 69-9.

Guitton, D. E. and Patel, R. P. (1969). An experimental study of the thermal wake interference between closely spaced wires of a X-type hot-wire probe. McGill University, Mechanical Engineering Research Laboratory Report 69-7.

Günkel, A. A., Patel, R. P., and Weber, M. E. (1971). A shielded hot-wire probe for highly turbulent flows and rapidly reversing flows. *Ind. Engr. Chem. Fundam.*, 10, 627–631.

Gupta, A. K. and Srivastava, A. (1979). Feasibility study of a reverse flow sensing probe. *J. Phys. E.: Sci. Instr.*, 12, 1029–1030.

Haghgooie, M., Kent, J. C., and Tabaczynski, R. J. (1984). Intake valve cylinder boundary flow characteristics in an internal combustion engine. *Combust. Sci. and Techn.*, 38, 49–57.

Haghgooie, M., Kent, J. C., and Tabaczynski, R. J. (1986). Verification of LDA and seed generator performance. *Exp. in Fluids*, 4, 27–32.

Hah, C. and Lakshminarayana, B. (1978). Effect of rotation on a rotating hot-wire sensor. *J. Phys. E.: Sci. Instr.*, 11, 999–1001.

Hah, C. and Lakshminarayana, B. (1981). Freestream turbulence effects on the development of a rotor wake. *AIAA J.*, 19, 724–730.

Hall, M. J. and Bracco, F. V. (1987). A study of velocities and turbulence intensities measured in firing and motored engines. SAE Paper No. 870453.

Hall, M. J., Bracco, F. V., and Santavicca, D. A. (1986). Cycle-resolved velocity and turbulence measurements in an I.C. engine with combustion. SAE Paper No. 860320.

Hall, R. M., Obara, C. J., Carraway, D. L., Johnson, C. B., Wright, R. E., Covell, P. F., and Azzazy, M. (1991). Comparisons of boundary-layer transition measurement techniques at supersonic Mach number. *AIAA J.*, 29, 865–871.

Hallbäck, M., Groth, J. and Johansson, A. V. (1991). Anisotropic dissipation rate – Implications for Reynolds stress models. In *Advances in Turbulence* 3, (ed. A. V. Johansson and P. H. Alfredsson). Springer, Berlin.

Handford, P. M. and Bradshaw, P. (1989). The pulsed-wire anemometer. *Exp. in Fluids*, 7, 125–132.

Hardin, L. W., Carta, F. O., and Verdon, J. M. (1987). Unsteady aerodynamic measurements on a rotating compressor blade row at low Mach number. *ASME, J. Turbomachinery*, 109, 499–507.

Haritonidis, J. H. (1989). The measurement of wall shear stress. In *Advances in Fluid Mechanics Measurements*, Lecture Notes in Engineering, Vol. 45 (ed. M. Gad-el-Hak), pp. 229–261. Springer, Berlin.

Harris, F. J. (1978). On the use of windows for harmonic analysis with the discrete Fourier transform. *Proc. IEEE*, 66, 51–83.

Hartmann, U. (1982). Wall interference effects on hot-wire probes a nominally two-dimensional highly curved wall jet. *J. Phys. E.: Sci. Instr.*, 15, 725–729.

Hartmann, U. and Dengel, P. (1989). On turbulence measurements with a split-film probe. *Advances in Turbulence* 2, (ed. H. H. Fernholz and H. E. Fiedler), pp. 262–266. Springer, Berlin.

Hatano, S. and Hotta, T. (1983). Turbulence measurements around a body in a re-circulating water channel by a hot film probe. *Exp. in Fluids*, **1**, 57–62.

Hatton, A. P., James, D. D., and Swire, H. W. (1970). Combined forced and natural convection with low-speed air flow over horizontal cylinders. *J. Fluid Mech.*, **42**, 17–31.

Haughdahl, J. and Lienhard, J. H. (1988). A low-cost, high-performance DC cold-wire bridge. *J. Phys. E.: Sci. Instr.*, **21**, 167–170.

Haw, R. C. and Foss, J. F. (1990). A facility for low speed calibration. *ASME, Proc. Symp. Heuristics of Thermal Anemometry, FED*, Vol. 97, University of Toronto, pp. 29–33.

Haw, R. C., Foss, J. K., and Foss, J. F. (1989). Vorticity based intermittency measurements in a single stream shear layer. In *Advances in Turbulence* 2, (ed, H. H. Fernholz and H. E. Fiedler), pp. 90–95. Springer, Berlin.

Haworth, D. C., EL Tahry, S. H., Huebler, M. S., and Chang, S. (1990). Multidimensional port-and-cylinder flow calculations for two- and four-valve-per cylinder engines: Influence of intake configuration on flow structure. SAE, Paper No. 900255.

Head, M. R. and Bandyopadhyay, P. (1981). New aspects of turbulent boundary-layer structures. *J. Fluid Mech.*, **107**, 297–338.

Head, M. R. and Rechenberg, I. (1962). The Preston tube as a means of measuring skin friction. *J. Fluid Mech.*, **14**, 1–17.

Head, M. R. and Vasanta Ram, V. (1971). Simplified presentation of Preston tube calibration. *Aero. Quart.*, **22**, 295–300.

Hebbar, K. S. (1980). Wall proximity corrections for hot-wire readings in turbulent flows. *DISA Info.*, No. 25, 15–16.

Hedley, T. B. and Keller, J. F. (1974). Turbulent/non-turbulent decisions in an intermittent flow. *J. Fluid Mech.*, **64**, 625–644.

Heikal, M., Antoniou, A., and Cowell, T. A. (1988). A rig for the static calibration of constant-temperature hot wires at very low velocities. *Exp. Thermal and Fluid Sci.*, **1**, 221–223.

Herringe, R. A. and Davis, M. R. (1974). Detection of instantaneous phase changes in gas-liquid mixtures. *J. Phys. E.: Sci. Instr.*, **7**, 807–812.

Heskestad, G. (1965). Hot-wire measurements in a plane turbulent jet. *ASME, J. Appl. Mech.*, **32**, 721–734.

Hetsroni, G., Cuttler, J. M., and Sokolov, M. (1969). Measurements of velocity and droplets concentration in two-phase flows. *ASME, J. Appl. Mech.*, **36**, 334–335.

Hetsroni, G. and Sokolov, M. (1971). Distribution of mass, velocity and intensity of turbulence in a two-phase turbulent jet. *ASME, J. Appl. Mech.*, **38**, 315–327.

Heywood, J. B. (1987). Fluid motion within the cylinder of internal combustion engines – The 1986 Freeman Scholar Lecture. *ASME, J. Fluids Eng.*, **109**, 3–35.

Heywood, J. B. (1989). *International Combustion Engine Fundamentals*. ISBN 0-07-100499-8.

Hilton, A. D. M., Roberts, J. B., and Hadded, O. (1991). Autocorrelation based analysis of ensemble averaged LDA engine data for bias-free turbulence estimates: A unified approach. SAE Paper No. 910479.

Hinze, J. O. (1959). *Turbulence. An introduction to its mechanism and theory.* McGraw-Hill, New York.

Hinze, J. O. (1975). *Turbulence*, (2nd ed.) McGraw-Hill, New York.

Hirano, S., Matsumoto, S., Tanaka, Y., and Yamamoto, T. (1989). Calibration of hot-film probes in water at low velocities. *DANTEC Info.*, No. 8, 11-12.

Hirota, M., Fujita, H., and Yokosawa, H. (1988). Influences of velocity gradient on hot-wire anemometry with an X-wire probe. *J. Phys. E.: Sci. Instr.*, **21**, 1077-1084.

Hirsch, Ch. and Kool, P. (1977). Measurement of the three-dimensional flow field behind an axial compressor stage. *ASME, J. Eng. for Power*, **99**, 168-180.

Hishida, M. and Nagano, Y. (1978). Simultaneous measurements of velocity and temperature in nonisothermal flows. *ASME, J. Heat Transf.*, **100**, 340-345.

Hishida, M. and Nagano, Y. (1988*a*). Turbulence measurements with symmetrically bent V-shaped hot-wires, Part 1: Principles of operation. *ASME, J. Fluids Eng.*, **110**,264-269.

Hishida, M. and Nagano, Y. (1988*b*). Turbulence measurements with symmetrically bent V-shaped hot-wires, Part 2: Measuring velocity components and turbulent shear stresses. *ASME, J. Fluids Eng.*, **110**, 270-274.

Ho, C. M. (1982). Response of a split film probe under electrical perturbations. *Rev. Sci. Instr.*, **53**, 1240-1245.

Ho, C. M. and Huang, L. S. (1982). Subharmonics and vortex merging in mixing layers. *J. Fluid Mech.*, **119**, 443-473.

Ho, C. H. and Huerre, P. (1984). Perturbed free shear layers. *Ann. Rev. Fluid Mech.*, **16**, 365-424.

Hodson, H. P. (1984). Boundary layer and loss measurements on the rotor of an axial-flow turbine. *ASME, J. Eng. for Gas Turbines and Power*, **106**, 391-399.

Hodson, H. P. (1985). Measurements of wake-generated unsteadiness in the rotor passages of axial flow turbines. *ASME, J. Eng. for Gas Turbines and Power*, **107**, 467-476.

Hodson, H. P. and Addison, J. S. (1989). Wake-boundary layer interactions in an axial flow turbine rotor at off-design conditions. *ASME, J. Turbomachinery*, **111**, 181-192.

Hoffmeister, M. (1972). Using a single hot-wire probe in three-dimensional turbulent flow fields. *DISA Info.*, No. 13, 26-28.

Højstrup, J., Rasmussen, K., and Larsen, S. E. (1976). Dynamic calibration of temperature wires in still air. *DISA Info.*, No. 20, 22-30.

Højstrup, J., Rasmussen, K., and Larsen, S. E. (1977). Dynamic calibration of temperature wires in moving air. *DISA Info.*, No. 21, 33.

Holmes, B. J., Croom, C. C., Gall, P. D., Manuel, G. S., and Carraway, D. C. (1986). Advanced transition measurement methods for flight applications. AIAA Paper No. 86-9786.

Hoole, B. J. and Calvert, J. R. (1967). The use of a hot-wire anemometer in turbulent flow. *J. Roy. Aero. Soc.*, **71**, 511-512.

Hooper, C. L. and Westphal, R. V. (1991). Hybrid approach to data reduction for multi-sensor hot wires. *Exp. in Fluids*, **11**, 398-400.

Hooper, J. D. (1980). Fully developed turbulent flow through a rod cluster. Ph.D. Thesis, New South Wales University.

Horstman, C. C. and Owen, F. K. (1974). New diagnostic technique for the study of turbulent boundary-layer separation. *AIAA J.*, **12**, 1436-1438.

Horstman, C. C. and Rose, W. C. (1977). Hot-wire anemometry in transonic flow. *AIAA J.*, **15**, 395-401.

Hotta, T. (1986). Flow measurement by hot-film probes. *Proc. Symp. on Circulating Water Channel*, Taipei, China, 187-197.

Houdeville, R. and Juillen, J. C. (1989). Skin friction measurements with hot elements. VKI Lecture Series 05.

Houdeville, R., Juillen, J. C., and Cousteix, J. (1983). Skin friction measurements with hot-element gages. *Proc. 20th Colloque d'Aerodynamique Appliquée (AAAF)*, Toulouse.

Hsu, Y. Y., Simon, F. F., and Graham, R. W. (1963). Application of hot-wire anemometry for two-phase flow measurements such as void fraction and slip velocity. *ASME, Multiphase Flow Symp.*, Philadelphia, pp. 26–34.

Hu, Z., Vafidis, C., Whitelaw, J. H., Chapman, J., and Head, R. A. (1992). Correlation between in-cylinder flow, performance and emissions characteristics of a Rover pentroof four-valve engine. *I. Mech. E.* Conf. Paper C448/026.

Huebner, K. H and McDonald, A. T. (1970). A dynamic model and measurement technique for studying induction air swirl in an engine cylinder. *ASME, J. Eng. for Power*, **92**, 189–197.

Huffman, G. D. (1980). Calibration of triaxial, hot-wire probes using a numerical search algorithm. *J. Phys. E.: Sci. Instr.*, **13**, 1177–1182.

Huffman, G. D., Rabe, D. C., and Poti, N. D. (1980). Flow direction probes from a theoretical and experimental point of view. *J. Phys. E.: Sci. Instr.*, **13,** 751–760.

Hunter, I. H. and Cumpsty, N. A. (1982). Casing wall boundary-layer development through an isolated compressor rotor. *ASME, J. Eng. for Power*, **104**, 805–818.

Hussain, A. K. M. F. (1983). Coherent structures — reality and myth. *Phys. Fluids*, **26**, 2816–2850.

Hussain, A. K. M. F. (1986). Coherent structures and turbulence. *J. Fluid Mech.*, **173**, 303–356.

Hussain, A. K. M. F. and Zaman, K. B. M. Q. (1980). Vortex pairing in a circular jet under controlled exitation, Part 2, Coherent structure dynamics. *J. Fluid Mech.*, **101**, 493–544.

Hussain, A. K. M. F. and Zaman, K. B. M. Q. (1981). The 'preferred mode' of the axisymmetric jet. *J. Fluid Mech.*, **110**, 39–71.

Hussein, H. J. (1990). Measurements of turbulent flows with flying hot-wire anemometry, *ASME, Proc. Symp. on the Heuristics of Thermal Anemometry, FED*, Vol. 97, University of Toronto, pp. 77–80.

Hussein, H. J. and George, W. K. (1989). Measurement of small scale turbulence in an axisymmetric jet using moving hot-wires. *Proc. Seventh Symp. on Turbulent Shear Flow*, Stanford University, 30.2.1–30.2.6.

Hussein, H. J. and George, W. K. (1990). Influence of wire spacing on derivative measurement with parallel hot-wire probes. *ASME, Forum on Turbulent Flow, FED*, Vol. 94, pp. 121–124.

Ikegami, M., Shioji, M., and Nishimoto, K. (1987). Turbulence intensity and spatial integral scale during compression and expansion strokes in a four-cycle reciprocating engine. SAE Paper No. 870372.

Inoue, M., Kuroumaru, M., Iwamoto, T., and Ando, Y. (1991). Detection of a rotating stall precursor in isolated axial flow compressor rotors. *ASME, J. Turbomachinery*, **113**, 281–289.

Jackson, A. D. (1987). Stall cell development in an axial compressor. *ASME, J. Turbomachinery*, **109**, 492–498.

Jackson, T. W. and Lilley, D. G. (1986). Accuracy and directional sensitivity of the single-wire technique. *AIAA J.*, **24**, 451–458.

Jackson, T. W. and Yen, H. H. (1971). Combining forced and free convective equations to represent combined heat transfer coefficients for a horizontal cylinder. *ASME, J. Heat Transf.*, **93**, 247–248.

Jacobsen, R. A. (1977). Hot-wire anemometry for in-flight measurement of aircraft wake vortices. *DISA Info.*, No. 21, 21–27.

Jaju, A. A. R. (1987). Development of a flying hot-wire probe system. M.Phil. Thesis, Department of Mechanical and Manufacturing Engineering, University of Bradford.

Janjua, S. I., McLaughlin, D. K., Jackson, T. W., and Lilley, D. G. (1983). Turbulence measurements in confined jets using a rotating single-wire probe technique. *AIAA J.*, **21**, 1609–1610.

Janke, G. (1987). Hot wire in wall proximity. In *Advances in Turbulence*, (ed. G. Comte-Bellot and J. Mathieu), pp. 488–498. Springer, Berlin.

Jaroch, M. (1985). Development and testing of pulsed-wire probes for measuring fluctuating quantities in highly turbulent flows. *Exp. in Fluids*, **3**, 315–322.

Jaroch, M. and Dahm, A. (1988). A new pulsed-wire probe for measuring the Reynolds stresses in the plane containing the main shear direction of a turbulent shear flow. *J. Phys. E.: Sci. Instr.*, **21**, 1085–1094.

Jaroch, M. P. and Fernholz, H. H. (1989). The three-dimensional character of a nominally two-dimensional separated turbulent shear flow, *J. Fluid Mech.*, **205**, 523–552.

Jenkins, G. M. and Watts, D. G. (1968). *Spectral Analysis and its Applications.* Holden-Day, San Francisco.

Jerome, F. E., Guitton, D. E., and Patel, R. P. (1971). Experimental study of the thermal wake interference between closely spaced wires of a X-type hot-wire probe. *Aero. Quart.*, **22**, 119–126.

Jimenez, J., Martinez-Val, R. and Rebollo, M. (1981). Hot-film sensors calibration drift in water. *J. Phys. E.: Sci. Instr.*, **14**, 569–572.

Johansson, A. V. and Alfredsson, P. H. (1982). On the structure of turbulent channel flow. *J. Fluid Mech.*, **122**, 295–314.

Johansson, A. V., Alfredsson, P. H., and Eckelmann, H. (1987a). On the evolution of shear-layer structures in near-wall turbulence. In *Advances in Turbulence*, (ed. G. Comte-Bellot and J. Mathiew), pp. 383–390. Springer, Berlin.

Johansson, A. V., Her, J. Y., and Haritonidis, J. H. (1987b). On the generation of high-amplitude wall-pressure peaks in turbulent boundary layers and spots. *J. Fluid Mech.*, **175**, 119–142.

Johansson, B. (1993). Influence of the velocity near the spark plug on early flame development. *SAE* Paper No. 930481.

Johnson, C. B. and Carraway, D. L. (1989). A transition detection study at Mach 1.5, 2.0 and 2.5 using a micro-thin hot-film system. *ICIASF '89 Record, Proc. Int. Congr. on Instrumentation in Aerospace Simulation Facilities*, pp. 82–94.

Johnson, C. B., Carraway, D. L., Stainback, P. C., and Fancher, M. F. (1987). A transition detection study using a cryogenic hot-film study in the Langley 0.3-meter transonic cryogenic tunnel. *AIAA* paper no. 87-0049, presented at the AIAA 25th Aerospace Sciences Meeting, Reno, Nevada.

Johnson, D. A., Modarress, D., and Owen, F. K. (1984). An experimental verification of laser-velocimeter sampling bias and its correction. *ASME, J. Fluids Eng.*, **106**, 5–12.

Johnson, F. D. and Eckelmann, H. (1983). Has a small-scale structure in turbulence been experimentally verified? *Phys. Fluids*, **26**, 2408–2414.

Johnson, F. D. and Eckelmann, H. (1984). A variable angle method of calibration for X-probes applied to wall-bounded turbulent shear flow. *Exp. in Fluids*, **2**, 121–130.

Johnson, M. C. and Greitzer, E. M. (1987). Effects of slotted hub and casing treatments on compressor endwall flow fields. *ASME, J. Turbomachinery*, **109**, 380–387.

Johnston, A. J., Grant, I., and Halliwell, A. R. (1983). Calibration of film probes using a submerged jet. *Inst. Eng. Aust. Civ. Eng. Trans.*, **25**, 143–147.

Johnston, J. P. (1970). Measurements in a three-dimensional turbulent boundary layer induced by a swept, forward-facing step. *J. Fluid Mech.*, **42**, 823–844.

Jones, O. C. and Zuber, N. (1978). Use of a cylindrical hot-film anemometer for measurement of two-phase void and volume flux profiles in a narrow rectangular channel. *AIChE Symp. Series*, **74**, No. 174, 191–204.

Jones, W. P. and Launder, B. E. (1972). The prediction of laminarization with a two-equation model of turbulence. *Int. J. Heat Mass Transf.*, **15**, 301–314.

Jones, W. P. and Launder, B. E. (1973). The calculation of low-Reynolds-number phenomena with a two-equation model of turbulence. *Int. J. Heat Mass Transf.*, **16**, 1119–1130.

Jørgensen, F. E. (1971*a*). Directional sensitivity of wire and fibre-film probes. *DISA Info.*, No. 11, 31–37.

Jørgensen, F. E. (1971*b*). DISA triaxial probe; measurements in a three-dimensional flow. *DISA Report*, FEJ/338.

Jørgensen, F. E. (1982). Characteristics and calibration of a triple-split probe for reversing flows. *DISA Info.*, No. 27, 15–22.

Joslyn, H. D., Dring, R. P., and Sharma, O. P. (1983). Unsteady three-dimensional turbine aerodynamics. *ASME, J. Eng. for Power*, **105**, 322–331.

Judd, A. M. (1975). Calibration of a five tube probe for measuring wind speed and direction. *J. Phys. E.: Sci. Instr.*, **8**, 115–116.

Kanevče, G. and Oka, S. (1973). Correcting hot-wire readings for influence of fluid temperature variations. *DISA Info.*, No. 15, 21–24.

Kaplan, R. E. and Laufer, J. (1968). The intermittently turbulent region of the boundary layer. *Proc. 12th Int. Cong. Applied Mechanics*, Stanford University., pp. 236–245.

Kassab, S. Z. (1993). A Preston tube calibration chart. *Rev. Sci. Instr.*, **64**, 253–256.

Kastrinakis, E. G. (1976). An experimental investigation of the fluctuations of the streamwise components of the velocity and vorticity vectors in a fully developed turbulent channel flow. Dissertation, Georg-August Universität zu Göttingen.

Kastrinakis, E. G. and Eckelmann, H. (1983). Measurement of streamwise vorticity fluctuations in a turbulent channel flow. *J. Fluid Mech.*, **137**, 165–186.

Kastrinakis, E. G., Wallace, J. M., Willmarth, W. W., Ghorashi, B., and Brodkey, R.S. (1977). On the mechanism of bounded turbulent shear flows. In *Lecture Notes in Physics.*, **75**, 175–189.

Kastrinakis, E. G., Eckelmann, H., and Willmarth, W. W. (1979). Influence of the flow velocity on a Kovasznay type vorticity probe. *Rev. Sci. Instr.*, **50**, 759–767.

Kastrinakis, E. G., Nychas, S.G., and Eckelman, H. (1983). Some streamwise vorticity characteristics of coherent structures. *Proc. IUTAM Symp. on Structure of Complex Turbulent Shear Flow*, (ed. R. Dumas and L. Fulachier), pp. 31–40. Springer, Berlin.

Kawall, J. G., Shokr, M. and Keffer, J. F. (1983). A digital technique for the simultaneous measurement of streamwise and lateral velocities in turbulent flows. *J. Fluid Mech.*, **133**, 83–112.

Keffer, J. F., Budny, R. S., and Kawall, J. G. (1978). Digital technique for the simultaneous measurement of velocity and temperature. *Rev. Sci. Instr.*, **49**, 1343–1346.

Keith, W. L. and Bennett, J. C. (1991). Low-frequency spectra of the wall shear stress and wall pressure in a turbulent boundary layer. *AIAA J.*, **29**, 526–530.

Khalighi, B. (1990). Intake-generated swirl and tumble motions in a four-valve engine with various intake configurations − flow visualization and particle tracking velocimetry, SAE Paper No. 90059.

Khalighi, B., El Tahry, S. H., and Kuziak, W. R. (1986). Measured steady flow velocity distributions around a valve/seat annulus. SAE Paper No. 860462.

Khan, M. A. (1991). Leakage flow in labyrinth seals. Ph.D. Thesis, Department of Mechanical and Manufacturing Engineering, University of Bradford.

Khan, M. A. and Bruun, H. H. (1990). Signal analysis of X hot-wire probes using a look-up table technique. *ASME, Proc. on the Heuristics of Thermal Anemometry, FED*, Vol. 97, University of Toronto, pp. 65–69.

Khan, M. K. (1989). Gas mixing processes in nuclear AGR boilers. Ph.D. Thesis, Department of Mechanical and Manufacturing Engineering, University of Bradford.

Khan, M. K., MacKenzie, K.A., and Bruun, H. H. (1987). The effects of blockage correction in hot-wire probe calibration facilities. *J. Phys. E.: Sci. Instr.*, **20**, 1031–1035.

Kibens, V., Kovasznay, L. S. G., and Oswald, L. J. (1974). Turbulent-nonturbulent interface detector. *Rev. Sci. Instr.*, **45**, 1138–1144.

Kidron, I. (1966). Measurement of the transfer function of hot-wire and hot-film turbulence transducers. *IEEE Trans. Instr. Meas.*, **IM-15**, 76–81.

Kidron, I. (1967). The signal-to-noise ratios of constant-current and constant-temperature hot-wire anemometers. *IEEE Trans. Instr. Meas.*, **IM-16**, 68–73.

Kim, H. T., Kline, S. J., and Reynolds, W. C. (1971). The production turbulence near a smooth wall in a turbulent boundary layer. *J. Fluid Mech.*, **50**, 133–160.

Kim, J., Moin, P., and Moser, R. (1987). Turbulence statistics in fully developed channel flow at low Reynolds numbers. *J. Fluid Mech.*, **177**, 133–166.

Kim, J. H. and Fiedler, H. E. (1989). Vorticity measurements in a turbulent mixing layer. In *Advances in Turbulence* 2, (ed. H. H. Fernholz and H. E. Fiedler), pp. 267–271. Springer, Berlin.

King, C. F. (1979). Some studies of vortex devices − vortex amplifier performance behaviour. Ph.D. Thesis. University College of Wales, Cardiff, Wales.

King, L. V. (1914). On the convection of heat from small cylinders in a stream of fluid: Determination of the convection constants of small platinum wires with applications to hot-wire anemometry. *Phil. Trans. Roy. Soc.*, **A214**, 373–432.

Kinns, R. (1973). Calibration of a hot-wire anemometer for velocity perturbation measurement. *J. Phys. E.: Sci. Instr.*, **6**, 253–6.

Kirchhoff, R. H. and Safarik, R. R. (1974). Turbulence calibration of a hot wire anemometer. *AIAA J.*, **12**, 710–11.

Kistler, A. L. (1952). The vorticity meter. M.Sc. Thesis, The Johns Hopkins University.

Kistler, A. L. (1959). Fluctuation measurements in a supersonic turbulent boundary layer. *Phys. Fluids*, **2**, 290–296.

Kiya, M. and Sasaki, K. (1983). Structure of a turbulent separation bubble. *J. Fluid Mech.*, **137**, 83–113.

Kjellström, B. and Hedberg, S. (1970). Calibration of a DISA hot-wire anemometer and measurements in a circular channel for confirmation of the calibration. *DISA Info.*, No. 9, 8–21.

Kjörk, A. and Löfdahl, L. (1989). Hot-wire measurements inside a centrifugal fan impeller. *ASME, J. Fluids Eng.*, **111**, 363–368.

Klebanoff, P. S. (1954). Characteristics of turbulence in a boundary layer with zero pressure gradient. NACA Technical Note No. 3178.

Klewicki, J. C. and Falco, R. E. (1990). On accurately measuring statistics associated with small-scale structure in turbulent boundary layers using hot-wire probes. *J. Fluid Mech.*, **219**, 119–142.

Kline, S. J. (1985). The purpose of uncertainty analysis. *ASME, J. Fluids Eng.*, **107**, 153–164.

Kline, S. J., Reynolds, W. C., Schraub, F. A., and Rundstadler, P. W. (1967). The structure of turbulent boundary layers. *J. Fluid Mech.*, **30**, 741–773.

Ko, C. L., McLaughlin, D. K., and Troutt, T. R. (1978). Supersonic hot-wire fluctation data analysis with a conduction end-loss correction. *J. Phys. E.: Sci. Instr.*, **11**, 488–493.

Ko, N. W. M. and Davies, P. O. A. L. (1971). Interference effect of hot-wires. *IEEE Trans. Instr. Meas.*, **IM-20**, 76–78.

Koch, F. A. and Gartshore, I. S. (1972). Temperature effects on hot wire anemometer calibrations. *J. Phys. E.: Sci. Instr.*, **5**, 58–61.

Kolmogorov, A. N. (1942). Equations of turbulent motion of an incompressible fluid. *Izv. Akad. Nauk SSR*, Seria Fiz. **VI**, No. 1-2, 56–58.

Komori, S. and Ueda, H. (1985). The large-scale coherent structure in the intermittent region of the self-preserving round free jet. *J. Fluid Mech.*, **152**, 337–359.

Konstantinov, N. I. and Dragnysh, G. L. (1960). The measurement of friction stress on a surface. (English translation), DSIR RTS 1499.

Koppius, A. M. and Trines, G. R. M. (1976). The dependence of hot-wire calibration on gas temperature at low Reynolds numbers. *Int. J. Heat Mass Transf.*, **19**, 967–974.

Kornberger, M. and Feyzi, F. (1990). Transitionsbestimmung mit multisensorheissfilmmesstechnik im windkanal und freiflug. DGLR-Bericht 90–06, 76–80.

Kostka, M. and Vasanta Ram, V. (1992). On the effects of fluid temperature on hot-wire characteristics, Part 1, Results of experiments. *Exp. in Fluids*, **13**, 155–162.

Kotidis, P. A. and Epstein, A. H. (1991). Unsteady radial transport in a transonic compressor stage. *ASME, J. Turbomachinery*, **113**, 207–218.

Kovasznay, L. S. G. (1950a). The hot-wire anemometer in supersonic flow. *J. Aero. Sci.*, **17**, 565–572, 584.

Kovasznay, L. S. G. (1950b). *Quart Prog Rep. of Aero. Dept.* Contract NORD–8036–JHB–39, The Johns Hopkins University.

Kovasznay, L. S. G. (1953). Turbulence in supersonic flow. *J. Aero. Sci.*, **20**, 657–674.

Kovasznay, L. S. G. (1954). Hot-wire method. In *Physical Measurements in Gasdynamics and Combustion*, (ed. R. W. Ladenburg *et al.*), Vol. 9, Section F2, pp. 219–285. Princeton University Press.

Kovasznay, L. S. G., Kibens, V., and Blackwelder, R. F. (1970). Large-scale motion in the intermittent region of a turbulent boundary layer. *J. Fluid Mech.*, **41**, 283–325.

Kramers, H. (1946). Heat transfer from spheres to flowing media. *Physica*, **12**, 61–80.

Kreplin, H. P. and Eckelmann, H. (1979a). Propagation of perturbations in the viscous sublayer and adjacent wall region. *J. Fluid Mech.*, **95**, 305-322.

Kreplin, H. P. and Eckelmann, H. (1979b). Behaviour of the three fluctuating velocity components in the wall region of a turbulent channel flow. *Phys. Fluids*, **22**, 1233-1239.

Krishnamoorthy, L. V. and Antonia, R. A. (1987). Temperature-dissipation measurements in a turbulent boundary layer. *J. Fluid Mech.*, **176**, 265-281.

Krishnamoorthy, L. V., Wood, D. H., Antonia, R. A., and Chambers, A. J. (1985). Effect of wire diameter and overheat ratio near a conducting wall. *Exp. in Fluids*, **3**, 121-127.

Kühn, W. and Dressler, B. (1985). Experimental investigations on the dynamic behaviour of hot-wire probes. *J. Phys. E.: Sci. Instr.*, **18**, 614-622.

Kuppa, S., Sarma, G. R., and Mangalam, S. M. (1993). Effect of thermal inertia on the frequency response of constant voltage hot wire anemometer and its compensation. *ASME*, Fluid Engineering Conference, Washington DC, June.

Kuroumaru, M., Inoue, M., Higaki, T., Abd-Elkhalek, F. A. E., and Ikui, T. (1982). Measurements of three dimensional flow field behind an impeller by means of periodic multi-sampling with a slanted hot wire. *Bull. JSME*, **25**, 1674-1681.

Kutler, P. (1985). A perspective of theoretical and applied computational fluid dynamics. *AIAA J.*, **23**, 328-341.

Kwok, F. T., Andrew, P. L., Ng, W. F., and Schetz, J. A. (1991). Experimental investigation of a supersonic shear layer with slot injection of helium. *AIAA J.*, **29**, 1426-1435.

Laderman, A. J. (1976). New measurements of turbulent shear stresses in hypersonic boundary layers. *AIAA J.*, **14**, 1286-1291.

Laderman, A. J. and Demetriades, A. (1973). Hot-wire measurements of hypersonic boundary layer turbulence. *Phys. Fluids*, **16**, 179-181.

Laderman, A. J. and Demetriades, A. (1974). Mean and fluctuating flow measurements in the hypersonic boundary layer over a cooled wall. *J. Fluid Mech.*, **63**, 121-144.

Laderman, A. J. and Demetriades, A. (1979). Turbulent shear stresses in compressible boundary layers. *AIAA J.*, **17**, 736-744.

Laksminarayana, B. (1980). An axial flow research compressor facility designed for flow measurement in rotor passages. *ASME, J. Fluids Eng.*, **102**, 402-411.

Lakshminarayana, B. (1981). Techniques for aerodynamic and turbulence measurements in turbomachinery rotors. *ASME, J. Eng. for Power*, **103**, 374-392.

Lakshminarayana, B. (1982a). Three sensor hot wire/film technique for three dimensional mean and turbulence flow field measurement. *TSI Quart.*, **VIII**, 3-13.

Lakshminarayana, B. (1982b). Fluid dynamics of inducers – A review. *ASME, J. Fluids Eng.*, **104**, 411-427.

Lakshminarayana, B. and Davino, R. (1980). Mean velocity and decay characteristics of the guidevane and stator blade wake of an axial flow compressor. *ASME, J. Eng. for Power*, **102**, 50-60.

Lakshminarayana, B. and Davino, R. (1988). Sensitivity of three sensor hot wire probe to yaw and pitch angle variation. *ASME, J. Fluids Eng.*, **110**, 120-122.

Lakshminarayana, B. and Pandya, A. (1984). Tip clearance flow in a compressor rotor passage at design and off-design conditions. *ASME, J. Eng. for Gas Turbines and Power*, **106**, 570-577.

Lakshminarayana, B. and Poncet, A. (1974). A method of measuring three-dimensional rotating wakes behind turbomachinery rotors. *ASME, J. Fluids Eng.*, **96**, 87–91.

Lakshminarayana, B., Pouagare, M., and Davino, R. (1982). Three-dimensional flow field in the tip region of a compressor rotor passage, Parts 1 and 2. *ASME, J. Eng. for Power*, **104**, 760–781.

Lam, C. K. G. and Bremhorst, K. (1981). A modified form of the k-ε model for predicting wall turbulence. *ASME, J. Fluids Eng.*, **103**, 456–460.

Lancaster, D. R. (1976). Effects of engine variables on turbulence in a spark-ignition engine. SAE Paper No. 760159.

Lance, M. (1979). Contribution à l'étude de la turbulence dans la phase liquide des écoulements à bulles. Thèse de Docteur-Ingénieur, Université Claude Bernard, Lyon, France.

Lance, M. and Bataille, J. (1983). Turbulence in the liquid phase of a bubbly air-water flow. *Advances in Two-Phase Flow and Heat Transfer*, (ed. Kakac and M. Ishii), Vol. 1, pp. 403–427. Martinus Nijhoff Publishers.

Lance, M. and Bataille, J. (1991). Turbulence in the liquid phase of a uniform bubbly air-water flow. *J. Fluid Mech.*, **222**, 95–118.

Larguier, R. (1981). Experimental analysis methods for unsteady flows in turbomachines. *ASME, J. Eng. for Power*, **103**, 415–423.

Larsen, S. and Højstrup, J. (1982). Spatial and temporal resolution of a thin-wire resistance thermometer. *J. Phys. E.: Sci. Instr.*, **15**, 471–477.

Larsen, S.E., Højstrup, J., and Fairall, C. W. (1986). Mixed and dynamic response of hot wires and cold wires and measurements of turbulence statistics. *J. Atmospheric Oceanic Techn.*, **3**, 236–247.

LaRue, J. C. (1974). Detection of the turbulent-nonturbulent interface in slightly heated turbulent shear flows. *Phys. Fluids*, **17**, 1513–1517.

LaRue, J. C. and Libby, P. A. (1974). Temperature fluctuations in the plane turbulent wake. *Phys. Fluids*, **17**, 1956–1967.

LaRue, J. C., Deaton, T., and Gibson, C. H. (1975). Measurement of high-frequency turbulent temperature. *Rev. Sci. Instr.*, **46**, 757–764.

Latigo, B. O. (1989). Coherent structure interactions in a two-stream plane turbulent mixing layer with impulsive acoustic excitation. *Phys. Fluids*, **A1**, 1701–1715.

Lau, J. C. and Fisher, M. J. (1975). The vortex-street structure of 'turbulent' jets, Part 1. *J. Fluid Mech.*, **67**, 299–337.

Lau, S., Schulz, V., and Vasanta Ram, V. I. (1993). A computer operated traversing gear for three-dimensional flow surveys in channels. *Exp. in Fluids*, **14**, 475–476.

Laufer, J. (1954). The structure of turbulence in fully developed pipe flow. NACA Technical Note No. 2954.

Laufer, J. and McClellan, R. (1956). Measurement of heat transfer from fine wires in supersonic flow. *J. Fluid Mech.*, **1**, 276–289.

Laufer, J. and Badri Narayanan, M. A. (1971). Mean period of the turbulent production mechanism in a boundary layer. *Phys. Fluids*, **14**, 182–183.

Launder, B. E. (1984). Second-moment closure: methodology and practice. In *Turbulence Models and their Applications*, Vol. 2. Editions Eyrolles, Paris.

Launder, B. E. (1989). Review: second moment closure: present and future? *Int. J. Heat and Fluid Flow*, **10**, 282–300.

Launder, B. E. and Spalding, D. B. (1972). *Mathematical Models of Turbulence*. Academic Press, London.

Launder, B. E. and Spalding, D. B. (1974). The numerical computation of turbulent flows. *Comp. Methods in Applied Mech. and Engng.*, **3**, 269–289.

Launder, B. E., Reece, G. J., and Rodi, W. (1975). Progress in the development of a Reynolds-stress turbulent closure. *J. Fluid Mech.*, **68**, 537–566.

Laurence, J. C. and Sandborn, V. A. Heat transfer from cylinders. *ASME, Symp. on Measurements in Unsteady Flow*, pp. 36–43.

LeBoeuf, R. L. and George, W. K. (1990). The calibration and use of non-orthogonal cross-wire probes. *ASME, Proc. Symp. on the Heuristics of Thermal Anemometry*, FED, Vol. 97, University of Toronto, pp. 59–63.

Lecordier, J. C., Dupont, A., Gajan, P., and Paranthoen, P. (1984). Correction of temperature fluctuation measurements using cold wires. *J. Phys. E.: Sci. Instr.*, **17**, 307–311.

Lecordier, J. C., Moreau, D., and Paranthoen, P. (1985). Prong-wire thermal interaction effect induced harmonics for the thin-wire resistance thermometer. *J. Phys. E.: Sci. Instr.*, **18**, 571–572.

Le Coz, J. F. (1992). Cycle-to-cycle correlations between flow field and combustion initiation in a S. I. Engine. SAE Paper No. 920517.

Le Coz, J. F., Henriot, S., and Pinchon, P. (1990). An experimental and computational analysis of the flow field in a four-valve spark ignition engine – focus on cycle-resolved turbulence. SAE Paper No. 900056.

Lee, J. and Farrel, P. V. (1993). Intake valve measurements of an IC engine using particle image velocimetry. SAE Paper No. 930480.

Lee, S. J. (1982). The development of a digital data processing system for two-phase turbulence data. M.S. Dissertation, Rensselaer Polytechnic Institute, Troy, NY.

Lee, T. and Budwig, R. (1991). Two improved methods for low-speed hot-wire calibration. *Meas. Sci. Technol.*, **1**, 643–646.

Legg, B. J., Coppin, P. A., and Raupach, M. R. (1984). A three-hot-wire anemometer for measuring two velocity components in high intensity turbulent boundary layers. *J. Phys. E.: Sci. Instr.*, **17**, 970–976.

Leithem, J. J., Kulik, R. A., and Weinstein, H. (1969). Turbulence in the mixing region between ducted coaxial streams. NASA CR-1335.

Lekakis, I. C., Adrian, R. J., and Jones, B. G. (1989). Measurement of velocity vectors with orthogonal and non-orthogonal triple-sensor probes. *Exp. in Fluids*, **7**, 228–240.

Lemieux, G. P. and Oosthuizen, P. H. (1984). A simple approach to the compensation of constant temperature hot-wire anemometers for fluid temperature fluctuations. *Proc. 30th Int. Instrumentation Symp.*, Denver, AIAA, pp. 277–282.

Leschziner, M. A. (1990). Modelling engineering flows with Reynolds stress turbulence closure. *J. Wind Eng. and Industr. Aerodyn.*, **35**, 21–47.

Li, Y. S. and Cumpsty, N. A. (1991). Mixing in axial flow compressor, Part 1, Test facilities and measurements in a four-stage compressor. *ASME, J. Turbomachinery*, **113**, pp. 161–165.

Libby, P. A. (1977). Studies in variable-density and reacting turbulent shear flows. In *Studies in Convection*, (ed. B. E. Launder), Vol. 2, pp. 1–43. Academic Press, New York.

Libby, P. A. and LaRue, J. C. (1978). Hot-wire anemometry for turbulence measurements in helium-air mixtures. *Proc. Dynamic Flow Conf.*, Marseille, pp. 115–130.

Lienhard, J. H. and Helland, K. N. (1989). An experimental analysis of fluctuating temperature measurements using hot-wires at different overheats. *Exp. in Fluids*, **7**, 265–270.

Liepmann, H. W. and Skinner, G. T. (1954). Shearing-stress measurements by use of a heated element. NACA Technical Note 3268.

Ligrani, P. M. and Bradshaw, P. (1987*a*). Subminiature hot-wire sensors: development and use. *J. Phys. E.: Sci. Instr.*, **20**, 323–332.

Ligrani, P. M. and Bradshaw, P. (1987*b*). Spatial resolution and measurement of turbulence in the viscous sublayer using miniature hot-wire probes. *Exp. in Fluids*, **5**, 407–417.

Ligrani, P. M., Gyles, B. R., Mathioudakis, K., and Breugelmans, F. A. E. (1983). A sensor for flow measurements near the surface of a compressor blade. *J. Phys. E.: Sci. Instr.*, **16**, 431–437.

Ligrani, P. M., Singer, B. A., and Baun, L. R. (1989*a*). Miniature five-hole pressure probe for measurement of three mean velocity components in low speed flow. *J. Phys. E.: Sci. Instr.*, **22**, 868–876.

Ligrani, P. M., Singer, B. A., and Baun, L. R. (1989*b*). Spatial resolution and downwash velocity corrections for multiple-hole pressure probes in complex flows. *Exp. in Fluids*, **7**, 424–426.

Ligrani, P. M., Westphal, R. V., and Lemos, F. R. (1989*c*). Fabrication and testing of subminiature multi-sensor hot-wire probes. *J. Phys. E.: Sci. Instr.*, **22**, 262–268.

Lin, C. N. and Chang, L. F. W. (1989). Liquid-droplet size measurement with the hot-film anemometer: an alternative approach. *J. Chinese Inst. Eng.*, **12**, 123–130.

Ling, S. C. and Hubbard, P. G. (1956). The hot-anemometers: A new device for fluid mechanics research. *J. Aero. Sci.*, **23**, 890–891.

Liou, T. M. and Santavicca, D. A. (1982). Cycle resolved LDV measurements in a motored IC engine. In *Engineering Applications of Laser Velocimetry*, (ed. H. W. Goldman and P. A. Pfund) 33–37. ASME, New York.

Liou, T. M. and Santavicca, D. A. (1985). Cycle resolved LDV measurements in a motored IC engine. *ASME, J. Fluids Eng.*, **107**, 232–40.

Liou, T. M., Santavicca, D. A., and Bracco, F. V. (1984). Laser doppler velocimetry measurements in valved and ported engines. SAE Paper No. 840375.

Liu, J. T. (1989). Coherent structures in transitional and turbulent free shear flows. *Ann. Rev. Fluid Mech.*, **21**, 285–315.

Liu, T. J. (1993). Bubble size and entrance length effect on void development in a vertical channel. *Int. J. Multiphase Flow*, **19**, 93–113.

Liu, T. J. and Bankoff, S. G. (1990*a*). Structure of air-water bubbly flow in a vertical pipe, Part I, Liquid mean velocity and turbulence measurements. *ASME, Symp. on Gas-liquid Two-phase Flows*, Dallas, pp. 9–17.

Liu, T. J. and Bankoff, S. G. (1990*b*). Structure of air-water bubbly flow in a vertical pipe, Part II, Void fraction, bubble velocity and bubble size distribution. *ASME, Symp. on Gas-Liquid Two-phase Flows*, Dallas, pp. 19–26.

Liu, T. J. and Bankoff, S. G. (1993*a*). Structure of air-water bubbly flow in a vertical pipe, Part I, Liquid mean velocity and turbulence measurements. *Int. J. Heat Mass Transf.*, **36**, 1049–1060.

Liu, T. J. and Bankoff, S. G. (1993*b*). Structure of air-water bubbly flow in a vertical pipe, Part II, Void fraction, bubble velocity and bubble size distribution. *Int. J. Heat Mass Transf.*, **36**, 1061–1072.

Löfdahl, L. (1988). Traverse mechanisms for the determination of Reynolds stresses using hot-wire techniques. *Exp in Fluids*, **6**, 352–354.

Löfdahl, L. and Larsson, L. (1984). Turbulence measurements near the stern of a ship model. *J. Ship Research*, **28**, 186–201.

Lomas, C. G. (1986). *Fundamentals of Hot-wire Anemometry*. Cambridge University Press, Cambridge.

Lord, R. G. (1974). Hot-wire probe end-loss corrections in low density flows. *J. Phys. E.: Sci. Instr.*, **7**, 56–60.

Lord, R. G. (1981). The dynamic behaviour of hot-wire anemometers with conduction end losses. *J. Phys. E.: Sci. Instr.*, **14**, 573–578.

Lord, R. G. (1982). Transfer functions of hot-wire anemometers with conduction end losses. *J. Phys. E.: Sci. Instr.*, **15**, 1045–1048.

Lorenz, M. and Prescher, K. (1990). Cycle-resolved LDV measurements on a fired SI-engine at high data rates using a conventional modular LDV-system. *SAE Paper No. 900054.*

Lu, S. S. (1979). Dynamic characteristics of a simple constant-temperature hot-wire anemometer. *Rev. Sci. Instr.*, **50**, 772–775.

Lu, S. S. and Willmarth, W. W. (1973). Measurements of the structure of the Reynolds stress in a turbulent boundary layer. *J. Fluid Mech.*, **60**, 481–511.

Luchik, T. S. and Tiederman, W. G. (1987). Timescale and structure of ejections and bursts in turbulent channel flows. *J. Fluid Mech.*, **174**, 529–552.

Ludwieg, H. (1949). Ein gerät zur messung der wandschubspannung turbulenter reibungsschichten. *Ing. Arch.*, **17**, 207–218.

Ludwieg, H. (1950). Instrument for measuring the wall shearing stress of turbulent boundary layers. NACA TM 1284.

Lueptow, R. M., Breuer, K. S., and Haritonidis, J. H. (1988). Computer-aided calibration of X-probes using a look-up table. *Exp. in fluids*, **6**, 115–118.

Lumley, J. L. (1970). *Stochastic Tools in Turbulence*. Academic Press, New York.

Lumley, J. L. (1983). Turbulence modelling. *ASME, J. Appl. Mech.*, **50**, 1097–1103.

Madavan, N. K., Deutsch, S., and Merkle, C. L. (1985). Measurements of local skin friction in a microbubble-modified turbulent boundary layer. *J. Fluid Mech.*, **156**, 237–256.

Mahajan, R. L. and Gebhart, B. (1980). Hot-wire anemometor calibration in pressurised nitrogen at low velocities. *J. Phys. E.: Sci. Instr.*, **13**, 1110–1118.

Mahler, D. S. (1982). Bidirectional hot-wire anemometer. *Rev. Sci. Instr.*, **53**, 1465–1466.

Manca, O., Mastullo, R., and Mazzei, P. (1988). Calibration of hot-wire probes at low velocities in air with variable temperature. *DANTEC Info.*, No. 6, 6–8.

Mangalam, S. M., Sarma, G. R., Kuppa, S., and Kubendran, L. R. (1992). A new approach to high-speed flow measurements using constant voltage anemometry. *AIAA 17th Aerospace Ground Testing Conference*, Nashville, TN, *AIAA* paper No. 92-3957.

Marasli, B., Nguyen, P., and Wallace, J. M. (1993). A calibration technique for multiple-sensor hot-wire probes and its application to vorticity measurements in the wake of a circular cyclinder. *Exp. in Fluids*, **15**, 209–218.

Mateer, G. G., Brosh, A., and Viegas, J. R. (1976). An experimental and numerical investigation of normal shock-wave turbulence boundary layer interaction at $M_\infty = 1.5$. AIAA paper 76-161.

Mathews, J. and Poll, D. I. A. (1985). The theory and application of heated films for the measurement of skin friction. College of Aeronautics, (COA) Cranfield Institute of Technology, CoA Report No. 8515.

Mathioudakis, K. and Breugelmans, F. A. E. (1985). Use of triple hot wires to measure unsteady flows with large direction changes. *J. Phys. E.: Sci. Instr.*, **18**, 414–419.

Mattioli, E. (1956). Una nuova sonda a filo caldo per misure di turbolenza nello strato limite. *Atti della Academia delle Scienze di Torino*, I. Classe di Scienze Fisiche, Matematiche e. Naturali, **91**, 71.

Maye, J. P. (1970). Error due to thermal conduction between the sensing wire and its supports when measuring temperatures with a wire anemometer used as a resistance thermometer. *DISA Info.*, No. 9, 22–26.

Mayle, R. E. and Anderson, A. (1991). Velocity and temperature profiles for stagnation film cooling. *ASME, J. Turbomachinery*, **113**, 457–462.

McAllister, J. E., Pierce, F. J., and Tennant, M. H. (1982). Preston tube calibrations and direct force floating element measurements in a two-dimensional turbulent boundary layer. *ASME, J. Fluids Eng.*, **104**, 156–161.

McConachie, P. J. and Bullock, K. J. (1976). Measurement of turbulence close to a wall with temperature wake sensing probes. *J. Phys. E.: Sci. Instr.*, **9**, 862–868.

McDougall, N. M., Cumpsty, N. A., and Hynes, T. P. (1990). Stall inception in axial compressors. *ASME, J. Turbomachinery*, **112**, 116–132.

McLaughlin, D. K. and McColgan, C. J. (1974). Hot-wire measurements in a supersonic jet at low Reynolds numbers. *AIAA J.*, **12**, 1279–1281.

McQuaid, J. and Wright, W. (1973). The response of a hot-wire anemometer in flows of gas mixtures. *Int. J. Heat Mass Trans.*, **16**, 819–828.

McQuaid, J. and Wright, W. (1974). Turbulence measurements with hot-wire anemometry in non-homogeneous jets. *Int. J. Heat Mass Transf.*, **17**, 341–349.

McQuivey, R. S. and Richardson, E. V. (1969). Some turbulence measurements in open-channel flow. *J. Eng. Mech. Div., ASCE*, **95**, HY1, 209–222.

Menendez, A. N. and Ramaprian, B. R. (1985). The use of flush-mounted hot-film gauges to measure skin friction in unsteady boundary layers. *J. Fluid Mech.*, **161**, 139–159.

Merati, P. and Adrian, R. J. (1984). Directional sensitivity of single and multiple sensor probes. *TSI Quart.*, **X**, No. 2, 3–12.

Mestayer, P. and Chambaud, P. (1979). Some limitations to measurements of turbulence micro-structure with hot and cold wires. *Bound. Layer Met.*, **16**, 311–329.

Miller, I. S., Shah, D. A., and Antonia, R. A. (1987). A constant temperature hot-wire anemometer. *J. Phys. E.: Sci. Instr.*, **20**, 311–314.

Millon, F., Paranthoen, P., and Trinite, M. (1978). Influence des echanges termiques entre le capteur et ses supports sur la mesure des fluctuations de température dans un écoulement turbulent. *Int. J. Heat Mass Transf.*, **21**, 1–6.

Mirow, P., Weiser, N., Devliotis, A., and Nitsche, W. (1990). Freiflugexperimente zur transitionserkennung an laminarflügeln. DGLR-Bericht 90–06, 81–85.

Mitchell, R. A. (1987). Hot-wire fabrication and repair: an interim report. MSU-ENGR., 87-015-DSFL, Michigan State University.

Mobarak, A., Sedrak, M.F., and El Telbany, M. M. M. (1986). On the direction sensitivity of hot-wire probes. *DANTEC Info.*, No. 2, 7–9.

Moffat, R. J. (1982). Contributions to the theory of single-sample uncertainty analysis. *ASME, J. Fluids Eng.*, **104**, 250–260.

Moffat, R. J. (1985). Using uncertainty analysis in the planning of an experiment. *ASME, J. Fluids Eng.*, **107**, 173–178.

Moffat, R. J., Yavuzkurt, S., and Crawford, M. E. (1978). Real-time measurements of turbulence quantities with a triple hot-wire system. *Proc. Dynamic Flow Conf.*, Marseille, pp. 1013–1035.

Moin, P. and Spalart, P. R. (1989). Contributions of numerical simulation data bases to the physics, modelling and measurements of turbulence. In *Advances in Turbulence*, (ed. W. K. George and R. Arndt), pp. 11–18. Hemisphere, New York.

Mojola, O. O. (1974). A hot-wire method for three-dimensional shear flows. *DISA Info.*, No. 16, 11–14.

Mollenkopf, G. (1972). Measuring nonstationary periodical flow in the draft tube of a water-powered Francis model turbine. *DISA Info.*, No. 13, 11–15.

Moon, I. M. (1962). Direction sensitive hot wire anemometer for two dimensional flow study near a wall. *ASME, Symp. on Measurement in Unsteady Flow*, pp. 71–75.

Moore, C. J. (1977). The role of shear-layer instability waves in jet exhaust noise. *J. Fluid Mech.*, **80**, 321–367.

Morkovin, M. V. (1956). *Fluctuations and Hot-wire Anemometry in Compressible Flows.* AGARDograph No. 24.

Morkovin, M. V. (1962). Effects of compressibility on turbulent flows. *Int. Symp. on the Mechanics of Turbulence*, CNRS, Paris, France.

Morkovin, M. V. and Phinney, R. E. (1958). Extended applications of hot-wire anemometry to high speed turbulent boundary layers. Department of Aeronautics, The Johns Hopkins University, Report No. AFORS TN-58-469.

Morrison, D. F., Sheppard, L. M., and Williams, M. J. (1967). Hole size effect on hemisphere pressure distributions. *J. Roy Aero. Soc.*, **71**, 317–319.

Morrison, G. L. (1974). Effects of fluid property variations on the response of hot-wire anemometers. *J. Phys. E.: Sci. Instr.*, **7**, 434–436.

Morrison, G. L., Perry, A. E., and Samuel, A. E. (1972). Dynamic calibration of inclined and crossed hot wires. *J. Fluid Mech.*, **52**, 465–474.

Morrison, J. F., Tsai, H. M., and Bradshaw, P. (1989). Conditional-sampling schemes for turbulent flow, based on the variable-interval time averaging (VITA) algorithm. *Exp. in Fluids*, **7**, 173–189.

Morrison, J. F., Subramaniam, C. S., and Bradshaw, P. (1992). Bursts and the law of the wall in turbulent boundary layers. *J. Fluid Mech.*, **241**, 75–108.

Morrow, T. B. and Kline, S. J. (1971). The evaluation and use of hot-wire and hot-film anemometers in liquids. Report MD-25, Mechanical Engineering Department Stanford University.

Moum, J. N., Kawall, J. G., and Keffer, J. F. (1979). Structural features of the plane turbulent jet. *Phys. Fluids*, **22**, 1240–1244.

Mulhearn, P. J. and Finnigan, J. J. (1978). A simple device for dynamic testing of X-configuration hot-wire anemometer probes. *J. Phys. E.: Sci. Instr.*, **11**, 679–681.

Müller, U. R. (1982a). Measurements of the Reynolds stresses and the mean-flow field in a three-dimensional pressure-driven boundary layer. *J. Fluid Mech.*, **119**, 121–153.

Müller, U. R. (1982b). On accuracy of turbulence measurements with inclined hot wires. *J. Fluid Mech.*, **119**, 155–172.

Müller, U. R. (1982c). Comments on 'Measurements by means of triple-sensor probes' by M. Acrivlellis. *J. Phys. E.: Sci. Instr.*, **15**, 386.

Müller, U. R. (1983). A hot-wire method for high-intensity turbulent flows. *ICIASF*

'83 Record. Proc. Int. Congr. Instrum. in Aerospace Simulation Facilities, ISL, St. Louis, France, pp. 298–304.

Müller, U. R. (1987). Developments in measuring Reynolds stresses. In *Perspective in Turbulence Studies*, (ed. H. U. Meier and P. Bradshaw), pp. 300–335. Springer, Berlin.

Müller, U. R. (1992). Comparison of turbulence measurements with single, X and triple hot-wire probes. *Exp. in Fluids*, **13**, 208–216.

Müller, U. and Krause, E. (1979). Measurements of mean velocities and Reynolds stresses in an incompressible three-dimensional turbulent boundary layer. *Proc. 2nd Int. Symp. on Turbulent Shear Flows*, Imperial College, London, 15.36–15.41.

Murakami, A., Sakimoto, M., Arai, M., and Hiroyasu, H. (1990). Measurement of turbulent flow in the combustion chamber of a D.I. diesel engine. *SAE* Paper No. 900061.

Murlis, J., Tsai, H. M., and Bradshaw, P. (1982). The structure of turbulent boundary layers at low Reynolds numbers. *J. Fluid Mech.*, **122**, 13–56.

Murthy, V. S. and Rose, W. C. (1978). Wall shear stress measurements in a shock-wave boundary-layer interaction. *AIAA J.*, **16**, 667–670.

Nabhani, N. (1989). Hot-wire anemometry study of confined turbulent swirling flow. Ph.D. Thesis, Department of Mechanical and Manufacturing Engineering, University of Bradford.

Nabhani, N. and Bruun, H. H. (1990). Three component velocity measurements using multi-position single-slant hot-wire technique. *ASME, Proc. Symp. Heuristics of Thermal Anemometry, FED*, Vol. 97, University of Toronto, pp. 23–28.

Nagano, Y. and Hishida, M. (1987). Improved form of the k–ε model for wall turbulent shear flows. *ASME, J. Fluids Eng.*, **109**, 156–160.

Nagano, Y. and Kim, C. (1988). A two-equation model for heat transport in wall turbulent shear flows. *ASME, J. Heat Transf.*, **110**, 583–589.

Nagano, Y. and Tagawa, M. (1990). An improved k-ε model for boundary layer flows. *ASME, J. Fluids Eng.*, **112**, 33–39.

Nakagawa, H. and Nezu, I. (1981). Structure of space–time correlation of bursting phenomena in an open-channel flow. *J. Fluid Mech.*, **104**, 1–43.

Namekawa, S., Ryu, H., and Asanuma, T. (1988). LDA measurement of turbulent flow in a motored and firing spark ignition engine with a horizontal prechamber. SAE Paper No. 881636.

Nasser, S. H. and Castro, I. P. (1989). Spatial correlation measurement in turbulent flows using pulsed wire anemometry. *Exp. in Fluids*, **7**, 119–124.

Nelson, E. W. and Borgos, J. A. (1983). Dynamic response of conical and wedge type hot films: Comparison of experimental and theoretical results. *TSI Quart.*, **IX**, No. 1, 3–10.

Neuerburg, W. (1969). Directional hot-wire probe. *DISA Info.*, No. 7, 30–31.

Newland, D. E. (1984). *An Introduction to Random Vibrations and Spectral Analysis*. Longman, New York.

Ng, W. F. and Epstein, A. H. (1983). High-frequency temperature and pressure probe for unsteady compressible flows. *Rev. Sci. Instr.*, **54**, 1678–1683.

Ng, W. F. and Epstein, A. H. (1985). Unsteady losses in transonic compressors. *ASME, J. Eng. for Gas Turbine and Power*, **107**, 345–353.

Ninnemann, T. A. and Ng, W. F. (1992). A concentration probe for the study of mixing in supersonic shear flows. *Exp. in Fluids*, **13**, 98–104.

Nino, E., Gajdeczko, B. F., and Felton, P. G. (1993). Two-color particle image velocimetry in an engine with combustion. SAE Paper No. 930872.

Nishioka, M. and Asai, M. (1988). A new approximate expression for the response of a hot-wire anemometer. *J. Fluid Mech.*, **190**, 113–119.

Nowack, F. R. (1970). Improved calibration method for a five-hole spherical pitot probe. *J. Phys. E.: Sci. Inst.*, **3**, 21–26.

Nychas, S. G., Hershey, H. C., and Brodkey, R. S. (1973). A visual study of turbulent shear flow. *J. Fluid Mech.*, **61**, 513–540.

Nuttall, A. H. (1981). Some windows with very good sidelobe behaviour. *IEEE Trans. on Acoustics, Speech and Signal Processing*, **29**, 84–91.

O'Brian, J. E. and Capp, S. P. (1989). Two-component phase-averaged turbulence statistics downstream of a rotating spoked-wheel wake generator. *ASME, J. Turbomachinery*, **111**, 475–482.

Offen, G. R. and Kline, S. J. (1974). Combined dye-streak and hydrogen-bubble visual observations of a turbulent boundary layer. *J. Fluid Mech.*, **62**, 223–239.

Ohigashi, S., Hamamoto, Y., and Tanabe, S. (1971). Swirl — Its measurement and effect on combustion in a diesel engine. *IMechE J.*, **134**.

Oka, S. and Kostic, Z. (1972). Influence of wall proximity on hot-wire velocity measurements. *DISA Info.*, No. 13, 29–33.

Okiishi, T. H. and Schmidt, D. P. (1978). Measurement of the periodic variation of turbomachine flow fields. *Proc. Dynamic Flow Conf.*, Marseille, pp. 249–269.

Olin, J. G. and Kiland, R. B. (1970). Split-film anemometer sensors for three-dimensional velocity-vector measurement. *Proc. Symp. on Aircraft Wake Turbulence*, Seattle, Washington, pp. 57–79.

Olivari, D. (1976). Effects of the wire supports on the frequency response of hot wire anemometers. Technical Note 118, Von Kármán Institute for Fluid Dynamics, Belgium.

Orlando, A. (1974). Turbulent transport of heat and momentum in a boundary layer subject to deceleration, suction and variable wall temperature. PhD. Thesis, Stanford University.

Ossofsky, E. (1948). Constant temperature operation of the hot-wire anemometer at high frequency. *Rev. Sci. Instr.*, **19**, 881–889.

Oster, D. and Wygnanski, I. (1982). The forced mixing layer between parallel streams. *J. Fluid Mech.*, **123**, 91–130.

Owen, F. K., Horstman, C. C., and Kussoy, M. I. (1975). Mean and fluctuating flow measurements of a fully-developed, non-adiabatic, hypersonic boundary layer. *J. Fluid Mech.*, **70**, 393–413.

Pailhas, G. and Cousteix, J. (1986). Méthod d'exploitation des données d'une sonde anémométrique a quatre fils chauds. *La Recherche Aerospatiale*, **2**, 161–168.

Panchapakesan, N. R., and Lumley, J. L. (1993). Turbulence measurements in axisymmetric jets of air and helium, Part 1, Air jet. *J. Fluid Mech.*, **246**, 197–223.

Pandya, A. and Lakshminarayana, B. (1983). Investigation of the tip clearance flow inside and at the exit of a compressor rotor passage, Parts 1 and 2. *ASME, J. Eng. for Power*, **105**, 1–17.

Pao, Y. H. (1965). Structure of turbulent velocity and scalar fields at large wavenumbers. *Phys. Fluids*, **8**, 1063–1075.

Papavergos, P. G. and Hedley, A. B. (1979). A simple practical method for establishing turbulence characteristics by means of a single 45° slant hot-wire probe in a field of known mean flow direction. *J. Phys. E.: Sci. Instr.*, **12**, 761–765.

Paranthoen, P., Lecordier, J. C., and Petit, C. (1982*a*). Influence of dust contamination on frequency response of wire resistance thermometers. *DISA Info.*, No. 27, 36–37.

Paranthoen, P., Petit, C., and Lecordier, J. C. (1982*b*). The effect of the thermal prong-wire interaction on the response of a cold wire in gaseous flows (air, argon and helium). *J. Fluid Mech.*, **124**, 457–473.

Paranthoen, P., Lecordier, J. C., and Petit, C. (1983). Dynamic sensitivity of the constant-temperature hot-wire anemometer to temperature fluctuations. *TSI Quart.*, **IX**, No. 3, 3–8.

Paranthoen, P., Lecordier, J. C. and Petit, C. (1984). Test of the P31 DISA probes. *DISA Info.*, No. 29, 27–28.

Patel, V. C. (1965). Calibration of the Preston tube and limitations on its use in pressure gradients. *J. Fluid Mech.*, **23**, 185–208.

Patel, V. C., Rodi, W., and Scheuerer, G. (1981). Evaluation of turbulence models for near-wall and low-Reynolds number flows. *Proc. 3rd Int. Symp. on Turbulent Shear Flows*, University of California, Davis, 1.1–1.8.

Patel, V. C., Rodi, W., and Scheuerer, G. (1985). Turbulence models for near-wall and low Reynolds number flows: A review. *AIAA J.*, **23**, 1308–1319.

Paul, J. and Steimle, F. (1977). New results from measurements with hot-wire anemometers at low velocities and superimposed convection. IIF–IIR Commission Bl, B2, El, Belgrade, Yugoslavia.

Paulsen, L. (1983). Triple hot-wire technique for simultaneous measurements of instantaneous velocity components in turbulent flows. *J. Phys. E.: Sci. Instr.*, **16**, 554–562.

Payne, F. R. and Lumley, J. L. (1966). One-dimensional spectra derived from an airborne hot-wire anemometer. *Q. J. Roy. Met. Soc.*, **92**, 397–401.

Peattie, R. (1987). A simple, low-drift circuit for measuring temperatures in fluids. *J. Phys. E.: Sci. Instr.*, **20**, 565–567.

Perry, A. E. (1972). The aeroelastic behaviour of hot-wire anemometer filaments in an airstream. *J. Sound Vib.*, **22**, 41–58. [Also see Perry 1973.]

Perry A. E. (1973). Errata aeroelastic behaviour of hot-wires. *J. Sound Vib.*, **26**, 279.

Perry, A. E. (1977). The time response of an aspirating probe in gas sampling. *J. Phys. E.: Sci Instr.*, **10**, 898–902.

Perry, A. E. (1982). *Hot-wire Anemometry*. Clarendon Press, Oxford.

Perry, A. E. and Morrison, G. L. (1971*a*). A study of the constant-temperature hot-wire anemometer. *J. Fluid Mech.*, **47**, 577–599.

Perry, A. E. and Morrison G. L. (1971*b*). Vibration of hot-wire anemometer filaments. *J. Fluid. Mech.*, **50**, 815–825.

Perry, A. E. and Morrison, G. L. (1971*c*). Static and dynamic calibrations of constant-temperature hot-wire systems. *J. Fluid Mech.*, **47**, 765–777.

Perry, A. E. and Morrison, G. L. (1972). Errors caused by hot-wire filament vibration. *J. Phys. E.: Sci. Instr.*, **5**, 1004–1008.

Perry, A. E., Smits, A. J., and Chong, M. S. (1979). The effects of certain low frequency phenomena on the calibration of hot wires. *J. Fluid Mech.*, **90**, 415–431.

Persen, L. N. and Saetran, L. R. (1983). Hot-film measurements in a water tunnel. *J. Hydr. Res.*, **21**, 379–387.

Pessoni, D. H. and Chao, B. T. (1974). A simple technique for turbulence measurements in nonisothermal air flows. *Proc. 5th Int. Heat Transfer Conf.*, Tokyo, (ISME), pp. 278–282.

Petit, C., Paranthoen, P., Lecordier J. C., and Gajan, P. (1985). Dynamic behaviour of cold wires in heated airflows (300 < *T* < 600 K). *Exp. in Fluids*, **3**, 169–173.

Phillips, W. R. C. (1985). Higher-order response equations for an inclined hot-wire. *J. Phys., E.: Sci. Instr.*, **18**, 314–318.

Pichon, J. (1970). Comparison of some methods of calibrating hot-film probes in water. *DISA Info.*, No. 10, 15–21.

Pierce, F. J. and Ezekwe, C. I. (1976). Comparison of Reynolds stress diagnostics by fixed and rotating probes. *AIAA J.*, **14**, 412–414.

Piomelli, U., Balint, J. L., and Wallace, J. M. (1989). On the validity of Taylor's hypothesis for wall-bounded flows. *Phys. Fluids*, **A1**, 609–611.

Pitts, W. M. and McCaffrey, B. J. (1986). Response behaviour of hot wires and films to flows of different gases. *J. Fluid Mech.*, **169**, 465–512.

Pluister, J. W. and Nagib, H. M. (1975). Evaluation of a hot-film calibration tunnel for low-velocity water flows. *DISA Info.*, No. 17, 29–33.

Polyakov, A. F. and Shindin, S. A. (1978). Peculiarities of hot-wire measurements of mean velocity and temperature in the wall vicinity. *Lett. Heat Mass Transfer*, **5**, 53–58.

Polyakov, A. F., Troitsky, V. V., Shehter, Yu.L., and Jørgensen, F. E. (1988). Shear stress measurements on heat transfer surfaces. *DANTEC Info.*, No. 6, 12–15.

Pompeo, L. and Thomann, H. (1993). Quadruple hot-wire probes in a simulated wall flow. *Exp. in Fluids*, **14**, 145–152.

Pougare, M., Galmes, J. M., and Lakshminarayana, B. (1985). An experimental study of the compressor rotor blade boundary layer. *ASME, J. Eng. for Gas Turbines and Power*, **107**, 364–373.

Prahl, J. M. and Win, H. (1987). Calibration of constant-temperature hot-film anemometers at low velocities in water of varying temperature. *Proc. Symp. Thermal Anemometry*, Cincinnati, pp. 35–37.

Prandtl, L. (1925). Über die ausgebildete turbulenz. *ZAMM*, **5**, 136–139.

Prandtl, L. (1945). Über ein neues formel-system fur die ausgebildete turbulenz. *Nachr. Akad. Wiss. Göttingen. Math.-Phys. Kl.* 1945, p. 6.

Preston, J. H. (1954). The determination of turbulent skin friction by means of pitot tubes. *J. R. Aero. Soc.*, **58**, 109–121.

Pronchick, S. W. (1983). An experimental investigation of the structure of a turbulent reattaching flow behind a backward-facing step. Ph.D. Thesis, Deparment of Mechanical Engineering, Stanford University.

Ra, S. H., Chang, P. K., and Park, S. O. (1990*a*). A modified calibration technique for the split film sensor. *Meas. Sci. Technol.*, **1**, 1156–1161.

Ra, S. H., Chang, P. K., and Park, S. O. (1990*b*). Measurement of the forward-flow fraction using a split film sensor. *Exp. in Fluids*, **10**, 57–59.

Rabiner, L. R. and Gold, B. (1975). *Theory and Application of Digital Signal Processing*. Prentice-Hall, Englewood Cliffs, New Jersey.

Raichlen, F. (1967). Some turbulence measurements in water. *J. Eng. Mech. Div. ASCE*, **93**, No. EM2, 73–97.

Raj, D. and Swim, W. B. (1981). Measurements of the mean flow velocity and velocity fluctuations at the exit of an FC centrifugal fan rotor. *ASME, J. Eng. for Power*, **103**, 393–399.

Raj, R. and Lakshminarayana, B. (1976). Three dimensional characteristics of turbulent wakes behind rotors of axial flow turbomachinery. *ASME, J. Eng. for Power*, **98**, 218–228.

Rajagopalan, S. and Antonia, R. A. (1993). RMS spanwise vorticity measurements in a turbulent boundary layer. *Exp. in Fluids*, **14**, 142–144.

Ramaprian, B. R. and Tu, S. W. (1983). Calibration of a heat flux gage for skin friction measurement. *ASME, J. Fluids Eng.*, **105**, 455–457.

Randolph, M., Eckelmann, H., and Nychas, S. G. (1987). Identification of sweeps with help of the instantaneous velocity gradient dU/dy. In *Advances in Turbulence*, (eds. G. Comte-Bellot and J. Mathiew), pp. 408–415. Springer, Berlin.

Randolph, M., Eckelmann, H., and Nychas, S. G. (1989). On the relation between sweeps and streamwise vortices. In *Advances in Turbulence* 2, (ed. H. H. Fernholz and H. E. Fiedler), Springer, Berlin.

Rao, K. N., Narasimha, R. and Badri Narayanan, M. A. (1971). The 'bursting' phenomenon in a turbulent boundary layer. *J. Fluid Mech.*, **48**, 339–352.

Rask, R. B. (1979). Laser doppler anemometer measurements in an internal combustion engine. SAE Paper No. 790094.

Rask, R. B. (1981). Comparison of window, smoothed-ensemble and cycle-by-cycle data reduction techniques for laser doppler anemometer measurements of in-cylinder velocity. *Proc. Symp. on Fluid Mechanics of Combustion Systems*, (ed. T. Morel, R. P. Lohmann and J. M. Rackley). ASME, New York.

Rask, R. B. (1984). Laser doppler anemometer measurements of mean velocity and turbulence in internal combustion engines. *ICALEO '84 Conference Proceedings Inspection, Measurement and Control and Laser Diagnostics and Photochemistry*, Vol. 45 and 47. Laser Institute of America, Boston.

Rasmussen, C. G. (1967). The air bubble problem in water flow hot-film anemometry. *DISA Info.*, No. 5, 21–26.

Rehmke, K. (1978). Some remarks on the response of hot-wire and hot-film probes to passage through an air-water interface. *J. Phys. E.: Sci. Instr.*, **11**, 94–96.

Reichert, J. K. and Azad, R. S. (1977). Wall shear stress measurement with a hot film in a variable temperature flow. *Rev. Sci. Instr.*, **48**, 341–345.

Resch, F. J. (1970). Hot-film turbulence measurements in water flow. *J. Hydr. Div. ASCE*, **96**, No. HY3, 787–800.

Resch, F. J. (1973). Use of the dual-sensor hot-film probe in water flow. *DISA Info.*, No. 14, 5–10.

Resch, F. J. and Abel, R. (1975). Spectral analysis using Fourier transform techniques. *Int. J. Num. Methods Eng.*, **9**, 869–902.

Resch, F. J. and Leutheusser, H. J. (1972). Reynolds stress measurements in hydraulic jumps. *J. Hydraulic Research*, **10**, No. 4, 409–430.

Resch, F. J., Leutheusser, H. J., and Alemu, S. (1974). Bubbly two-phase flow in hydraulic jump, *J. Hyd. Div.*, *ASCE*, **100**, No. HYI, 137–149.

Reuss, D. L., Adrian, R. J., Landreth, C. C., French, D. T., and Fansler, T. D. (1989). Instantaneous planar measurements of velocity and large-scale vorticity and strain rate in an engine using particle-image velocimetry. SAE Paper No. 890616.

Reuss, D. L., Bardsley, M., Felton, P. G. Landreth, C. C., and Adrian, R. J. (1990). Velocity, vorticity, and strain rate ahead of a flame measured in an engine using particle image velocimetry. SAE Paper No. 900053.

Rey, C. (1973). Etude d'une sonde anémometrique a trois fils chauds parallèles. Thesis Dr. 3emè: University de Provence, Marseille, France.

Rey, C. and Beguier, C. (1977). On the use of a three parallel wire probe. *DISA Info.*, No. 21, 11–15.

Reynolds, W. C. (1976). Computation of turbulent flows. *Ann. Rev. Fluid Mech.*, **8**, 183–208.

Reynolds, W. C. (1980). Modelling of fluid motion in engines – an introductory overview. In *Combustion Modelling in Reciprocating Engines*, (ed. J. N. Mattavi and C. A. Amann), pp. 69–124. Plenum Press, New York.

Richard, P. H. and Johnson, C. G. (1988). Development of secondary flows in the stator of a model turbine. *Exp. in Fluids*, **6**, 2–10.

Richardson, E. V. and McQuivey, R. S. (1968). Measurement of turbulence in water. *J. Hydr. Div. ASCE*, **94**, No. HY2, 411–430.

Richardson, P. D. (1965). Convection from heated wires at moderate and low Reynolds Numbers. *AIAA J.*, **3**, 537–538.

Richter, H. (1985). Measurement of two-dimensional periodical flow behind turbine guide-vanes by means of a split-fiber probe. *DANTEC Info.*, No. 1, 10–12.

Ritsch, M. L. and Davidson, J. H. (1992). Phase discrimination in gas-particle flows using thermal anemometry. *ASME, J. Fluids Eng.*, **114**, 692–694.

Roberts, J. B. and Gaster, M. (1980). On the estimation of spectra from randomly sampled signals: a method of reducing variability. *Proc. Roy. Soc. Lond.*, **A371**, 235–258.

Roberts, R. A. and Mullis, C. T. (1987). *Digital Signal Processing*. Addison-Wesley.

Robinson, S. K. (1991). Coherent motions in the turbulent boundary layer. *Ann. Rev. Fluid Mech.*, **23**, 601–639.

Rodi, W. (1975). A new method of analysing hot-wire signals in highly turbulent flow, and its evaluation in a round jet. *DISA Info.*, No. 17, 9–17.

Rodi, W. (1979). Influence of buoyancy and rotation on equations for turbulent length scale. *Proc. 2nd Int. Symp on Turbulent Shear Flows*, Imperial College, London, 10.37–10.42.

Rodi, W. and Scheuerer, G. (1983). Calculation of curved shear layers with two-equation turbulence models. *Phys. Fluids*, **26**, 1422–1436.

Rodi, W. and Scheuerer, G. (1986). Scrutinizing the k-ε turbulence model under adverse pressure gradient conditions. *ASME, J. Fluids Eng.*, **108**, 174–179.

Rogallo, R. S. and Moin, P. (1984). Numerical simulation of turbulent flows. *Ann. Rev. Fluid Mech.*, **16**, 99–137.

Rogers, B. K. and Head, M. R. (1969). Measurement of three-dimensional boundary layers. *J. Roy. Aero. Soc.*, **73**, 796–797.

Rong, B. S., Tan, D. K. M. and Smits, A. J. (1985). Calibration of the constant temperature normal hot-wire anemometer in transonic flow. Princeton University Department of Mechanical and Aerospace Engineering, Report No. 1696.

Rose, W. C. and McDaid, E. P. (1976). Turbulence measurement in transonic flow. *Proc AIAA 9th Aerodynamic Testing Conf.*, Arlington, Texas.

Rose, W. G. (1962). Some corretions to the linearized response of a constant-temperature hot-wire anemometer operated in a low-speed flow. *ASME, J. Appl. Mech.*, **29**, 554–558 (erratum p. 758).

Roshko, A. (1954). On the development of turbulent wakes from vortex streets. *NACA Report*, 1191.

Rothrock, M. D., Jay, R. L., and Riffel, R. E. (1982). Time-variant aerodynamics of high-turning blade elements. *ASME, J. Eng. for Power*, **104**, 412–419.

Rubatto, G. (1970). Calibration of probes for flow velocity measurements in liquids, in the range 2-5 m/sec. *DISA Info.*, No. 9, 3-7.

Rubesin, M. W., Okuno, A. F., Mateer, G. G., and Brosh, A. (1975). A hot-wire surface gage for skin friction and separation detection measurements. NASA TM X-62, 465.

Ruck, B. (1991). Distortion of LDA fringe pattern by tracer particles. *Exp. in Fluids*, **10**, 349-354.

Ruderich, R. and Fernholz, H. H. (1986). An experimental investigation of a turbulent shear flow with separation, reverse flow, and reattachment. *J. Fluid Mech.*, **163**, 283-322.

Rundstadler, P. W., Kline, S. J., and Reynolds, W. C. (1963). An experimental investigation of the flow structure of the turbulent boundary layer. Stanford University, Department of Mechanical Engineering Report No. MD-8.

Russ, S. and Simon, T. W. (1990). Signal processing using the orthogonal triple-wire equations. *TSI Flow Lines*, Winter, 3-9.

Russ, S. and Simon, T. W. (1991). On the rotating, slanted, hot-wire technique. *Exp. in Fluids*, **12**, 76-80.

Sabot, J. and Comte-Bellot, G. (1976). Intermittency of coherent structures in the core region of fully developed turbulent pipe flow. *J. Fluid Mech.*, **74**, 767-796.

Sakao, F. (1973). Constant temperature hot wires for determining velocity fluctuation in an air flow accompanied by temperature fluctuations. *J. Phys. E.: Sci. Instr.*, **6**, 913-916.

Samet, M. and Einav, S. (1984). Directional pressure probe. *Rev. Sci. Instr.*, **55**, 582-588.

Samet, M. and Einav, S. (1985). Directional sensitivity of unplated normal-wire probes. *Rev. Sci. Instr.*, **56**, 2299-2305.

Samet, M. and Einav, S. (1987). A hot-wire technique for simultaneous measurement of instantaneous velocities in 3D flows. *J. Phys. E.: Sci.*, **20**, 683-690.

Sammler, B. and Kitzing, H. (1990). Interpretation of signals of triple-wire probes. *DANTEC Info.*, No. 9, 16-18.

Sampath, S., Ganesan, V., and Gowda, B. H. L. (1982). Improved method for the measurement of turbulence quantities. *AIAA J.*, **20**, 148-149.

Samways, A. L. and Bruun, H. H. (1992). Vertical up-flow of oil/water mixtures. Technical Report No. 200, Department of Mechanical and Manufacturing Engineering, University of Bradford.

Sandborn, V. A. (1972). *Resistance Temperature Transducers*. Metrology Press, Ft Collins, CO.

Sandborn, V. A. (1976). Effect of velocity gradients on measurements of turbulent shear stress. *AIAA J.*, **14**, 400-402.

Sandborn, V. A. (1979). Evaluation of the time dependent surface shear stress in turbulent flows. ASME Paper 79-WA/FE-17. ASME Winter Annual Meeting.

Sandborn, V. A. and Seegmiller, H. L. (1975). Evaluation of mean and turbulent velocity measurements in subsonic, accelerated boundary layers. NASA TM X-62, 488.

Saunders, L. J. and Lawrence, P. (1972). Calibration of hot-film anemometers. *Proc. Conf. on Fluid Dynamic Measurements in the Industrial and Medical Environment*, Leicester University, pp. 125-130.

Saxena, V. and Rask, R. B. (1987). Influence of inlet flows on the flow field in an engine. SAE Paper No. 870369.

Schewe J. and Ronneberger, D. (1990). Error-tolerant calibration of dual sensor probes used in a turbulent wall boundary layer. *Exp. in Fluids*, **9**, 285–289.

Schmidt, D. P. and Okiishi, T. H. (1977). Multistage axial-flow turbomachine wake production, transport and interaction. *AIAA J.*, **15**, 1138–1145.

Schraub, F. A. and Kline, S. J. (1965). A study of the structure of the turbulent boundary layer with and without longitudinal pressure gradients. Stanford University, Department of Mechanical Engineering, Report No. MD-12.

Schubauer, G. B. and Klebanoff, P. S. (1946). Theory and application of hot wire instruments in the investigation of turbulent boundary layers. NACA WR W-86.

Schulz, H. D., Gallus, H. E., and Lakshminarayana, B. (1990). Three- dimensional separated flow field in the endwall region of an annular compressor cascade in the presence of rotor-stator interaction, Parts 1 and 2. *ASME, J. Turbomachinery*, **112**, 669–688.

Schumann, U. and Friedrich, R. (1987). On direct and large eddy simulation of turbulence. In *Perspective in Turbulence Studies*, (ed. H. U. Meier and P. Bradshaw), pp. 88–104. Springer, Berlin.

Seed, W. A. (1969). Fabrication of thin-film microcircuits on curved substrates. *J. Phys. E.: Sci. Instr.*, **2**, 206.

Seed, W. A. and Thomas, I. R. (1972). The application of hot-film anemometry to the measurement of blood flow velocity in man. *Proc. Conf. on Fluid Dynamic Measurements in the Industrial and Medical Environments*, pp. 298–304. Leicester University.

Seed, W. A. and Wood, N. B. (1969). An apparatus for calibrating velocity probes in liquids. *J. Phys. E.: Sci. Instr.*, **2**, 896–898.

Seed, W. A. and Wood, N. B. (1970a). Development and evaluation of a hot-film velocity probe for cardio vascular studies. *Cardiovasc. Res.*, **4**, 253–263.

Seed, W. A. and Wood, N. B. (1970b). Use of a hot-film velocity probe for cardiovascular studies. *J. Phys. E.: Sci., Instr.*, **3**, 377–384.

Seifert, G. and Graichen, K. (1982). A calibration method for hot-wire probes including the low velocity range. *DISA Info.*, No. 27, 8–11.

Semenov, E. S. (1958). Studies of turbulent gas flow in piston engines. *Otedelinie Technicheskikh Nauk*, No. 8, (NASA Technical Translation F97).

Senda, M., Suzuki, K. and Sato, T. (1980). Turbulence structure related to heat transfer in a turbulent boundary layer with injection. *Proc. Turbulent Shear Flow Conf.*, Berlin, pp. 143–157.

Serizawa, A., Kataoka, I., and Michiyoshi, I. (1975a). Turbulence structure of air-water bubbly flow. Part I, Measuring techniques. *Int. J. Multiphase Flow*, **2**, 221–233.

Serizawa, A., Kataoka, I., and Michiyoshi, I. (1975b). Turbulence structure of air-water bubbly flow, Part II, Local properties. *Int. J. Multiphase Flow*, **2**, 235–246.

Serizawa, A., Kataoka, I., and Michiyoshi, I. (1975c). Turbulence structure of air-water bubbly flow, Part III, Transport properties. *Int. J. Multiphase Flow*, **2**, 247–259.

Serizawa, A., Tsuda, K., and Michiyoshi, I. (1983). Real-time measurement of two-phase flow turbulence using a dual-sensor anemometry. *Proc. Symp. on Measuring Techniques in Gas-Liquid Two-Phase Flows*, Nancy, France, pp. 495–523.

Shah, D. A. and Antonia, R. A. (1986). A comparison between measured and calculated thermal inertia of cold wires. *DANTEC Info.*, No. 2, 13–14.

Shah, D. A. and Antonia, R. A. (1989). Scaling of the 'bursting' period in turbulent boundary layer and duct flows. *Phys. Fluids*, **Al**, 318–325.

Shaukatullah, H. and Gebhart, B. (1977). Hot film anemometer calibration and use in fluids at varying background temperature. *Letters in Heat and Mass Transfer*, **4**, 309-317.

Shayesteh, M. V. and Bradshaw, P. (1987). Microcomputer-controlled traverse gear for three-dimensional flow explorations. *J. Phys. E.: Sci. Instr.*, **20**, 320-322.

Sheih, C. M., Tennekes, H., and Lumley, J. L. (1971). Airborne hot-wire measurements of the small-scale structure of atmospheric turbulence. *Phys. Fluids*, **14**, 201-215.

Sheng, Y. Y. and Irons, G. A. (1991). A combined laser doppler anemometry and electrical probe diagnostic for bubbly two-phase flow. *Int. J. Multiphase flow*, **17**, 585-598.

Shepherd, I. C. (1981). A four hole pressure probe for fluid flow measurements in three dimensions. *ASME, J. Fluids Eng.*, **103**, 590-594.

Sherif, S. A. and Pletcher, R. H. (1983). Prediction of the constant-temperature anemometer response in known flow fields. *ICIASF '83 Record, Proc. Int. Congr. Instrumentation in Aerospace Simulation Facilities*, St. Louis, France, pp. 45-48.

Sherif, S. A. and Pletcher, R. H. (1986). Temperature correction for the output response of a constant-temperature hot-film anemometer in nonisothermal flows with strong property temperature dependence. *Proc. 8th Intern. Heat Transfer Conf.*, San Francisco, pp. 549-554.

Sherif, S. A. and Pletcher, R. H. (1987). A normal-sensor hot-wire/film probe method for the analysis of highly three-dimensional flows. *ASME, Fluid Measurement and Instrumentation Forum, FED*, Vol. 49, pp. 19-22.

Sherif, S. A. and Pletcher, R. H. (1989). Measurements of the flow and turbulence characteristics of round jets in crossflow. *ASME, J. Fluids Eng.*, **111**, 165-171.

Sherlock, R. A. (1984). Frequency response optimisation of the constant temperature detector system—a detailed root-locus analysis. *J. Phys. E.: Sci. Instr.*, **17**, 386-393.

Shin, H. W. and Hu, Z. A. (1986). Measurement of swirling flow field using the single slanted hot-wire technique. *Int. J. Turbo and Jet Engines*, **3**, 139-145.

Shishov, E. V., Roganov, P. S., Klorikian, P. V., and Zabolotsky, V. P. (1987). Correction of constant current anemometer readings in high-frequency temperature fluctuation measurements. *DANTEC Info.*, No. 5, 21-23.

Shivaprasad, B. G. and Simpson, R. L. (1982). Evaluation of a wall-direction probe for measurements in separated flows. *ASME, J. Fluids Eng.*, **104**, 162-166.

Shook, M., Stock, D. E., and Bowen, A. J. (1990). Split-film anemometry. *ASME, Proc. Symp. on Heuristics of Thermal Anemometry, FED*, Vol. 97, University of Toronto, pp. 93-98.

Shreeve, R. P. and Neuhoff, F. (1984). Measurements of the flow from a high-speed compressor rotor using a dual probe digital sampling (DPDS) technique. *ASME, J. Eng. for Gas Turbines and Power*, **106**, 366-375.

Siddall, R. G. and Davies, T. W. (1972). An improved response equation for hot-wire anemometry. *Int. J. Heat Mass Transf.*, **15**, 367-368.

Simpson, R. L. and Wyatt, W. G. (1973). The behaviour of hot-film anemometers in gas mixtures. *J. Phys. E.: Sci. Instr.*, **6**, 981-987.

Sirivat, A. (1989). Measurement and interpretation of space-time correlation functions and derivative statistics from a rotating hot wire in a grid turbulence. *Exp. in Fluids*, **7**, 361-370.

Sitaram, N. and Treaster, A. L. (1985). A simplified method of using four-hole probes to measure three-dimensional flow fields. *ASME, J. Fluids Eng.*, **107**, 31–35.

Skinner, G. T. and Rae, W. J., (1984). Calibration of a three-element hot-wire anemometer. *Rev. Sci. Instr.*, **55**, 578–581.

Smith, G. D. J. and Cumpsty, N. A. (1984). Flow phenomena in compressor casing treatment. *ASME, J. Eng for Gas Turbines and Power*, **106**, 532–541.

Smits, A. J. (1990). An introduction to constant-temperature hot-wire anemometry in supersonic flows. *ASME, Proc. Symp. on the Heuristics of Thermal Anemometry, FED*, Vol. 97, University of Toronto, pp. 35–40.

Smits, A. J. and Dussauge, J. P. (1989). Hot-wire anemometry in supersonic flow, In *A Survey of Measurements and Measuring Techniques in Rapidly Distorted Compressible Turbulent Boundary Layer*, AGARDograph No. 315.

Smits, A. J. and Muck, K. C. (1984). Constant temperature hot-wire anemometer practice in supersonic flows, Part 2, The inclined wire. *Exp. in Fluids*, **2**, 33–41.

Smits, A. J. and Perry, A. E. (1980). The effect of varying resistance ratio on the behaviour of constant temperature hot-wire anemometers. *J. Phys. E.: Sci. Instr.*, **13**, 451–456.

Smits, A. J. and Perry, A. E. (1981). A note on hot-wire anemometer measurements of turbulence in the presence of temperature fluctuations. *J. Phys. E.: Sci. Instr.*, **14**, 311–312.

Smits, A. J., Perry, A. E., and Hoffmann, P. H. (1978). The response to temperature fluctuations of a constant-current hot-wire anemometer. *J. Phys. E.: Sci. Instr.*, **11**, 909–914.

Smits, A. J., Hayakawa, K., and Muck, K. C. (1983). Constant-temperature hot-wire anemometer practice in supersonic flows, Part 1, The normal wire. *Exp. in Fluids*, **1**, 83–92.

So, R. M. C. (1976). Shear effects in hot-wire measurements. *ASME, J. Fluids Eng.*, **98**, 771–773.

So, R. M. C., Lai, Y. G., Zhang, H. S., and Hwang, B. C. (1991). Second-order near-wall turbulence closures: A review. *AIAA J.*, **29**, 1819–1835.

Soria, J. and Norton, M. P. (1990). A study of the three-term hot-wire system equation and analog linearization based on it. *Exp. Thermal and Fluid Sci.*, **3**, 346–353.

Spalding, D. B. (1961). A single formula for the law of the wall. *ASME, J. Appl. Mech.*, **28**, 455–457.

Spencer, B. W. and Jones, B. G. (1971). Turbulence measurements with the split-film anemometer probe. *Proc. Symp. on Turbulence in Liquids*, University of Missouri, Rolla, pp. 7–15.

Spina, E. F., Donovan, J. F., and Smits, A. J. (1991). On the structure of high-Reynolds-number supersonic turbulent boundary layers. *J. Fluid Mech.*, **222**, 293–327.

Sreenivasan, K. R. and Tavoularis, S. (1980). On the skewness of the temperature derivative in turbulent flows. *J. Fluid Mech.*, **101**, 783–795.

Sreenivasan, K. R., Antonia, R. A., and Danh, H. Q. (1977). Temperature dissipation fluctuations in a turbulent boundary layer. *Phys. Fluids*, **20**, 1238–1249.

Stack, J. P., Mangalam, S. M., and Berry, S. A. (1987). A unique measurement technique to study laminar-separation bubble characteristics on an airfoil. AIAA paper 87–1271.

Stack, J. P., Mangalam, S. M., and Kalburgi, V. (1988). The phase reversal phenomenon at flow separation and reattachment. AIAA paper 88–0408.

Stanford, R. A. and Libby, P. A. (1974). Further applications of hot-wire anemometry to turbulence measurements in helium-air mixtures. *Phys. Fluids*, **17**, 1353-1361.

Stevenson, W. H. and Thompson, H. D. and Craig, R. R. (1984). Laser velocimeter measurements in highly turbulent recirculating flows. *ASME, J. Fluids Eng.*, **106**, 173-180.

Stock, D. E. and Fadeff, K. G. (1983). Measuring particle transverse velocity using an LDA. *ASME, J. Fluids Eng.*, **105**, 458-460.

Stock, D. E. and Jaballa, T. M. (1985). Turbulence measurements using split-film anemometry. *Proc. Int. Symp. Refined Flow Modelling and Turbulence Measurements*, University of Iowa, H15 1-10.

Stock D. E., Wells, M. R., Barriga, A., and Crowe, C. T. (1977). Application of split-film anemometry to low-speed flows with high turbulence intensity and recirculation as found in electrostatic precipitators. *Proc. Fifth Biennial Symp. on Turbulence*, University of Missouri, Rolla, pp. 117-123.

Strohl, A. and Comte-Bellot, G. (1973). Aerodynamic effects due to configuration of X-wire anemometers. *ASME, J. Appl. Mech.*, **40**, 661-666.

Subramanian, C. S. and Antonia, R. A. (1981). Effect of Reynolds number on a slightly heated turbulent boundary layer. *Int. J. Heat Mass Transf*, **24**, 1833-1846.

Subramanian, C. S., Rajagopalan, S., Antonia, R. A., and Chambers, A. J. (1982). Comparison of conditional sampling and averaging techniques in a turbulent boundary layer. *J. Fluid Mech.*, **123**, 335-362.

Subramaniyam, S., Ganesan, V., and Srinivasa Rao, P. (1990). Turbulent flow inside the cylinder of a diesel engine — an experimental investigation using hot wire anemometer. *Exp. in Fluids*, **9**, 167-174.

Sugeng, F. and Fiedler, K. (1986). An experimental investigation into unsteady blade forces and blade losses in axial compressor blade cascade. *ASME, J. Eng. for Gas Turbines and Power*, **108**, 47-52.

Suzuki, Y. and Kasagi, N. (1990). Evaluation of hot-wire measurements in wall shear turbulence using a direct numerical simulation data base. *Exp. Thermal Fluid Sci.*, **5**, 69-77.

Swaminathan, M. K., Bacic, R., Rankin, G. W., and Sridhar, K., (1983). Improved calibration of hot-wire anemometers. *J. Phys. E.: Sci. Instr.*, **16**, 335-338.

Swaminathan, M. K., Kokate, P. Y., Rankin, G. W., and Sridhar, K. (1984a). An improved method of obtaining the yaw sensitivity of hot-wires. Technical Report of the Department of Mechanical Engineering, University of Windsor, Ontario, Canada.

Swaminathan, M. K., Rankin, G. W., and Sridhar, K. (1984b). Some studies on hot-wire calibration using Monte Carlo technique. *J. Phys. E.: Sci. Instr.*, **17**, 1148-1151.

Swaminathan, M. K., Rankin, G. W., and Sridhar, K. (1986a). A note on the response equations for hot-wire anemometry. *ASME, J. Fluids Eng.*, **108**, 115-118.

Swaminathan, M. K., Rankin, G. W., and Sridhar, K. (1986b). Evaluation of the basic systems of equation for turbulence measurements using the Monte Carlo technique. *J. Fluid Mech.*, **170**, 1-19.

Swoboda, M., Nitsche, W., and Suttan, J. (1991). Shock detection on airfoils by means of piezo foil- and hot-film arrays. *ICIASF '91 Record, Proc. Int. Congr. on Instrumentation in Aerospace Simulation Facilities*, pp. 284-294.

Tabaczynski, R. J. (1983). Turbulence measurements and modelling in reciprocating engines – an overview. IMechE C51/83.

Tabatabai, M., Pollard, A., and McPhail, A. (1986). A device for calibrating hot-wire probes at low velocities. *J. Phys. E.: Sci. Instr.*, **19**, 630-632.

Tagawa, M., Tsuji, T., and Nagano, Y. (1992). Evaluation of X-probe response to wire separation for wall turbulence measurements. *Exp. in Fluids*, **12**, 413-421.

Takagi, S. (1985). Hot-wire height gauge using a laser and photodiodes. *Exp. in Fluids*, **3**, 341-342.

Takagi, S. (1986). A hot-wire anemometer compensated for ambient temperature variations. *J. Phys. E.: Sci. Instr.*, **19**, 739-743.

Tan-atichat, J., Nagib, H. M., and Pluister, J. W. (1973). On the interpretation of the output of hot-film anemometers and a scheme of dynamic compensation for water temperature variation. *Proc. Third Symp. Turbulence in Liquids*, University of Missouri, Rolla, pp. 352-374.

Tavoularis, S. (1978). A circuit for the measurement of instantaneous temperature in heated turbulent flows. *J. Phys. E.: Sci. Instr.*, **11**, 21-23.

Tennekes, H. and Lumley, J. L. (1972). *A First Course in Turbulence*. MIT Press, Cambridge, Massachusets.

Tewari, S. S. and Jaluria, Y. (1990). Calibration of constant-temperature hot-wire anemometers for very low velocities in air. *Rev. Sci. Instr.*, **61**, 3834-3845.

Thomas, R. M. (1973). Conditional sampling and other measurements in a plane turbulent wake. *J. Fluid Mech.*, **57**, 549-582.

Thompson, B. E. (1984). The turbulent separating boundary layer and downstream wake. Ph.D. Thesis, Department of Mechanical Engineering, Imperial College, London.

Thompson, B. E. (1987). Appraisal of a flying hot-wire anemometer. *DANTEC Info.*, No. 4, 14-18.

Thompson, B. E. and Whitelaw, J. H. (1984). Flying hot-wire anemometry. *Exp. in Fluids*, **2**, 47-55.

Tindal, M. J., Williams, T. J., and Aldoory, M. (1982). The effect of inlet port design on cylinder gas motion in direct injection diesel engines. In *Flows in Internal Combustion Engines*, (ed. T. Uzkan), pp. 101-111. ASME, New York.

Tombach, I. H. (1969). Velocity measurements with a new probe in inhomogeneous turbulent jets. Ph.D. Dissertation, California Institute of Technology.

Tombach, I. H. (1973). An evaluation of the heat pulse anemometer for velocity measurement in inhomogeneous turbulent flow. *Rev. Sci. Instr.*, **44**, 144-148.

Toral, H. (1981). A study of the hot-wire anemometer for measuring void fraction in two phase flow. *J. Phys. E.: Sci. Instr.*, **14**, 822-7.

Townsend, A. A. (1947). Measurements in the turbulent wake of a cylinder. *Proc. Roy. Soc. London.* **A190**, 551-561.

Townsend, A. A. (1949). The fully developed turbulent wake of a circular cylinder. *Aust. J. Sci. Res.*, **2**, 451-468.

Towsend, A. A. (1956). *The structure of turbulent shear flow*. Cambridge University Press, Cambridge.

Treaster, A. L. and Yocum, A. M. (1979). The calibration and application of five-hole probes. *ISA Trans.*, **18**, 23-34.

Tropea, C. (1991). Digital Time Series Analysis. Advanced HWA course, UMIST, Manchester, England.

Tsanis, I. K. (1987). Calibration of hot-wire anemometers at very low velocities. *DANTEC Info.*, No. 4, 13–14.

TSI split sensor. Calibration and application. Thermo-Systems Inc. Technical Bulletin TB20.

Tsinober, A., Kit, E., and Dracos, T. (1992). Experimental investigation of the field of velocity gradients in turbulent flows. *J. Fluid Mech.*, **242**, 169–192.

Tsiolakis, E. P., Krause, E., and Müller, U. R. (1983). Turbulent boundary layer-wake interaction. *Proc. 4th Int. Symp. on Turbulent Shear Flow*, Karlsruhe, 5.19–5.24.

Tsuji, T. and Nagano, Y. (1989). An anemometry technique for turbulence measurements at low velocity. *Exp in Fluids*, **7**, 547–559.

Tsuji, T., Nagano, Y., and Tagawa, M. (1992). Frequency response and instantaneous temperature profile of cold-wire sensors for fluid temperature fluctuation measurements. *Exp. in Fluids*, **13**, 171–178.

Tsuji, Y. and Morikawa, Y. (1982). LDV measurements of an air-solid two-phase flow in a horizontal pipe. *J. Fluid Mech.*, **126**, 385–409.

Tubergen, R. G. and Tiederman, W. G. (1993). Evaluation of ejection detection schemes in turbulent wall flows. *Exp. in Fluids.*, **15**, 255–262.

Tuckey, P. R., Morsi, Y. S., and Clayton, B. R. (1984). The experimental analysis of three-dimensional flow fields. *Int. J. Mech. Eng. Educ.*, **12**, 149–166.

Turan, Ö. and Azad, R. S. (1989). Effect of hot-wire probe defects on a new method of evaluating turbulence dissipation. *J. Phys. E.: Sci. Instr.*, **22**, 254–261.

Turan, Ö., Azad, R. S., and Atamanchuk, T. M. (1987). Wall effect on the hot-wire signal without flow. *J. Phys. E.: Sci. Instr.*, **20**, 1278–1280.

Tureaud, T. F., Szewcyk, A. A., Nee, V. W., and Yang, K. T. (1988). Some results from the calibration and use of a three-wire probe in a highly stratified wake. *Proc. 1st World Conf. on Experimental Heat Transfer, Fluid Mechanics and Thermodynamics*, Dubrovnik, Yugoslavia, pp. 269–273.

Tutu, N. K. and Chevray, R. (1975). Cross-wire anemometry in high intensity turbulence. *J. Fluid Mech.*, **71**, 785–800.

Uberoi, M. S. and Corrsin, S. (1951). Progress report on the propagation of turbulence into a non-turbulent flow. NACA Contract NAW 5504, The Johns Hopkins University.

Uberoi, M. S. and Kovasznay, L. S. G. (1953). On mapping and measurement of random fields. *Q. Appl. Math.*, **10**, 375–393.

Vagt, J. D. (1979). Hot-wire probes in low speed flow. In *Progress in Aerospace Sciences*, (ed. J. A. Bagleyand and P. J. Finley), Vol. 18, pp. 271–323. Pergamon, Oxford.

Vagt, J. D. and Fernholz, H. H. (1973). Use of surface fences to measure wall shear stress in three-dimensional boundary layers. *Aero. Q.*, **XXIV**, 2, 87–91.

Vagt, J. D. and Fernholz, H. H. (1979). A discussion of probe effects and improved measuring techniques in the near-wall region of an imcompressible three-dimensional turbulent boundary layer. *AGARD Conf. Proc.*, **271**, 10.1–10.17.

Van Atta, C. W. (1974) Sampling techniques in turbulence measurements. *Ann. Rev. Fluid Mech.*, **6**, 75–91.

Van der Hegge Zijnen, B. G. (1924). Measurement of the velocity distribution in the boundary layer along a plane surface. Thesis, University of Delft.

Van der Hegge Zijnen, B. G. (1956). Modified correlation formulae for the heat transfers by natural and by forced convection from horizontal cylinders. *Appl. Sci. Res.*, **A6**, 129–140.

Van Thinh, N. (1969). On some measurements made by means of a hot wire in a turbulent flow near a wall. *DISA Info.*, No. 7, 13–18.

Venkateswaran, S. (1991). Experimental study of casing boundary layers in a multistage axial compressor. *ASME, J. Fluids Eng.*, 113, 240–244.

Vonnegut, B. and Neubauer, R. (1952). Detection and measurement of aerosol particles. *Anal. Chem.*, 24, 1000–1005.

Vrebalovich, T. (1962a). Heat loss from hot wires in transonic flow. Research Summary No. 36–14, Jet Propulsion Laboratory, Pasadena, California.

Vrebalovich, T. (1962b). Applications of hot-wire techniques, in unsteady compressible flows. Research Summary No. 32–229, Jet Propulsion Laboratory, Pasadena, California.

Vukoslavčević, P. and Wallace, J. M. (1981). Influence of velocity gradients on measurements of velocity and streamwise vorticity with hot-wire X-array probes. *Rev. Sci. Instr.*, 52, 869–879.

Vukoslavčević, P., Balint, J. L., and Wallace, J. M. (1989). A multi-sensor hot-wire probe to measure vorticity and velocity in turbulent flows. *ASME, J. Fluids Eng.*, 111, 220–224.

Vukoslavčević, P., Wallace, J. M. and Balint, J. L. (1991). The velocity and vorticity vector fields of a turbulent boundary layer, Part 1, Simultaneous measurement by hot-wire anemometry. *J. Fluid Mech.*, 228, 25–51.

Wacker, D. (1985). Die numerische lösung der Navier-Stokesschen gleichungen für einen beheizten zylinder in wandnähe mit berücksichtigung des wärmeübergangs. Dissertation, Technical University Berlin.

Wadcock, A. (1978). Flying-hot-wire study of two-dimensional turbulent separation on an NACA 4412 airfoil at maximum lift. Ph.D. Thesis, California Institute of Technology.

Wagner, P. M. (1991). The use of near-wall hot-wire probes for time resolved skin-friction measurements. In *Advances in Turbulence* 3, (ed. A. V. Johansson and P. H. Alfredsson), pp. 524–529. Springer, Berlin.

Wagner, T. C. and Kent, J. C. (1988a). On the directional sensitivity of hot-wires: a new look at an old phenomenon. *Exp. in Fluids*, 6, 553–560.

Wagner, T. C. and Kent, J. C. (1988b). Measurement of intake valve/cylinder boundary, flows using a multiple orientation hot-wire technique. *ASME, J. Fluids Eng.*, 110, 361–366.

Wakisaka, T., Hamamoto, Y., and Shimamoto, Y. (1982). Turbulence structure of air swirl in reciprocating engine cylinders. In *Flows in Internal Combustion Engines*, (ed. T. Uzkan). ASME, New York.

Wakuri, Y., Kido, H., Murase, E. and Wang, Z. (1983). Variations of turbulence scales in an engine cylinder and a fine structure model of engine turbulence. *Memoirs of the Faculty of Engineering, Kyushu University*, Vol. 43, pp. 179–191.

Walker, D. A. and Walker, M. D. (1990). Method for fast sine-wave calibration of hot-wire frequency response. *Rev. Sci. Instr.*, 61, 1131–1135.

Walker, D. A., Ng, W. F., and Walker, M. D. (1989). Experimental comparison of two hot-wire techniques in supersonic flow. *AIAA J.*, 27, 1074–1080.

Walker, M. D. and Maxey, M. R. (1985). A whirling hot-wire anemometer with optical data transmission. *J. Phys. E.: Sci. Instr.*, 18, 516–521.

Walker, T. B. and Bullock, K. J. (1972). Measurement of longitudinal and normal velocity fluctuations by sensing the temperature downstream of a hot wire. *J. Phys. E.: Sci. Instr.*, 5, 1173–1178.

Wallace, J. M. (1986). Methods of measuring vorticity in turbulent flows. A Review. *Exp. in Fluids*, **4**, 61–71.

Wallace, J. M., Eckelmann, H., and Brodkey, R. S. (1972). The wall region in turbulent shear flow. *J. Fluid Mech.*, **54**, 39–48.

Wang, S. K., Lee, S. J., Jones, O. C., and Lahey, R. T. (1986). Three-dimensional conical probe measurements in turbulent air/water two-phase pipe flow. *TSI Flow Lines*, Fall, 10–16.

Wang, S. K., Lee, S. J., Jones, O. C., and Lahey, R. T. (1987). 3-D turbulence structure and phase distribution measurements in bubbly two-phase flows. *Int. J. Multiphase Flow*, **13**, 327–343.

Wang, S. K., Lee, S. J., Jones, O. C., and Lahey, R. T. (1990). Statistical analysis of turbulent two-phase pipe flow. *ASME, J. Fluids Eng.*, **112**, 89–95.

Wark, C. E., Nagib, H. M., and Jennings, M. J. (1990). A rotating hot-wire technique for spatial sampling of disturbed and manipulated duct flows. *Exp. in Fluids*, **9**, 191–196.

Warschauer, K. A., Vijge, J. B. A., and Boschloo, G. A. (1974). Some experiences and considerations on measuring turbulence in water with hot films. *Appl. Sci. Res.*, **29**, 81–98.

Wasan, D. T. and Baid, K. M. (1971). Measurement of velocity in gas mixtures: Hot-wire and hot-film anemometry. *AIChE J.*, **17**, 729–731.

Wasan, D. T., Davis, R. M., and Wilke, C. R. (1968). Measurement of the velocity of gases with variable fluid properties. *AIChE J.*, **14**, 227–234.

Wassmann, W. W. and Wallace, J. M. (1979). Measurement of vorticity in turbulent shear flow. *Bull. Am. Phys. Soc.*, **24**, 1142.

Watmuff, J. H., Perry, A. E., and Chong, M. S. (1983). A flying hot-wire system. *Exp. in Fluids*, **1**, 63–71.

Way, J. and Libby, P. A. (1970). Hot-wire probes for measuring velocity and concentration in helium-air mixtures. *AIAA J.*, **8**, 976–978.

Way, J. and Libby, P. A. (1971). Application of hot-wire anemometry and digital techniques to measurements in a turbulent helium jet. *AIAA J.*, **9**, 1567–1573.

Webster, C. A. G. (1962). A note on the sensitivity to yaw of a hot-wire anemometer. *J. Fluid Mech.*, **13**, 307–312.

Weeks, A. R., Beck, J. K., and Joshi, M. L. (1988). Response and compensation of temperature sensors. *J. Phys. E.: Sci. Instr.*, **21**, 989–993.

Weir, A. D. and Bradshaw, P. (1977). The interaction of two parallel free shear layers. *Proc. Symp. Turbulent Shear Flows*. Penn. State University, USA, 1.9–1.15.

Weir, A. D., Wood, D. H., and Bradshaw, P. (1981). Interacting turbulent shear layers in a plane jet. *J. Fluid Mech.*, **107**, 237–260.

Weisbrot, I. and Wygnanski, I. (1988) On coherent structures in a highly excited mixing layer. *J. Fluid Mech.*, **195**, 137–159.

Welch, P. D. (1967). The use of Fast Fourier Transform for the estimation of power spectra: A method based on time averaging over short, modified periodograms. *IEEE Trans. on Audio and Electroacoustics*, **AU-15**, 70–73.

Wentz, W. H. and Seetharam, H. C. (1977). Split-film anemometer measurements on an airfoil with turbulent separated flow. *Proc. Symp. on Turbulence in Fluids*, University of Missouri, Rolla, pp. 31–39.

Weske, J. R. (1943). A hot-wire circuit with very small time lag. NACA Technical Note No. 881.

Westphal, R. V. and Johnston, J. P. (1984). Effect of initial conditions on turbulent reattachment downstream of a backward-facing step. *AIAA J.*, **22**, 1727–1732.

Westphal, R. V. and Mehta, R. D. (1984). Crossed hot-wire data acquisition and reduction system. NASA TM-85871.

Westphal, R. V., Eaton, J. K., and Johnston, J. P. (1981). A new probe for measurement of velocity and wall shear in unsteady, reversing flow. *ASME, J. Fluids Eng.*, **103**, 478–482.

Weyer, H. (1980). Optical methods of flow measurement and visualization in rotors. *Proc. Symp. on Measurement Methods in Rotating Components of Turbomachinery*, (ed. B. Lakshminarayana and R. Runstadler), pp. 99–110. ASME, New York.

Whitfield, C. E., Kelly, J. C., and Barry, B. (1972). A three-dimensional analysis of rotor wakes. *Aero. Q.*, **23**, 285–300.

Whittington, R. B. and Clapp, D. M. (1958). The measurement by optical reflection of water levels in hydraulic models. *Civil Engineering and Public Works Review*, **53**, 160–162.

Williams, D. A. (1981). *The Analysis of Random Data*. AGARDograph No. 160, Vol. 14.

Willmarth, W. W. (1975). Structure of turbulence in boundary layers. *Adv. Appl. Mech.*, **15**, 159–254.

Willmarth, W. W. and Bogar, T. J. (1977). Survey and new measurements of turbulent structure near the wall. *Phys. Fluids*, **20**, S9–S21.

Wilmarth, W. W. and Lu, S. S. (1972). Structure of the Reynolds stress near the wall. *J. Fluid Mech.*, **55**, 65–92.

Willmarth, W. W. and Sharma, L. K. (1984). Study of turbulent structure with hot wires smaller than the viscous length. *J. Fluid Mech.*, **142**, 121–149.

Wills, J. A. B. (1962). The correction of hot-wire readings for proximity to a solid boundary. *J. Fluid Mech.*, **12**, 388–396.

Wilson, D. J. and Netterville, D. D. J. (1981). A fast-response heated element concentration detector for wind tunnel applications. *J. Wind Eng. and Industr. Aerodyn.*, **7**, 55–64.

Winant, C. D. and Browand, F. K. (1974). Vortex pairing: the mechanism of turbulent mixing-layer growth at moderate Reynolds number. *J. Fluid Mech.*, **63**, 237–255.

Winter, A. R., Graham, L. J. W., and Bremhorst, K. (1991*a*). Velocity bias associated with laser doppler anemometer controlled processors. *ASME, J. Fluids Eng.*, **113**, 250–255.

Winter, A. R., Graham, L. J. W., and Bremhorst, K. (1991*b*). Effects of time scales on velocity bias in LDA measurements using sample and hold processing. *Exp. in Fluids*, **11**, 147–152.

Winter, K. G. (1977). An outline of the techniques available for the measurement of skin friction in turbulent boundary layers. *Prog. Aero. Sci.*, **18**, 1–57.

Wisler, D. C., Bauer, R. C., and Okiishi, T. H. (1987). Secondary flow, turbulent diffusion, and mixing in axial flow compressors. *ASME, J. Turbomachinery*, **109**, 455–483.

Witze, P. O. (1977). Measurements of the spatial distribution and engine speed dependence of turbulent air motion in an I.C. engine. SAE Paper No. 770220.

Witze, P. O. (1980). A critical comparison of hot-wire anemometry and laser doppler velocimetry for I.C. engine applications. SAE Paper No. 800132.

Witze, P. O. and Mendes-Lopes, J. M. C. (1986). Direct measurement of the turbulent burning velocity in a homogenous-charge engine. SAE Paper No. 861531.

Witze, P. O., Martin, J. K., and Borgnakke, C. (1984). Conditionally-sampled velocity and turbulence measurements in a spark ignition engine. *Combust. Sci. and Techn.*, **36**, 301–317.

Witze, P. O., Hall, M. J., and Bennett, M. J. (1989). Are gas velocities in a motored piston engine representative of the pre-ignition fluid motion in a fired engine? *Exp. in Fluids*, **8**, 103–107.

Witze, P. O., Hall, M. J., and Bennett, M. J. (1990). Cycle-resolved measurements of flame kernel growth and motion related with combustion duration. SAE Paper No. 900023.

Wood, N. B. (1975). A method for determination and control of the frequency response of the constant-temperature hot-wire anemometer. *J. Fluid Mech.*, **67**, 769–786.

Wu, S. and Bose, N. (1993). Calibration of a wedge-shaped vee hot-film probe in a towing tank. *Meas. Sci. Technol*, **4**, 101–108.

Wyatt, L. A. (1953). A technique for cleaning hot-wires used in anemometry. *J. Phys. E.: Sci. Instr.*, **30**, 13–14.

Wyngaard, J. C. (1968). Measurement of small-scale turbulence structure with hot wires. *J. Phys. E.: Sci.: Instr.*, **1**, 1105–1108.

Wyngaard, J. C. (1969). Spatial resolution of the vorticity meter and other hot-wire arrays. *J. Phys. E.: Sci. Instr.*, **2**, 983–987.

Wyngaard, J. C. (1971*a*). The effect of velocity sensitivity on temperature derivative statistics in isotropic turbulence. *J. Fluid Mech.*, **48**, 763–769.

Wyngaard, J. C. (1971*b*). Spatial resolution of a resistance wire temperature sensor. *Phys. Fluid*, **14**, 2052–2054.

Wyngaard, J. C. and Lumley, J. L. (1967). A constant temperature hot-wire anemometer. *J. Sci. Instr.*, **44**, 363–365.

Wyngaard, J. C. and Sheih, C. M. (1968). Further studies of the constant temperature hot-wire anemometer. *J. Phys. E.: Sci. Instr.*, **1**, 61–62.

Wygnanski, I. and Fiedler, H. E. (1970). The two-dimensional mixing region. *J. Fluid Mech.*, **41**, 327–361.

Wygnanski, I. and Ho, C. (1978). Note on the prong configuration of an X-array hot-wire probe. *Rev. Sci. Instr.*, **49**, 865–866.

Wygnanski, I. J. and Petersen, R. A. (1987). Coherent motion in excited free shear flows. *AIAA J.*, **25**, 201–213.

Wygnanski, I. and Weisbrot, I. (1988). On the pairing process in an excited plane turbulent mixing layer. *J. Fluid Mech.*, **195**, 161–173.

Yavuzkurt, S. (1984). A guide to uncertainty analysis of hot-wire data. *ASME, J. Fluids Eng.*, **106**, 181–186.

Yavuzkurt, S., Crawford, M. E., and Moffat, R. J. (1977). Real-time hot-wire measurements in three-dimensional flows. *Proc. 5th Biennial Symp. on Turbulence*, University of Missouri, Rolla, pp. 265–279.

Yeh, T. T. and Van Atta, C. W. (1973). Spectral transfer of scalar and velocity fields in heated-grid turbulence. *J. Fluid Mech.*, **58**, 233–261.

Yianneskis, M., Tindal, M. J., and Suen, K. O. (1989). A comparison of cycle-resolved and ensemble-averaged velocity variations in a diesel engine. SAE Paper No. 890617.

Yoshino, F., Waka, R., and Hayashi, T. (1989). Hot-wire direction-error response equations in two-dimensional flow. *J. Phys. E.: Sci. Instr.*, **22**, 480–490.

Young, M. F. (1976). Calibration of hot-wires and hot-films for velocity fluctuations. Report TMC-3. Thermosciences division, Department of Mechanical Engineering, Stanford University.

Yowakim, F. M. and Kind, R. J. (1988). Mean flow and turbulence measurements of annular swirling flows. *ASME, J. Fluids Eng.*, **110**, 257–263.

Yule, A. J. (1978). Large-scale structure in the mixing layer of a round jet. *J. Fluid Mech.*, **89**, 413–432.

Zabat, M., Browand, F. K., and Plocher, D. (1992). In-situ swinging arm calibration for hot-film anemometers. *Exp. in Fluids*, **12**, 223–228.

Zaman, K. B. M. Q., and Hussain, A. K. M. F. (1980). Vortex pairing in a circular jet under controlled exitation, Part 1, General jet response. *J. Fluid Mech.*, **101**, 449–491.

Zemskaya, A. S., Levitskiy, V. N., Repik, Ye.U., and Sosedko, Yu.P. (1979). Effect of the proximity of the wall on hot-wire readings in laminar and turbulent boundary layers. *Fluid Mech.—Sov. Res.*, **8**, 133–141.

Zhu, Y. and Antonia, R. A. (1993). Temperature dissipation measurements in a fully developed turbulent channel flow. *Exp. in Fluids*, **15**, 191–199.

Zilberman, M. (1981). On the interaction of transitional spots and generation of a synthetic turbulent boundary layer. Ph.D. Thesis, Tel-Aviv University, Israel.

Zilberman, M, Wynanski, I., and Kaplan, R. E. (1977). Transitional boundary layer spot in a fully turbulent environment. *Phys. Fluids*, **20**, S258–S271.

Zilliac, G. G. (1993). Modelling, calibration and error analysis of seven-hole pressure probe. *Exp. in Fluids*, **14**, 104–120.

zur Loye, A. O., Siebers, D. L., Mckinley, T. L., Ng, H. K., and Primus, R. J. (1989). Cycle-resolved LDV measurements in a motored diesel engine and comparison with k–ε model predictions. SAE paper No. 890618.

INDEX

Printed in the United States
136673LV00003B/16/A

9 780198 563426